D1337524

How

Things Work

The Universal Encyclopedia of Machines

Volume Two

Paladin

Granada Publishing Limited
Published in 1974 by Paladin
Frogmore, St Albans, Herts AL2 2NF

First published in Great Britain by George Allen & Unwin Ltd 1971
German original first published by Bibliographisches Institut
in 1967 under the title *Und Wie Funktioniert Dies?*
This edition translated and adapted from the German by
C. van Amerongen, M.Sc., A.M.I.C.E.
Translation copyright © George Allen & Unwin Ltd 1971
Printed in Great Britain by Fletcher & Son Ltd, Norwich

INTRODUCTION

BY THE RIGHT HON. THE LORD RITCHIE-CALDER

*Kalinga Prize winner for promoting
the common understanding of science*

"The world will never starve for want of wonders but
only for the want of wonder…"—G. K. CHESTERTON

So MUCH has happened since the end of World War II—the Atomic Age, the Computer Age and the Space Age, epochs concentrated into a brief lifetime—that wonder is liable to become numbed and wonders taken for granted. We flick a switch, press a button, or put a coin in the slot and expect things to happen…without asking "How?" That is sad, because as Sir Edward Appleton, the Nobel Prize-winning scientist, said, "It is such fun finding out."

And we cannot all be like my Arctic guide, Luke the Inquisitive Eskimo, who had a direct way of finding out. If he wanted to know how something worked—a wristwatch, for instance—he would just take it to pieces and memorize the mechanism. I found this childlike curiosity amusing until, when we were holed up in an igloo by a blizzard, I woke up one morning to find him dismantling my tape recorder.

Or one can do what I have done most of my life as a science writer: go to the chap who discovered something and ask him, "How does it work?" That is how, over the years, I learned about a lot of things in this book.

In going through the first volume of *The Way Things Work*, however, I was pleased (and sometimes professionally mortified) to find explanations of things which had slipped into the commonplace of my experience and about which I had never asked "How?" I was given one of the first practical ball-point pens in 1938 but forgot—all this time—to ask just exactly how it worked. It is simple, like most good ideas. For years I have been using a Polaroid camera. To photograph and develop immediately is like practicing a conjuring trick; it fascinates children and grownups alike; and sometimes it spoils the relish of a trick when one knows exactly how it is done. So I went on conjuring without asking! Now I know that the "How" is just as fascinating as the "Hey presto!"

Away back in 140 A.D., Hero of Alexandria invented the first coin-in-slot vending machine for dispensing holy water. It is interesting to have

5

it brought up to date in the modern mechanisms of the automats and the refined tortures of the jukebox. When they were first invented, I wrote about transistors, lasers, the magnetron, xerography, jet engines, hovercraft and computers, and it is nice to see them, grown-up, as it were, in the family album.

The bits and pieces of technology, however, are not enough. We would still be mystified, left with unanswered questions, if we did not see how they fitted into the patterns and the systems. If this and its preceding volume were just expanded glossaries of gadgets, they might suffice for a quiz game of "Ask me another." But this encyclopedia is designed to be more than that—it deals with principles and is grouped in subjects so that the picture of technology emerges and embraces the items.

I do not know who thought of *The Way Things Work*, nor how it was compiled, but I know it is a valuable piece of communication (to use an overworked word) to the general public and, speaking professionally, a useful crib for science writers and for scientists and technologists as well. It is simple without being condescending.

A FOREWORD
FROM THE PUBLISHERS

WHEN WE first saw the original German edition of *Wie Funktioniert Das?*,* some five years ago, it seemed to us the answer to the modern man's prayer: a guide and an encyclopedia that explained modern technology—from simple, everyday mechanical items to complex processes —in terms that anyone could understand.

It seemed to us that the book would be both useful and successful, and we were delighted to have it on our list. What we did *not* expect was a nationwide explosion of enthusiasm that almost instantly made the English translation into an institution and a best seller. Our first and relatively modest printing vanished in a week, as did successive printings, until, today, close to *one million* copies of *The Way Things Work* (Volume One) have been sold in various editions.

Not many books sell a million copies, and it is natural that we should have explored the possibility of a further volume, one not in any way overlapping the first, but providing, quite simply, several hundred subjects that couldn't be covered in one volume—more of the processes, machines, principles and conveniences of the mod-

*British title, *The Universal Encyclopedia of Machines;* American title, *The Way Things Work*.

ern industrial age, concisely explained and made crystal-clear by a brilliant two-color schematic drawing opposite each page of text, as well as a section of full-color illustrations.

We hope that readers of Volume One will derive an equal pleasure from Volume Two. Certainly we have, and have found that we are now in a position, with two volumes, of being able to answer twice as many questions from our children as we were before. However, there is no real need to start with Volume One. If you don't already have it, you can plunge into Volume Two, and indeed start on any page with any topic. The key to enjoying *The Way Things Work* is to browse for pleasure. When you have something specific to look up, there is a full index to help you find whatever it is you (or your child) want to know about.

As for us, we are now launched on a new enterprise. Since, like most modern businesses, our company uses a computer for accounting and information storage, we have decided to find out how it works, and why it sometimes seems not to, and what makes it tick (or rather, hum). So we have started work on a third volume in this series, to be called *The Way Things Work Book of the Computer*, a volume that will explain the principles, the mechanism, the uses, the theory and the functions of computers of all kinds to every layman who has ever wanted to know more about the central technological innovation of our times. It may take us a year or more to have this book ready, but we think you'll find it interesting.

When Volume One of *The Way Things Work* was published, *Time* Magazine said of it:

"In this technological age, the world has long been divided into two camps. There are those who want to know how the complex machines that are so much a part of their lives really work; and there are those who couldn't care less, as long as the machines somehow keep functioning. For those who nourish any technical curiosity at all, this lucid book provides a thorough collection of answers."

That's what we had in mind when we began. The purpose of these books is to help you *understand* the world you live in. In the words of Isaac Asimov, one of America's most distinguished scientific writers:

Once upon a time, there were primitive priesthoods of magic, and members of those priesthoods cast spells, muttered runes, made intricate diagrams on the floor with powders of arcane composition. All this was intended to make the future clear, or bring the rain, or ward off evil, or lure an enemy to an agonizing end. Onlookers, when there were any, would watch with awe and no little fear, believing utterly in the efficacy of all this and in the existence of powerful and dangerous forces they could not themselves control.

Nowadays, there is a modern priesthood of science that calls on the power of expanding steam, of shifting electrons or drifting neutrons, of exploding gasoline or uranium, and does so without spells, runes, powders or even any visible change of expression. In response, onlookers are without awe, for, indeed, they seem to participate in the magic. By moving a lever, interrupting a light beam, or closing a contact, they can accelerate a car, open a door, or blow up a city.

Yet have we the right to sneer at the centuries of "ignorance" when witches were feared and gloat at our own age of "enlightenment" when scientists are respected? Do we understand more about the scientific principles and technological details of today than our ancestors did of the witches' spells of yester-century?

What it amounts to is that we live in a world of black boxes—all of us do, however well we may be educated. We drive automobiles that are complete mysteries to us, and, for that matter, the lock on our front door may be equally mysterious.

It is to this problem that these two volumes address themselves. Their aim is to take any device you can think of, from the ball-point pen, through lightning conductors and gunpowder, to gearbox transmission and the nuclear bomb, and to explain how each of them works.

The mechanics of explanation are as follows: Each double-page spread is devoted to one particular item, without regard to intricacy. The electric bell gets one double-page spread, for instance, and the electron microscope also gets one double-page spread. (However, it is usual for several successive spreads to deal with related subjects, as, for instance, in the case of seven spreads on fundamental electrical matters.)

On the right-hand side of each spread is a diagram or a series of diagrams in red and black, drawn and labeled neatly and clearly. On the left-hand side of the spread is descriptive matter....

The book is exhaustive, and has an excellent index. There is little you can wonder about and not find, for there is mention of things as disparate as fuses, dry ice, television cameras, power looms, multistage rockets and eyeglasses....

For instance, I use an electric typewriter with a typing head instead of a keyboard. It had never seriously occurred to me to wonder how the keys control the typing head. With the book in my hand, it did occur to me to be curious and I turned to the two double-page spreads devǒted to typewriters (one to mechanical and electrical keyboard typewriters and the second to the typing head). Well, I know now—in general, anyway—even though I still wouldn't dream of trying to repair the machine myself.*

*From the review of *The Way Things Work* by Isaac Asimov, *New York Times Book Review,* November 19, 1967; © 1967 by The New York Times Company. Reprinted by permission.

CONTENTS

Introduction by The Right Hon. The Lord Ritchie-Calder . . 5

A Foreword from the Publishers 7

Aerosols 20

Nuclear Fusion 22

Nuclear Reactors 26

    Pressurized-water Reactor 26

    Pressure-tube Reactor 26

    Boiling-water Reactor 28

    Superheated-steam Reactor 28

    Gas-cooled Reactor 30

    Advanced Gas-cooled Reactor 30

    High-temperature Reactor 32

    Sodium Reactor 36

    Breeder Reactors 38

Synthetic Ammonia Process 40

Synthesis Gas and Methanol Synthesis 42

Town-gas Detoxication (Conversion of Carbon Monoxide) . . 44

Phosphorus 46

Sulphuric Acid 48

Nitric Acid 50

Foam Plastics 52

Tank Trucks 54

Plastics Processing 56

Prospecting for Minerals 60

Deep-drilling Engineering 62

Mines and Mining 64

Shaft Sinking 68

Winding 70

Mine Ventilation 72

Precautions against Mine Gas 74

Mechanization and Automation in Mining 76

Preparation of Ores 78

    Lead 84

    Copper 86

    Zinc 88

    Aluminum 92

Roasting of Ores and Concentrates 94
Ingot Casting 96
Continuous Casting 98
Metal Casting 102
    Sand Casting 102
    Shell Molding 104
Precision Casting 106
Gravity Die Casting 108
Pressure Die Casting 110
Centrifugal Casting 112
Composite Casting 114
Powder Metallurgy 116
Forging 118
Cold Extrusion of Metals 122
Hot Extrusion of Metals 124
Cutting and Machining of Metals 126
Pressure Welding 130
Fusion Welding 134
Soldering 138
Sheet-metal Work 140
Galvanizing 144
Metal Spraying 146
Electroforming 148
Electroplating 150
Power-operated Valves 152
Screw Threads 156
Screw Cutting 158
Thread Milling 160
Gears 162
Gear Cutting 166
Ball and Roller Bearings 170
Lathe 178
Milling Machine 186
Couplings 188
Clutches 194
Mechanical Power Transmission 198
Linkages 200
Gear Mechanism 206
Friction Drives 208

Pulley Drives 210
Cam Mechanism 212
Locking and Arresting Mechanisms 214
Guns 218
Projectiles 226
Fuses 230
Mines 234
Torpedoes 236
Tanks 240
Road Intersections and Junctions 248
Operating an Automobile 256
Improving Engine Performance: Principles . . . 260
Increasing the Engine's Cubic Capacity 264
Increasing Engine Efficiency: Mean Effective Pressure . . 266
Improving Volumetric Efficiency 272
Resonance and Engine Efficiency 274
Valve Timing and Engine Efficiency 276
Fuel Injection and Supercharging 278
Improving the Engine's Mechanical Efficiency . . . 280
Improving Engine Speed 282
Color Television 288
Cartography 296
Engine Tuning 304
Generator (Dynamo) 322
    Principles 322
    Direct-current Generator 322
    Regulator 324
    Three-phase Generator 326
Present-day Methods of Aircraft Construction . . . 328
Wing Geometry 330
Airfoils and Airflow Phenomena 332
Wind Tunnel 336
Hydraulic Power Systems in Aircraft 340
Ramjet Propulsion 344
Vertical-Takeoff-and-Landing Aircraft (VTOL) . . . 346
Yachts 348
Submarines 352
Ship Stabilizing 358
Liquid-propellant Rocket Systems 362

Solid-propellant Rocket Systems 364
Ion-drive Rocket Propulsion Systems 366
Multistage Rockets 368
Reentry and Ablation 370
Space Suits and Space Capsules 372
Inertial Navigation 374
Space-flight Orbits and Trajectories 376
Landing the First Men on the Moon 382
Satellites and Space Probes 386
    Space Probes 388
    Communications Satellites 390
Planetarium 392
Medical Applications of the Betatron 398
Hearing Aids 400
High-speed Turbine Drill 402
Incubator and Oxygen Tent 404
Gas Mask 406
Artificial Heart 408
Artificial Kidney 418
Packaging Technology 426
Silos and Silage 430
The Modern Cattle Barn (Cow Shed) 432
Rotary Haymaking Machine 436
Forage Harvesters 438
Self-contained Underwater Breathing Apparatus (SCUBA) . 442
Safety Binding for Skis 444
Pianoforte 448
Piano Tone and Tuning 450
Accordion 452
Electronic Organ 454
Electronic Music 456
High Fidelity 468
Cloud Chamber and Bubble Chamber 476
Linear Accelerator 478
Electric-resistance Strain Gauge 480
Slide Rule 484
Straight-line Link Mechanisms 488
Measurement of Pressure 490
Measurement of Flow of Fluids 494

Mixing of Materials 498
Electromagnets 502
Direct-current Machines 506
Alternating-current Machines 510
Converters 516
Ward-Leonard Control 518
Overhead Transmission Lines 520
Wave Guides 526
Displacement Current 528
Basic Electronic Circuits 530
Carrier Transmission 534
VHF Stereophonic Broadcasting 538
Time Division Multiplexing 542
Telephony and Telephone-exchange Techniques . . . 544
Telephone Cables 552
Automatic Message Accounting and Circuit Testing . . . 556
Current Supply for Telecommunication Systems . . . 558
Video Tape Recording 560
Teaching Machines 566

Index 575

THE WAY THINGS WORK

AEROSOLS

An aerosol is a colloidal system consisting of very finely divided liquid or solid particles dispersed in and surrounded by a gas. In recent years aerosols have become familiar as products discharged from spray dispensers, and the term "aerosol" has, in popular speech, also come to mean the dispenser itself—i.e., a pressurized container made of metal or glass and provided with a discharge valve, which may be a spray valve or a foam valve. It is filled with the product to be sprayed and the propellant gas under pressure (Fig. 4).

The product that is to be dispersed as an aerosol may have the liquefied propellant mixed with it in the form of a solution (Fig. 1). Alternatively, the propellant may be present as a separate gaseous phase in the dispenser, in which case it does not mingle with the product (two-phase system: Fig. 2).

An example of the first type is afforded by hair spray. The spray (or lacquer), usually dissolved in alcohol, is completely miscible with the liquefied propellant. When the valve button on the dispenser is pressed, the propellant vaporizes immediately, and its pressure forces the liquid out of the nozzle. The liquid (i.e., the lacquer solution) is discharged as a fine mist. The most commonly employed propellants are chlorinated hydrocarbons, butane, propane, isobutane, vinyl chloride and nitrogen. Nitrogen is used particularly for products that must on no account be contaminated in flavor or smell: e.g., toothpaste packaged in aerosol dispensers.

Aerosol toothpaste is an example of the second category of aerosol systems—namely, the two-phase system—in which the propellant gas forms a separate layer over the product to be discharged. The dispenser is usually about half filled with nitrogen (or some other suitable gas) and half with the product. The pressure in the dispenser is about 6 to 8 atm. (90 to 120 lb./in.^{2}). Nitrogen is also used as the propellant for foods packaged in aerosol form—e.g., cheese spreads, malt extracts, vitamin preparations, syrups, pudding sauces, whipped cream.

Filling an aerosol dispenser at the factory is a simple operation (Fig. 3: stages 1 to 5). First, the product is introduced into the dispenser (1). This is done by a pneumatic filling machine. Then the aerosol valve is placed on the dispenser (2). In the next stage the valve is force-fitted under high pressure (about $\frac{3}{4}$ ton) into the neck of the dispenser, so that a strong gastight seal is formed between the latter and the valve unit (3). The propellant gas is now forced into the dispenser (4). Finally, the dispenser is immersed in water to test it for possible leakage, which is manifested by escaping bubbles of gas (5).

Aerosols are coming into increasingly widespread use in industry, too. For instance, they are used for the disinfection of milk tanks. For this purpose a spraying device is used which draws the disinfectant solution by suction from a container and disperses it as an aerosol by means of two atomizing discs. These discs rotate, and their centrifugal action sets up a suction which draws the disinfectant forward through the hollow shaft of the motor. To make the aerosol mist flow in the desired direction, a second air stream is needed. A turbine installed behind the motor sucks in air, which flows along the motor and emerges from the annular aperture around the atomizing discs. This stream of air carries along the aerosol particles of disinfectant (Fig. 5).

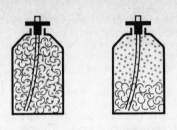

FIG. 1   FIG. 2

PRINCIPLE OF AEROSOL PRESSURIZED PACKAGING

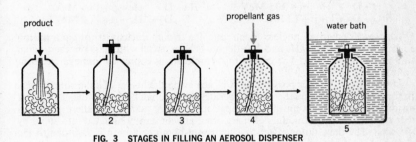

product

propellant gas

water bath

1   2   3   4   5

FIG. 3   STAGES IN FILLING AN AEROSOL DISPENSER

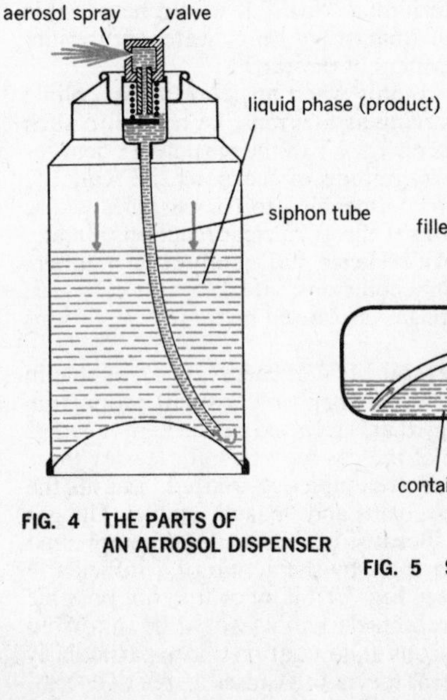

aerosol spray   valve

liquid phase (product)

siphon tube

filler opening   valve

container   air inlet   motor   atomizing discs

**FIG. 4   THE PARTS OF
AN AEROSOL DISPENSER**

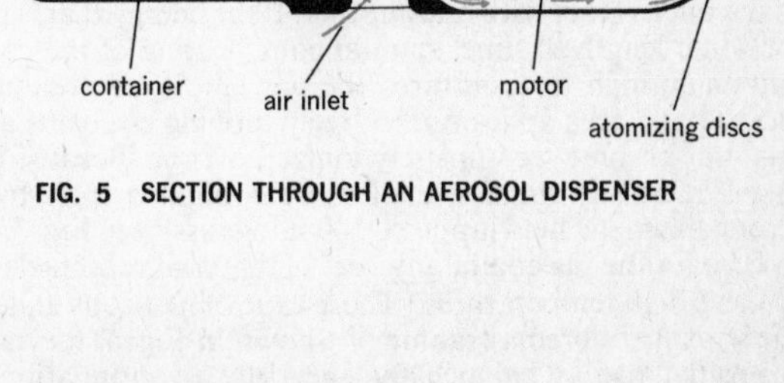

FIG. 5   SECTION THROUGH AN AEROSOL DISPENSER

NUCLEAR FUSION

Energy that is released as a result of fission of heavy atom nuclei (e.g., uranium) in atomic reactors has been utilized already for a good many years. Another possible method of nuclear-energy production is by the fusion of the nuclei of the lightest chemical elements. Fusion of this kind—called nuclear (or thermonuclear) fusion— may occur, for example, when in a mixture of the two gases deuterium and tritium (which are heavy hydrogen isotopes indicated by the symbols ^{2}H and ^{3}H respectively) the atom nuclei collide with one another with sufficiently high energy (i.e., with sufficiently high relative velocity). In such circumstances the electrostatic repulsion operating between the nuclei (which are positively charged) can be overcome (see Fig. 1), so that the colliding nuclei "fuse" together. This phenomenon is accompanied by the release of individual nuclear components with high kinetic energy.

The principal processes that may occur in a mixture of deuterium and tritium are:

$$D + D \rightarrow {}^3He + n + 3.25 \text{ MeV} \qquad {}^3He + D \rightarrow {}^4He + p + 18.3 \text{ MeV}$$
$$D + D \rightarrow T + p + 4 \text{ MeV} \qquad T + D \rightarrow {}^4He + n + 17.6 \text{ MeV}$$

D denotes a deuterium nucleus (deuteron), T a tritium nucleus (triton), p a proton and n a neutron, while ^{3}He and ^{4}He denote helium nuclei with mass numbers equal to 3 and 4, respectively. The energy that is released is expressed in mega-electron-volts (1 MeV = 4.45×10^{-20} kWh).

There are many other possible ways of achieving the fusion of light atom nuclei. In the sun and other stars there occur, besides the fusion of hydrogen nuclei, complex fusion processes involving the participation of heavy nuclei. It is these processes that produce the vast amounts of energy that are constantly radiated into space by the stars. The devastating effect of a hydrogen-bomb explosion is also due to the energy released by fusion processes of this type: with an atomic bomb as a detonator, an explosive—or uncontrolled—chain reaction of nuclear fusion processes is initiated. "Taming the hydrogen bomb" in the sense of bringing these reactions under control in a so-called fusion reactor would place a tremendous new source of energy at man's disposal. Using pure deuterium as "fuel," it would be possible to produce about 10^{25} kWh of energy from the quantity of heavy water (deuterium oxide) estimated to be present in the natural waters of the earth.

As already stated, fusion processes take place only when nuclei of atoms collide at high velocities. To achieve these, the gas serving as fuel must be heated to such extremely high temperatures T that the average energy kT of the particles (k denotes the Boltzmann constant) is of the order of magnitude of the potential wall U_0. Since the potential wall is to a certain extent "permeable" to the particles whose energy is less than U_0 (quantum-mechanical tunnel effect), nuclear reactions already take place at kinetic-energy values of 10 to 100 keV. Hence, sufficiently large numbers of nuclei can react with one another in "thermal collisions," if kT $\approx$ 100 keV—i.e., T $\approx$ 100 million degrees centigrade. Nuclear reactions caused by thermal collisions are generally called *thermonuclear reactions*.

Since each fusion results in the release of several MeV of energy, it is possible in principle to achieve a positive energy balance even when only a small proportion of the nuclei react with one another. If the energy that is released can be kept together for some length of time, spontaneous heating of the gas may be initiated. At these extremely high temperatures the gas employed is completely ionized: i.e., all the atoms have been split up into freely moving electrons and "naked" nuclei. The gas has thus become a completely ionized plasma. Because of this, it can be compressed into a relatively small space and be contained there by the action of a sufficiently strong magnetic field (approx. 100 kilogauss); see Fig. 3. (It is, of course, not possible to contain the plasma in any sort of material receptacle, as this would be vaporized at these high temperatures.) The escape of neutrons and radiation (more particularly the so-called "bremsstrahlung," shown in Fig. 4) inevitably causes losses of energy. A positive energy balance and therefore a continuation of the fusion processes can

(more)

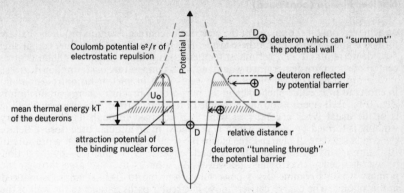

FIG. 1 POTENTIAL DISTRIBUTION FOR THE INTERACTION OF CHARGED NUCLEI (DEUTERONS)

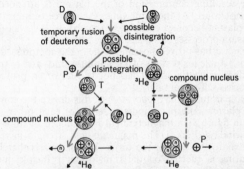

FIG. 2 SCHEMATIC REPRESENTATION OF THE PRINCIPAL NUCLEAR-FUSION PROCESSES IN A DEUTERIUM-TRITIUM MIXTURE

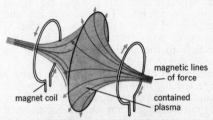

FIG. 3 CONTAINMENT OF A PLASMA BY A MAGNETIC FIELD; THE CURRENTS IN THE COILS PRODUCING THE FIELD FLOW IN OPPOSITE DIRECTIONS

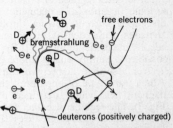

FIG. 4 FULLY IONIZED DEUTERIUM PLASMA; AN ELECTRON EMITS "BREMSSTRAHLUNG" RADIATION AS SOON AS ITS PATH CEASES TO BE LINEAR, IN CONSEQUENCE OF INTERACTION WITH OTHER PARTICLES

be achieved only if the energy of the electrically charged reaction products that are formed is able to make up for these losses. This is the case with a mixture comprising 50% deuterium and 50% tritium at temperatures of more than 50 million degrees centigrade (for pure deuterium it would require a temperature of 400 million degrees).

One of the best-known pieces of apparatus used in nuclear-fusion research and in efforts to utilize this phenomenon for purposes of practical energy production is *Zeta* (Fig. 5). It operates on the principle of the transformer. The primary winding is of the usual type; a condenser bank is discharged through it. The secondary winding is formed by plasma which is produced in an annular tube (torus). Before the condensers are discharged, the gas in the torus (e.g., deuterium) at a pressure of 10^{-4} mm is slightly ionized—that is, made electrically conductive—by the action of high-frequency electromagnetic waves. As soon as the discharge through the primary winding commences, a powerful current (up to 200,000 amps.) is induced in this plasma. The charge carriers move in circular paths parallel to the wall of the torus. These "current filaments," like all electric currents flowing in parallel paths, attract one another: The ring of plasma, which initially fills the entire space within the torus, contracts and becomes detached from the wall. This phenomenon is called "pinch effect." The resulting compression of the plasma is attended with a considerable rise in temperature. At the same time, the degree of ionization increases so greatly that the plasma becomes completely ionized. In this way the conditions in which nuclear-fusion processes can occur are established.

With the aid of Zeta it has proved possible to keep the compressed plasma tube, the so-called "pinch," stable for periods of a few milliseconds and to reach temperatures of 5 million degrees centigrade. It has not yet, however, proved possible to make the fusion process self-sustaining.

Another type of apparatus is the *Stellarator*. In this device the containment and the heating of the plasma take place independently of each other. In a torus of double-loop shape (a figure 8), a magnetic field is produced by an electric current flowing through a winding (Fig. 6). The magnetic-field strength increases with the distance from the axis to the wall of the torus. The plasma is thereby kept away from the wall. Heating is effected on the transformer principle, just as in Zeta (electrical resistance produces heat in the plasma functioning as the secondary winding). However, this phenomenon generates a temperature of only about 1 million degrees centigrade, because at elevated temperatures the electrical conductivity of the plasma increases and the resistance therefore decreases.

Yet another method of raising the temperature to very high values is based on periodic variation of a magnetic field by means of a booster coil (see Fig. 6). An alternating current flowing through this coil produces periodic increases and decreases in the density of the magnetic lines of force in the torus (this effect is called "magnetic pumping"). By choosing an appropriate frequency it is thus possible to supply energy more particularly to the nuclei (as distinct from the electrons), so that the bremsstrahlung losses due to the electrons are kept low.

In a third type of apparatus, the containment of the plasma is achieved by a magnetic field which is stronger at the (externally open) ends than in the middle (Fig. 7). The regions of higher field strength act as "magnetic mirrors": They are able to reflect plasma particles. This type of configuration for the containment of a plasma in controlled-thermonuclear-reaction experiments is referred to as a "magnetic bottle." The initially cold plasma is compressed and heated by rapid intensification of the magnetic field. Temperatures exceeding 10 million degrees centigrade are thus attained in very small regions of the plasma.

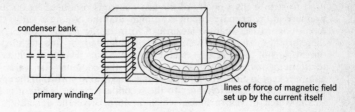

FIG. 5 ZETA (SCHEMATIC)

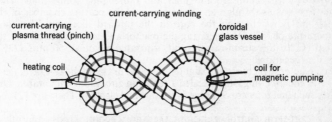

FIG. 6 DOUBLE-LOOP STELLARATOR TORUS; THE GUIDING MAGNETIC FIELD IS PRODUCED BY MAGNET COILS (HERE REPRESENTED SCHEMATICALLY BY THE WINDING); STABILIZATION OF THE PINCH IS ACHIEVED BY MEANS OF ADDITIONAL WINDINGS

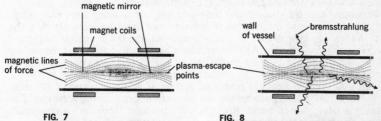

**FIG. 7
PRINCIPLE OF A MAGNETIC BOTTLE WITH CONTAINED PLASMA**

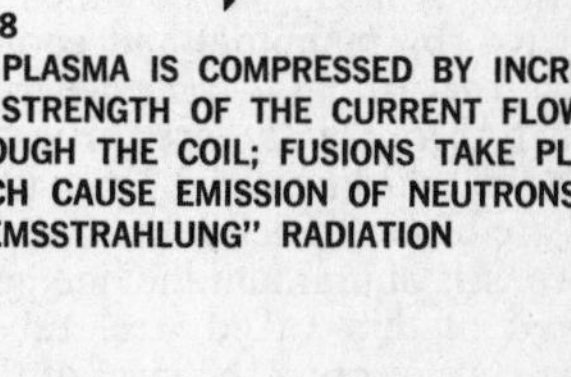

**FIG. 8
THE PLASMA IS COMPRESSED BY INCREASING THE STRENGTH OF THE CURRENT FLOWING THROUGH THE COIL; FUSIONS TAKE PLACE WHICH CAUSE EMISSION OF NEUTRONS AND "BREMSSTRAHLUNG" RADIATION**

Pressurized-water reactor: This is the simplest form of thermal reactor, in which water serves as the coolant and also as the moderator (i.e., the substance that is used to reduce the velocity of the fast neutrons produced by nuclear fission). The pressure in the primary circuit is so high, and the boiling point of the water consequently so raised, that no steam can form in the reactor core. The pressure, and therefore the attainable temperature, is limited by the technically practicable dimensions of the reactor vessel. Ordinary water (H_2O) as well as "heavy" water (deuterium oxide, D_2O) may be used as the coolant. The water in the primary circuit is kept in circulation by pumping. The heat absorbed in the core is transferred by means of a heat exchanger to the secondary circuit, where it is utilized to raise steam which drives turbines, which in turn drive the generators for producing electricity.

An example of this reactor type is the KWO pressurized water reactor built at Obrigheim on the River Neckar, in Germany, which has a thermal output of 907.5 MW and an electric power output of 283 MW (Fig. 1). The reactor core is accommodated in a pressure vessel with an internal diameter of 3.27 m (10 ft. 9 in.), which is provided with cooling-water inlet and outlet pipe connections. The inflowing water first passes through an annular gap in the bottom chamber of the pressure vessel and flows through the core in the upward direction. The coolant has a temperature of 283° C on entering the reactor and is heated to an exit temperature of 310° C. In the steam-raising unit, saturated steam of 50 atm. (735 lb./in.^{2}) at 263° C is generated.

The reactor fuel consists of slightly enriched uranium dioxide (average 3% U 235), which is enclosed in gastight sealed zircaloy (a zirconium alloy) tubes. One hundred eighty such fuel rods are combined into one fuel element. There are 121 fuel elements in the reactor core.

For short-term control operations, there are 27 control rods uniformly distributed over the core, which are inserted into it from above. To compensate for the initial reactivity necessary for high consumption rates or for slow control operations, the boron concentration in the coolant can be altered.

Pressure-tube reactor: A very interesting and promising form of construction of the pressurized-water reactor is the pressure-tube type. Instead of the pressure vessel there is now a set of parallel pressure tubes through which the coolant flows. The fuel elements are inside these tubes. Externally the latter are surrounded by the moderator. Since the coolant and the moderator are separate from each other, the moderator can be at a much lower pressure and lower temperature than the coolant. Almost any type of coolant can be employed, and the temperatures thereof can be freely chosen. Thus, the AKB project for a heavy-water pressure-tube reactor (100 MW electric-power output) developed by the German engineering firm of Siemens uses carbon dioxide as the coolant. The exit temperature of this substance is so high that the live steam is generated at 530° C and a pressure of 105 atm. (1545 lb./in.^{2}). This makes possible the use of modern steam turbogenerators. The moderator is heavy water, which surrounds the pressure tubes and is constantly circulated (by pumping) and cooled (Fig. 2). In contrast with the Canadian and French reactors of the same general type, in this German reactor the cooling tubes, of which there are 351, are vertical. In the fission zone they consist of zircaloy and have a wall thickness of 2.7 mm (0.106 in.). Above and below this zone the cooling tubes are of steel. Each tube contains a fuel element comprising 19 fuel rods. These rods consist of uranium dioxide pellets, made with very slightly enriched uranium, enclosed in thin-walled steel tubes. The pressure-tube reactor is controlled by raising and lowering the level of the moderator. For this purpose the heavy water is kept under a low-pressure helium atmosphere.

(more)

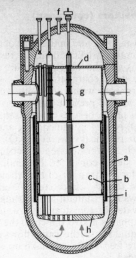

a   pressure vessel
b   core-supporting
     framework
c   core cage
d   top supporting
     plate
e   bundle of fuel
     elements
f   control-rod drive
g   control rod
h   bottom supporting
     plate
i   thermal shield

**FIG. 1   SECTION THROUGH THE KWO
PRESSURIZED-WATER REACTOR
AT OBRIGHEIM, GERMANY**

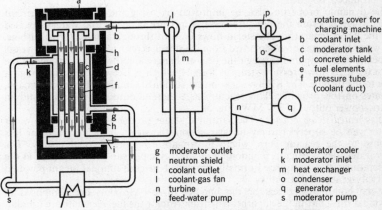

a   rotating cover for
     charging machine
b   coolant inlet
c   moderator tank
d   concrete shield
e   fuel elements
f   pressure tube
     (coolant duct)

g   moderator outlet   r   moderator cooler
h   neutron shield   k   moderator inlet
i   coolant outlet   m   heat exchanger
l   coolant-gas fan   o   condenser
n   turbine   q   generator
p   feed-water pump   s   moderator pump

**FIG. 2   CIRCUIT DIAGRAM OF A PRESSURE-TUBE REACTOR
WITH SCHEMATIC SECTION THROUGH THE REACTOR
(AKB PROJECT FOR NIEDERAICHBACH, GERMANY)**

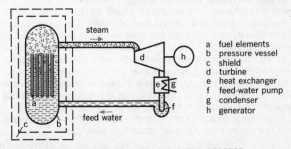

steam

a   fuel elements
b   pressure vessel
c   shield
d   turbine
e   heat exchanger
f   feed-water pump
g   condenser
h   generator

feed water

FIG. 3   DIRECT CIRCUIT OF A BOILING-WATER REACTOR

Boiling-water reactor: In this type of reactor the coolant also serves as the moderator. Its general construction closely resembles that of the pressurized-water reactor (see page 26), and the fuel elements are also very similar. As a rule, the reactor vessel comprises a steam chamber above the water level. The object of this steam chamber, which is provided with various internal fittings, is to promote the separation of the phases and to prevent carry-over of large quantities of water from the liquid to the vapor phase, besides compensating for minor variations in pressure. Ordinary water (H_2O) is used as the coolant and moderator; it is, however, also possible in principle to use heavy water (D_2O). The reactor normally generates saturated steam, which is either passed direct to the turbine or used to generate secondary steam in a steam converter. Depending on whether or not there is a main heat exchanger for the transfer of the whole of the energy between the reactor and the turbine, the terms "direct circuit" and "indirect circuit" are applicable. Fig. 3 (p. 27) shows an installation operating on the former principle: inside the reactor core, the water or the steam-and-water mixture flows upwards. After separation of the two phases above the core, the saturated steam flows direct to the turbine, while the liquid phase (the water) flows downwards in an externally disposed annular space, where feed water, to make up for the quantity of saturated steam that has been produced, is added.

The dynamic properties can be improved by using the so-called two-circuit system (Fig. 1). From the pressure vessel a mixture of steam and water is passed into the water separator. The steam flows onward through a valve to the turbine. The water, which has been separated from the steam, is returned to the reactor vessel by means of a pump and through the heat exchanger. On the secondary side of the heat exchanger low-pressure steam is formed, which is likewise passed through a valve to the turbine. The KRB nuclear power station at Gundremmingen, Germany, operates on this principle. This reactor has a thermal output of 801 MW and an electric-power output of 237 MW (Fig. 2).

Superheated-steam reactor: The saturated steam produced in the boiling-water reactor can be superheated by nuclear action, the steam being returned for this purpose to the reactor core, where it can absorb more thermal energy. The superheated-steam reactor, functioning on this principle, can generate steam at the temperatures and pressures normally employed in conventional steam-powered electricity-generating stations.

In the boiling-water reactor the bundles of fuel elements are composed of individual rods, but in the superheated-steam reactor the fuel elements are tubular. For example, in the nuclear power station at Grosswelzheim, Germany, the fuel-element assembly comprising 8×8 tubes is accommodated in a container, just as in the boiling-water reactor. These tubes are cooled externally by boiling water, so that saturated steam is produced. The saturated steam, which collects in the space above the water surface, is returned to the fuel elements for superheating. The saturated steam flows downwards through the interior of the tubular fuel elements in its first pass through the core of the reactor; in the course of this journey it cools the elements internally and becomes superheated. Below the core, the direction of flow of the steam is reversed: it now flows upwards in its second pass through other fuel elements and so becomes further superheated. This cycle is shown schematically in Fig. 3, while Fig. 4 is a vertical section through this reactor. The latter is of relatively low capacity, and for this reason the steam actually makes four passes through the core. With large superheated steam reactors, however, a double pass will suffice to attain the desired superheated-steam temperature.

(more)

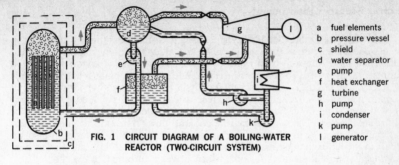

FIG. 1 CIRCUIT DIAGRAM OF A BOILING-WATER REACTOR (TWO-CIRCUIT SYSTEM)

a fuel elements
b pressure vessel
c shield
d water separator
e pump
f heat exchanger
g turbine
h pump
i condenser
k pump
l generator

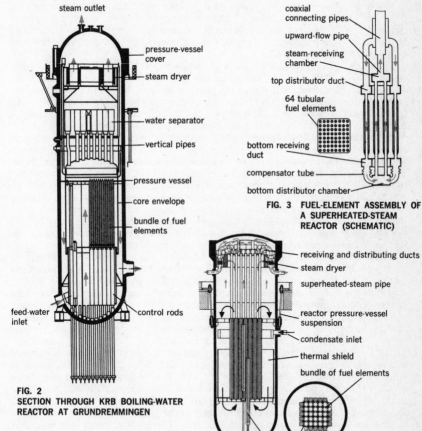

FIG. 2
SECTION THROUGH KRB BOILING-WATER REACTOR AT GRUNDREMMINGEN

steam outlet
pressure-vessel cover
steam dryer
water separator
vertical pipes
pressure vessel
core envelope
bundle of fuel elements
feed-water inlet
control rods

FIG. 3 FUEL-ELEMENT ASSEMBLY OF A SUPERHEATED-STEAM REACTOR (SCHEMATIC)

coaxial connecting pipes
upward-flow pipe
steam-receiving chamber
top distributor duct
64 tubular fuel elements
bottom receiving duct
compensator tube
bottom distributor chamber

FIG. 4
SECTION THROUGH THE SUPERHEATED-STEAM REACTOR AT GROSSWELZHEIM, GERMANY

receiving and distributing ducts
steam dryer
superheated-steam pipe
reactor pressure-vessel suspension
condensate inlet
thermal shield
bundle of fuel elements
control rods

→ saturated steam 285° C
---→ superheated steam 400° C and superheated steam 500° C
→ water circulation

29

Gas-cooled reactor (Calder Hall type): Along with the pressurized water reactor, the gas-cooled reactor of the type installed at Calder Hall power station in Great Britain is one of the well-tried types of nuclear reactor. This reactor is operated with natural uranium. The moderator is graphite and the coolant is carbon dioxide gas (CO_2). The circuit is shown schematically in Fig. 1. A large cylindrical block of graphite pierced by channels forms the reactor core. The rod-shaped fuel elements are so disposed in these channels that a gap remains between each element and the wall of its channel. Carbon dioxide (the coolant) flows through this gap. The heat that the gas absorbs in the core is transferred to the secondary water-and-steam circuit in a heat exchanger installed under the reactor pressure vessel. In this heat exchanger saturated steam is formed, which drives a turbine (see Fig. 2).

The gas-cooled reactor has the advantage over the liquid-cooled reactor that there are no corrosion problems. A disadvantage, however, is the high power consumption of the fans for circulating the gaseous coolant. This can be reduced by increasing the pressure of the gas, but the construction of the pressure vessel imposes limits in this respect.

The Calder Hall reactor has a thermal output of 180 MW and an electric-power output of 34.5 MW. The pressure vessel has an internal diameter of 11.3 m (37 ft.) and is 28 m (70 ft.) high, with a wall thickness of about 50 mm (2 in.). It encloses the core, which consists of 58,000 blocks of graphite stacked on a grid structure. The core contains 1696 vertical channels for the fuel elements. The latter consist of natural uranium clad ("canned") in magnesium sheaths provided with spiral fins. The gaseous coolant flows upwards along the elements in the channels. Each uranium rod is about 30 mm (1.18 in.) in diameter and 1 m (3 ft. $3\frac{1}{2}$ in.) long. Control of the reactor is effected by means of 160 rods which are inserted into the core from above. The carbon dioxide coolant is used at a pressure of 6.8 atm. (100 lb./in.^{2}). On entering the reactor the gas has a temperature of 140° C; its exit temperature is 345° C. After Calder Hall, ten more nuclear power stations operating on the same principle have been built in Britain. The gas pressure employed has progressively been increased to 28 atm. (about 400 lb./in.^{2}) and the gas exit temperature to 414° C. At the two latest power stations (Oldbury and Wylfa) the pressure vessels are of pre-stressed concrete, not steel, which was previously employed.

Advanced gas-cooled reactor: Derived from the Calder Hall type, this reactor represents the next stage of British nuclear-reactor development. A characteristic feature is that the coolant temperature has been raised from 414° C to over 600° C. It is thus possible to attain steam conditions and efficiencies comparable to those of conventional thermal power stations. The attainment of these high temperatures is made possible by the use of sintered UO_2 as the fuel and of corrosion-resisting steel as the cladding material. The Dungeness nuclear power station has an electrical output of 1200 MW, supplied by two reactors and two turbogenerating sets (Fig. 1, p. 33). Each reactor, with its core, cooling gas fan and heat exchanger, is enclosed in a cylindrical pressure vessel of prestressed concrete (Fig. 2). The core is similar in construction to that of the Calder Hall reactor. It is enclosed within a steel dome which is installed coaxially in the pressure vessel. This dome subdivides the interior of the vessel into a "cold gas" and a "hot gas" space. In the annular space between the dome and the cylindrical concrete shell of the pressure vessel are four steam-raising units and their respective coolant fans. The graphite moderator assembly is supported by a steel structure which is anchored into the concrete bottom of the pressure vessel.

(more)

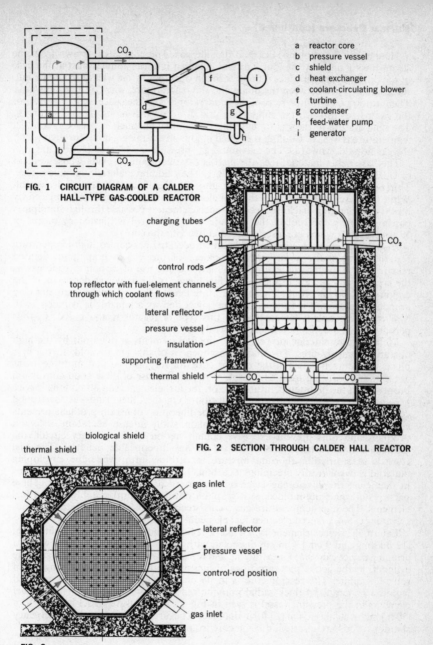

a reactor core
b pressure vessel
c shield
d heat exchanger
e coolant-circulating blower
f turbine
g condenser
h feed-water pump
i generator

FIG. 1 CIRCUIT DIAGRAM OF A CALDER HALL–TYPE GAS-COOLED REACTOR

charging tubes

CO_2

CO_2

control rods

top reflector with fuel-element channels through which coolant flows

lateral reflector

pressure vessel

insulation

supporting framework

thermal shield

CO_2

CO_2

FIG. 2 SECTION THROUGH CALDER HALL REACTOR

thermal shield

biological shield

gas inlet

lateral reflector

pressure vessel

control-rod position

gas inlet

FIG. 3

Here again the coolant is carbon dioxide gas. The cold gas is drawn from the steam-raising units by the four fans and is forced into the annular duct at the base of the dome. From this duct it flows through openings in the wall of the dome into the annular space between the dome and the reactor core, where it flows upwards. Then it flows downwards to cool the moderator. From the space under the reactor the gas passes through the supporting grid and upwards along the fuel elements. The temperature of the gas on entering the cooling channels is 318° C; its outlet temperature is 675° C. Cooling is effected in four steam-raising units, through which the gas flows downwards. The maximum gas pressure is 34.3 atm. (500 lb./in.^{2}).

The fissionable material (i.e., the fuel) is ceramic uranium oxide whose content of U 235 has been enriched to 1.47–1.76%. The cladding material is stainless steel. Thirty-six fuel rods enclosed within a graphite sheath constitute a fuel element. While the reactor is in service, the fuel elements can be removed and replaced by means of a charging machine. Access to each element is possible through standpipes passing through the roof of the pressure vessel. Control is effected by means of 53 control rods which are inserted from above into the core.

High-temperature reactor: The graphite-moderated gas-cooled high-temperature reactor is essentially a further development of the British graphite-moderated reactors. A characteristic feature of the latter is the use of carbon dioxide gas as the coolant and the use of metal-clad fuel elements. Their power density is in the region of 1 MW/m^3. In the high-temperature reactor the fuel elements are not clad in metal, the coolant is helium, and the power density attainable is as much as 10 MW/m^3. The coolant temperature in this type of reactor is above 700° C, thus permitting the use of modern turbogenerators.

Metal as the material for "canning" the fuel elements is ruled out by the high operating temperatures. The main consequences of this are twofold. In the first place, the neutron losses in the reactor core are very low, so that more neutrons are left for producing fresh fissionable material. Because of these good conversion properties of the high-temperature reactor, the fuel elements can attain long service lives and thus ensure efficient fuel utilization. On the other hand—as the second consequence—with unclad fuel elements the liberation of fission products presents a special problem. In order to obviate inadmissibly intense liberation of fission products and thus prevent excessive contamination of the primary circuit, the individual fuel particles are enveloped in an impermeable material. The fuel consists of a uranium-thorium mixture in carbide form, while the enveloping material is pyrolytically precipitated carbon. The carbide particles are 200 microns in diameter; the enveloping layer on each particle is 100 microns thick. These particles are pressure-molded with graphite powder to form suitably shaped fuel elements. The high-temperature gas-cooled reactor built at Peach Bottom, Pennsylvania (U.S.A.), has rod-shaped fuel elements (Fig. 1, p. 35).

Externally, a fuel element looks like a solid cylinder of graphite, about 3.7 m (12 ft.) long and 9 cm (3½ in.) in diameter, with a graphite mushroom head which a gripping device can seize for extracting and replacing the element. Inside the graphite cylinder, in the active part, are rings of fuel material loosely threaded on a solid rod of graphite. The reactor core is made of 804 such rods, which are enclosed within a 60 cm (2 ft.) thick radial graphite reflector. The arrangement of the fuel elements in the pressure vessel is seen in Fig. 2. The pressure vessel is about 9 m (30 ft.) high and about 4 m (13 ft.) in diameter. The 36 control rods and 19 emergency shutdown rods are inserted into the core from below.

(more)

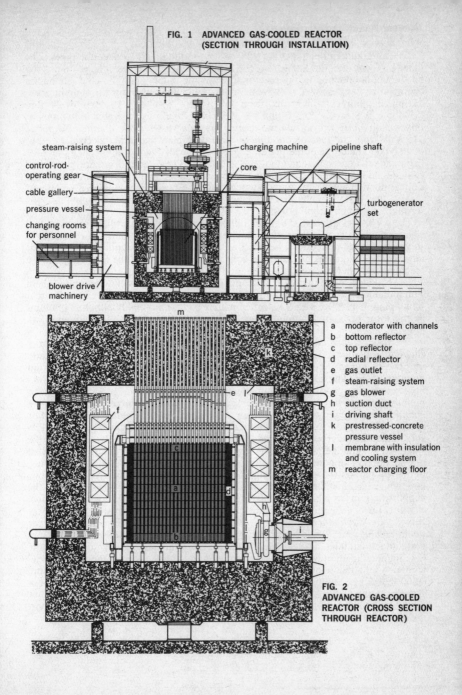

FIG. 1 ADVANCED GAS-COOLED REACTOR
(SECTION THROUGH INSTALLATION)

steam-raising system

control-rod-
operating gear

cable gallery

pressure vessel

changing rooms
for personnel

blower drive
machinery

charging machine

pipeline shaft

core

turbogenerator
set

m

k

e l

f

c

a

d

h

b

g

i

a moderator with channels
b bottom reflector
c top reflector
d radial reflector
e gas outlet
f steam-raising system
g gas blower
h suction duct
i driving shaft
k prestressed-concrete
   pressure vessel
l membrane with insulation
   and cooling system
m reactor charging floor

FIG. 2
ADVANCED GAS-COOLED
REACTOR (CROSS SECTION
THROUGH REACTOR)

33

The inlet and outlet pipes for the coolant (helium) are concentric pipes. The inner pipe conveys the hot gas to the heat exchanger, and the cold gas delivered by the fan is returned to the reactor through the outer pipe. The cold gas passes downwards along the wall of the reactor vessel and flows upwards through the core; during this journey it is heated from $350°$ C to $720°$ C at a pressure of 24 atm. (350 lb./in.^{2}). Steam at $538°$ C and 100 atm. (1470 lb./in.^{2}) is generated in two steam-raising units installed outside the reactor pressure vessel. For a thermal output of 115 MW from the reactor, the power-generating plant has a net electrical output of 40 MW. The cover of the pressure vessel is provided with a number of ports for the removal and replacement of fuel elements. These operations are performed by a special manipulating machine, which is introduced into the vessel through the central port while the reactor is switched off. This machine is able to reach and manipulate every fuel element and to replace it by a fresh element, which is inserted through one of the lateral ports.

The AVR nuclear reactor built at Jülich, Germany (Fig. 4), utilizes spherical fuel elements, which enable the fuel to be inserted and removed while the reactor is in service. Thus the reactor can be fed continuously with fuel; significant operational advantages are thus gained, and the conversion properties of the high-temperature reactor are further improved.

The fuel element of the AVR reactor (Fig. 3) consists of a hollow graphite sphere with a diameter of 6 cm ($2\frac{3}{8}$ in.) and a shell thickness of 1 cm (0.4 in.), closed with a screw plug. It is filled with a so-called matrix containing highly enriched uranium in the form of coated uranium–thorium carbide particles mixed with graphite powder.

The reactor core consists of about 100,000 of these spherical elements and is enclosed in a 50 cm (20 in.) thick graphite reflector. The spheres are fed pneumatically into the core from above through five tubes and are removed through a bottom discharge tube. The space occupied by the charge in the core is 3 m (9 ft. 10 in.) in diameter and 3 m high. From the graphite reflector four projections extend radially into the core over the entire height. Each of these projections contains a channel for the emergency rods, which are inserted into the core from below. For reasons of thermal insulation and radiation shielding, the graphite reflector is surrounded by a carbon-block blanket and a thermal shield. The reactor is of the single-vessel type: i.e., all the components of the primary circuit are accommodated in one vessel. The core is in the bottom part of the pressure vessel, above which, in the top part, is the steam-raising unit. The gaseous coolant is circulated by two fans which are installed in the so-called fan dome. The gas flows through the supporting grid on which the reactor is seated and into the mass of spherical fuel elements through channels in the bottom reflector. After passing through the spaces between the fuel elements, the gas flows through slots in the carbon bridge into the steam-raising unit. On leaving the latter, it flows along the wall of the pressure vessel to the fan inlet. The gas used as the coolant is helium at a pressure of 10 atm. (147 lb./in.^{2}). In the reactor core the temperature of the gas is raised from $175°$ C to $850°$ C. In the steam-raising unit, superheated steam at $505°$ C and 75 atm. (1100 lb./in.^{2}) is produced. The thermal output of the reactor is 46 MW and the electrical output is 15 MW.

(more)

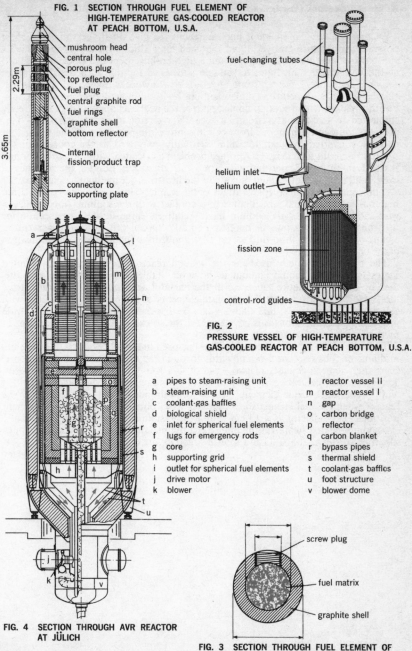

FIG. 1 SECTION THROUGH FUEL ELEMENT OF HIGH-TEMPERATURE GAS-COOLED REACTOR AT PEACH BOTTOM, U.S.A.

2.29m
3.65m

mushroom head
central hole
porous plug
top reflector
fuel plug
central graphite rod
fuel rings
graphite shell
bottom reflector

internal
fission-product trap

connector to
supporting plate

fuel-changing tubes

helium inlet
helium outlet

fission zone

control-rod guides

**FIG. 2
PRESSURE VESSEL OF HIGH-TEMPERATURE
GAS-COOLED REACTOR AT PEACH BOTTOM, U.S.A.**

a	pipes to steam-raising unit	l	reactor vessel II
b	steam-raising unit	m	reactor vessel I
c	coolant-gas baffles	n	gap
d	biological shield	o	carbon bridge
e	inlet for spherical fuel elements	p	reflector
f	lugs for emergency rods	q	carbon blanket
g	core	r	bypass pipes
h	supporting grid	s	thermal shield
i	outlet for spherical fuel elements	t	coolant-gas baffles
j	drive motor	u	foot structure
k	blower	v	blower dome

screw plug

fuel matrix

graphite shell

**FIG. 4 SECTION THROUGH AVR REACTOR
AT JÜLICH**

**FIG. 3 SECTION THROUGH FUEL ELEMENT OF
AVR REACTOR AT JÜLICH, GERMANY**

Sodium reactor: To enable a nuclear reactor to give off its heat at the highest possible temperature and yet avoid the need for a thick-walled pressure vessel, a substance with a low melting point and a high boiling point can efficiently be used as the heat-transfer medium. A suitable substance for the purpose is the metal sodium. There are, however, some unavoidable drawbacks associated with its use. Bombardment with neutrons makes sodium highly radioactive within the reactor. For this reason, with a sodium-cooled reactor the heat exchanger cannot be directly connected to the primary circuit; a secondary circuit must be interposed. This prevents radioactive material from coming into close proximity to the water that is to be converted to steam. Sodium is usually employed as the coolant for the secondary circuit also (Fig. 1). Another problem associated with the use of sodium is its reactivity with water and with atmospheric oxygen. Besides, the presence of even small amounts of sodium dioxide in the heat-transfer medium (coolant) causes a significant increase in corrosive attack of the stainless steel used as the construction material for those parts which come into contact with the sodium. In comparison with moderator materials, sodium has a relatively large initial cross section for neutrons; for this reason it is necessary to take special precautions to prevent an escape of the sodium from the reactor core and thus avoid a sudden intensification of the chain reaction.

The moderator chiefly used for the sodium reactors developed in the United States is graphite. Liquid sodium is, however, liable to penetrate into graphite, thereby causing a marked increase in the harmful absorption of neutrons in the moderator. For this reason the graphite elements of the moderator are clad with zirconium. Alternatively, this cladding can be dispensed with if zirconium hydride is used as the moderator (as is envisaged for the so-called KNK reactor developed by the Interatom undertaking).

The core of the KNK reactor (Fig. 2) is enclosed in a steel vessel 1.9 m (6 ft. 3 in.) in diameter. The coolant flows upwards through the 66 fuel elements of the core. Each fuel element consists of two rows of fuel rods in an annular arrangement. The cylindrical central cavity as well as the annular intermediate cavity is filled with zirconium hydride elements, which form the moderator and remain in the reactor when the fuel elements are changed. The control rods occupy fuel element positions. The reflector is a 13 cm (5 in.) thick stainless steel casing. Each of the two sodium-heated steam-raising units has a thermal output of 29 MW and supplies superheated steam at 510° C and 85 atm. (1250 lb./in.²).

(more)

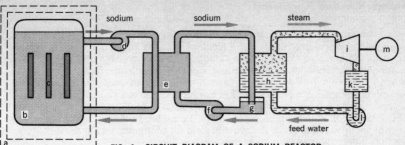

FIG. 1 CIRCUIT DIAGRAM OF A SODIUM REACTOR

a shield
b reactor tank
c fuel element
d primary circulation pump
e primary heat exchanger
f secondary circulation pump

g compensating tank
h steam-raising unit (boiler)
i turbine
k condenser
l feed-water pump
m generator

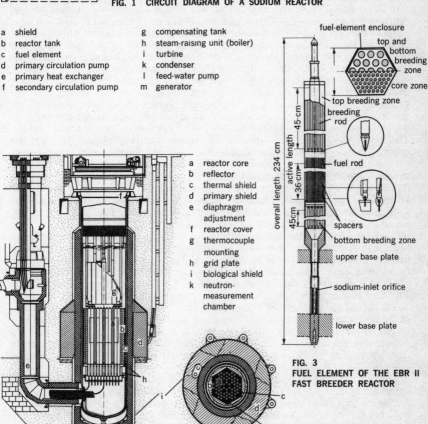

a reactor core
b reflector
c thermal shield
d primary shield
e diaphragm
   adjustment
f reactor cover
g thermocouple
   mounting
h grid plate
i biological shield
k neutron-
   measurement
   chamber

fuel-element enclosure
top and bottom breeding zone
core zone
top breeding zone
breeding rod
fuel rod
spacers
bottom breeding zone
upper base plate
sodium-inlet orifice
lower base plate

overall length 234 cm
active length
45 cm
36 cm
45 cm

**FIG. 3
FUEL ELEMENT OF THE EBR II
FAST BREEDER REACTOR**

reserve position (external)
● control rods
● fuel elements

**FIG. 2 KNK REACTOR (LONGITUDINAL SECTION
AND CROSS SECTION)**

Breeder reactors: In nuclear fission each neutron that causes fission releases more than one new neutron. It is this fact that not only enables a chain reaction to be sustained but also, under certain conditions, makes possible the "breeding" of fissionable material—i.e., the procedure whereby the chain reaction is made to produce more fissionable material than is consumed in generating energy. The isotopes uranium 238 and thorium 232 are suitable for this process. The breeding process based on uranium extends from the fission of U 235 to the transmutation of U 238 into fissionable plutonium 239 (U-Pu cycle). In this cycle the use of fast neutrons is advantageous, as the neutron yield for plutonium 239 in the fast energy range is higher than at thermal energy levels.

In the so-called *fast breeder reactor*, the moderator needed for slowing down the neutrons is dispensed with. This in turn results in a more compact core, which now contains only the requisite fuel elements and the coolant. In contrast with the U-Pu cycle, the thorium-uranium cycle (Th-U cycle) operates more favorably in the thermal energy range. In this process the fission of U 235 likewise results in the transmutation of Th 232 into fissionable U 233. In comparison with the fast breeder reactor, the so-called *thermal breeder reactor* is characterized by a lower breeding gain, a lower fissionable-material requirement, and the absence of special safety problems. A significant feature is that with the breeder reactor, as compared with the ordinary nuclear reactor, it is possible to attain fifty times better utilization of the available uranium resources when the U-Pu cycle is employed, and that with the Th-U cycle the large natural deposits of thorium can be exploited. Whereas the thermal breeder reactors were progressively evolved from existing types (pressurized- or boiling-water reactor with D_2O coolant; high-temperature reactor), the fast breeder reactors represent what is very largely an entirely new technical development. The first experimental reactors of this kind were built in the United States, Great Britain and the U.S.S.R. All use fuel elements in the form of rods and sodium as the coolant. A typical fuel element is illustrated in Fig. 3, p. 37. The core zone consists of a number of sheathing tubes containing the fuel (U 235), usually in metallic form. Above and below this zone are breeding zones in which the neutrons escaping from the reactor are trapped. These zones likewise comprise a number of sheathing tubes, here containing U 238 as the fertile material. The whole assembly is accommodated in a fuel-element container made of stainless steel.

Fig. 4 is a cutaway drawing of the EBR II, an American fast breeder reactor. The core has an approximately hexagonal cross section, with a height of about 36 cm (14 in.) and a volume of 73 dm^3 (4500 $in.^3$). The coolant is fed into the reactor at the bottom and flows upwards through the core and the surrounding "blanket." At full output, the inlet and outlet temperatures of the sodium are 371° and 473° C respectively. The reactor is enclosed in a tank into which the control rods are inserted from above. The whole assembly is surrounded laterally by a shield of boronated graphite.

The first large power reactor of this type is the Enrico Fermi reactor, which has a thermal output of 300 MW and an electrical output of 100 MW. Fig. 5 is a cutaway drawing of this American reactor. The core, surrounded by a blanket of uranium, is 77 cm (30 in.) in height and has a volume of 380 dm^3 (23,000 $in.^3$). The reactor inlet and outlet temperatures of the sodium are 288° C and 427° C respectively. The coolant flows upwards through the rod-shaped fuel elements. Above the reactor are arranged the control-rod drive equipment and the fuel-charging devices. The whole installation is surrounded by primary shield tanks, and the reactor itself is accommodated in a reactor vessel.

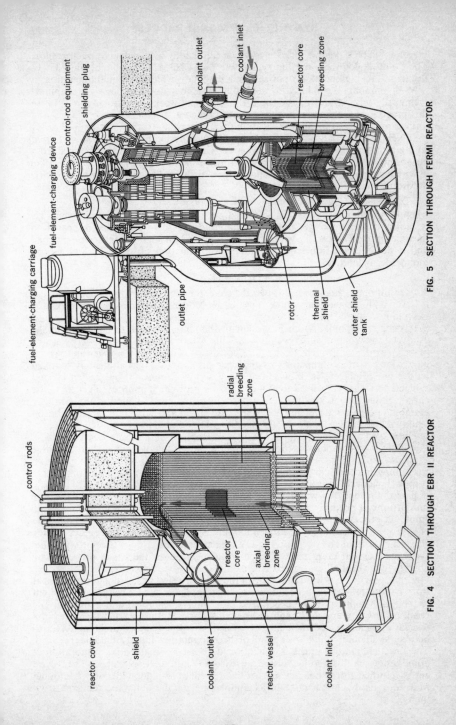

fuel-element-charging carriage

fuel-element-charging device

control-rod equipment

shielding plug

coolant outlet

coolant inlet

reactor core

breeding zone

outlet pipe

rotor

thermal shield

outer shield tank

FIG. 5 SECTION THROUGH FERMI REACTOR

control rods

radial breeding zone

reactor core

axial breeding zone

reactor cover

shield

coolant outlet

reactor vessel

coolant inlet

FIG. 4 SECTION THROUGH EBR II REACTOR

39

SYNTHETIC AMMONIA PROCESS

Ammonia (NH_3) is a colorless gas which can readily be liquefied by compression. The liquid is colorless and highly refractive and has a boiling point of $-33°$ C. Because of its high heat of vaporization, ammonia is widely used in refrigeration.

Most ammonia is produced by the Haber, or Haber-Bosch, process—first applied technically on a large scale by the BASF chemical works in Germany in 1913. In this process, hydrogen and nitrogen are combined at a temperature of $500°$ C and a pressure of 200 atm. (approx. 3000 lb./in.^{2}) according to the reaction $3 H_2 + N_2 \rightarrow 2 NH_3$, which takes place in contact with a catalyst consisting of iron stabilized with aluminum oxide and potassium oxide. The initial materials are obtained from air and water, the oxygen being removed by means of carbon (in the form of coke) as a reducing agent. At elevated temperature, air is converted in a gas producer (1) according to the following reaction:

$$4 N_2 + O_2 + 2 C \rightarrow 4 N_2 + 2 CO$$

whereby so-called producer gas is formed. As a result of this reaction, in which air is blown through the gas producer, the coke charge in the latter is heated to incandescence. Then water vapor (steam) is passed through the incandescent coke and is decomposed into so-called water gas, a mixture of hydrogen and carbon monoxide, as follows:

$$C + H_2O \rightarrow H_2 + CO$$

As a result of passing steam through the coke, the temperature in the gas producer is lowered. The cycle can then be repeated by blowing air again, to yield producer gas and heat up the coke, and so on.

Hydrogen sulphide, which is present in the coke, is liable to contaminate the catalyst and therefore has to be removed. This is done in the gas washer (2) by means of the alkazide method. The gas mixture, consisting of nitrogen, hydrogen and carbon monoxide (i.e., the combination of producer gas and water gas alternately supplied by the gas producer), is drawn from the gas storage tank (3) by the fan (4) and subjected to what is referred to as a conversion process (5) in order to get rid of the carbon monoxide, as this gas too is harmful to the catalyst used in the synthetic ammonia process. Conversion is effected with steam and catalytic contact with iron oxide and chromium oxide at $500°$ C: $CO + H_2O \rightarrow CO_2 + H_2$.

The cooled converted gas (N_2, H_2, CO_2 and traces of CO) is compressed to 25 atm. (370 lb./in.^{2}) by compressors (6) and passed to the carbon dioxide washer (7), where about 99% of the carbon dioxide is removed by means of water under pressure. Remaining traces of carbon monoxide and dioxide are removed with ammoniacal copper (I) chloride solution. The gas mixture, compressed to 100 atm. (1470 lb./in.^{2}), is passed through the washing tower (8, 9). With correct adjustment of the mix proportions of producer gas and water gas, the right gas mixture for the synthetic ammonia process can be obtained, consisting of three parts of nitrogen to one part of hydrogen. This nitrogen-hydrogen mixture is passed through the contact reactor (10), which contains a system of heat-exchanger tubes and contact tubes in which the synthesis reaction takes place. On entering the reactor, the gas is preheated by absorbing the heat of reaction evolved from the gas mixture that has already reacted. Once the reaction has been started with the aid of an electric heating device, no heat from an outside source is needed to sustain it: it produces its own heat. Eleven percent of the gas introduced into the synthetic process is transformed into ammonia. The resulting gas mixture (NH_3 and unreacted N_2 and H_2) is cooled with water in a tubular cooler (11) and then further cooled in a low-temperature cooler (12) to between $-20°$ and $-30°$ C. As a result of this refrigeration, liquid ammonia is formed. The low temperature in the last-mentioned cooler is obtained by evaporation of liquid ammonia diverted for this purpose from the production process. The recycle gas is returned to the contact reactor by a circulation pump (13), with additional fresh nitrogen-hydrogen mixture to compensate for the ammonia removed from the system.

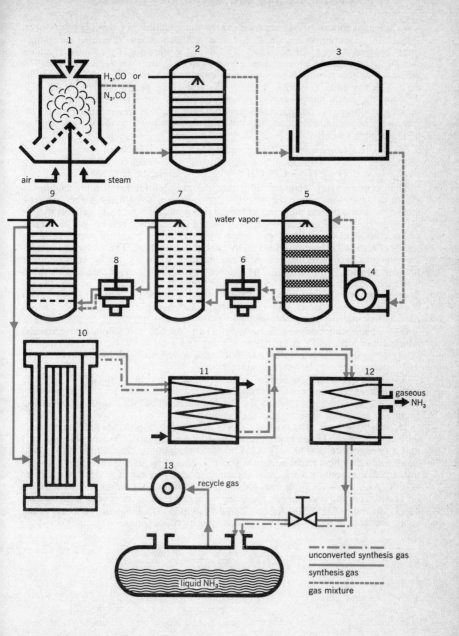

H₂, CO or
N₂, CO

air   steam

water vapor

gaseous
NH₃

recycle gas

liquid NH₃

unconverted synthesis gas

synthesis gas

gas mixture

SYNTHESIS GAS AND METHANOL SYNTHESIS

Various gas mixtures that are used for the manufacture of chemical compounds —mixtures of ammonia, methanol, ethylene, or hydrocarbons generally—are sometimes referred to as synthesis gases. The term "synthesis gas" is, however, more specifically applicable to a mixture of nitrogen (N_2) and hydrogen (H_2) which is employed in the synthetic production of ammonia ($N_2 + 3 H_2 \rightarrow 2 NH_3$). In the early 1960s over 80% of this mixture was still being produced from solid fuels (cf. synthetic ammonia process, p. 40). At the present time, however, processes for making synthesis gas from natural gas and hydrocarbons are gaining ground. Of increasing importance, too, are gas mixtures that are used for the manufacture of hydrocarbons and alcohols:

$$n\ CO + 2\ n\ H_2 \rightarrow (CH_2)_n + n\ H_2O \qquad \text{(n = number of}$$
$$\qquad\qquad\qquad\qquad\qquad\qquad\text{molecules or}$$
$$CO + 2\ H_2 \rightarrow CH_3OH \qquad\qquad \text{molecule groups)}$$

The accompanying diagrams illustrate the possibilities for the production of synthesis gases. As contrasted with the processing of solid fuels in the synthetic ammonia process, these diagrams show the processing of liquid and gaseous hydrocarbons. Crude oil, light gasoline (petrol), liquid gas (butane), refinery gas or natural gas is converted to synthesis gas either with the aid of platinum-nickel catalysts (no combustion being employed) or by partial combustion. The preheated hydrocarbons and preheated oxygen enter the reactor. Synthesis gas for the ammonia process requires additional air. The hot decomposition gases are passed through a waste-heat boiler. The steam formed in the boiler may be supplied to the reactor. Soot sludge is removed in a pressure-washing installation. The synthesis gas is cooled in a spray cooler and then undergoes cleaning or aftertreatment. More particularly, the products obtained by gasification of solid fuels contain tar, dust and organic compounds (such as naphthalene, phenol and sulphur compounds) which are objectionable in further processing and must therefore be removed. The liquid or gaseous fuels employed usually have a very low sulphur content or are desulphurized before the reaction takes place.

Another type of aftertreatment is re-forming: the transformation of methane with water vapor into carbon monoxide and hydrogen. If the ratio of carbon monoxide to hydrogen is not suited for the synthesis to be performed, a conversion process is applied (cf. town-gas detoxication, p. 44).

Methanol synthesis: The synthesis gas in the correct proportions reacts in contact with chromium oxide and zinc oxide catalysts at a pressure of 300 atm. (4400 lb./in.^{2}) and a temperature of $330°-370°$ C. The gas containing methanol (methyl alcohol) flows through a heat exchanger in which it is cooled by so-called recycle gas—i.e., unreacted synthesis gas which is returned to the process. Then follows further cooling with water, causing methanol to condense. In a separator the methanol is separated from the residual gas, the latter being returned to the reactor (recycle gas). The crude methanol is distilled (methanol boils at $64.5°$ C) in order to free it of the small amounts of water and by-products that it still contains.

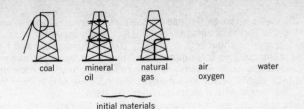

coal   mineral   natural   air   water
        oil   gas   oxygen

initial materials

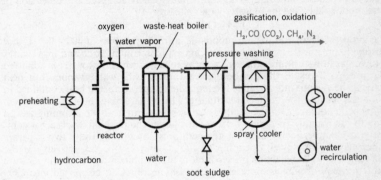

gasification, oxidation

$H_2, CO (CO_2), CH_4, N_2$

oxygen   waste-heat boiler

water vapor

pressure washing

preheating

cooler

reactor

spray cooler

hydrocarbon

water

water recirculation

soot sludge

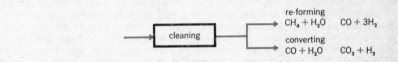

cleaning

re-forming
$CH_4 + H_2O \quad CO + 3H_2$

converting
$CO + H_2O \quad CO_2 + H_2$

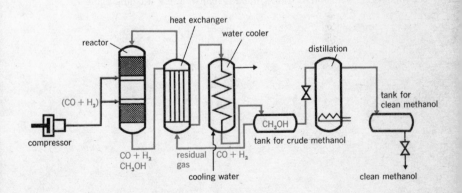

heat exchanger

water cooler

distillation

reactor

$(CO + H_2)$

tank for clean methanol

compressor

CH_3OH

$CO + H_2$
CH_3OH

residual gas

$CO + H_2$

tank for crude methanol

cooling water

clean methanol

43

TOWN-GAS DETOXICATION (Conversion of Carbon Monoxide)

"Town gas" is the name given to the gas with which many Americans cook. Town gas that is produced by the gasification of solid fuels contains—after removal of tar, ammonia and benzene—about 50% hydrogen, 20 to 30% methane and 6 to 17% carbon monoxide (percentages by volume). In addition, the gas contains some carbon dioxide, nitrogen and other impurities. Detoxication consists in removing the carbon monoxide, a highly poisonous colorless and odorless gas. It does not sustain combustion, but burns with a characteristic blue flame.

In order to obtain a nontoxic gas, it is necessary to reduce the carbon monoxide content to between 1 and 1.5%. This can be fairly easily achieved by conversion of carbon monoxide by means of water vapor (steam):

$$CO + H_2O \rightarrow CO_2 + H_2$$

In this process, steam and carbon monoxide, at a temperature of 400°–480° C and atmospheric pressure, are passed over a catalyst (chromium oxide and iron oxide). The town gas flows upwards through a spray tower in which hot water is admitted at the top. As a result, the gas becomes saturated with water vapor. In a heat exchanger the gas mixture acquires the requisite temperature, the heat contained in the hot gas leaving the contact reactor (in which the above-mentioned reaction takes place in contact with the catalyst) being utilized to heat the incoming gas on its way to the reactor. To obtain better control over the reaction, the reactor comprises two chambers with different temperatures; or alternatively, two separate reactors may be employed, operating at 400° C and 480° C respectively. At the higher temperature the reaction proceeds more rapidly, but at the lower temperature the reaction equilibrium is more favorable: i.e., more CO is converted into CO_2 than at the higher temperature. In the two-stage process envisaged here, the greater part of the CO is rapidly converted at the higher temperature, and the reaction is then "finalized" at the lower temperature. The hot gases are first cooled in the heat exchanger (as already mentioned) and then further cooled in an injection condenser. The conversion process is associated with a final cleaning treatment: virtually all ammonia, hydrocyanic acid and nitric oxide are removed as well. To the detoxicated town gas obtained in this way certain odorous components may then be added in order to give it a characteristic smell, so that a leakage or escape of gas can be detected speedily. Conversion is also applied in the synthetic ammonia process (cf. p. 40) to get rid of the carbon monoxide, which would otherwise adversely affect the process. The carbon dioxide formed is removed in a washing plant.

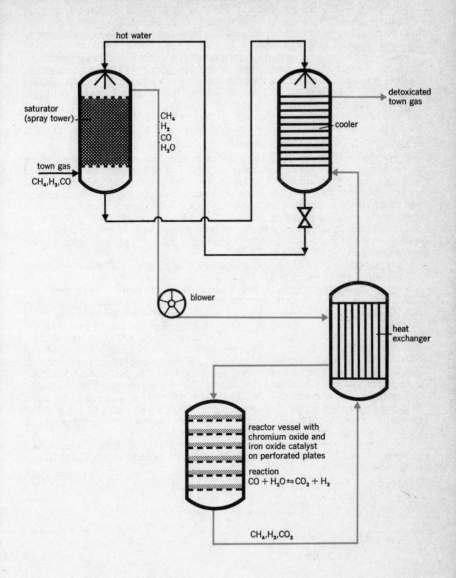

hot water

saturator
(spray tower)

town gas
CH_4, H_2, CO

CH_4
H_2
CO
H_2O

detoxicated
town gas

cooler

blower

heat
exchanger

reactor vessel with
chromium oxide and
iron oxide catalyst
on perforated plates

reaction
$CO + H_2O \rightleftharpoons CO_2 + H_2$

CH_4, H_2, CO_2

PHOSPHORUS

Phosphorus occurs in nature only in the form of the salts of phosphoric acid. From these it is obtained by reduction. There are three so-called allotropic forms of elementary phosphorus: white, red and black. Of these, black phosphorus is the form that is most stable at room temperature; it is obtained from the white form by the application of high pressures.

Black phosphorus is mainly of scientific interest, whereas the other two allotropic forms are technically important. White phosphorus melts at 44.1° C and, when finely divided, reacts with atmospheric oxygen even at room temperature. Red phosphorus is obtained from white phosphorus by heating the latter in a closed container (exclusion of air).

Phosphorus is prepared by heating calcium phosphate with carbon and silica (SiO_2) in an electric furnace. The reaction is represented by the following equation:

$$2\ Ca_3(PO_4)_2 + 6\ SiO_2 + 10\ C \rightarrow 6\ CaSiO_3 + P_4 + 10\ CO$$

The phosphate is fed to the furnace in lump form. To make them suitable for processing in this way, finely granular phosphates first have to be agglomerated by pelletizing. Agents used for binding the particles together into pellets are soda water glass (sodium silicate), Cottrell dust (from electrostatic precipitation processes) and other admixtures, which are intimately mixed with the phosphate grains in a screw mixer and then pelletized in a revolving pan. The pellets are transformed to firm, hard balls (nodules) by sintering at high temperatures in rotary kilns or on special sintering grates. The nodules, mixed with coke and silica pebbles, are fed to the furnace.

The three-phase electric furnace consists of a steel tank of which the bottom part is lined with hard-burned carbon blocks and the top part with fireclay bricks. At the bottom are two tapholes for tapping the ferrophosphorus and the slag respectively. The furnace cover is provided with openings for the three electrodes, the feed pipe and the gas outlet. The electrodes are made of carbon and are fed from above at a rate sufficient to compensate for loss by burning. The gas discharged from the furnace, consisting of phosphorus and CO, is passed through Cottrell-type electrostatic dust precipitators (dust filters) in which the dust in the gas is trapped and collected. These precipitators are heated to prevent condensation of phosphorus inside them. The dust is returned to the sintering plant. The exit gases, which have a temperature of 250°–350° C, are passed to Ströder washers in which the phosphorus is condensed. The white phosphorus obtained in this process is stored under water to prevent the spontaneous combustion that results from contact with air. The CO gas that is discharged from the washers is utilized for heating the sintering plant and steam boilers or is burned at the top of high flare stacks.

Phosphorus is used in the manufacture of detergents. The plastics industry uses phosphorus-based plasticizers. Red phosphorus is employed in the friction striking surfaces on matchboxes.

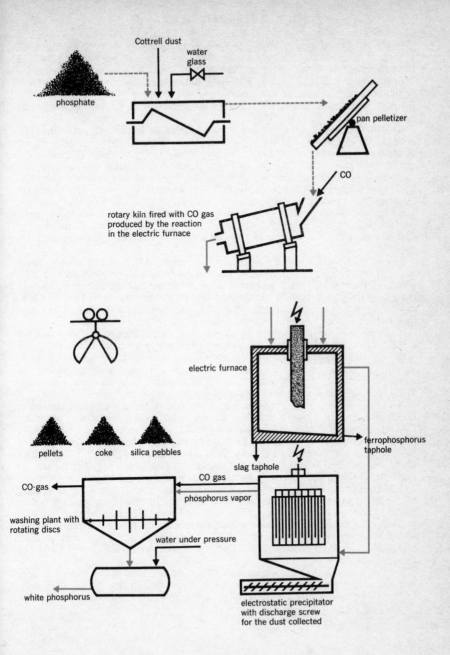

Cottrell dust

water glass

phosphate

pan pelletizer

CO

rotary kiln fired with CO gas
produced by the reaction
in the electric furnace

electric furnace

pellets   coke   silica pebbles

ferrophosphorus
taphole

slag taphole

CO gas

CO-gas

phosphorus vapor

washing plant with
rotating discs

water under pressure

white phosphorus

electrostatic precipitator
with discharge screw
for the dust collected

SULPHURIC ACID

In the pure state, sulphuric acid is a clear, colorless, oily liquid. One-hundred-percent H_2SO_4 has its melting point at $10°$ C; when heated, it gives off SO_3 until the concentration of the acid has fallen to 98.5%, and it then boils at a constant temperature of $338°$ C. Considerable evolution of heat occurs when concentrated sulphuric acid is diluted with water. Substantial amounts of SO_3 can dissolve in the acid. The resulting solution is known commercially as fuming sulphuric acid (oleum).

Sulphuric acid does not occur as a free acid in nature. It is found only in the form of its salts (sulphates): gypsum ($CaSO_4 \cdot 2H_2O$), Epsom salts ($MgSO_4 \cdot 7H_2O$), barite ($BaSO_4$) and Glauber's salt ($Na_2SO_4 \cdot 1OH_2O$). Up to about the eighteenth century, sulphuric acid was made by heating alum (aluminum potassium sulphate) or iron vitriol (hydrous ferrous sulphate). This method was superseded by the burning of natural sulphur with saltpeter, which eventually evolved into the so-called lead-chamber process, first introduced in the early part of the nineteenth century and still used. At the present time, however, most sulphuric acid is produced by the contact process, which has the advantage that the acid can be obtained in any desired concentration, whereas the highest attainable concentration with the lead-chamber process is 78%.

The contact process is as follows. Sulphur dioxide (SO_2) is obtained by roasting iron pyrites (FeS_2) in a rotary kiln, shelved roasting kiln or fluidized bed kiln. Which of these kiln types is employed depends on the particle size and nature of the pyrites to be processed. When the gases from the roasting process have cooled—in gas ducts, by radiation of heat—from $1000°$ C to about $400°$–$500°$ C, the dust they contain is removed in electrostatic precipitators ("electric filters").

Next, the SO_2 gas is passed through a washing tower, where constituents that are present in vapor form (mainly compounds of arsenic, selenium and chlorine) are removed with sulphuric acid serving as the washing liquid. Remaining traces of impurities present as very fine suspended droplets ("fog") are removed in an irrigated electrostatic precipitator ("wet" precipitator). Then the gas is dried by being brought into contact with concentrated (98%) sulphuric acid.

A blower draws in the cold dried SO_2 gas and delivers it into the converter, which is a tank or tower in which a suitable catalyst—e.g., vanadium pentoxide (V_2O_5)—is placed in layers on shelves or arranged in some other appropriate manner to ensure thorough contact with the gas. The reaction whereby SO_2 is converted to SO_3 by oxidation ($2SO_2 + O_2 \rightarrow 2\ SO_3$) takes place at $430°$ to $550°$ C. A heat exchanger installed before the converter serves to cool the gas discharged from the converter and at the same time preheats the incoming gas flowing to the converter.

The hot SO_3 gas that comes out of the converter is cooled with air to around $120°$–$150°$ C in tubular coolers and is then passed to absorbers, which are steel towers lined with ceramic materials and containing specially shaped packing bodies called Raschig rings. These towers are rather similar to the washing towers. The SO_3 is absorbed by concentrated (98%) sulphuric acid. Water is then added to obtain the desired concentration. The heat evolved by the dilution is dissipated by spray coolers.

Sulphuric acid is the most important of all acids. It is used in the manufacture of superphosphate and ammonium sulphate fertilizers, petroleum refining, the manufacture of explosives and synthetic fibers, and a host of other industrial processes.

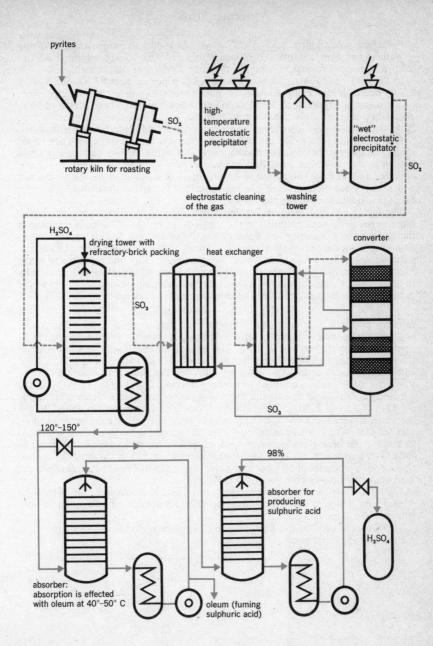

pyrites

SO₂

rotary kiln for roasting

high-temperature electrostatic precipitator

electrostatic cleaning of the gas

washing tower

"wet" electrostatic precipitator

SO₂

H₂SO₄

drying tower with refractory-brick packing

heat exchanger

converter

SO₂

SO₃

120°–150°

98%

absorber for producing sulphuric acid

H₂SO₄

absorber: absorption is effected with oleum at 40°–50° C

oleum (fuming sulphuric acid)

For a long time all nitric acid (HNO_3) used to be produced from nitrates occurring in nature. For example, nitric acid can be obtained by adding sulphuric acid to saltpeter (potassium nitrate), whereby the latter undergoes decomposition.

In the pure state the acid is a colorless liquid which boils at 87° C. The boiling point rises with increasing dilution. As pure nitric acid will always undergo some decomposition when left to stand, especially when exposed to the action of light, acid with a concentration exceeding 90% almost invariably contains some dissolved NO_2. This decomposition process can be stopped by diluting the acid with water.

The various processes for the commercial manufacture of nitric acid are based on any of three principles: the decomposition of nitrates (more particularly Chile saltpeter) with sulphuric acid (this method is now virtually obsolete); the direct synthesis of NO from nitrogen and oxygen in an electric arc, followed by the last two reactions of the ammonia process (described in further detail below); or the catalytic oxidation of ammonia.

The last-mentioned method (ammonia process) is now most widely employed, more particularly with platinum as the catalyst. (In certain variants of the process the catalyst may, however, be Fe_2O_3, Mn_2O_3 or Bi_2O_3). In this process liquid ammonia is vaporized in an evaporator and is mixed with air. The gas mixture (10% ammonia and 90% air) makes its way through a filter and a preheater to the contact reactor, in which it passes over platinum gauze (the catalyst) heated initially to about 900° C. Heat liberated in the reaction maintains the temperature of the catalyst. In the reactor about 90% of the ammonia is oxidized to nitric oxide:

$$4\,NH_3 + 5\,O_2 \rightarrow 4\,NO + 6\,H_2O$$

Directly after the platinum gauze are filtering agents which trap the unstable platinum compounds present in the gas discharged from the reactor: the platinum precipitated in this way is recovered. This gas, of which $97\frac{1}{2}$% is NO, is cooled in the gas cooler. On leaving the cooler, it is passed through absorption towers, filled with rings made of ceramic material, in which two reactions take place. In the first tower the nitric oxide is oxidized to nitrogen dioxide: $2\,NO + O_2 \rightarrow 2\,NO_2$. In the following towers (four in all) the dioxide reacts with water to form nitric acid: $3\,NO_2 + H_2O \rightarrow 2\,HNO_3$. The requisite water is added in the last tower. In the preceding towers the NO_2 is brought into contact not with water, but with nitric acid solution, which is circulated by pumps and passed through cooling apparatus to remove the heat evolved in the reaction, as low temperatures are favorable to the absorption reactions. In a degasifying tower air is blown through the acid to remove such amounts of NO gas as are still present in it. The exhaust gas from the final absorption tower contains 0.3 to 0.4% NO.

The installation described here produces nitric acid in a concentration of between 40 and 60%. A higher concentration (up to 99.5%) is obtainable by distillation with concentrated sulphuric acid.

Nitric acid is used in the manufacture of fertilizers, explosives, lacquers, dyes, plastics and synthetic fibers.

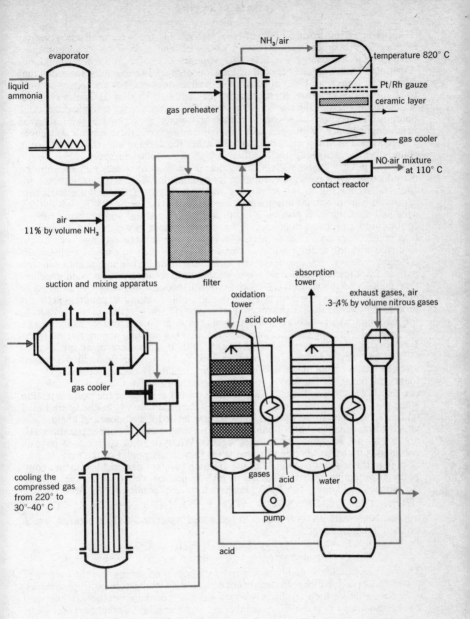

evaporator

liquid ammonia

gas preheater

NH₃/air

temperature 820° C
Pt/Rh gauze
ceramic layer
gas cooler
NO-air mixture at 110° C
contact reactor

air 11% by volume NH₃

suction and mixing apparatus

filter

gas cooler

cooling the compressed gas from 220° to 30°-40° C

oxidation tower

acid cooler

absorption tower

exhaust gases, air .3-4% by volume nitrous gases

gases   acid   water

pump

acid

FOAM PLASTICS

A characteristic feature of foam-type materials is the structural configuration of the cells. Absorbent cotton (cotton wool), felt and glass wool, for example, do not belong to this category of materials; sponges and cork, on the other hand, do. A distinction can be made between true and false foams. In a true foam the individual cells are not mere relatively thick-walled cavities or pores, but are separated only by thin partitions and are interdependent for their stability (Fig. 1). The mechanical strength is highest in the case of foams with closed (nonintercommunicating) cells. Since no convection is possible in such materials, they possess good thermal insulating capacity. With open (intercommunicating) cells the mechanical strength and thermal insulation are lower; on the other hand, these materials have a high sound-absorbing capacity and are therefore good acoustic insulators.

Artificial foam materials, including more particularly foam plastics, can be manufactured by three different methods: by churning (Fig. 2), by expansion with chemical agents (Fig. 4) and by physical methods (Fig. 3). The initial materials that can suitably be processed into foams include polyvinyl chloride (PVC), polystyrene, urea and formaldehyde condensation products, and natural and synthetic rubber. In the churning process of producing foam rubber, latex to which fillers, vulcanization accelerators and foaming agents (surface-active substances) have been added is stirred with air to form a foam, which sets and is then vulcanized with hot air (Dunlop process). Urea-formaldehyde foams are made by foaming a soap solution with an incompletely condensed water-soluble resin solution and air in an impeller-type high-speed mixer. Further condensation is brought about by the addition of acid. In the process based on physical methods, the foaming (expanding) action is produced by gases such as nitrogen, carbon dioxide or pentane. Gas dissolved in the material under pressure is liberated from the solution and thus forms bubbles in the material when the pressure is reduced; this is the foaming action. For example, PVC pastes are processed with carbon dioxide at a pressure of about 20 atm. (300 lb./in.^{2}) and a temperature between $-5°$ and $0°$ C. The fluid mass is passed into the heating zone of the installation. Here the dissolved CO_2 escapes and thus foams the material. The foam sets at a temperature of $150°$ C and is solidified by cooling. Polystyrene is foamed with pentane, which is added at the polymerization stage (e.g., in the manufacture of Styropor). Chemical foaming methods are based on the fact that certain substances will, on being heated, decompose and liberate gas, which forms small bubbles (foam cells). Azo compounds, N-nitroso compounds and azides are employed as foaming agents. What all these compounds have in common is that they liberate nitrogen when they decompose. For the manufacture of polyurethane foam plastics (e.g., the German product named Moltopren), compounds containing hydroxyl groups of high molecular weight are mixed with di-isocyanates and water. The foam plastic is formed according to the equation

$$HO \cdot R_1 \cdot OH + OCN \cdot R_2 \cdot NCO \rightarrow \ldots CO_2 \cdot R_1 CO_2 \cdot NH \cdot R_2 NH \cdot CO_2 \cdot R_1 \cdot CO_2 \ldots$$

Surplus isocyanate groups react with the added water and CO_2 is evolved, which acts as a foaming agent:

$$\underset{\text{isocyanate}}{R-NCO} + H_2O \rightarrow \underset{\text{amine}}{R-NH_2} + CO_2$$

The reaction mixture is cast in molds in which both the foaming and the hardening process take place. Blocks of foam are cut up into slabs or sheets by cutting machines.

In the building industry foam plastics have achieved importance as heat and sound insulating materials. They are also used for a number of other purposes—e.g., as paddings, packing materials, materials for the manufacture of sponges, and bath mats.

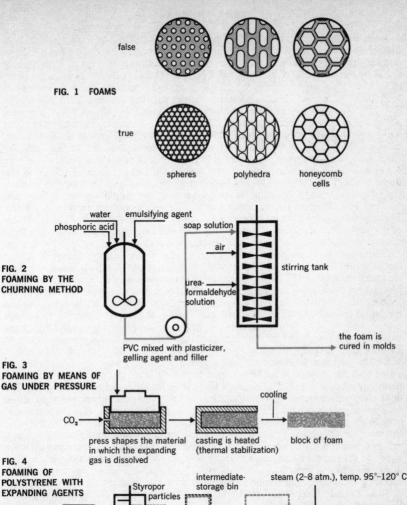

false

FIG. 1 FOAMS

true

spheres polyhedra honeycomb cells

**FIG. 2
FOAMING BY THE
CHURNING METHOD**

phosphoric acid

water emulsifying agent

soap solution

air

urea-
formaldehyde
solution

stirring tank

the foam is
cured in molds

PVC mixed with plasticizer,
gelling agent and filler

**FIG. 3
FOAMING BY MEANS OF
GAS UNDER PRESSURE**

CO_2

press shapes the material
in which the expanding
gas is dissolved

casting is heated
(thermal stabilization)

cooling

block of foam

**FIG. 4
FOAMING OF
POLYSTYRENE WITH
EXPANDING AGENTS**

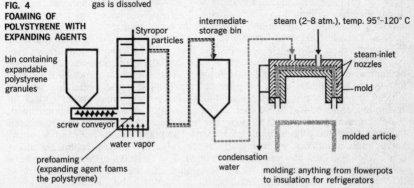

bin containing
expandable
polystyrene
granules

Styropor
particles

intermediate-
storage bin

steam (2–8 atm.), temp. 95°–120° C

steam-inlet
nozzles

mold

screw conveyor

water vapor

prefoaming
(expanding agent foams
the polystyrene)

condensation
water

molded article

molding: anything from flowerpots
to insulation for refrigerators

53

Tank trucks (road tank vehicles) are used for supplying gasoline (petrol) and diesel fuel to roadside filling stations and fuel oil to schools, hospitals, office buildings and residential buildings. Fig. 1 shows three types of tank vehicles most frequently found on the roads: (a) truck (lorry) with removable tank, capacity approx. 2.5–6 m^3 (90–210 ft.^{3}); (b) tank vehicle, capacity approx. 6–22 m^3 (210–780 ft.^{3}), sometimes with tank trailer; (c) articulated tank vehicle, capacity approx. 22–35 m^3 (780–1240 ft.^{3}). Removable tanks are usually not subdivided, but larger tanks comprise a number of compartments (up to six) for reasons of stability. Each compartment has its own bottom valve and is connected by pipes to the equipment and instrument cubicle. The latter can be dispensed with if the entire contents of the tank are delivered to one customer. The tank, or the individual chambers, are sealed, and the driver is issued a certificate showing the quantity of fuel put into his vehicle at the filling point. Fig. 2 shows the arrangement of pipelines and fittings, the volume-metering equipment (operating on the oval-gear or the oscillating-piston principle) and the hose drum in the equipment cubicle of a two-compartment tank vehicle. The various possible connection combinations are listed in a table. Since the contents of the tank are a commercial commodity, the metering equipment has to conform to certain standards of accuracy and reliability laid down by the public authorities. An important requirement is that the liquid must be free of bubbles on discharge from the tank. Since a volume meter will measure gaseous as well as liquid volumes, it is necessary to install a gas separator or, in some cases, an arresting device to prevent gas from being carried along into the meter. The gas separator is connected to the tank or compartment by a pipe through which flows a certain proportion of the liquid together with the separated gas. The liquid being metered has to be kept under constant observation through a sight glass. If bubbles appear, the flow must be throttled down or cut off. On the other hand, the gas-arresting device functions automatically (Fig. 3).

Before metering commences, the metering apparatus must be filled with the liquid and the air in it must be removed by actuation of the air-release valve. The metering apparatus is full of liquid when the latter has reached the level (1). The float lever and the control-valve push rod are now not in contact: the connection between supply pipe and return-flow pipe is cut off. The pointers on the meter dial may be set to zero, if necessary, and then the three-way valve can be opened. When the pump lever is moved to and fro, oil is delivered from the pump casing through valves 1 and 2 into the supply pipe and to the control piston of the three-way valve until the oil pressure compels the latter to open. Valve 1 remains closed, valve 2 open; metering commences. The gas that flows into the arresting device during metering—and more particularly on changing over from one tank compartment to another—slowly forces the liquid level down, so that the float likewise descends. When the liquid reaches the level (3), the lever attached to the float comes into contact with the push rod of the control valve, and the latter is pushed up and thus opened when the liquid level falls still lower. At the level (4) the control valve is fully open, and the oil under pressure can now flow into the return-flow pipe, in which there is no previously existing pressure. The control valve is relieved, and the spring-loaded three-way valve closes. The metering operation is thus interrupted. Valve 2 is now also closed. If metering has to be stopped earlier, this can be effected by moving the pump lever briefly backwards. The pump piston thus ascends, opens valve 3 and equalizes the pressure. To resume the metering operation, the initial procedure described above must be repeated. The metering apparatus on a tank vehicle may also be provided with a printing mechanism which issues a printed delivery note, stating the metered quantity delivered to the customer.

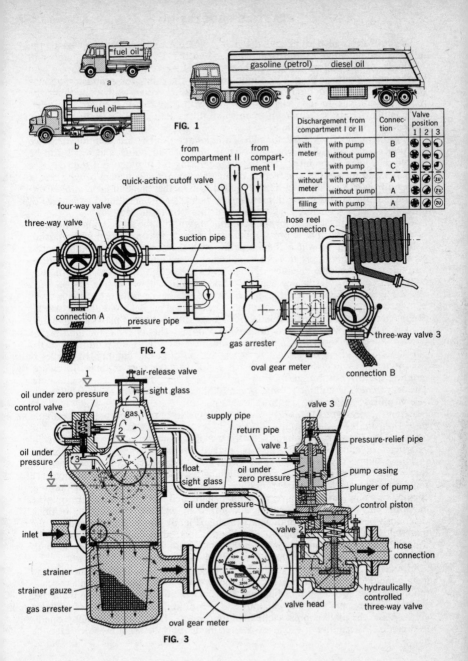

FIG. 1

a — fuel oil
b — fuel oil
c — gasoline (petrol)   diesel oil

Dischargement from compartment I or II		Connection	Valve position 1	2	3
with meter	with pump	B			
	without pump	B			
	with pump	C			
without meter	with pump	A			
	without pump	A			
filling	with pump	A			

FIG. 2

from compartment II
from compartment I
quick-action cutoff valve
four-way valve
three-way valve
suction pipe
connection A
pressure pipe
gas arrester
oval gear meter
hose reel connection C
three-way valve 3
connection B

FIG. 3

air-release valve
sight glass
gas
oil under zero pressure
control valve
oil under pressure
float
sight glass
supply pipe
return pipe
valve 1
oil under zero pressure
oil under pressure
valve 3
pressure-relief pipe
pump casing
plunger of pump
control piston
valve 2
hose connection
hydraulically controlled three-way valve
valve head
oval gear meter
gas arrester
strainer gauze
strainer
inlet

The properties of plastics and the many different requirements applied to the finished products made from them have led to the development of a number of methods for shaping and molding these materials. From the manufacturers who synthetically produce plastics for industrial use the fabricating industry obtains the specified initial materials, i.e., the appropriate polymers with or without the requisite additives. In the latter case the user will have to add auxiliary materials such as plasticizers, stabilizers, pigments and fillers. Batch mixing of the powdered ingredients is performed in agitators or mixing drums. Alternatively, kneaders or mixing rolls (Fig. 1) are used for plastifiable materials. The last-mentioned device comprises a pair of rollers which revolve in opposite directions and which can be heated or cooled as required. The material entering the gap between the rollers is squeezed and mixed. On completion of this treatment the so-called rough sheet is stripped from the rollers (Fig. 1) and passed to a further stage of processing. Continuous mixing is performed in extruders, which offer the additional advantage of filtering the plastics before they undergo further processing (Fig. 4).

The shaping of plastic articles and components without the application of pressure is effected by casting. The simplest method of shaping in conjunction with pressure is by molding (Fig. 2), which is suitable for both thermosetting and thermoplastic compositions. (Thermoplastics can be softened by the application of heat; thermosetting plastics undergo chemical change under the action of heat and are thereby converted to infusible masses which cannot be softened by subsequent heating.) The material is fed into the mold in the form of powder or pellets. For the process known as compression molding the mold is heated; for impact molding the material itself is preheated. Paper or textile fabrics for making laminated plastics (laminates) are impregnated with thermosetting compositions; this is done on multiplaten presses. Such presses are also used for the manufacture of fiberboard (phenolic plastic with wood chips as filler). Another method of producing molded articles is by injection molding (Fig. 5), which has the advantage over ordinary (pressure) molding that preheating, plasticizing and shaping are done by the same machine. The only materials suitable for injection molding are thermoplastics of high fluidity. The granules are introduced through a hopper into the cylinder, in which they are heated—by means of a heating jacket—to above their softening point. A moving piston plasticizes the material and forces it through a nozzle into the mold. The plasticizing action can be enhanced by the use of a screw instead of a piston (Fig. 6).

Articles or components can also be shaped by the machining of semifinished products—films, sheets, rods or tubes. Machining is more particularly employed in cases where the articles are of complex shape or where only a small number are required. Whereas thermosetting plastics can be shaped only by machining (milling, turning, cutting, drilling) once they have hardened, semifinished thermoplastic materials can be shaped by heating and joined by welding. Hot shaping of thick sheets can be effected by bending or drawing (Fig. 3). In the drawing process the material to be shaped is gripped, heated and deformed to the desired shape. If the wall thickness must remain constant, the sheet must be resiliently gripped; with so-called stretch forming a reduction in wall thickness occurs.

In recent years shaping by the vacuum process has gained importance. In the female-mold (or negative-mold) method, the heated plastic sheet is laid on a concave mold and subjected to further heating. Air is extracted through holes in the mold, so that the sheet is drawn (by suction) into the mold. For the molding of complex components the plate is prestretched before the actual "negative"-molding operation begins. Alternatively, a convex master model may be used, in which case the

(more)

feeding mixing stripping

FIG. 1 MIXING ROLLERS

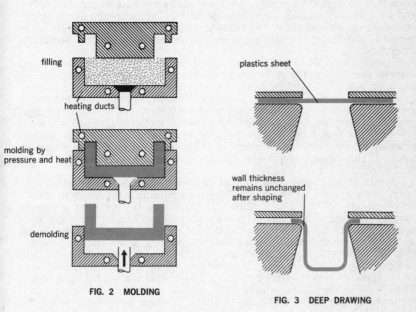

filling

heating ducts

molding by pressure and heat

demolding

FIG. 2 MOLDING

plastics sheet

wall thickness remains unchanged after shaping

FIG. 3 DEEP DRAWING

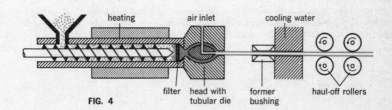

heating   air inlet   cooling water

filter   head with tubular die   former bushing   haul-off rollers

FIG. 4

process is known as the male-mold (or positive-mold) method. The preheated sheet is placed over the master model and preformed. When the air is evacuated, the desired shape is obtained. These molding techniques are schematically illustrated in Figs. 8 and 9.

"Endless" products such as sections, sheet, strip and film are produced by extruders (Figs. 4 and 6). Extrusion consists in forcing a plastic material through a suitably shaped die to produce the desired cross-section shape. The extruding force may be exerted by a piston or ram (ram extrusion) or by a rotating screw (screw extrusion) which operates within a cylinder in which the material is heated and plasticized and from which it is then extruded through the die in a continuous flow. Different kinds of die are used to produce different products—e.g., blown film (formed by blow head for blown extrusions), sheet and strip (slot dies) and hollow and solid sections (circular dies). Wires and cables can be sheathed with plastics extruded from oblique heads. The extruded material is cooled and is taken off by means of suitable devices which are so designed as to prevent any subsequent deformation.

For the manufacture of large quantities of film or thin sheet, the so-called sheeting calender is employed (Fig. 7). The rough sheet from the two-roll mill is fed into the gap of the calender, a machine comprising a number of heatable parallel cylindrical rollers which rotate in opposite directions and spread out the plastics and stretch the material to the required thickness. The last roller smoothes the sheet or film thus produced. If the sheet is required to have a textured surface (e.g., to resemble wood graining), the final roller is provided with an appropriate embossing pattern; alternatively, the sheet may be reheated and then passed through an embossing calender. The calender is followed by one or more cooling drums. Finally, the finished sheet or film is reeled up.

Another field of application consists in coating a supporting material—e.g., textile fabrics, paper, cardboard, metals, various building materials—with plastics for the purpose of electrical insulation, protection against corrosion, protection against the action of moisture or chemicals, providing impermeability to gases and liquids, or increasing the mechanical strength. Coatings are applied to textiles, foil and other sheet materials by continuously operating spread-coating machines (Fig. 10). A coating knife, also known as a "doctor knife," ensures uniform spreading of the coating materials (in the form of solutions, emulsions or dispersions in water or an organic medium) on the supporting material, which is moved along by rollers. The coating is then dried. Alternatively, the coating applied to the supporting material may take the form of a film of plastic, in which case the process is called laminating.

Metal articles of complex shape can be coated with plastics by means of the whirl sintering process. The articles, heated to above the melting point of the plastics, are introduced into a fluidized bed of powdered plastics (a rising stream of air in which the powder particles are held in suspension), whereby a firmly adhering coating is deposited on the metal by sintering.

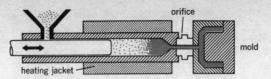

FIG. 5 INJECTION MOLDING WITH RAM EXTRUSION

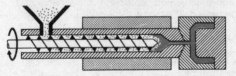

FIG. 6 INJECTION MOLDING WITH SCREW EXTRUSION

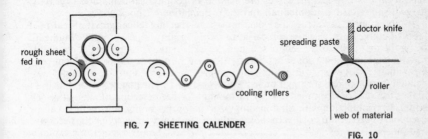

FIG. 7 SHEETING CALENDER

FIG. 10

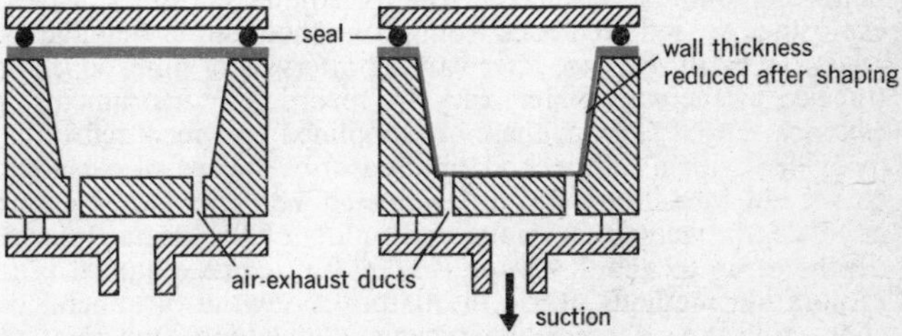

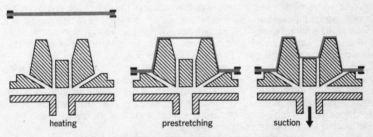

FIG. 8 VACUUM PROCESS: NEGATIVE-MOLD METHOD

FIG. 9 VACUUM PROCESS: POSITIVE-MOLD METHOD

PROSPECTING FOR MINERALS

The increasing need for mineral raw materials of all kinds has led to the development of methods for the detection of deposits concealed underground or for ascertaining the extent of known deposits. For thousands of years man had to rely on features visible at the surface of the ground as indications of what might be hidden underneath: e.g., outcrops of mineral-bearing veins or strata, or the presence of oil or salt springs. In addition, he had recourse to the services of diviners. Indeed, the divining rod is sometimes still used in the search for underground water or other minerals.

Systematic geological mapping and a better understanding of how mineral deposits are formed (genesis) and of the history of the earth's crust (paleogeography, sedimentary rocks and the fossils they contain, the life conditions of the fossil organisms, differentiation of strata by means of "key fossils," etc.) have provided valuable clues in the quest for mineral wealth. By the beginning of the present century, applied geophysics had placed underground exploration on a scientific basis. Various physical principles are utilized in geophysical exploration: gravity, magnetism, electric equipotential, and the propagation of shock waves.

In gravimetric surveying, the deviations from the theoretical gravitational field (as predictable from the geographical latitude) of a locality is measured, the "milligal" being the unit of measurement of the gradient of gravity. This gradient corresponds to a change in gravitational acceleration by an amount equal to 0.001 cm/sec.[2] Gravimetric surveyors use equipment comprising a pendulum or a weight suspended from a helical spring or quartz filament (Fig. 1). The magnetometric method makes use of a device called a local variometer for the detection of magnetic rock strata. In conjunction with photogeology and aerotopography, this method can be used for the exploration of inaccessible territories from aircraft carrying the necessary equipment. The most widely used geophysical method, however, is seismic exploration, which is employed especially in prospecting for oil or natural gas. High-explosive charges are electrically detonated in shallow boreholes located at grid points set out in the terrain. The "earthquake" shock waves generated by the explosions are reflected back from pronounced strata boundaries underground and reach the earth's surface after varying intervals of time, depending on the distance traveled by the waves. Here they are recorded by instruments called pickups. The electrical impulses from these are amplified and recorded on a moving film strip (provided with a millisecond time base) by means of oscillographs (Fig. 2). The curves obtained in this way are evaluated with reference to empirically known data as to depth, inclination, faults and uplifts of the strata. Information on strata at depths of up to about 4000 m (13,000 ft.) can be obtained in this way. Electrical exploration methods utilize the natural potential of mineral compounds. In this connection the differences in oxygen and metal ion concentration in the pore water contained in the relevant strata, which are associated with an oxidation and a reduction zone, are especially important. In consequence of the difference in concentration, galvanic contact potential differences are set up between separate points on the surface of the mineral deposit. A metallic ore deposit acts as an electric cell in which the ore constitutes the electrode material, while the pore water is the electrolyte whose concentration is subject to local variations (Fig. 3).

Quite often several of the above-mentioned methods are successively used on the same prospecting survey. It is important always to ascertain the "strike" and the "dip" of the deposit (Fig. 4). Before any mining operations are started, a deposit is usually further investigated by means of exploratory boreholes located at closely spaced grid points. The cores (specimens of rock or soil) obtained from the holes are subjected to petrographic examination.

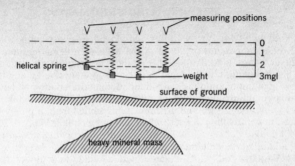

measuring positions

helical spring

weight

0
1
2
3mgl

surface of ground

heavy mineral mass

FIG. 1   GRAVIMETRIC EXPLORATION

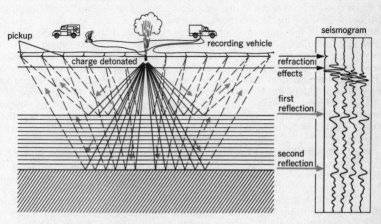

pickup

recording vehicle

charge detonated

seismogram

refraction
effects

first
reflection

second
reflection

FIG. 2   SEISMIC EXPLORATION

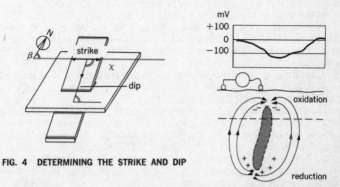

N

β

strike

χ

dip

FIG. 4   DETERMINING THE STRIKE AND DIP

mV

+100
0
−100

oxidation

reduction

FIG. 3   NATURAL-POTENTIAL EXPLORATION

Deep drill holes (or boreholes) are employed in prospecting for minerals and —where these are gaseous or liquid in character—for bringing them to the surface. The high demands placed upon this branch of engineering (attainment of great depths in conjunction with rapid drilling progress) have caused the traditional percussion drilling techniques, with their various drawbacks, to be superseded by continuous rotary drilling. Depths of more than 8000 m (26,000 ft.) have been reached by this technique. The drill rod is suspended from a pulley block within a lattice steel tower (drilling derrick) which may be as much as 60 m (200 ft.) high and designed for loads of up to 600 tons. (For drilling operations on a more limited scale, a jackknife-type collapsible mast on a mobile chassis may more conveniently be used.) A square rod, which engages with a socket in a power-driven turntable, transmits the rotary motion to the drill rod and thus to the drill bit attached (by means of a screw thread) to the rod (Fig. 1). For drilling in hard rock a so-called roller bit is employed (Fig. 2), which comprises three toothed conical steel elements with welded-on hard-metal (tungsten carbide) tips. The drill rod is hollow; during drilling, a flushing liquid is pumped down through the rod and then rises to the surface through the annular space between the rod and the wall of the drill hole. This liquid, which consists of water to which certain substances (which are held in suspension) have been added to increase its specific gravity (1.2 to 1.6) and is referred to as "drilling mud," is kept in circulation by pumping. It serves to cool the drill bit and to keep the drill hole clean and free of obstructing matter; it washes away the debris produced by the drill and carries it to the surface. On emerging at the surface, the liquid is passed through vibrating screens to remove the debris. Additives to the liquid serve to consolidate the wall of the drill hole, preventing its collapse. Another function of the liquid is to counteract, by its high specific gravity, any gas or oil pressure that may build up in the hole.

With rotary drilling it is possible to obtain rock specimens (cores) for examination. For this purpose a core barrel provided with an annular bit (likewise tipped with hard metal) is used instead of an ordinary bit. A cylindrical specimen of rock is thus cut from the bottom of the hole and can be brought to the surface.

The drill rod is assembled from units up to 32 m (105 ft.) in length. On the working platform in the tower is the control panel for operating the machinery. Below and beside the platform are the electric motors (up to 2500 hp) for driving the winch, turntable, pumps, etc.

Another form of rotary drilling is carried out with the turbodrill. In this type of equipment the drill bit is rotated by an axial turbine power unit near the bottom of the hole, so that the long transmission distance for the rotary motion from the turntable through the drill rod is eliminated and a very considerable saving in power is effected. The turbodrill is driven by the circulating liquid with which the drill hole is flushed and which is pumped at pressures of up to 150 atm. (2200 lb./in.²). Such drills rotate at speeds ranging from 400 to 900 rpm, and drilling progress rates of 10 to 20 m/hour (about 33 to 66 ft./hour) are attained, depending on the hardness of the rock encountered by the drill.

The exploration and exploitation of mineral deposits, especially those of petroleum and natural gas, in the relatively shallow coastal seas (continental shelf) have led to the extensive use of offshore drilling. The installations employed are designed for drilling to depths of about 6000 m (20,000 ft.). The essential thing is to provide a steady base for the drilling tower. There are various systems, illustrated in Figs. 3 and 4. Thus the drilling platform may be a floating barge, or be supported by a sinkable barge or pontoon resting on the seabed, or it may take the form of a spud-leg pontoon which is provided with vertical columns ("spuds") which are lowered to the seabed and serve as the legs of a huge table. So far, the maximum working depth of water for such installations has been 42 m (140 ft.).

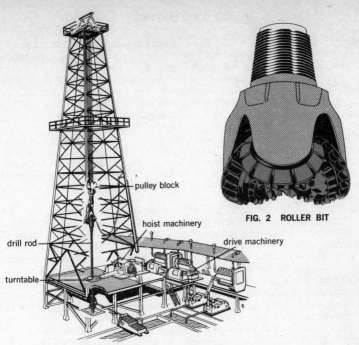

pulley block

hoist machinery

drive machinery

drill rod

turntable

FIG. 2   ROLLER BIT

FIG. 1   ROTARY-DRILLING INSTALLATION

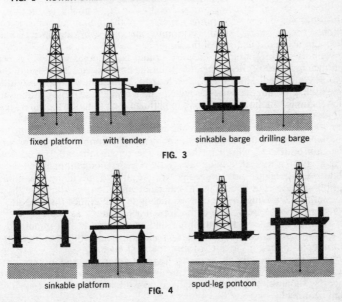

fixed platform   with tender   sinkable barge   drilling barge

FIG. 3

sinkable platform   **FIG. 4**   spud-leg pontoon

The exploitation of mineral wealth dates from prehistoric times. For thousands of years the mining operations consisted mainly in the excavation of outcropping material or tunneling more or less horizontally into mountainsides rather than descending vertically into the ground. Mining from deep shafts became possible only when better control of rock pressure was achieved (reliable supports for tunnels, etc.), along with drainage, ventilation and the use of mechanical appliances. This progress has been closely bound up with the discovery of steam and electric power and the use of compressed air for driving tools and machinery.

Before a mining operation is undertaken, elaborate calculations are made, which relate not only to geological and technical aspects of working the mineral deposit, but also to commercial and many other matters: financing, prospects for marketing the newly mined mineral, location of the mine with regard to accessibility and transport facilities, electric-power supply, local population (availability of manpower), etc.

Although methods for the mining of mineral deposits may differ considerably from one another because of differences in geological conditions and location, all mines have certain features in common. Access to the deposit is gained either by means of a horizontal tunnel driven into a mountainside or by means of a vertical shaft. The deposit is divided into sections of suitable size and shape for mining. Only after this extensive preparatory work has been done does the actual mining commence. The various horizontal tunnels (called "levels" or "roads") extending from the shaft may be situated at vertical intervals of as much as 300 ft. and more in deposits extending to great depths (coal, potash); on the other hand, in the working of lodes or veins the levels may be at intervals of only about 100 ft. (ores, fluorite). Individual levels or sublevels are interconnected by blind shafts.

Every underground deposit is characterized by its "dip" (inclination of the strata in relation to the horizontal: the angle of dip is measured in the direction of maximum slope) and its "strike" (the horizontal direction at right angles to the dip). These two directions very largely determine the extent of the preliminary and preparatory operations (access shafts, subdivision into sections, working levels) and thus establishing the overall features of the mine.

The supports and lining to be installed in the shafts and tunnels will depend not only on the condition of the rock but also on the anticipated service life of the parts of the mine concerned. Thus, main haulage roads, which may have to last as long as the mine itself, are from the outset provided with stronger and more durable supports than, for example, headings which have to perform their function for only a relatively short time.

For every mineral deposit a systematically planned mining program must be prepared and adhered to. Although this may entail a certain amount of tunneling through unproductive rock (i.e., other than the actual mineral), the advantage of working according to a strictly applied system is that long straight haulage roads are obtained, which are advantageous with regard to mechanical handling and ventilation and are thus conducive to efficiency and safety in the mine. Especially with modern fast-moving or continuous means of conveyance (loco-hauled trains, belt conveyors), it is essential to eliminate curves as much as possible.

There are various methods of underground or deep mining for mineral deposits. The following may be used for metal ores, rock salt, etc. (methods more particularly applied in coal mining are dealt with on the next page):

(a) Stoping: levels are driven at fairly frequent intervals, and rooms ("stopes") are excavated to remove as much of the ore as possible; pillars of ore are left standing as supports; sometimes the rooms are filled up with stone waste, so that the pillars can then be mined.

(b) Caving: small sections of the ore are successively mined, and the overlying rock

(more)

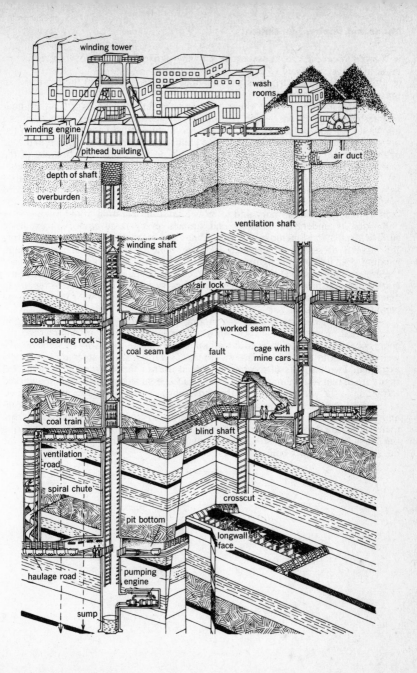

winding tower

wash rooms

winding engine

pithead building

air duct

depth of shaft

overburden

ventilation shaft

winding shaft

air lock

worked seam

coal-bearing rock

coal seam

fault

cage with mine cars

coal train

blind shaft

ventilation road

spiral chute

crosscut

pit bottom

longwall face

haulage road

pumping engine

sump

is allowed to cave (i.e., collapse) each time; alternatively, portions of the ore may be undermined and allowed to cave.

(c) Pillaring: levels are driven in a rectangular pattern throughout the ore body, so that pillars are left standing.

Viewed as a whole, the working of a modern mine comprises a large number of interlinked operations which require careful advance planning, with adequate allowance for compensation and latitude at critical points in the system. Even the disposal of the large quantities of gangue (valueless minerals encountered in a lode or vein) presents a separate and often difficult problem, not least because of the manpower and cost it involves. In mining, wages account for something like 40 to 70% of the total prime cost of the mineral. It is therefore essential to utilize labor as carefully and efficiently as possible. At the same time the mining must, perhaps more than any other industry, concern itself with matters of safety to protect the men from accident and injury and to ensure smooth operation of the mine.

Besides its underground shafts and tunnels, every mine comprises extensive surface installations, including large buildings and other structures accommodating the mineral dressing and dispatch facilities, boiler and power-generating plant, winding gear, lamps and lighting equipment, dressing rooms and washrooms for the miners, stores, surveyors' office, management offices, etc. In addition, there are workshops that have to be fully equipped to carry out a wide variety of repairs to machinery and tools which are liable to suffer damage in the rough working conditions of the mine.

A well-established method for the mining of sedimentary deposits occurring in seams—coal, in particular—is known as "longwall working." In this method the coal is obtained from a continuous wall up to 200 yds. long, usually by removal of a web of coal about 5 ft. wide. In this way areas of several hundred acres are completely extracted. Two variants of the method are in use: in the "retreating" system, the roads are driven to the boundaries of the area to be mined, and the faces are worked retreating to the winding shaft; in the "advancing" system, the faces are opened up at the shaft and then advanced to the boundaries. In the United States the "room-and-pillar" method of coal mining is extensively used: roads are first made and from them rooms are driven; between 30 and 50% of the coal is mined in this way during the "first working"; subsequently the pillars of coal remaining are mined in the retreat, or "second working."

SEAM ROAD EXTENDING LONGITUDINALLY

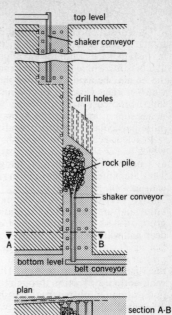

top level

shaker conveyor

drill holes

rock pile

shaker conveyor

A   B

bottom level   belt conveyor

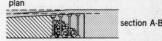

plan

section A-B

LONGWALL WORKING

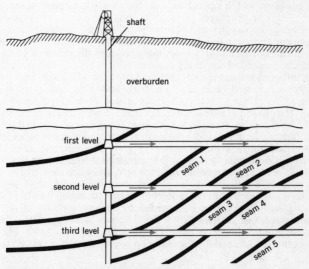

shaft

overburden

first level

seam 1   seam 2

second level

seam 3   seam 4

third level

seam 5

**CROSSCUTS (AT RIGHT ANGLES TO STRIKE
DIRECTION) OPEN UP A WHOLE GROUP OF SEAMS**

SHAFT SINKING

The construction of large and deep shafts is carried out by specialist firms, not by the mining company itself. In stable dry ground the excavation work is done by manual methods, with the aid of pneumatic picks and spades, augmented by drilling and blasting where hard rock is encountered. A multiblade grab or other mechanical device may be used for loading the loosened soil or rock into a heavy steel bucket, which is then winched to the surface (Fig. 1). Normally, shaft sinking and shaft lining proceed together, sinking being interrupted at intervals in order to line the newly sunk portion. Brick or concrete is used as a lining material, brickwork being more particularly used for round shafts in Britain and on the Continent.

In water-bearing strata the shaft lining is usually constructed from "tubbing," which consists of cast-steel segments (Fig. 3). The latter are about 1.3 m (4$\frac{1}{2}$ ft.) high and provided with ribs and flanges for strengthening and interconnecting them. They are bolted together to form rings which are installed one above the other, with lead gaskets in the joints, so that a closed watertight lining is obtained which is able to withstand pressure acting on it from the outside. The space between the lining and the wall of the excavation is filled with concrete. In loose water-bearing ground a shaft-boring technique may be employed, whereby it is possible to construct deep and wide shafts, e.g., up to about 500 m (1600 ft.) depth and 5 m (16 ft.) diameter. Boring is carried out with the aid of drilling mud, which helps to withstand the pressures (from the ground and from the underground water) which tend to collapse the shaft during construction. The cuttings from the drill are removed from the shaft bottom by suction through the hollow drill rod by means of special pneumatic pumps (mammoth pumps). The shaft lining, a cylinder fabricated from steel plate, is lowered into the shaft as excavation proceeds and is cemented in.

In the drop-shaft method of construction (Fig. 2), the brickwork or concrete lining is built up at the surface and sinks into the shaft—under its own weight, assisted by ballast loading—as excavation at the bottom proceeds. The bottom of the lining is provided with a cutting edge of steel to assist its penetration. Another technique that may be employed in water-bearing ground is cementation, which consists in sealing the cavities and fissures with "grout" (cement slurry—a fluid mixture of cement and water) which is injected under pressure through holes drilled into the strata concerned. The grout solidifies and stops the inflow of water, so that shaft sinking can be carried out in the ordinary way.

In soft waterlogged ground the so-called freezing process is sometimes employed. It is an expensive method because of the fairly elaborate equipment it requires (refrigerating plant, cooling tank, etc.). Pipes spaced about 1 yd. apart are sunk vertically into the ground at a distance of 2 or 3 yds. from the edge of the shaft to be excavated. Pipes of smaller diameter are installed concentrically inside these vertical pipes, and a freezing liquid (brine) is circulated through the inner pipes and flows back to the surface through the outer ones (Fig. 4). The temperature of the brine is about −20° C. Ammonia is used as the refrigerating agent. A solid cylinder of frozen ground is gradually formed around each pipe, and when these frozen cylinders unite to form a solid ring round the ground to be excavated, shaft sinking can proceed in the usual way within the protection of this encircling "wall" of solidified ground. Sometimes the freezing is preceded by cement grouting if the ground contains wide fissures or cavities. When the shaft has been completed, freezing is stopped and the ground allowed to thaw. The pipes are then withdrawn. The freezing process has been used successfully for shafts up to about 600 m (2000 ft.) deep.

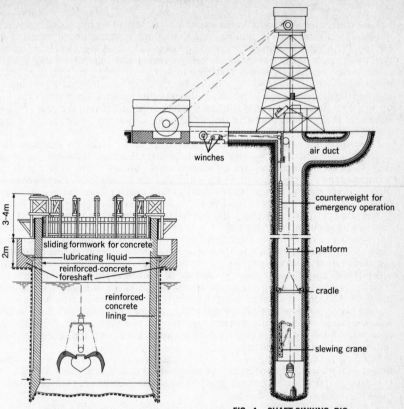

winches

air duct

counterweight for
emergency operation

platform

cradle

slewing crane

FIG. 1   SHAFT-SINKING RIG
FOR A MAIN SHAFT

3-4m

2m

sliding formwork for concrete

lubricating liquid

reinforced-concrete
foreshaft

reinforced-
concrete
lining

FIG. 2   DROP-SHAFT METHOD

FIG. 3   SHAFT LINING WITH
CAST-STEEL SEGMENTS

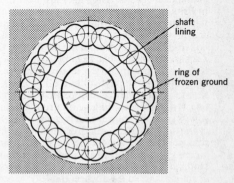

shaft
lining

ring of
frozen ground

FIG. 4   FREEZING PROCESS

WINDING

The term "winding" refers to the operations associated with hoisting the mined mineral to the surface. In modern mines winding is automated (controlled by electrical contacts) in conjunction with acoustic and visual signaling and various devices to ensure safety when men are being raised or lowered. All winding systems embody the counterweight principle, with two cages (or skips) moving in opposite directions—one ascending while the other descends. The basic features of the winding gear are the winding engine or motor, the headframe (usually a lattice steel structure, up to 200 ft. high, over the shaft), the winding rope and drum, the cages (or the skips), the intermediate gear whereby the latter are connected to the rope, and cage guides in the shaft.

Large-diameter shafts—of circular section up to 7 m (23 ft.) diameter—are normally equipped with two sets of winding gear; there are thus four hoistways (Fig. 1), in each of which a cage (or a skip) moves up and down. Steam-powered winding engines or electric winders are employed, the latter usually being driven by direct-current electric motors with Ward-Leonard control. Such motors may have power ratings of as high as 12,000 kW and hoist coal or other minerals at a rate of 10,000 tons per day from shafts 800 m (2600 ft.) in depth. The hoisting speed in deep shafts is about 22 m/sec. (70–75 ft./sec.) with a 30-ton payload. The winding operations are controlled by a device called a winding-speed regulator. It determines the hoisting speed in relation to the distance traveled and more particularly limits the acceleration of ascent and descent. In addition, every winding system includes a depth indicator, which consists essentially of a screw spindle along which a nut travels; the position of the nut on the spindle indicates the position of the cage in the shaft. This device is linked to an overwind-prevention device, which actuates a second brake (drop-weight brake) if the cage ascends too high and the counterbalancing cage consequently descends too low. In the event of overwinding, the cage is slowed down and braked by such means as thickening the cage-guide rods in the top part of the headframe.

Winding ropes are composed of several strands which in turn consist of cold-drawn steel wires of 2.5 mm (0.1 in.) diameter, with a tensile strength of 200 kg/mm^2 (127 tons/in.^{2}) (Fig. 4). Such ropes may be as much as 100 mm (4 in.) thick and have breaking loads of around 700 tons.

The cages that are raised and lowered in the shaft are multideck structures into which the tubs or mine cars are pushed and from which they are removed by mechanical means. The cages are also used for raising and lowering the miners. In Britain, cage winding is still predominant in coal mining. However, skip winding is progressively being introduced; it is already extensively used in the United States and in various Continental countries. A skip is a guided steel or aluminum-alloy box (Fig. 5) which is automatically filled at the bottom of the shaft and automatically discharged when it has been hoisted aboveground. The advantage of skip winding is that a skip can carry a relatively larger payload than a cage with tubs or cars; the proportion of payload to total load can be raised by nearly 40% by substituting skip for cage winding, and loading and unloading are faster. In the main, there are two systems of winding gear: drum winding and Koepe winding. Of these, drum winding is the more widely employed. The drum may be variously shaped: cylindrical, conical, cylindroconical. The object of the conical shape (Fig. 3) is to equalize the driving torque by using a smaller diameter of drum when exerting maximum rope pull to lift and accelerate the cage (or the skip). In Germany and Holland the Koepe system (Fig. 2) is generally preferred. In this system the massive drum is replaced by a wheel (pulley) with one peripheral groove. A single winding rope lies in this groove, and all controlling forces transmitted through the rope depend on the friction of the rope in the groove. The cages (or the skips) are suspended one on each end of the rope, which passes over the Koepe wheel. The system offers certain advantages and is more particularly suitable for winding heavy loads from deep levels.

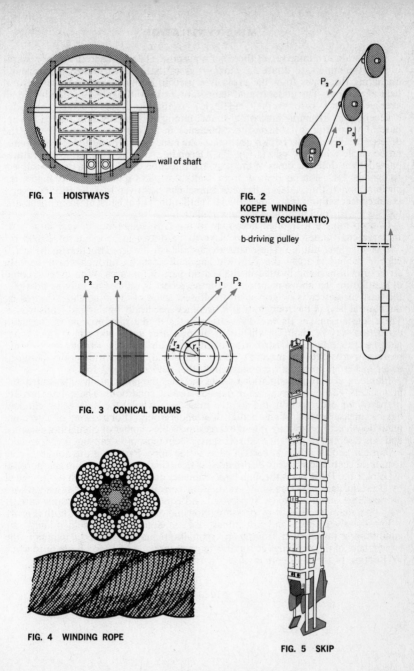

FIG. 1 HOISTWAYS

wall of shaft

FIG. 2
KOEPE WINDING
SYSTEM (SCHEMATIC)

b-driving pulley

P_1 P_2 P_3

FIG. 3 CONICAL DRUMS

P_2 P_1 P_1 P_2 r_2 r_1

FIG. 4 WINDING ROPE

FIG. 5 SKIP

MINE VENTILATION

Ventilation in a mine serves three main purposes: to provide fresh air for respiration by the miners, to dilute any noxious gases that may be formed underground (including the fumes from the explosives used in blasting), and to lower natural heat of the rock. The underground temperature rises with increasing depth—on an average, about $1°$ C for every 30 m (100 ft.)—so that the deeper the mine, the hotter it generally is. In simple horizontal-tunnel mining it is usually sufficient to rely on natural ventilation by utilizing the difference in air pressure associated with the difference in level between two openings—the mine entrance and the top of a ventilation shaft (chimney effect, Fig. 1). Depending on the external temperatures prevailing at different times of the year, the direction of flow of the draft is subject to change. Diffusion (as distinct from draft) also plays some part in changing the air in a tunnel; thus, a large-diameter tunnel can be driven to a distance of several hundred feet without requiring artificial ventilation. The exhaust air from pneumatic tools is also helpful in promoting air circulation.

In deep mining it is often necessary to use fans, sometimes of very large size. In large coal mines fresh air may have to be drawn in at a rate of 20,000 m³ (700,000 ft.³) per minute. These fans are installed at the air-extraction shafts at the edge of the mined area—the main winding shaft or shafts, through which the fresh air is drawn in, being located in the central part of this area. With every method of ventilation the above-mentioned chimney effect is utilized as fully as possible: the fresh air descends by gravity to the lowest levels of the mine and is heated by the natural heat of the rock, so that it becomes specifically lighter and tends to rise. The rising air makes its way by various paths to the suction zone of the main extraction way or shaft, in which suction pressures up to 400 mm (17 in.) water gauge are maintained. Distribution of the fresh air over the various levels, main roadways, crosscuts, rooms and workings is assisted by ventilation doors (designed as air locks), stoppings, air crossings and other devices (Figs. 2 and 3).

Planning of a mine-ventilation system includes the preparation of so-called air-flow sheets—diagrams comprising data on airflow conditions. These diagrams are prepared for each section and for the mine as a whole, the data being checked against measurements of the actual flow underground. For reasons of safety the main flow has to be split up into the largest possible number of circulating currents, and it is essential to prevent "short circuits"—circumstances causing the air to take a shortcut and thus bypass certain parts of the mine. Parts that are not accessible to natural ventilation have to be provided with auxiliary ventilation. For this purpose air is piped to those parts through large-diameter ducts through which it is impelled by powerful fans. This auxiliary ventilation constitutes a separate system whose proper functioning has to be supervised and controlled with considerable care. It may operate by suction or by pressure (blowing) or a combination of both (Fig. 4).

Particularly in deep and hot mines—e.g., in South African gold mining—air conditioning (as distinct from mere ventilation) may be used to maintain the atmosphere of the workings at suitable temperature and humidity for men to work in. Because of the high cost involved, it is seldom used, however.

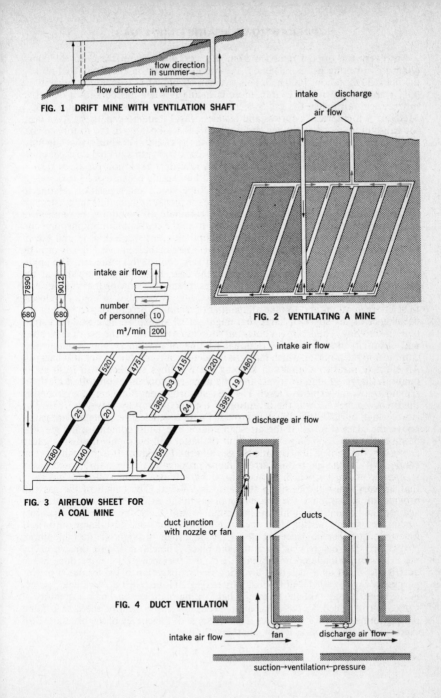

FIG. 1 DRIFT MINE WITH VENTILATION SHAFT

flow direction in summer
flow direction in winter

intake   discharge
air flow

FIG. 2 VENTILATING A MINE

intake air flow

number of personnel ⑩

m³/min 200

FIG. 3 AIRFLOW SHEET FOR A COAL MINE

intake air flow

discharge air flow

duct junction with nozzle or fan

ducts

FIG. 4 DUCT VENTILATION

intake air flow   fan   discharge air flow

suction→ventilation←pressure

PRECAUTIONS AGAINST MINE GAS

For every ton of coal removed from the working face, anything up to 200 m³ (7000 ft.³) of mine gas is released; the average amount in the Ruhr coal-mining district is 30 m³ (about 1000 ft.³). This gas consists of hydrocarbons, chiefly methane (CH_4), and is known as firedamp, more particularly when it occurs in an explosive mixture with air (when the gas concentration in the air is between 4.5 and 14.5%). Methane is a colorless, odorless and tasteless gas. Firedamp explosions have been the cause of many catastrophes and a vast number of deaths in the history of coal mining. The initial explosion is liable to ignite the cloud of coal dust that the blast disperses. When the air contains between 70 and 1000 grams of coal dust per cubic meter, it constitutes an explosive mixture. A coal-dust explosion produces carbon monoxide (CO), which is a serious danger to human life because of its toxicity.

Accordingly, the main object of research, supervision and legislation relating to safety in coal mines is to develop and improve the precautions against the occurrence of such mishaps. Essentially, the aim is to eliminate all possibility of ignition of such explosive mixtures. The emission of gas from the coal is a natural phenomenon associated with the constitution of the coal and adjacent strata (Fig. 1), and except in some rare circumstances where methane gas can be removed in advance by suction, it is not possible to reduce or effectively control this emission. The application of special procedures in extracting the coal, high speed of advance at the coal face, complete sealing off of old workings by suitable stowing (backfilling), etc., may have a favorable effect. It nevertheless remains essential to conform closely to safety regulations and to take all manner of technical precautions to prevent the occurrence of any spark or flame that might set off an explosion, including a strict ban on smoking. There remain potential sources of ignition in the use of explosives and electricity underground. Frictional heating of machinery and spontaneous-combustion phenomena, which may arise under certain conditions, are also hazards. An effective measure consists in keeping the workings supplied with fresh air in quantities large enough to ensure that the mine-gas concentration will at all times remain below the explosive level. The first important development in overcoming the firedamp menace was the invention of the safety lamp by Davey in 1816. The lamp, which burns a liquid fuel, is provided with an enclosure of metal-wire gauze above the glass (Fig. 2). If the air surrounding the lamp contains mine gas, the flame of the lamp will ignite the gas, but the latter will burn only inside the gauze enclosure; the flame will not ignite the gas all round the lamp. It is the high thermal conductivity of the gauze that arrests flame propagation. The lamp is used today as a detector for gas, which burns with a characteristic flame called a "gas cap" that appears when the flame in the lamp is lowered. The length of the gas cap provides an indication of the percentage of gas in the air (Fig. 3). A countermeasure against the formation of highly inflammable and therefore explosive coal-dust clouds consists in "dusting": i.e., specified quantities of stone dust (finely pulverized limestone or other nonsiliceous stone) are deposited throughout the mine, more particularly just before blasting is to take place. The cloud of dust thrown up by the explosion is rendered nonflammable by the presence of the stone dust. Stone dust is also used as a means of arresting the propagation of explosions. For this purpose a device called a stone-dust barrier (Fig. 4) is installed at "strategic" points in the mine. It may take the form of a light tilting platform on which a quantity of stone dust is placed. In the event of an explosion in the vicinity, the dust is flung off and forms a dense cloud which absorbs some of the energy of the blast and also exercises a cooling action which smothers the flame.

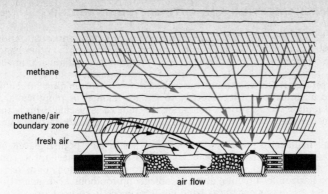

methane

methane/air
boundary zone

fresh air

air flow

FIG. 1   FLOW OF MINE GAS THROUGH STRATA

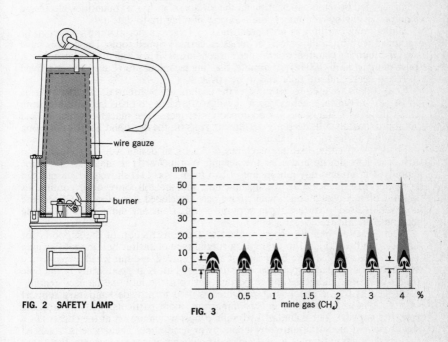

wire gauze

burner

FIG. 2   SAFETY LAMP

mm

0   0.5   1   1.5   2   3   4   %
mine gas (CH₄)

FIG. 3

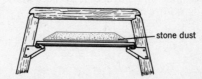

stone dust

FIG. 4   STONE-DUST BARRIER

Because of the high proportion of wages in the cost of production, mechanization is more necessary in mining than in almost any other industry. On the other hand, because of the often difficult conditions in which the product—the mineral mined underground—must be obtained, the scope for mechanization of the actual "production" process is limited. Among the earliest developments in mine mechanization was the introduction of locomotives to replace manual or pony haulage. In present-day mining the operations of loading the mine cars, placing them in the cage at the winding shaft, discharging the cars at the surface, etc., are all performed automatically.

Underground haulage in mine cars and the like has in part been replaced by continuous conveying systems of various kinds (belt conveyors, steel-apron conveyors, chain conveyors, etc.), which can handle up to 600 or 700 tons of material per hour in the horizontal direction and at gradients up to about 12 degrees. Coordination of the various mechanical handling appliances is ensured by control centers operating with automatic interlocking optical and acoustic signaling systems in conjunction with measuring and monitoring equipment, overall control being assisted by remote indication (at the surface) of measured quantities, electronic data processing by computers, and a variety of other up-to-date aids.

Underground drilling is now performed by large crawler-mounted power drills equipped with high-alloy-steel- or tungsten carbide-tipped tools. These tools have swivel mountings and hydraulic feed mechanisms. They can drill blastholes at rates of up to 30 m (100 ft.) per minute. Blasting techniques too have been improved in recent years and are safer and more efficient.

Coal cutters are used to get most of the coal mined by longwall or other systems in Britain. A typical machine for longwall work has a jib provided with a cutting chain fitted with tungsten carbide-tipped cutter picks. The machine, which makes an approximately 5 ft.-deep horizontal cut, rests on the floor and hauls itself along by winding a wire rope on a drum.

Loaders and cutter-loading machines of American design were introduced into Britain chiefly during and after the Second World War. The majority of loaders employed in present-day mining are of two main types: (1) shovels which pick up and empty in sequence; (2) gathering machines with integral conveyors in continuous motion. Shovel-type mobile loaders have been developed since about 1925 and are now widely used in certain types of mining—e.g., ironstone mining. Some loading machines are illustrated in Figs. 1–3.

Mechanization has also been applied to the construction of the supports for underground workings. For example, a special type of anchor bolt for roof lagging is secured in a drilled hole by means of a screw thread and nut which expands the two halves of a tapered split bushing, so that the bolt is gripped firmly in the hole (Fig. 4). A more elaborate and important device is the fully mechanized remote-controlled hydraulic chock (Fig. 5), which is used to provide temporary support in situations where stability is very important, more particularly at the working face. It comprises four columns (hydraulic props) which are set at the corners of a steel base and are surmounted by a canopy or headframe. The device is advanced to a fresh position by means of a hydraulic cylinder whose movements, like those of the four props, are controlled by a system of valves.

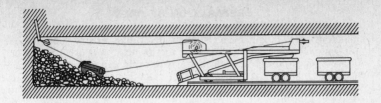

FIG. 1   SCRAPER-LOADER

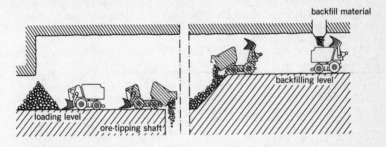

backfill material

backfilling level

loading level
ore-tipping shaft

FIG. 2   MOBILE SHOVEL LOADER

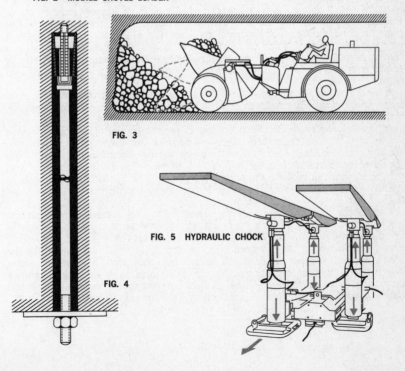

FIG. 3

FIG. 5   HYDRAULIC CHOCK

FIG. 4

Metalliferous ores straight from the mine are seldom directly suited for metal smelting. Quite often their metal content is too low (e.g., only 0.8% copper or 5% lead) for processing in the furnace, or they may be composed of minerals containing different metals requiring different kinds of metallurgical treatment. For these reasons, most ores have to undergo various preparatory processes, collectively referred to as "dressing," for the removal or separation of waste matter or other minerals, so that the concentration of the desired mineral is increased. The principal mineral-dressing processes are presented systematically on page 79. Some of the processes used more particularly in ore dressing are sorting (by hand); comminution (crushing and grinding); sizing (by screening); classifying (e.g., the grading of finely divided material by rates of settling); separation (e.g., by magnetism, electrical conductivity, specific gravity, etc.).

The preparation plants are usually located at the mines, so that only the processed ore, free of waste matter, has to be transported to the smelting works. In most cases preparation starts with crushing and grinding (Fig. 1). The degree of comminution (size reduction) to be applied will depend on the size of the ore lumps and on the requirements of the subsequent treatment to be applied. Sizing and classifying— i.e., grading the comminuted material according to particle size—are important operations in ore dressing. To relieve the crushing and grinding machines of unnecessary load, particles that have been sufficiently reduced in size are removed by screening (Fig. 5). Sizing of relatively coarse particles can most efficiently be performed by screening, and screens of many kinds are used for the purpose. Small particles (below about 1 mm in size) can usually be more suitably sized by classification based on different rates of settling of different particle sizes in water. For example, the Hardinge countercurrent classifier (Fig. 6) is a slowly rotating drum on the inner surface of which are located spiral flanges. As the classifier rotates, the coarser particles are settled out, moved forward by the spiral flanges, and repeatedly turned over in a forward motion, releasing any finely divided material mixed with them.

It is not possible, within the scope of this article, to describe all the many processes and types of equipment employed in the preparation of metalliferous ores. However, three important methods of treatment will be dealt with.

In *wet-mill concentration* the differences in specific gravity of different minerals are utilized for separating them. The metalliferous ores—sulphides and oxides— are as a rule specifically heavier than the waste material. Separation of the ore from the waste (or separation of different ores) may be effected in a settling classifier with the aid of water in motion, the underlying principle being that the differences in specific gravity are associated with different rates of settlement of the particles (Figs. 2 and 7).

Coarser particles, ranging in size from about 0.5 to 30 mm, can be settled out in a machine called a jig in which a horizontal stream of water is subjected to a rhythmical up-and-down motion. Finer-grained materials, approx. 0.3 to 0.5 mm in size, may be treated on a table concentrator—a slightly inclined plate on which the lighter material is separated from the heavier by a thin, shallow stream of water. This hydromechanical separating action may be augmented by the action of gravity developed by oscillating or jolting motions applied to the table (shaking and bumping tables, Fig. 4). These concentration processes are referred to as "tabling."

(more)

MINERAL DRESSING
(Survey of methods and processes for the preparation of minerals*)

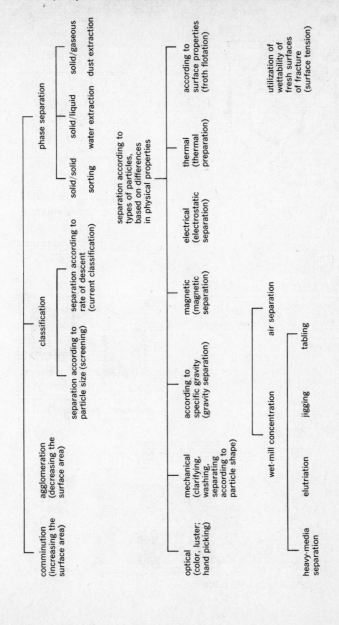

comminution (increasing the surface area)

agglomeration (decreasing the surface area)

classification

phase separation

separation according to particle size (screening)

separation according to rate of descent (current classification)

solid/solid — sorting

solid/liquid — water extraction

solid/gaseous — dust extraction

separation according to types of particles, based on differences in physical properties

optical (color, luster; hand picking)

mechanical (clarifying, washing, separating according to particle shape)

according to specific gravity (gravity separation)

magnetic (magnetic separation)

electrical (electrostatic separation)

thermal (thermal preparation)

according to surface properties (froth flotation)

utilization of wettability of fresh surfaces of fracture (surface tension)

wet-mill concentration

air separation

heavy-media separation

elutriation

jigging

tabling

——————
* according to Prof. Dr.-Ing. Gründer

79

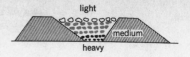

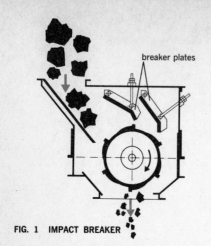

FIG. 1 IMPACT BREAKER

FIG. 2 PRINCIPLE OF
WET-MILL CONCENTRATION

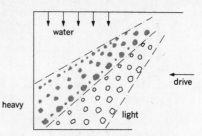

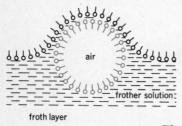

FIG. 3

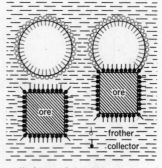

attachment of air bubble
to ore particle

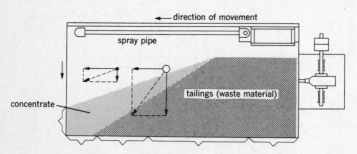

FIG. 4 SHAKING TABLE

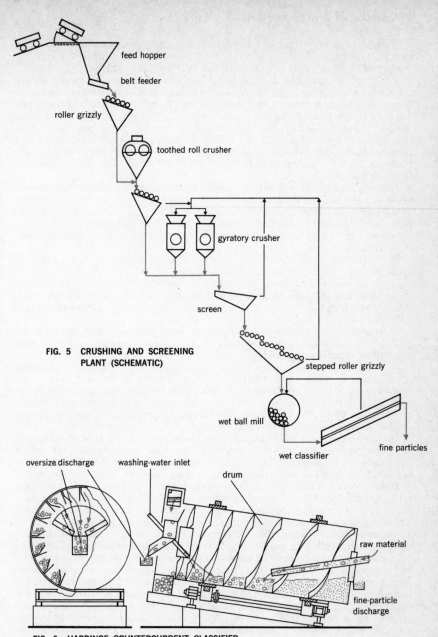

feed hopper

belt feeder

roller grizzly

toothed roll crusher

gyratory crusher

screen

FIG. 5 CRUSHING AND SCREENING PLANT (SCHEMATIC)

stepped roller grizzly

wet ball mill

wet classifier

fine particles

oversize discharge

washing-water inlet

drum

raw material

fine-particle discharge

FIG. 6 HARDINGE COUNTERCURRENT CLASSIFIER

Flotation is an important and widely used separation process which is based on the fact that some of the components in the comminuted minerals are wettable (hydrophilic), whereas others are water-repellent (hydrophobic). See Figs. 3 and 8. The hydrophobic particles have an ability to hold air bubbles by surface action, the nature of the film on the outside of the particles being the controlling factor. Finely divided air which is introduced into the "pulp," the mixture of solids and water in which flotation is performed, adheres in the form of bubbles to these particles, more particularly the metalliferous components of the pulp, and causes them to rise to the surface. Here they collect in a mass of froth and are removed by a skimmer device. The hydrophilic components remain behind in the pulp. As a rule, these are the worthless minerals (gangue), which are removed as tailings from the flotation machine.

The sulphides of heavy metals are readily floatable, and flotation is therefore an important method for the concentrating of copper, lead and zinc ores. A further development has been the selective flotation of two or more useful minerals, particularly the ores of different metals, which can thus be collected as separated "concentrates." This principle is, for example, applied to the preparation of sulphidic lead-zinc ores.

The floatability of minerals can be controlled by certain chemical additives called flotation agents. These are of various kinds:

Frothers, whose function is to produce froth by combining the air bubbles (introduced into the pulp by stirring or by the injection of compressed air) into a stable froth which will buoy up the ore particles. Oils and allied substances are used as frothers.

Collectors are substances that increase the water repellency and make the ore particularly receptive to the attachment of air bubbles. Collectors usually consist of synthetic organic compounds.

Other flotation agents help to regulate the process. So-called depressors can make hydrophobic minerals temporarily hydrophilic and can in this way help in the selective separation of one mineral from another by depressing one, thereby inhibiting its flotation. The "depressed" mineral can subsequently be made hydrophobic again by an activating agent. The various agents for regulating the flotation process in this manner are inorganic compounds, mostly salts.

Magnetic separation: If a comminuted and classified ore is brought into a magnetic field, the magnetic components (generally the useful metalliferous ore) can be extracted and thus separated from the nonmagnetic residual material. The treatment is carried out with the aid of magnetic separators, of which there are many kinds. A drum separator is illustrated in Fig. 9. This method of separation plays an important part, for example, in the concentrating of certain iron and manganese ores. Those substances which are attracted by a magnetic field are called paramagnetic. These are subdivided into strongly magnetic (ferromagnetic) and weakly magnetic substances. Various techniques have been devised for the separation of both categories of materials. In general, magnetic fields of greater intensity (high-intensity magnetic separators) have to be employed for dealing with weakly magnetic ores. In both categories "wet" and "dry" processes are employed, depending on whether or not water is used as an aid in the process. The magnetic properties of certain ferrous minerals can be enhanced by suitable preliminary heat treatment. For instance, in the case of siderite (a particular kind of iron ore, which is a carbonate— $FeCO_3$) the carbon dioxide can be expelled by heating the ore in a kiln. As a result, the carbonate is converted into the strongly magnetic compound named ferrosoferric oxide (Fe_3O_4), which can readily be separated by magnetic action.

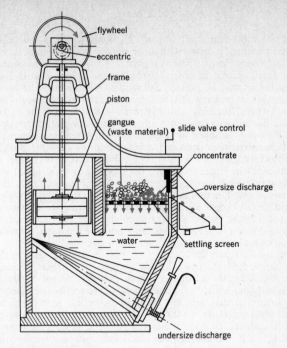

FIG. 7   SETTLING CLASSIFIER

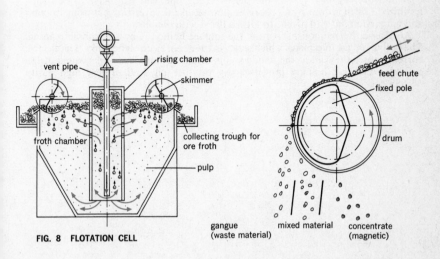

FIG. 8   FLOTATION CELL

FIG. 9   ELECTROMAGNETIC DRUM SEPARATOR

The lead ore most commonly mined is galena, which is the sulphide of lead (PbS). It occurs intimately mixed with other metalliferous minerals, such as sphalerite (zinc sulphide), copper pyrites and iron pyrites. The ore has to be concentrated, e.g., by flotation (see page 83), in order to separate the galena from the sphalerite and other minerals that may be present. Subsequent treatment of the concentrate thus obtained consists in roasting followed by reduction in a vertical-shaft furnace, a form of blast furnace. Roasting is usually performed by heating the lead ores, blended with suitable fluxing minerals, on a traveling endless grate through which air is sucked. In this way the material is sintered—converted into lumps (called sinter) which are then mixed with coke and charged into the shaft furnace (Fig. 2). Air is forced into the furnace at the bottom. The coke, which serves as fuel and reducing agent, reacts with the sinter to reduce the oxides and yield liquid lead, which is, however, contaminated with other metals—silver, copper, zinc, tin, antimony, bismuth, arsenic, etc. The nonreduced components form a liquid slag which floats on the liquid metal. Preparing the charge and operating the furnace call for great skill. In particular, the charge must contain the correct proportions of iron, lime and silica to produce a liquid slag that can readily be separated from the metal; it is also essential to maintain the proper balance of coke and sinter.

When the impure liquid lead (known as "bullion") cools, some of the impurities, especially copper, separate out as drosses, which are further processed to extract the copper. Further removal of the copper may be effected by treatment of the bullion with sulphur. Antimony, tin and arsenic are removed by elective oxidation in a reverberatory furnace or by treatment of the bullion with chemical reagents to separate out these metals in the form of salt-type compounds. Desilverizing of the liquid lead is achieved by adding metallic zinc and raising the temperature sufficiently to dissolve it. On cooling, the zinc forms a dross or crust which contains nearly all the silver and other metallic impurities. The dross is skimmed off, and the silver is recovered from it in a separate process. The zinc is distilled off and used over and over again (Fig. 3). After desilverizing, the lead may have to be debismuthized, which is done by a process somewhat like desilverizing but using calcium and magnesium instead of zinc to form a dross with the bismuth.

An alternative method of treating the impure bullion is by electrolytic refining. The bullion is cast into plates which serve as anodes in electrolytic tanks. The electric current causes the lead at the anode to dissolve, and pure lead is deposited at the cathode. All these refining processes can produce pig lead of very high purity (99.999%!).

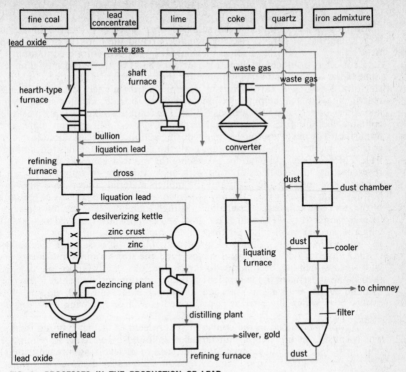

| fine coal | lead concentrate | lime | coke | quartz | iron admixture |

FIG. 1 PROCESSES IN THE PRODUCTION OF LEAD

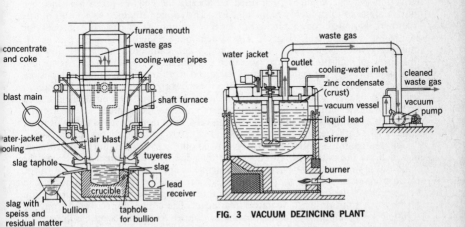

FIG. 2 SHAFT FURNACE FOR LEAD SMELTING

FIG. 3 VACUUM DEZINCING PLANT

COPPER

Most copper is obtained from sulphide ores. These are found admixed with large quantities of gangue (worthless material), and the initial content of copper may be very low (1 or 2%, sometimes only 0.7%). A concentrate containing 15 to 35% copper is produced by flotation (see page 83). For the purpose of eliminating some of the sulphur and certain impurities, the concentrate is usually roasted before smelting. Roasting is carried out in multiple-hearth furnaces, in which oxidizing reactions take place: sulphur is eliminated as the dioxide; metallic sulphides (iron and some copper) remain behind as oxides. The resulting mixture, known as calcine, contains the sulphides of iron and copper, together with gangue material and impurities. The next step consists in producing a molten artificial sulphide of copper and iron, known as matte, containing all the copper and the desired amount of iron. The smelting operation for producing the matte is generally carried out in a reverberatory furnace (Fig. 2), fired with oil, natural gas or pulverized coal. The charge is fed through the roof, and the molten material collects in a pool at the bottom. Slag, which rises to the top, is tapped off. The matte collects at the bottom of the pool and is likewise discharged through a taphole. The molten matte is fed to a converter (Fig. 3) in which the iron and sulphur are removed by oxidation, which is effected by blowing air through the molten mass and is based on the fact that copper has a lower affinity for oxygen than has iron or sulphur. The reactions in the converter occur in several stages. First the iron oxidizes and forms a slag with silica, which has been added to the charge; this slag is tapped off, the copper then being present as the sulphide. Further oxidation results in the formation of metallic copper with a small amount of copper oxide and other impurities. The converter in which the process is performed is a large revolving refractory-lined drum.

The copper obtained in the converter is subjected to further refining treatment, which consists in fire refining (in furnaces) followed by electrolytic refining. Fire refining is done in small reverberatory furnaces or in revolving furnaces similar to the copper converter. Air is blown through the molten material to oxidize all impurities; the oxides rise to the surface and are skimmed off. Then follows a reduction process which is performed by forcing the ends of green logs into the molten metal to form highly reducing gases. The copper obtained as a result of this treatment is called tough pitch. For further refining, it is cast into anodes for electrolytic refining cells (Fig. 4). The system most widely used is known as the multiple system, comprising separate anodes and cathodes; the latter consist of thin sheets of high-purity copper (so-called starting sheets). When an electric current is passed through the cells, copper is dissolved from the anodes and is deposited in a very pure form on the cathodes. When these have grown to a thickness of about $\frac{1}{2}$ inch they are replaced by fresh starting sheets. About four starting sheets are used for each anode, until the latter has been completely consumed. An electrolytic refinery may comprise hundreds of refining cells, each using perhaps thirty or more anodes. The cathodes are melted down and cast into suitable shapes for market.

Ore treatment may, alternatively, be carried out by hydrometallurgical processes in which the ore is treated with a solvent that dissolves the copper and leaves the undesirable material unaffected. This principle is applied more particularly to the oxide ores of copper, or to sulphide ores after suitable roasting, sulphuric acid being used as the leaching solvent. Elaborate washing, filtration and purification of the leach solution are associated treatments. The copper is recovered from the solution by precipitation (displacement of copper from the copper sulphate by metallic iron that is added to the solution) or by electrolysis (using insoluble anodes and copper cathodes, which are high-purity starting sheets).

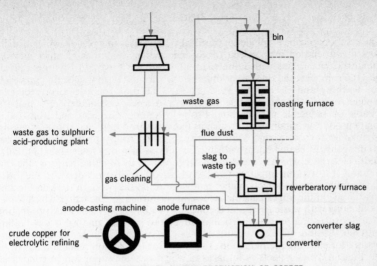

bin

waste gas

roasting furnace

waste gas to sulphuric
acid–producing plant

flue dust

slag to
waste tip

gas cleaning

reverberatory furnace

anode-casting machine   anode furnace

converter slag

crude copper for
electrolytic refining

converter

FIG. 1   PROCESSES IN THE PRODUCTION OF COPPER

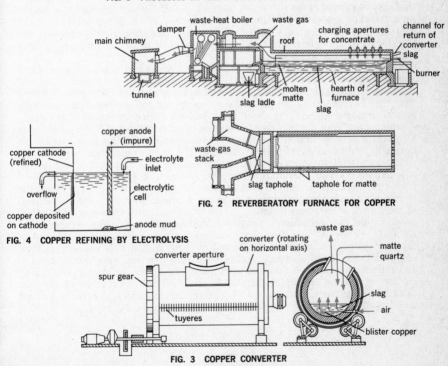

main chimney

damper

waste-heat boiler

waste gas

roof

charging apertures
for concentrate

channel for
return of
converter
slag

burner

tunnel

slag ladle

molten
matte

hearth of
furnace

slag

copper anode
(impure)

waste-gas
stack

copper cathode
(refined)

electrolyte
inlet

electrolytic
cell

overflow

slag taphole   taphole for matte

copper deposited
on cathode

anode mud

FIG. 2   REVERBERATORY FURNACE FOR COPPER

FIG. 4   COPPER REFINING BY ELECTROLYSIS

waste gas

converter (rotating
on horizontal axis)

matte
quartz

converter aperture

spur gear

slag

air

tuyeres

blister copper

FIG. 3   COPPER CONVERTER

ZINC

The chief ore of zinc is sphalerite (ZnS), also known as zinc blende, which usually occurs in association with galena (PbS) and smaller quantities of other metallic sulphides. Concentrates with more than 50% zinc are produced by flotation (see page 83). From these the zinc can be obtained by various thermal reduction processes or by leaching and electrolysis. In every case the process must be preceded by complete roasting to convert zinc sulphide into zinc oxide and thereby make it leachable or reducible with carbon. From low-grade zinc ores and from intermediate products containing zinc, such as flue dusts from lead and copper smelting, it is possible to obtain zinc oxide by the so-called rotary process: the material containing zinc is heated under reducing conditions in a long rotary kiln (a tubular cylindrical furnace which is inclined and revolves on its longitudinal axis; the raw material is fed in at the upper end and gradually makes its way down to the lower end, acquiring a progressively higher temperature on the way). In the kiln the zinc is volatilized; in the upper part the zinc vapor is burned to zinc oxide, which leaves the kiln along with waste gases and is collected in a bag-filter plant. Zinc production by "dry" processing presents particular difficulties because zinc has a low boiling point (906° C) and therefore occurs only in the gaseous form at the temperature necessary for effecting the reduction (1300° C). For this reason reduction has to be carried out in closed vessels or furnaces in which the zinc vapor can be condensed in the absence of air. For thermal reduction the concentrate is roasted—a treatment that may be carried out in two stages (multiple-hearth furnace followed by sintering on a traveling grate) or in a single-stage operation in which a proportion of the roasted material is fed back to the sintering machine. Sintering is necessary for transforming the material into suitable lumps to allow air to flow through it during the subsequent reduction process.

Reduction of the zinc oxide can be done by various methods. In the so-called standard process, reduction is effected in horizontal retorts in a retort furnace or distilling furnace (Fig. 2), which comprises a lower part containing regenerative chambers for preheating the gas and combustion air and an upper part in which retorts, arranged in tiers one above the other, are heated by the hot-flame gases. A retort of this kind is a rectangular distilling vessel, about 6 or 7 ft. long and about 1 ft. square in cross section. It is made of fireclay and has only a short service life, having to be renewed every four to six weeks. The retorts are charged with crushed sintered material, intermediate products containing zinc, and coke breeze (as the reducing agent). This mixture is fed into the retorts by high-speed belt conveyors; hand charging is still employed in some zinc-processing plants, however. When the retorts have been charged, their mouths are provided with so-called condensers, likewise made of refractory material, and the furnace is heated up to about 1300° C, at which temperature reduction of the zinc oxide takes place according to the reaction $ZnO + C \rightarrow Zn + CO$. The zinc vapor escapes from the retorts and is collected in the condensers as liquid metal. At the end of about 20 hours the process has been completed; the liquid zinc and zinc dust are then removed from the condensers, and the residual matter is removed from the retorts by special machines. Cleaning out the retorts and recharging them takes about 4 hours, so that the whole cycle can be repeated every 24 hours. In a more recent development of the process a single large condensing chamber is used instead of individual condensers.

(more)

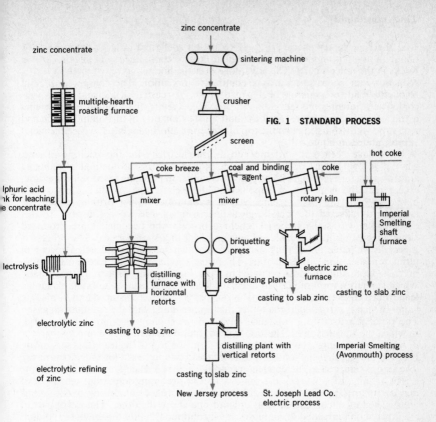

FIG. 1 STANDARD PROCESS

zinc concentrate

zinc concentrate

sintering machine

crusher

multiple-hearth roasting furnace

screen

hot coke

coke breeze

coal and binding agent

coke

lphuric acid
nk for leaching
e concentrate

mixer

mixer

rotary kiln

Imperial Smelting shaft furnace

lectrolysis

distilling furnace with horizontal retorts

briquetting press

carbonizing plant

electric zinc furnace

casting to slab zinc

electrolytic zinc

casting to slab zinc

casting to slab zinc

electrolytic refining of zinc

distilling plant with vertical retorts

Imperial Smelting (Avonmouth) process

casting to slab zinc

New Jersey process

St. Joseph Lead Co. electric process

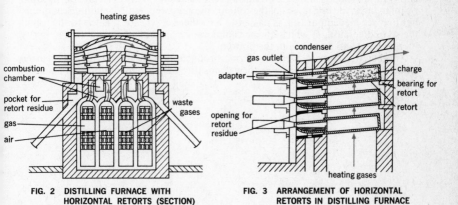

heating gases

combustion chamber

pocket for retort residue

gas

air

waste gases

condenser

gas outlet

adapter

charge

bearing for retort

retort

opening for retort residue

heating gases

FIG. 2 DISTILLING FURNACE WITH HORIZONTAL RETORTS (SECTION)

FIG. 3 ARRANGEMENT OF HORIZONTAL RETORTS IN DISTILLING FURNACE

In the New Jersey process (Fig. 4), reduction is effected in large vertical retorts about 14 m (45 ft.) high and of rectangular cross section, lined with silicon-carbide bricks in the hottest parts. The advantage over the horizontal retort method is that reduction can be performed as a continuous operation. The charge consists of briquettes made of a mixture of zinc oxide material (roasted blende) and bituminous coal. The briquettes are fed automatically to the retort, the residual matter being extracted by a screw conveyor at the bottom. The mixture of carbon monoxide and zinc vapor is discharged from the top of the retort into a condenser, where the metal is precipitated in liquid form.

There have been many attempts to utilize electricity for the smelting of zinc. The St. Joseph Lead Co. (U.S.A.) has developed a successful method of reducing zinc in an arc furnace (Fig. 5). The process is similar in principle to the retort method, except that the heat is now supplied by the electric arc. Another fairly recent development is the method employed at Avonmouth (Great Britain), which has significantly affected the metallurgical processing of lead as well as zinc. In this method, known as the Imperial Smelting process, the two metals are produced simultaneously from the oxides of zinc and lead in a shaft furnace. The process is especially valuable for dealing with ores in which sphalerite and galena occur in intimate association with each other. The furnace is charged with coke and a mixture of roasted lead and zinc ores, prepared by a pressure sintering process. In the furnace, which is really a form of blast furnace, the lead oxide is reduced to molten metallic lead, which collects at the bottom. The zinc oxide is likewise reduced and forms zinc vapor, which is extracted at the top of the furnace along with the combustion gases. The vapor is passed to a condenser in which the cooling medium is molten lead, in which the zinc dissolves. The zinc-in-lead solution is then passed into a separator in which, on cooling, a layer of liquid zinc forms on top of the lead (this separation is due to the fact that the solubility of zinc in lead diminishes at lower temperatures). The lead is returned to the condenser, and the zinc is further processed by refining.

The refining of zinc—i.e., the removal of the remaining impurities (chiefly lead and cadmium)—is effected by redistillation in a furnace comprising two columns constructed of silicon-carbide trays placed one above the other. The lower part of the first column is heated. Impure zinc is fed continuously into the top of the column and is vaporized as it flows down through the heated trays. After further purification by refluxing in the upper part of the column, the zinc vapor (still containing cadmium, but free of other impurities) is passed to a condenser, whence it is fed to the top of the second column, in which all the cadmium is driven off. Zinc of 99.995% purity is condensed and drawn from the bottom.

Another widely used method of zinc production is by the leaching of roasted zinc concentrates with acid and then depositing the zinc by electrolysis from the solution thus obtained.

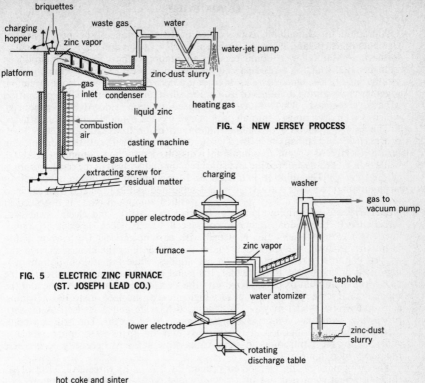

FIG. 4   NEW JERSEY PROCESS

FIG. 5   ELECTRIC ZINC FURNACE
(ST. JOSEPH LEAD CO.)

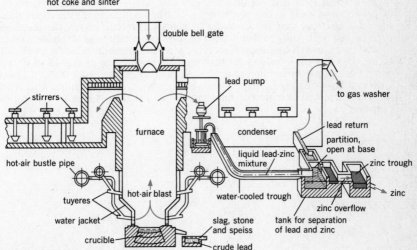

FIG. 6   IMPERIAL SMELTING PLANT

ALUMINUM

Aluminum (or aluminium, as it is called in Britain) is one of the most abundant elements on earth, and its oxide is present in clay, kaolin and many other mineral formations. Mainly for economic reasons, aluminum is almost exclusively produced from bauxite, which is a residual clay formed in tropical regions by the chemical weathering of basic igneous rocks. It contains 55 to 65% aluminum oxide (alumina) together with varying amounts of iron oxide, silica and titanium oxide. Preparation of bauxite is carried out in two stages: first, pure aluminum oxide (Al_2O_3) is produced, which is then decomposed into aluminum and oxygen by an electrolytic treatment.

The principal method of making aluminum oxide from bauxite is the Bayer process (Fig. 1). The bauxite is dried, ground and treated with caustic soda solution in an autoclave. As a result, the aluminum is dissolved as sodium aluminate ($NaAlO_2$), while iron oxide, titanium oxide and silica remain undissolved in the residue (known as "red mud"). The solution is filtered, and the aluminum is precipitated from it as aluminum hydroxide $Al(OH)_3$, which is separated by filtration and then calcined to aluminum oxide in a rotary kiln. The purified aluminum oxide is dissolved in molten cryolite—a sodium-aluminum fluoride (Na_3AlF_6)—and electrolyzed with direct current. This is done in an electrolytic cell (Fig. 2), which is essentially a tank lined with carbon bricks and provided with carbon anodes. The carbon lining forms the negative pole (cathode). Under the influence of the electric current the oxygen of the Al_2O_3 is deposited on the anodes, while the molten aluminum is deposited on the lining. In particular, the metal accumulates at the bottom of the cell. More aluminum oxide is stirred into the electrolyte from time to time and the molten metal removed. Currents of very high intensity are used (up to 100,000 amps. at 5 or 6 volts). A cell may be 20 ft. long, 6 ft. wide and 3 ft. deep. A modern processing plant may comprise a large number of such cells. The ordinary commercial aluminum obtained in this process may be up to 99.9% pure, which is sufficient for most purposes. In some cases, however, it is necessary to increase the purity by refining.

The principal refining method in present-day use is by three-layer electrolysis (Fig. 3), which is carried out in a cell provided with a carbon-lined bottom and magnesite-lined walls. In this type of cell the carbon bottom forms the anode, while a graphite electrode forms the cathode. To increase its specific gravity, the aluminum to be refined is first alloyed with copper or some other metal and is introduced into the cell in the molten condition. Over it is a layer of molten salt which is specifically lighter than this alloy, but heavier than pure aluminum. The passage of an electric current causes pure aluminum to go to the cathode, with the result that it accumulates as a layer floating on the molten salt. This aluminum, which has a purity of 99.99%, is removed from time to time and cast into suitable shapes for commercial purposes.

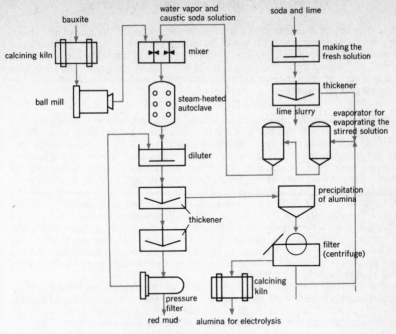

FIG. 1 FLOW SHEET OF THE BAYER PROCESS

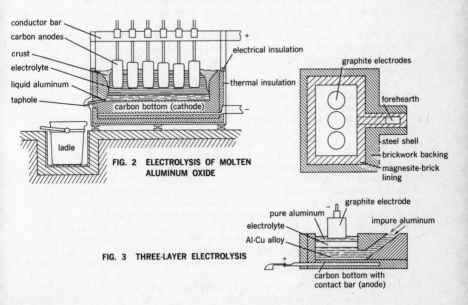

FIG. 2 ELECTROLYSIS OF MOLTEN ALUMINUM OXIDE

FIG. 3 THREE-LAYER ELECTROLYSIS

93

The ores or the concentrates of the heavy metals copper, lead, zinc and nickel (but not iron and tin) consist mainly of the sulphides of those metals. To enable the sulphides to be reduced with carbon or dissolved with dilute acids, total or partial removal of the sulphur is necessary. This is achieved by roasting, which is a heat treatment carried out in an oxidizing atmosphere and which conforms to the following general reaction:

$$MeS + 1\tfrac{1}{2}O_2 \rightarrow MeO + SO_2$$

(where Me denotes any bivalent metal). Thus, roasting results in the formation of the metallic oxide and sulphur dioxide gas, which may be processed to sulphuric acid. Once the roasting process has been initiated by ignition, it produces its own heat and generally requires no additional fuel. Arsenic and antimony, when present, are likewise removed by roasting. In a wider sense, roasting may also denote the process of driving off the carbon dioxide from carbonate ores; this process is more particularly referred to as calcining.

Depending on the further treatment that the roasted material will undergo, roasting is so controlled that all the metal is transformed into its oxide (dead roasting) or that, alternatively, only a proportion of the metal is oxidized while the rest remains combined with sulphur (partial roasting). The roasting process can, moreover, be regulated by varying the temperature and the air-supply rate in such a way that the sulphate of the metal is formed; or its chloride can be formed by the addition of chloride salts or chlorine gas.

There are many different types and varieties of roasting furnace. The multiple-hearth furnace (Herreshoff furnace, Fig. 1) has a number of annular-shaped hearths mounted one above the other. Each hearth has rabble arms driven from a common center shaft. The material for roasting is charged at the periphery of the top hearth; the arms push it inwards to the center, where it falls to the next hearth; here it is moved to the periphery and falls to the next hearth; and so on.

In the process known as flash roasting, the finely pulverized sulphide material is roasted in a special kind of fluidized-solids reactor (Fig. 2). Sulphur dioxide and the metallic oxide are formed. The material for roasting is introduced into the top of the combustion chamber along with a stream of preheated air. A swirling motion is imparted to the gases in the chamber, so that the larger particles settle in the hopper bottom.

The same general principle of roasting the finely divided material while it is in suspension in an air stream is applied in the fluosolids process (Fig. 3). Air is blown in through a bottom grate, and the finely pulverized material in the reactor is thus kept in a state of turbulent suspension. The material in this condition forms a "fluidized bed." The roasted material is discharged through an overflow pipe. Some of it is carried along as dust with the exhaust gas and is collected in a dust filter.

A widely employed machine for the roasting or calcining of a variety of materials is the rotary kiln (Fig. 4), which consists of a slightly inclined cylindrical steel tube which rotates on its longitudinal axis and is lined with refractory material. The charge to be roasted or calcined is fed in at the upper end and is progressively heated in the course of its journey down to the other end of the kiln. The kiln is fired from the lower end, where an oil, gas or pulverized-coal burner is installed.

The roasted material obtained from the above-mentioned processes is in the form of a powder. For further treatment in the shaft furnace, however, a lumpy charge material is required. The powder can be agglomerated by a process called sintering, whereby the particles become caked together by partial fusion (in addition to undergoing roasting), so that a porous mass called sinter is formed. This can be done on a sintering belt (such as the Dwight-Lloyd machine illustrated in Fig. 5) which comprises a traveling endless grate on which a bed of ore or concentrate is placed. Air is sucked or blown through the bed. Roasting is initiated by a gas, oil or pulverized-coal flame in an ignition hood under which the grate passes.

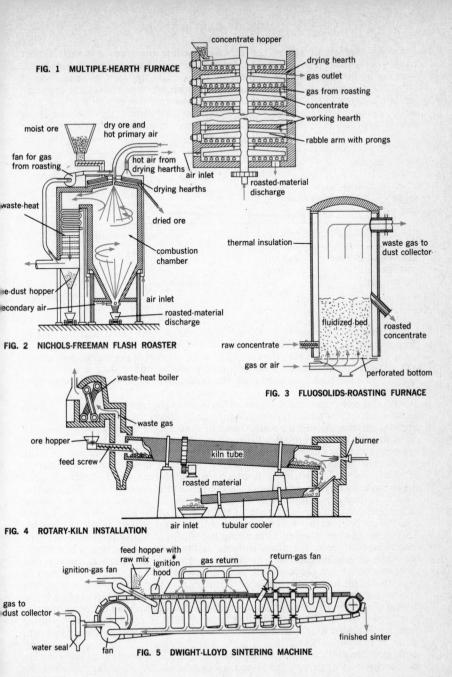

FIG. 1 MULTIPLE-HEARTH FURNACE

concentrate hopper
drying hearth
gas outlet
gas from roasting
concentrate
working hearth
rabble arm with prongs

moist ore
dry ore and hot primary air
fan for gas from roasting
hot air from drying hearths
air inlet
roasted-material discharge
drying hearths
dried ore
waste-heat
combustion chamber
e-dust hopper
econdary air
air inlet
roasted-material discharge

FIG. 2 NICHOLS-FREEMAN FLASH ROASTER

thermal insulation
waste gas to dust collector
fluidized bed
roasted concentrate
raw concentrate
gas or air
perforated bottom

FIG. 3 FLUOSOLIDS-ROASTING FURNACE

waste-heat boiler
waste gas
ore hopper
feed screw
kiln tube
burner
roasted material
air inlet
tubular cooler

FIG. 4 ROTARY-KILN INSTALLATION

feed hopper with raw mix
ignition hood
gas return
return-gas fan
ignition-gas fan
gas to dust collector
water seal
fan
finished sinter

FIG. 5 DWIGHT-LLOYD SINTERING MACHINE

95

INGOT CASTING

An intermediate operation between the smelting of metals and their further treatment is casting. The metal may be cast in the form of large ingots for subsequent working by forging, rolling, stamping, etc., or the castings may be smaller ingots or billets—especially for nonferrous metals—which are more convenient to handle and are afterwards recast or are eventually formed into castings for machine parts or other components that require only a small amount of finishing treatment of one kind or another. The principal casting processes in industrial use are as follows:

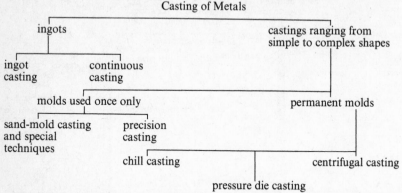

Casting of Metals

ingots · castings ranging from simple to complex shapes

ingot casting · continuous casting

molds used once only · permanent molds

sand-mold casting and special techniques · precision casting

chill casting · centrifugal casting

pressure die casting

The choice of method will depend upon the melting point and the casting properties of the metal or alloy concerned and, in the case of final castings, also upon the purpose for which the casting is to be employed, the desired mechanical and technological properties, the requisite precision and surface condition, the number of castings to be produced, the question of economy, and other considerations.

Steel and nonferrous-metal ingots for further working are normally cast in ingot molds made of cast iron. These molds are usually filled from the top, either directly from the furnace through a pouring spout or from a ladle. In such cases the molten metal may be poured through a runner, a kind of long funnel, which is withdrawn before the ingot solidifies. Alternatively, bottom (or uphill) pouring may be used, to reduce turbulence during pouring. This is illustrated in Fig. 1, where the arrangement known as group casting is presented—i.e., several ingot molds are filled simultaneously through one runner. In the procedure shown in Fig. 2 the molten metal is poured direct from the ladle into the mold. The latter is at first held almost in the horizontal (tilted-over) position and is gradually swung back to the upright position as pouring proceeds, while the ladle is progressively tilted farther over. This technique too is aimed at reducing turbulence and thus producing a better-quality ingot. In order to achieve uniform quality and not too coarse a structure of the metal on solidification, and moreover to obviate such defects as segregations, pipes and blowholes, the ingots are generally cooled with water. The water may be circulated in the walls of the molds, or the molds may be sprayed with water, or they may be immersed bodily in water. This last-mentioned principle is applied in the casting of nonferrous-metal ingots to produce a fine-grained and uniform structure. In this process an electrically heated jacket, which is provided with a pouring aperture, is placed over the thin-walled ingot. When the mold has been filled with metal, it is slowly lowered into a tank of cooling water. A development of this method is the multiple-immersion process (Fig. 3), in which several ingot molds, each enclosed by a heated jacket, stand in a casting pit. The metal is poured into the molds through distributor channels and long runners. The latter are withdrawn when the molds have been filled, and the heated jackets are then slowly removed and the pit is flooded with water.

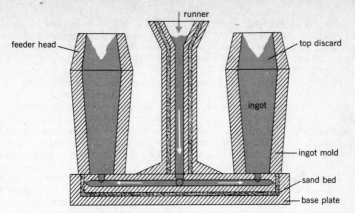

**FIG. 1 GROUP CASTING BY THE
BOTTOM-POURING METHOD**

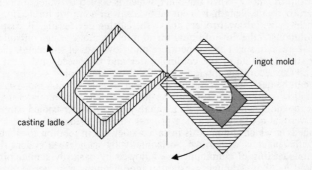

FIG. 2 TILTING-MOLD CASTING

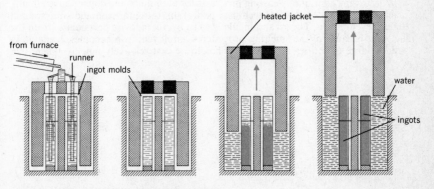

FIG. 3 MULTIPLE-IMMERSION CASTING

CONTINUOUS CASTING

More than a century ago Bessemer patented a proposed method of obviating the intermediate stages between the molten crude metal and the semifinished product. According to his invention, the steel would be cast between two water-cooled rollers and be pulled out in the form of a solid plate (Fig. 1). Since those days there have been a large number of proposals based on this fundamental idea, but its technical realization presented great difficulties which proved insurmountable until the first major breakthrough was achieved, not long before the Second World War. The first metal for which continuous casting was successfully employed was aluminum, which, because of its low melting point and its favorable casting and solidification properties, presented the least difficulty. At the present time, vertical continuous casting is in fact the most commonly used method for the production of ingots, tubes and other sections of this metal and its alloys. The principle of the aluminum-casting process was developed by Junghans: the metal is molded into a billet (or other desired barlike casting) in a mold that is open at the top and bottom and is of the appropriate square, round or rectangular cross-sectional shape (Fig. 2, page 101). When casting commences, the bottom of the mold, which is made of copper or aluminum and is water-cooled, is at first closed by a retractable table. When the billet has solidified in the mold, the table is lowered and the billet can now be slowly withdrawn from the mold. When the billet, which is cooled by water spraying, has reached a certain desired length, it is cut off by a saw or a cutting torch. If the billet is suitably held by withdrawal rollers above the cutting device (Fig. 3), casting can proceed continuously, the flow of liquid metal into the mold being controlled by a float-operated mechanism. The temperature and flow rate of the metal, the mold temperature, the flow rate of the cooling water and the speed of descent of the billet must all be accurately interadjusted. With several molds side by side, the installation can be designed to produce a number of billets simultaneously. Tubes and other hollow sections can likewise be produced by continuous casting. For this purpose the mold is provided with a suitably shaped water-cooled core.

From the vertical continuous-casting process various horizontal processes have been developed, in which the mold is installed sideways in relation to the holding furnace and the billet is withdrawn horizontally by withdrawal rollers (Fig. 4). The continuous casting of aluminum was followed by basically similar processes for copper and its alloys, though substantial modifications were needed: graphite molds cooled at the base are employed, and casting is carried out under protective gas shielding.

Whereas the continuous-casting processes for aluminum and copper have already reached an advanced stage of technical perfection, those for iron and steel are still under development. The great difficulties to be overcome arise particularly from the fact that because of the smelting procedures employed, iron becomes available in sudden large quantities and therefore has to be cast relatively quickly. There is a

(more)

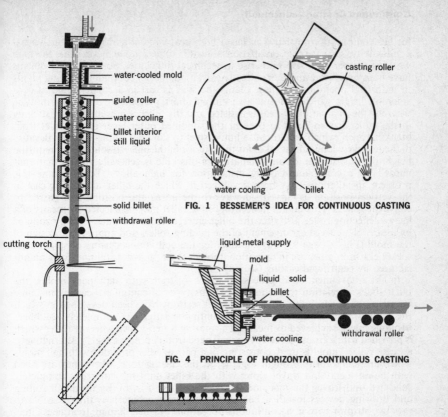

water-cooled mold

guide roller

water cooling

billet interior
still liquid

casting roller

water cooling — billet

FIG. 1   BESSEMER'S IDEA FOR CONTINUOUS CASTING

solid billet

withdrawal roller

cutting torch

liquid-metal supply

mold

liquid   solid

billet

water cooling   withdrawal roller

FIG. 4   PRINCIPLE OF HORIZONTAL CONTINUOUS CASTING

FIG. 3   VERTICAL CONTINUOUS CASTING

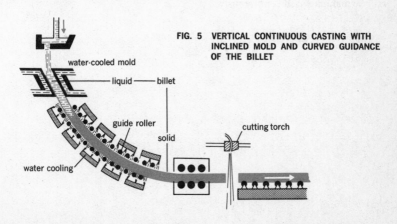

**FIG. 5   VERTICAL CONTINUOUS CASTING WITH
INCLINED MOLD AND CURVED GUIDANCE
OF THE BILLET**

water-cooled mold

liquid —— billet

guide roller

solid

cutting torch

water cooling

further complication in that iron has a high heat content and fairly poor thermal conductivity, so that rapid dissipation of heat is difficult to achieve, while the high working temperature also gives rise to technical difficulties. Some of the problems have been overcome, and there already exist a number of continuous-casting plants in industrial operation. In these plants the steel is cast in a copper mold of considerable height which performs a very rapid short, periodic up-and-down motion, whereby the dissipation of heat is assisted and the billet is prevented from sticking in the mold. It also helps to prevent the solidified outer layer of the billet from breaking open, which would cause liquid steel to flow out of the interior. Because of the relatively long time it takes for the interior of the billet to solidify, the requisite height of construction for an installation of this kind is generally greater than that needed for a continuous-casting installation for light alloys. To overcome this problem, installations have been developed in which the billet, after being cast in the vertical position, is curved sideways to the horizontal position and passed to conveyor rollers. This lateral deflection of the billet may be effected by means of a heavy deflecting roller just after the billet emerges from the water-cooling zone or by means of a curved arrangement of the guiding rollers and appropriate design of the mold (Fig. 5, page 99). Installations for the continuous casting of spigot-and-socket cast-iron pipes are in operation, though these pipes are still unusually manufactured by centrifugal casting (see page 112).

In the above-mentioned processes, which operate with stationary molds, the billet slides in relation to the mold. Various types of continuous-casting machine for strips, bars, wires and other small steel sections have been developed in which the mold travels along with the billet until the latter has completely solidified. Molding may be achieved by means of a rotating water-cooled wheel or roller which is provided with a groove closed by a metal band (rotary process, Fig. 7). Alternatively, a "caterpillar" mold is used, consisting of a pair of belts provided with half molds which are temporarily brought together in the casting zone, where they are filled with liquid metal and travel along with the billet until the latter has completely solidified and leaves the machine (Hazelett process, Fig. 6). By means of rolling and drawing devices installed behind the casting machine it is thus possible to produce strip or wire in a continuous operation. Further shaping treatments may be added: e.g., stamping or deep-drawing devices, whereby sheet-steel products can be manufactured continuously.

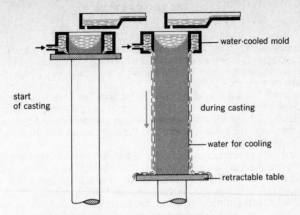

FIG. 2   SEMICONTINUOUS CASTING OF ALUMINUM

start
of casting

water-cooled mold

during casting

water for cooling

retractable table

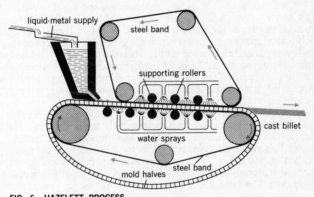

liquid-metal supply

steel band

supporting rollers

water sprays

cast billet

mold halves

steel band

FIG. 6   HAZELETT PROCESS

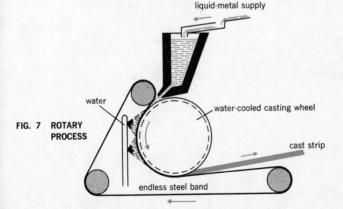

liquid-metal supply

water

water-cooled casting wheel

**FIG. 7   ROTARY
PROCESS**

cast strip

endless steel band

Sand Casting

The numerous processes for the casting of metals can be divided into two main groups: casting with expendable molds (sand casting, shell molding, etc.) and casting with permanent molds which can be reused a large number of times (chill casting, pressure die casting, etc.). In either case it is necessary to make a model of the casting to be produced. Such a model is called a "pattern" in founding. The mold is then produced from the pattern. Wood, plaster, metal and plastics are materials used for pattern making. Except for very simple castings, the pattern will generally comprise two or more parts: the actual pattern and the cores which will form the cavities and recesses in the casting.

In casting with expendable molds, the individual pattern parts are first made by hand or by mechanical means and then assembled. The molding materials—i.e., those used for constructing the actual molds in which the metal will be cast—are usually mineral substances such as sand, cement, fireclay, plaster, etc., in conjunction with bonding agents (sulphite solution, oil, water glass, synthetic resins, etc.) which give the molds the necessary strength and dimensional accuracy. The bonding action may be achieved by drying or by chemical consolidation (curing). In dry-sand molding the mold is baked; in green-sand molding the mold is used with sand in the damp ("green") condition. The metal is poured from above into an open mold. Closed molds, which are the more usual kind, are filled through a special system of channels (called runners or gates) which are generally so contrived that the metal enters at a low point and rises in the mold. When the metal has solidified and cooled, the casting is removed from the mold, and the runners and risers (which ensure that the mold is properly filled and compensated for shrinkage) are detached from the casting. The latter is then cleaned up by abrasive blasting, tumbling, grinding and cutting.

Sand casting can be used for all the common metals. There are many sand-casting processes and special processes derived from this method. These are known by various names such as box molding, open sand molding, pit molding, template molding, etc. The most widely employed method for making comparatively small castings is box molding (Fig. 1). In this method the pattern is embedded in the sand (or other mold material) within a molding box which usually comprises an upper and a lower part, the sand being compacted by ramming, pressure or vibration. Then the box is opened, the pattern removed, the cores inserted, the box closed again and casting carried out. For the casting of very large, heavy and intricate components the pit molding process is employed. Here the mold is built up in a casting pit. To give the sand greater strength when used as a mold material for large castings, cement may be added to it (cement-sand method). For symmetrically shaped castings the mold is sometimes formed by means of a template, a metal plate cut to the desired profile for producing a certain shape when it is moved along— e.g., slid along a guide track or rotated on a pivot (as in Fig. 2).

(more)

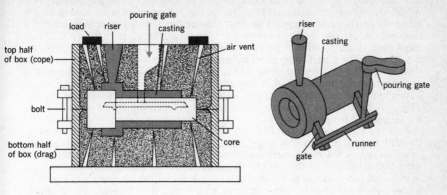

FIG. 1   BOX MOLDING

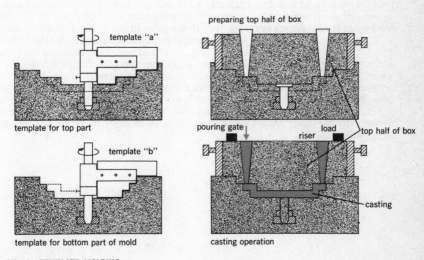

FIG. 2   TEMPLATE MOLDING

Shell Molding

A substantial saving of mold material may be effected by using a molding box whose shape roughly corresponds to that of the pattern it encloses, so that only a relatively thin layer of material is needed. The molding sand (or other mold material) is introduced into the box by a blowing device and is compacted to form a shell-like mold around the pattern. In plaster molding, the pattern made of metal or plastic is surrounded by a paste of gypsum plaster which is removed when it has set and is then assembled to form the mold which will receive the metal. With plaster it is possible to make molds of high precision, but this material has the disadvantage of having only low permeability to gas, which may give rise to difficulties in casting. This drawback can be overcome by the addition of foaming agents which increase the porosity of the plaster.

An important advance was achieved with the shell molding process, invented by Croning and patented in 1944 (Figs. 3 and 4). Its principle consists in making a thin "shell"—only a few millimeters thick—around a pattern and then assembling the parts of the shell to form the mold for the metal. A suitable mold material is a mixture of 95% fine quartz sand and 5% synthetic resin and a hardening agent. Alternatively, a sand whose grains have been precoated with a resin may be used, the advantage of this technique being that segregation of the material is thereby prevented. The metal parts that make up the pattern are mounted on a metal base plate, and together they are heated to about 250° C. Mold material is heaped on the pattern or deposited on it by a blowing device. The temperature causes the mold material to form a thin adherent coating ("shell") on the pattern by melting the resin constituent in immediate contact with the pattern. The rest of the mold material (in which the resin has not melted, so that the material is still of a loose granular constitution) is then removed, and the shell is cured by heating for a short time to 450° C. With the help of the hardening agent the shell thus attains the necessary strength to serve as a mold to receive the metal and is detached from the pattern (Fig. 3). The hollow core is produced by a similar method (Fig. 4). The halves of the shell and the cores are assembled to form the mold and are gripped in special holding devices or are glued together with a special adhesive. For casting, a number of molds may be arranged in stacks or installed side by side in a box, the voids between the molds being packed with steel balls to hold them firmly in position. When the casting metal is poured, the resin in the shell is burned away by the heat released from the metal so that only the sand remains, which can afterwards be easily shaken or knocked off the solidified casting.

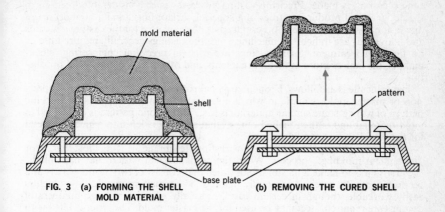

FIG. 3 (a) FORMING THE SHELL MOLD MATERIAL

(b) REMOVING THE CURED SHELL

Croning's shell-molding process

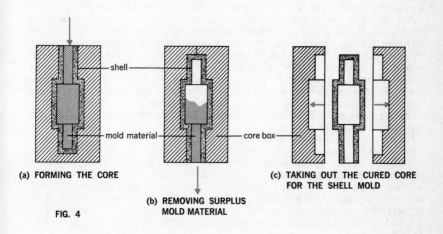

(a) FORMING THE CORE

(b) REMOVING SURPLUS MOLD MATERIAL

FIG. 4

(c) TAKING OUT THE CURED CORE FOR THE SHELL MOLD

PRECISION CASTING

The processes under the heading of "precision casting" differ from sand casting and shell molding in that the molds they employ consist of only one part—i.e., are not assembled from two or more parts—while the pattern itself is expendable each time a casting is made. Precision casting processes offer considerable freedom to the designer and produce castings of a superior surface finish and a high degree of dimensional accuracy. Among other purposes, they are used for the casting of metals and alloys that are difficult to machine, since the castings generally require little or no finishing treatment. Such castings are used in precision engineering, clock-making, the manufacture of metal ornaments, and other fields of industrial production.

The principle is as follows. From an original pattern (usually of metal) an impression or master mold is made in which a second pattern (the expendable working pattern of wax or some similar material) is cast. This second pattern is embedded in mold material and is then melted out, so that a cavity is left into which the metal for the actual casting can be poured.

The most widely used precision casting technique is the "lost-wax," or "investment molding," process. When the original metal pattern has been made, the first step is to make a master mold, which may consist of two or more parts and be provided with cores. It is usually made of a low-melting metal alloy which is easily workable, though in certain cases, especially where very large numbers of castings are required, steel may be used for the master mold. The latter is filled with molten wax, which is allowed to solidify and is then removed. Thus a wax pattern similar to the original is obtained. In some cases a number of wax patterns may be joined together in a treelike assembly, so that a corresponding number of castings can be produced in one operation. This is represented in the accompanying illustrations. The wax pattern (or "tree" of patterns) is immersed in a wet slurry or paste consisting of a fine-grained refractory mold material and a bonding agent, so that the wax pattern becomes coated with this mixture. The wax pattern is now taken out of the paste bath and the coating is built up to a greater thickness by having strewn on grains of a coarser mold material. The coating, which closely envelops the wax pattern and reproduces every detail of its shape, is called the "investment." The pattern thus "invested" is placed, with the pouring gate downwards, in a special box, which is filled up with more mold material. The complete mold is then heated, causing the wax to melt and run out. This is followed by baking the mold at about 1000° C for several hours, so that it becomes hard and strong. The metal is cast in the mold while the latter is still hot. Filling the mold may be done by gravity, pressure, or centrifugal means. Finally, when the metal has solidified, the mold is broken up.

There are many variants of the lost-wax process. For instance, plastics may be used instead of wax; or the wax may be removed with solvents instead of being melted out. In the Mercast process (used in the United States), mercury is used for filling the master mold at normal temperature. Then this mold is cooled to −40° C, causing the mercury to solidify. The "frozen" mercury pattern is removed from the master mold and dipped a number of times in a special investment mixture, so that it receives a multilayer coating which forms the final mold. The temperature is allowed to rise, and the mercury liquefies and is retrieved from the mold, which is then baked.

original pattern

injecting wax into the
master mold

wax pattern

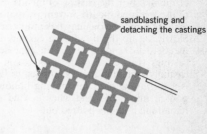

multiple wax pattern
("tree")

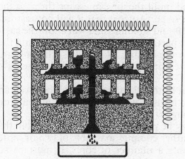

wax pattern immersed
in investment mixture

filling up the box
with mold sand

melting out the wax and
baking the mold

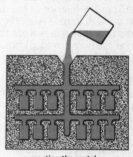

casting the metal

sandblasting and
detaching the castings

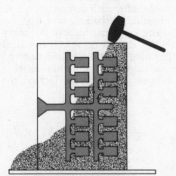

destroying the mold
to release the casting

finished castings

GRAVITY DIE CASTING

With sand casting and shell-mold casting the mold has to be broken up after each casting operation. On the other hand, in the process known as die casting, the mold, called a "die," is made of metal and is used a large number of times. It is, of course, more expensive to make than an expendable "once-only" mold. An intermediate technique is the use of semipermanent molds made of fireclay or gypsum plaster, from which a limited number of castings can be obtained. The most widely used die-making materials are steel, cast iron and heat-resisting alloys of iron. For particular purposes other materials are sometimes employed for the dies, e.g., copper, aluminum or graphite. A die can produce castings with a smooth and clean surface and high dimensional accuracy, requiring little or no final machining or other finishing treatment. The service life of a die, in terms of the number of castings that can be produced from it, depends on such factors as the thermal shock resistance of the die material, the casting material, the temperature at which it is poured, and the casting method employed.

A great many details have to be taken into consideration in designing the pattern from which the die is made. Thus, in designing the pouring-gate system and risers it must be borne in mind that the walls of the mold exert a quenching (sudden cooling) action upon the molten metal so that this solidifies much more rapidly than in sand casting. In addition, the die must be provided with fine channels at the joints and with air-vent holes and thus enable the air displaced by the casting metal to escape from the interior of the die. (In a sand mold the air can escape through the porous mold material.) Also, the die must be so constructed that it will not restrain the shrinkage that occurs when the metal cools and solidifies and will allow the casting to be easily removed. Shrinkage presents particular difficulties in designing the cores which form the cavities and recesses in the casting. Normally such cores are made of steel or special alloys. Sometimes compressible sand or shell cores are used, however.

To prevent the casting metal from sticking to the die, the latter may be given an internal coating of clay, chalk or bone ash with water glass as a binder, this mixture being applied to the die by brushing, spraying, or immersion.

In the case of simple castings the metal may be poured into the open die from the top (Fig. 3). Usually the die is a closed and rather complex assembly of two or more parts, however (Fig. 1), sometimes comprising a number of cores. It must be so designed that the molten metal will flow quickly, without turbulence, into all parts of the die. For the casting of metals with a low melting point it may be necessary to use a heated die (to prevent premature solidification), and for metals with a high melting point the die may have to be artificially cooled after each casting operation. Slowly tilting the die during casting, in order to reduce turbulence and help the metal to flow smoothly, is an expedient that is employed particularly for heavy castings (Fig. 2). For the production of awkwardly shaped or very thin-walled castings a vacuum may be applied to facilitate the filling of the die. "Slush casting" is a technique used for making hollow ornamental castings: the molten metal is poured into a die, and when a solid shell of sufficient thickness has formed, the remaining liquid is poured out.

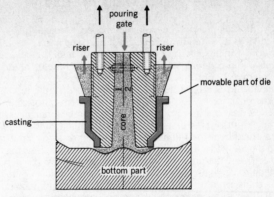

FIG. 1   GRAVITY DIE CASTING WITH MULTIPART DIE

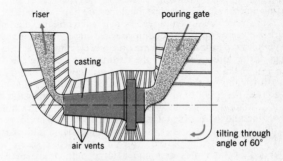

FIG. 2   TILTING DIE-CASTING TECHNIQUE

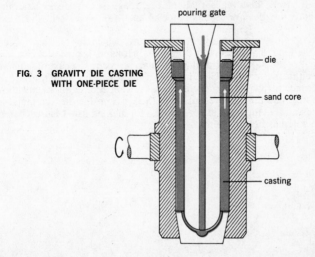

FIG. 3   GRAVITY DIE CASTING WITH ONE-PIECE DIE

PRESSURE DIE CASTING

Pressure die casting makes possible the economical quantity production of intricate castings at a rapid rate. Such castings, which may comprise various holes, recesses, screw threads, etc., are characterized by high dimensional accuracy, good surface finish, and economy of metal; they require little or no final machining. The principle on which pressure die-casting processes are based consists in forcing molten metal into a mold (the "die") under considerable pressure. The machines used for the purpose operate on one of two systems: hot-chamber machines (for metals with a low melting point) and cold-chamber machines (for metals with a high melting point). In the first-mentioned type of machine (Fig. 1) the metal is kept liquid in a crucible inside the machine, and the pressure chamber that delivers the metal into the die is located in the metal bath. Such a machine may be pneumatically operated or, more frequently, develop the pressure by the action of a ram. The casting metal for a cold-chamber machine is kept liquid in a holding furnace from which it is transferred to the pressure chamber (which may be integral with the die or separate from it) by means of a scoop or a special automatic device and is forced into the die by means of a ram. Depending on the arrangement of the ram, a distinction can be made between "vertical" (Fig. 2) and "horizontal" (Fig. 3) pressure die-casting machines. The simplest machines are hand-operated, but fully automatic machines are more usually employed, which make possible high rates of production. A complete cycle of operations comprises closing the die, forcing the molten metal into it, withdrawing the cores, opening the die, ejecting the casting and (if necessary) shearing off the sprue, deburring the casting, and cleaning the die. The number of cycles per hour that a casting machine can attain will depend on the size and shape of the castings and on the casting metal used. Thus, with zinc alloys it is possible, within a given period of time, to produce about seven times the number of castings that can be made with brass. There are fully automatic machines which can turn out small zinc-alloy castings at a rate of more than 1500 cycles per hour. In pressure die casting, precision-made dies, sometimes of intricate multipart design and therefore very expensive, are employed, which are exposed to severe working conditions characterized by high pressures and a large number of successive variations in temperature. For the production of zinc and zinc-alloy castings the dies may be made of unalloyed steel; however, for aluminum, magnesium, copper and the alloys of these metals the dies are usually made of hot-work tool steel, which has greater durability.

A more recent development is vacuum die casting, which has hitherto been applied chiefly in the United States, Great Britain and Holland. It produces castings which have an even better surface finish than ordinary pressure die castings. There are two systems: either the die is enclosed within a hood which evacuates the air, or the holding furnace (in which the casting metal is kept molten) is so installed under the casting machine that on evacuation of the air from the die, the metal is sucked into the die and is compacted in it. A process for making iron castings based on this latter principle has been developed.

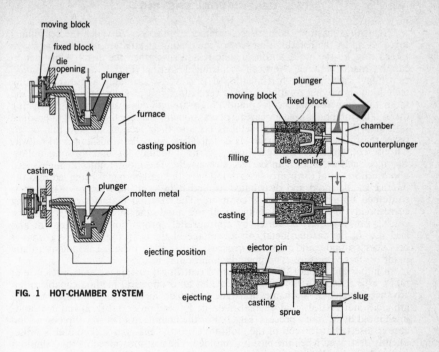

FIG. 1 HOT-CHAMBER SYSTEM

FIG. 2 VERTICAL COLD-CHAMBER SYSTEM

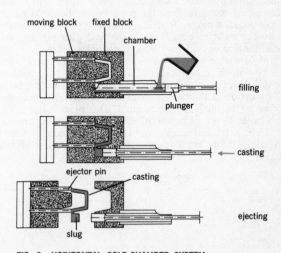

FIG. 3 HORIZONTAL COLD-CHAMBER SYSTEM

CENTRIFUGAL CASTING

"Centrifugal casting" comprises a number of processes in which the centrifugal force set up by the rotation of a part of the casting installation is utilized to shape the casting, fill the mold, and help solidify and strengthen the metal. A distinction can be made between "vertical" centrifugal casting (Fig. 1) and "horizontal" centrifugal casting. The first-mentioned process is essentially a pressure-casting technique employing rotation about a vertical axis; it produces good filling of the mold, high dimensional accuracy, and a high-strength dense structure of the casting metal. This method is used for casting of components that are too difficult to produce satisfactorily by static casting methods because their sections are too thin or for other reasons: e.g., gears, piston rings, impellers, propellers, bushings and railway wheels. Horizontal centrifugal casting is used mainly for making long hollow castings, such as pipes, gun barrels, sleeves, etc. The mold rotates at high speed about a horizontal (or nearly horizontal) axis, the molten metal being fed into the interior of the mold and distributed around it by centrifugal action. Rotation is continued until solidification is complete. The external diameter of the casting corresponds to the internal diameter of the mold; the internal diameter of the casting can, however, be varied by appropriately proportioning the amount and feed rate of the casting metal. An advantage of the centrifugal process is that it produces a sounder and more uniform casting than static means. The mold is usually made of steel or cast iron; nonmetallic linings may be used.

An important application of horizontal centrifugal casting is the manufacture of pipes, especially cast-iron pipes. It provides an economical method capable of an advanced degree of mechanization. The two main methods of centrifugal casting are in a water-cooled mold by the Briede–de Lavaud process (Fig. 2) and in a sand-lined mold by Moore's process (Fig. 3). For the manufacture of spigot pipes, a sand core is inserted at the end of the mold (to form the enlarged socket) and is subsequently destroyed when the pipe is demolded. The first-mentioned method employs a slightly inclined mold which can move longitudinally. The molten iron is introduced into the mold through a long duct from a tilting ladle containing the correct amount of casting metal to form the pipe. When the mold has reached a certain speed of rotation, the molten iron is admitted to it, and the mold is moved slowly forwards (to the right in Fig. 2) while the feed duct remains stationary, so that uniform distribution of the metal along the mold is achieved. Moore's method uses a rotating mold with a sand lining, which protects the metal shell of the mold so that water cooling is not necessary. The sand itself is applied to the mold wall and compacted by centrifugal action. The inlet duct is short because, with a sand lining, solidification of the casting takes a relatively long time (no rapid cooling); proper filling of the mold is thus ensured. This process has the advantage of not requiring a wide range of molds of different diameters, since any desired pipe diameter can be produced simply by varying the thickness of the sand lining.

Centrifugal methods are also used for the production of composite castings (see page 114).

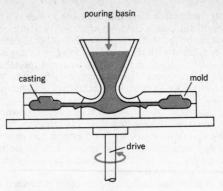

FIG. 1   CENTRIFUGAL CASTING

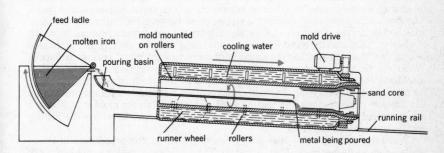

FIG. 2   BRIEDE–DE LAVAUD PROCESS

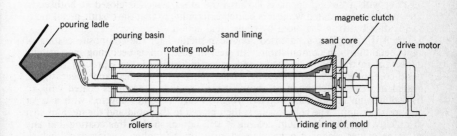

FIG. 3   MOORE'S PROCESS

COMPOSITE CASTING

"Composite casting" refers to a large number of processes in which molten metal is cast on to a solid metal component, so that the two subsequently form one integral unit. In this way the favorable mechanical or other technological properties of one metal can be combined with those of another. For example, bearings may be made from an outer shell of strong metal with a lining of special low-friction metal cast inside it, or a metal possessing high strength may be given a cast-on covering of corrosion-resistant metal, or an expensive metal may be combined with a cheaper metal for economy.

The mechanical connection of the two metals can be promoted by means of interlocking devices such as dovetailing, grooves, recesses, etc., which form a physical "key," or by shrink fitting or by bond established as a result of diffusion at the interface of the two metals so that local interpenetration occurs. The bond may be further strengthened by heat treatment (annealing) or by the interposition of special bonding layers of metal at the junction.

In the simple case of a bearing comprising an outer steel shell and a metal-alloy lining, the shell is preheated and the molten lining metal (with a lower melting point than steel) is cast inside it. In the integral-casting process developed by Mascher (Fig. 1) the steel shell is first given a coating of zinc and then, while it is still hot, is filled with molten lining metal (lead bronze) in a compound iron-and-sand mold. The zinc coating serves to establish a strong bond between the lining and the shell. Another technique is represented by the salt-bath displacement process, in which the bearing shell and the mold for forming the lining within it are together preheated in a salt bath, which also has the function of protecting the metal against oxidation. When the required temperature has been reached, the bearing metal is poured, displacing the salt. In the immersion process (Fig. 2), which is used, for example, for making bushings provided with bearing metal on both sides or for the production of relatively large castings, the bearing shell enclosed within the mold is heated in a salt bath and is then lowered into a bath of molten bronze in which the bronze displaces the salt in the mold.

Centrifugal casting techniques are extensively used for the production of thick-walled composite bearings (Figs. 3 and 4). The steel outer shell is heated in a salt bath or is zinc-coated; it is then gripped in a centrifugal casting machine, and when the latter has reached the requisite speed of rotation, the molten metal is introduced into it. In the case of a composite bearing built up from different metals in concentric layers, an intermediate bonding layer of tin is applied to the first layer of metal after it has solidified. Then the second layer is cast. In another technique for the casting of bearings the appropriate quantity of casting metal in the form of chips or granules, together with a fluxing agent, is fed into the mold, which is closed at both ends. The mold is then rotated while it is heated externally, so that the casting metal melts and is distributed in the mold by centrifugal action.

The Al-Fin process, a patented American method for the composite casting of light metal alloys in combination with steel or cast iron, is becoming increasingly important in the motor industry and other industries. In this process the steel basic component, whose surface has been thoroughly cleaned and degreased, is heated in molten aluminum at about 750° C, so that it becomes coated with an iron-aluminum compound (Fe_2Al_5) which in turn becomes covered with a layer of pure aluminum. The component is then placed in the mold and the metal is cast so quickly that the aluminum coating is still liquid, so that this coating and any adhering oxides are, as it were, washed away by the molten casting metal, which can now bond itself firmly to the underlying layer of Fe_2Al_5.

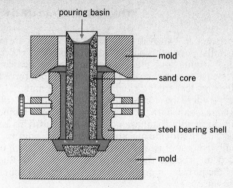

FIG. 1   MASCHER'S INTEGRAL-CASTING PROCESS

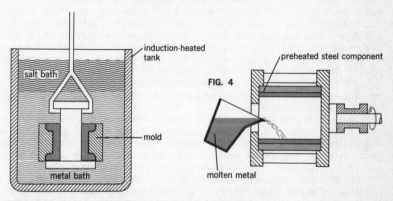

FIG. 2   IMMERSION PROCESS (MOLTEN BRONZE
WITH OVERLYING SALT BATH)

FIGS. 3 and 4
CENTRIFUGAL COMPOSITE CASTING

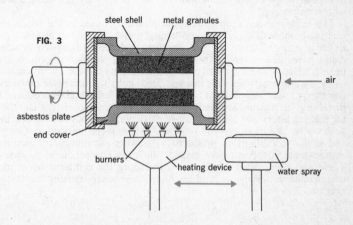

POWDER METALLURGY

This process, in which articles and components are produced by agglomeration of fine metallic powder, is employed in cases where other methods of shaping – such as casting, forging and machining – are impracticable or where special material properties have to be achieved. By way of example, three techniques for the extrusion of metallic powders are indicated in Fig. 2. The materials used in powder metallurgy (metallic powders or, for some purposes, mixtures of metallic and nonmetallic powders) are shaped by cold pressing at room temperature between steel dies, which produces initial adhesion of the particles. This is followed by heating of the compacts in a nonoxidizing atmosphere (sintering) to obtain final cohesion. The dies usually consist of two parts thrusting against each other, and each part may be subdivided to produce the required shape (Fig. 4). Another technique is isostatic pressing: the powder is pressed in a closed flexible container (rubber, plastic) under liquid pressure.

The function of the sintering treatment is to bond the powder particles of the compact into a coherent mass. As a rule, the sintering temperature is somewhat below the melting point of the powder, or the temperature may be so controlled that fusion of certain constituents of the powder mixture is achieved. Sintering as a subsequent separate treatment may be dispensed with by pressing of the powder at elevated temperature or by subjection of cold-pressed compacts to hot shaping —e.g., by drop forging, rolling or extrusion. In certain cases it is advantageous to process the powder in a protective metal envelope which provides mechanical strength and/or protection against oxidation (Fig. 2c). To prevent oxidation, hot pressing or sintering is usually carried out under the protection of a shielding gas or in a reducing atmosphere.

Shaping of the powder is generally done by the application of pressure. However, in the so-called slip casting process, a technique adopted from the ceramics industry (Fig. 1), the powder is mixed with a suitable liquid suspension medium to form a "slip" (a thick suspension), which is put into a mold (a, Fig. 1). The liquid is absorbed by the walls of the mold, usually consisting of gypsum plaster (b). Then the shaped component is removed from the mold, dried and sintered (c). The powders used in powder metallurgy are produced by comminution of solid materials, by atomizing of molten materials in a stream of gas or water (Fig. 3), or by chemical processing. It is essential to obtain particles that are suitably graded in size and are of regular shape and surface condition, so that they interlock and adhere properly when compressed.

The technique has numerous applications. It is used in the production of high-melting-point metals such as tungsten and molybdenum. For instance, pure tungstic oxide is prepared from the ore and then reduced to tungsten powder, which is cold-pressed and sintered. Another important application of powder metallurgy is the manufacture of hard-metal cutting and working tools in which cemented carbides —e.g., tungsten carbide—are incorporated: cobalt and carbide powders are mixed together, pressed and sintered, so that the cobalt fuses. Metals produced by powder-metallurgy techniques are characterized by their fine porosity, a fact that is utilized for making filters and bearings, more particularly porous bronze bearings that can soak up oil like a sponge and require no subsequent lubrication. Copper-tungsten and similar combinations produced by powder metallurgy are used as electrical-contact materials. Permanent-magnet alloys are also produced by such techniques. There are many other applications, including the combination of metallic with nonmetallic materials to produce high-temperature-resisting materials (so-called "cermets").

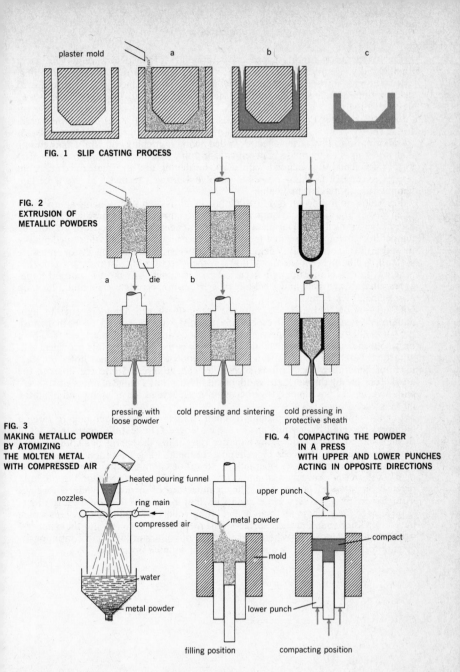

plaster mold

a b c

FIG. 1 SLIP CASTING PROCESS

FIG. 2 EXTRUSION OF METALLIC POWDERS

die

a b c

pressing with loose powder

cold pressing and sintering

cold pressing in protective sheath

FIG. 3 MAKING METALLIC POWDER BY ATOMIZING THE MOLTEN METAL WITH COMPRESSED AIR

FIG. 4 COMPACTING THE POWDER IN A PRESS WITH UPPER AND LOWER PUNCHES ACTING IN OPPOSITE DIRECTIONS

heated pouring funnel

nozzles

ring main

compressed air

water

metal powder

upper punch

metal powder

mold

lower punch

compact

filling position

compacting position

FORGING

One of the most important properties of metals is their deformability. The term "malleability" denotes the ability to be mechanically deformed by forging, rolling, extrusion, etc., without rupture and without significant increase in resistance to deformation. Metals such as lead and tin are malleable at ordinary temperatures, whereas others, such as iron, become malleable only when heated. The term "ductility" denotes the ability of metals to be mechanically deformed when cold. In the course of such deformation most metals become progressively more resistant to deformation; this latter effect is called work hardening (or strain hardening). A distinction is to be made between cold-forming and hot-forming processes. The former are usually associated with work hardening and are performed at room temperature. Hot-forming processes involve heating the metal above a certain temperature to make it malleable.

Forging is an important hot-forming process. It is used in producing components of all shapes and sizes, from quite small items to large units weighing several tons—e.g., heavy crankshafts. The metal, which is preheated to the appropriate forging temperature in a forge fire, in a forging furnace or by induction, is deformed mainly by upsetting (compressive deformation) between impact surfaces or pressure surfaces. In the process the metal flows in the direction of least resistance, so that generally lateral elongation will occur unless restrained. The most important forgeable materials are steel and steel alloys. Certain nonferrous metals and alloys are also shaped by forging.

Hand forging tools (Fig. 1) comprise variously shaped hammers, such as light and heavy sledgehammers (respectively wielded with one hand or both hands), the square flatter, the cross-peen hammer, various auxiliary hammers, etc. The base on which the work is supported during forging is the anvil, which is provided with a hardened steel face and terminates at one end or both ends in a horn, or beak, used for bending work. Various accessories can be inserted into the holes in the anvil. For holding the work, the smith has at his disposal a range of tongs and pincers with a variety of jaw shapes, together with other devices for gripping and handling larger pieces.

For the semimechanized forging of small to medium-sized components, forging hammers powered by various means are employed. The feature common to all of them is that, like the hand forging hammer, they utilize the energy of a falling weight to develop the pressure needed for shaping the metal. Larger components are forged by means of forging presses operated by steam or compressed air or by hydraulic or electric power. Largely automatic forging machines are used for the quantity production of engineering parts. The manufacture of intricately shaped forgings from bar material in very large quantities may be carried out by forging rollers (Fig. 5, page 121). These are matched rotating rollers or segments of rollers which have impressions sunk in their surfaces. The metal blank is rolled into these impressions as the rollers turn. Whereas the rollers of rolling mills rotate continuously, forging rollers perform only one revolution per shaping operation.

(more)

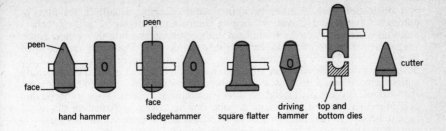

peen

peen

face

face

hand hammer   sledgehammer   square flatter   driving hammer   top and bottom dies   cutter

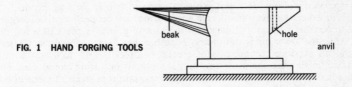

beak   hole

FIG. 1 HAND FORGING TOOLS   anvil

FIG. 2 HAMMER-FORGING OPERATIONS

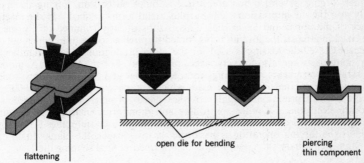

flattening   open die for bending   piercing thin component

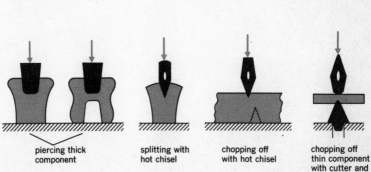

piercing thick component   splitting with hot chisel   chopping off with hot chisel   chopping off thin component with cutter and hand hammer

A distinction may be made between open-die forging, usually in the form of hammer forging, and closed-die forging. In hammer forging (Fig. 2), which is essentially derived from the traditional craft of the blacksmith, the component is shaped by hammer blows aided by relatively simple tools. These may include open dies—i.e., dies that do not completely enclose the metal to be shaped. One of the basic operations of hammer forging is the elongation of a piece of metal by stretching with hammer blows (Fig. 3), causing it to become thinner and longer. In hand forging the work piece is usually turned 90° after each blow, in order to forge it thoroughly and prevent its lateral expansion. A tube can be forged by flattening the metal longitudinally or tangentially around a mandrel. The opposite of elongation is upsetting, which produces compressive shortening. For example, the diameter of a bar can be increased uniformly or locally by heating and hammering axially.

More important is closed-die forging (Fig. 4), very widely used for mass production in industry, in which the metal blank is shaped by pressing between a pair of forging dies. The upper die is usually attached to the ram of a forging press or a forging hammer, while the lower die is stationary. Together they form a closed die. Closed-die forging can produce components of greater complexity and accuracy, with a better surface finish, than the more traditional methods not using closed dies. The dies are made of special heat-resistant and wear-resistant tool steels. A piece of hot metal sufficient to slightly overfill the die shape is placed in the bottom die, and the top die is forced against it, so that the metal takes the internal shape of the die. In hammer forging, several blows are struck in quick succession, forcing the metal evenly into the die impressions. The surplus metal forms a "flash" at the meeting surface of the upper and lower dies. This is subsequently trimmed off by special tools fixed in a press, the forging being forced through a hollow tool which cuts off the flash. Closed-die forging is used for the rapid production of large numbers of fairly small parts and also for very large components. For the latter—e.g., modern jet-aircraft components (including complete wing and airframe units)—giant hydraulically operated presses are used, which can develop forces of 50,000 tons and more. Such presses are highly complex pieces of machinery, equipped with elaborate electronic and other controlling and monitoring instruments. Forgings produced in closed dies are known as "drop forgings" or "stampings." For some purposes the forging operation is performed in two stages, the blanks first being treated in preliminary shaping dies and then formed in final shaping dies.

FIG. 3 DRAWING OUT METAL BY HAMMERING

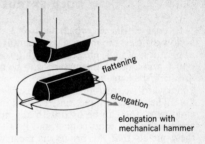

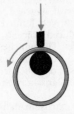

elongation with peen of hammer
flattening
elongation
elongation with mechanical hammer
longitunal and tangential flattening of a tube around a mandrel

FIG. 4 CLOSED-DIE FORGING

upper die
metal blank
lower die
in the die
forging completed
forged article
trimming off the flash
finished product

FIG. 5 FORGING ROLLS

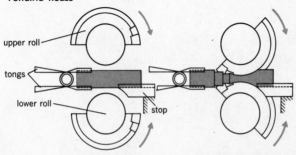

upper roll
tongs
lower roll
stop
inserting the workpiece
forging the workpiece

COLD EXTRUSION OF METALS

In cold extrusion, which is used for the manufacture of special sections and hollow articles, the material is generally made to flow in the cold condition by the application of high pressure, which forces it through the cavity enclosed between a punch and a die. A distinction is to be made between "forward" extrusion (Fig. 2), in which the extruded metal flows in the direction of movement of the punch, and "backward" extrusion (Fig. 1), which is characterized by the opposite direction of flow. The two techniques may be used in simultaneous combination (Figs. 3 and 4). Cold extrusion can be used with any material that possesses adequate cold workability—e.g., tin, zinc, copper and its alloys, aluminum and its alloys—and it is for these metals that the process is most widely adopted. Low-carbon soft-annealed steel can also be cold-extruded. If the product cannot be fully shaped in a single operation, the extrusion process may be performed in several stages. The solid or hollow products that can be made by cold extrusion are relatively limited in size.

The initial stock from which cold extrusions are produced consists of round blanks, lengths cut from bars, or specially preformed blanks. The punches and dies used in cold extrusion are subject to severe working conditions and are made of wear-resistant tool steels—e.g., high-alloy chromium steels. To reduce friction, the tool surfaces are polished. In the cold extrusion of steel the blank may additionally be given a phosphate coating to minimize friction.

A widely used special cold-extrusion method is impact extrusion. It is used for making collapsible tubes of lead, tin, zinc and aluminum—e.g., toothpaste tubes. Zinc battery cans are also produced in this way. A pointed punch descends swiftly on to a disc-shaped blank arranged in a die, causing the metal to flow upwards and around the punch.

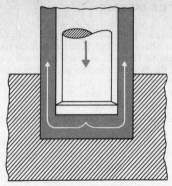

FIG. 1   BACKWARD EXTRUSION

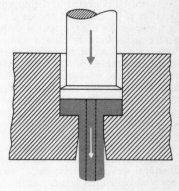

FIG. 2   FORWARD EXTRUSION

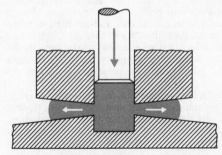

FIG. 3   EXTRUSION TRANSVERSELY TO DIRECTION
OF PUNCH MOTION

FIG. 4   COMBINATION OF FORWARD
AND BACKWARD EXTRUSION

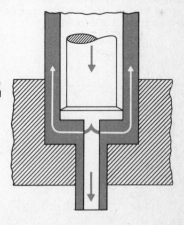

HOT EXTRUSION OF METALS

Extrusion is a hot-working process which, like forging, rolling, etc., uses the good deformability of heated metallic materials for shaping them. The most important aspect of the process is that it enables considerable changes of shape to be achieved in a single operation and provides a means of dealing with metals and alloys whose physical structure renders them unsuitable for shaping by other methods. Besides, with extrusion it is possible to form complex sections that cannot be produced in other ways. Finally, extrusion also offers economic advantages in that the dies are relatively inexpensive and are interchangeable, so that one extrusion machine can be used for the production of a wide variety of sections.

A metal billet heated to the appropriate temperature is fed into the cylindrical container of the extrusion press and is forced by the action of a ram through a steel die whose orifice has the desired shape to produce the solid or hollow section. The metal emerges from the die as a continuous bar, which is cut to the required lengths. Extrusion products are therefore essentially "linear" in character, in the sense that shaping is confined to the cross section only. The process is therefore eminently suitable for the production of barlike and tubular objects. A distinction is to be made between "direct" extrusion (Figs. 1 and 2, showing production of solid and hollow sections respectively) and "inverted" extrusion (Fig. 3), in which the extruded metal flows in the opposite direction to the movement of the ram, the extrusion die being in the ram itself.

Most metals and alloys can be shaped by extrusion. At first the process was confined to nonferrous metals and has now in fact largely superseded other methods for the shaping of such metals. Cable sheathing, lead pipe and aluminum-alloy structural sections are typical of such extrusion products. The extrusion of steel presented difficulties because of the heavy wear on the dies and the high working temperatures and stresses. However, these difficulties have been overcome, and extrusion is used, for example, in the production of stainless steel tubes. In the Ungine-Séjournet method the steel billet is coated with glass powder, which melts and forms a viscous heat-insulating and lubricating layer between the die and the extruded metal.

For making tubular sections, a mandrel is arranged in the die orifice (Fig. 2), and during extrusion the metal flows through the annular space so formed. Hollow billets are used for tubes, or solid billets are first pierced in the extrusion operation. Extrusion machines are generally hydraulic presses, with capacities ranging from about 500 tons to about 7500 tons. Graphite grease is commonly used for lubrication between metal and tools.

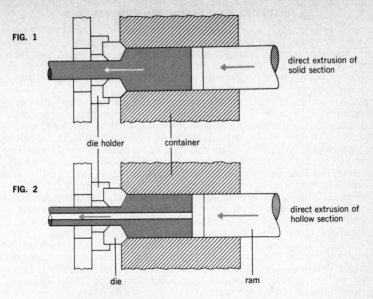

FIG. 1 — direct extrusion of solid section

FIG. 2 — direct extrusion of hollow section

die holder

container

die

ram

FIG. 3 — indirect extrusion with hollow ram

ram

examples of extruded sections

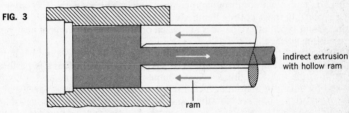

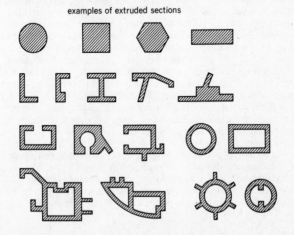

Forging, rolling, extrusion, etc., are metal-shaping processes that do not involve the removal of metal by means of cutting tools. On the other hand, many important shaping processes are based on cutting and similar operations. The tools used are made of special steels (tool steels), hard metals (cemented carbide alloys), oxide ceramic materials, and diamonds.

In this article the principles of the various methods are briefly outlined, without detailed descriptions of the machines used for performing the shaping operations. For each of these methods a whole range of tools has been developed, each type of tool being employed for a particular purpose. In chiseling (Fig. 1), the cutting edge of the tool is driven into the surface of the workpiece by the action of blows. To ensure even and regular removal of the chips it is essential to hold the chisel correctly and take care that it does not slip on the metal surface or dig too deeply into it. Chiseling is used chiefly for cutting off and for the removal of edges, burrs, fins, etc.

Planing, shaping and slotting (Fig. 2) are machining operations comparable to chiseling, characterized by the removal of the chips in one direction, the tool being moved to and fro or up and down in relation to the workpiece. In sawing (Fig. 3) the removal of metal is effected by a series of saw teeth. With power-driven band saws and circular saws, cutting can be performed in a continuous operation. The shape, spacing and number of teeth vary greatly for different saws. Large-diameter circular saws may have interchangeable teeth or interchangeable segments comprising a number of teeth. Sawing is used mainly for cutting off and for cutting plate material of not too great a thickness. Thick pieces of metal can, while still hot from the furnace, be cut with hot sawing machines or with cutting discs. The latter achieve very high cutting rates, cutting being effected by melting of the metal due to frictional heating. Band-type cutting devices are based on the same principle of heat generated at the cutting surface.

Another important basic process is filing (Fig. 4). By using suitably shaped files it is possible to cut metal to any desired shape. The actual cutting is performed by the teeth of the file. Roughing out the shape is first done by means of coarse files, followed by finishing with finer files. Files are available in a great variety of shapes, sizes and grades. They are classified and named according to sectional shape (e.g., half-round, square, triangular, round), length, and the relative fineness of cut of the teeth. With regard to fineness, the following classes of file are distinguished: bastard, second-cut, smooth, dead smooth. If there is only one series of parallel teeth, the file is known as a single-cut file; if the first series is intersected by a second and finer series, so as to form diamond-shaped teeth, the file is a so-called double-cut.

A broach (Fig. 5) is a tapered tool provided with a series of cutting teeth which are lower at one end of the tool than they are at the other. Broaching is mainly employed for machining out holes or other internal surfaces, but can also be used for external surfaces and for burnishing already-formed holes. The cut starts with the smaller teeth, which enter the hole, and finishes with the larger teeth, which bring the hole to the finished size. Fig. 5 shows an internal broach. Its cross-sectional shape may be round, rectangular, etc., depending on the desired shape of the hole. The broaching operation is performed by a machine that pulls or pushes the broach through the workpiece.

Turning (shown in Fig. 6) is one of the most important machining processes. It is the process of reducing the diameter of material held in a lathe. The workpiece is attached to a driven spindle and, while rotating, is brought into contact with a cutting tool. The position of the tool in relation to the axis of rotation can be varied so as to cut the workpiece to the desired shape. In longitudinal turning, the tool is moved

(more)

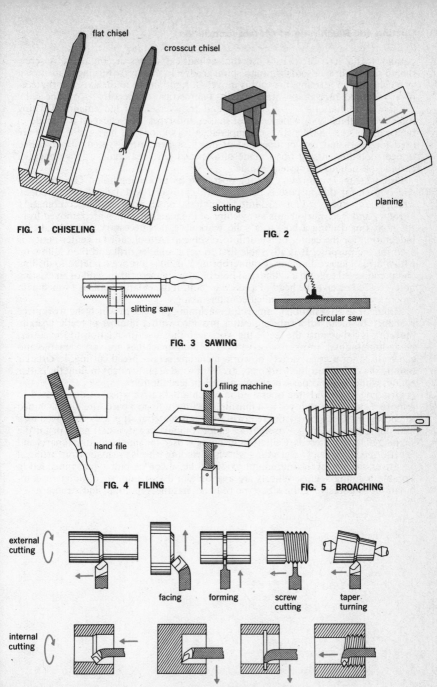

flat chisel

crosscut chisel

FIG. 1 CHISELING

slotting

planing

FIG. 2

slitting saw

circular saw

FIG. 3 SAWING

hand file

filing machine

FIG. 4 FILING

FIG. 5 BROACHING

external cutting

facing forming screw cutting taper turning

internal cutting

FIG. 6 TURNING

parallel to the axis of rotation, so that cylindrical shapes are obtained. A screw thread can be cut by a tool forming a spiral groove. In transverse turning (also known as facing) and in forming, the tool is moved at right angles to the axis. Workpieces of any desired tapered or other axially symmetrical shape can be produced by suitable combinations of longitudinal and transverse tool movements. Turning tools are available in a wide range of shapes and types. The cheapest are made of high-carbon steel, hardened and tempered. Alloys known as high-speed steels are used for tools that can be operated at much higher cutting speeds. In so-called tipped tools the cutting tip is made of a special hard material—e.g., a cemented carbide, particularly tungsten carbide.

Drilling (Fig. 7) is a rotary cutting operation for producing holes. The tool most widely used for the purpose is the twist drill, provided with helical cutting edges, which rotates and is fed forward into the material under pressure. The combination of rotary and feed motion cuts away chips of the material, which are removed from the hole. For drilling a hole in a solid workpiece, it is necessary first to make an indentation for the center of the drill to revolve in. A tool called a center punch is used for the purpose. It is advisable first to use a smaller drill and then follow up with a drill of larger diameter. Counterboring (Fig. 8) is a process related to drilling and is employed to form a cylindrical hole of large diameter at the end of an existing hole—e.g., to receive the head of a screw or bolt. If the enlarged hole is formed with tapered sides, the process is called countersinking.

Milling (Fig. 9) is another important machining process, in which the workpiece is shaped by means of a rotating cutter provided with a number of teeth. Usually the work is fed against the teeth, the work-feed direction (in relation to the cutter) being longitudinal, transverse or vertical. Milling machines are very versatile and can be used for a great variety of work, including screw-thread cutting. In circular milling the cutter and the workpiece are both rotated; in straight milling the cutter rotates while the workpiece performs a straight feed motion.

Grinding (Fig. 10) is the operation in which an abrasive wheel or disc is used to remove metal. It is employed as a finishing treatment to give parts already machined the necessary precision by the removal of excess material. It is also employed as a machining process in its own right—e.g., for roughly forged or cast parts or for the shaping of hard materials. Centerless grinding is used for small cylindrical parts and is performed between two grinding wheels. Grinding wheels are made from artificial abrasives, usually of the aluminum oxide or the silicon carbide type, embedded in suitable bonding agents. Wheels are available in a vast number of different combinations of abrasive, grain size, type of bond, hardness of bond, and structure.

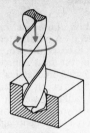

FIG. 7 DRILLING

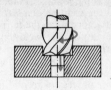

FIG. 8 COUNTERBORING

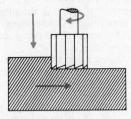

face milling

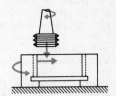

thread milling

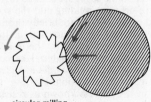

circular milling

straight milling

FIG. 9 MILLING

cylindrical grinding

face grinding

centerless grinding

FIG. 10 GRINDING

PRESSURE WELDING

Welding is the joining of metals by the application of heat and/or pressure, with or without the addition of a similar metal (filler metal). A welding technique may be designated according to the purpose for which it is used, or the procedure employed (manual or automatic welding), or the nature of the heat source (gas welding, electrical-resistance welding, arc welding, welding based on chemical reactions, etc.). As regards the purpose, a distinction is to be made between connective welding—i.e., the forming of joints and connections—and build-up welding (or surfacing), which is the process of reconditioning damaged or worn engineering components by the application of weld metal or the protection of components against corrosion or wear by the application of an armoring layer of more resistant metal (hard surfacing). As to the nature of the welding process itself, a distinction may be made between pressure welding and fusion welding.

In pressure welding, the parts to be joined are first locally heated at the place where the joint is to be formed and are then squeezed together in the plastic state so that they are united. In general, no filler metal is employed. Cold pressure welding makes use of high pressure, without the help of heat, to unite the parts. Related to this process are ultrasonic welding and explosion welding.

The oldest welding technique, still used in art ironwork and smith's work, is forge welding. This is the process of joining steel or iron parts by heating them in a forge until they reach a plastic state and are then united by hammering, by pressure, or by rolling.

In gas pressure welding (Fig. 1) the parts to be joined are heated by a gas-and-oxygen flame and are united by the exertion of continuous or of impact-type pressure. This principle is applied, for example, in the manufacture of small-diameter tubes from steel strip (Fretz-Moon process). By means of a special die, called a bell, and shaping rollers the continuous strip is formed into a tube. The edges of the strip are heated to welding temperature by gas burners, and the edges are pressed together and united by pressure rollers. In the case of arc pressure welding and the special techniques derived from it, the heat is generated by an electric arc briefly produced between the parts to be joined, which are then united by impact action. In the resistance pressure-welding process (Fig. 2) the heat is generated by the resistance encountered by an electric current which is passed through the material, especially the high resistance at the contact faces of the parts to be joined. The current is applied through electrodes or generated in the parts by induction. Heating based on electrical resistance is, for example, utilized in the process known as resistance butt welding (Fig. 2a): the two parts to be joined end to end are gripped, in contact with each other, in copper jaws which serve as electrodes for the passage of current across the joint. When the metal at the joint has reached a sufficiently high temperature, the current is switched off and the contact pressure is increased to unite the parts.

(more)

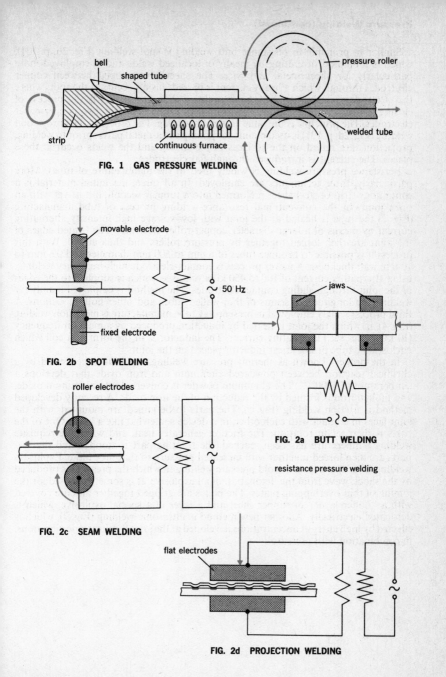

FIG. 1 GAS PRESSURE WELDING

bell

shaped tube

pressure roller

strip

continuous furnace

welded tube

FIG. 2b SPOT WELDING

movable electrode

fixed electrode

50 Hz

FIG. 2a BUTT WELDING

jaws

FIG. 2c SEAM WELDING

roller electrodes

resistance pressure welding

FIG. 2d PROJECTION WELDING

flat electrodes

Similar in principle to resistance butt welding is spot welding (Fig. 2b, p. 131), which is a method of uniting by means of localized welds and is employed more particularly for sheet metal and wire. The sheets are gripped between copper electrodes through which a heavy current is passed; fusion occurs at the spots where the electrodes are thus applied. Seam welding (Fig. 2c, p. 131) is the process of closing a seam by a continuous resistance weld formed between two copper-roller electrodes. The principle is the same as in spot welding. The process called projection welding (Fig. 2d, p. 131) is used mainly for joining sheet-metal parts: prior to welding, projections are raised on the surfaces of the sheets, and the welds occur at these places. The current is introduced through flat electrodes.

Resistance pressure welding is widely used in the manufacture of tubes. More particularly, three techniques are employed. In all three, the initial material is a continuous strip of steel sheet preformed into a tubular section, but as yet with an open joint. In the conventional resistance-welding process of tube manufacture (Fig. 3) the tube is heated at the joint with low-voltage high-intensity alternating current by means of a large-diameter copper-roller electrode. The heated edges of the joint are then forced together by pressure rollers and thus united. With this process it is possible to produce tubes of 6 mm to 500 mm diameter and 0.6 mm to 10 mm wall thickness. A newer process is contact-electrode high-frequency welding, using alternating current of 100 to 450 kilocycles/sec. which is supplied to the edges of the joint through sliding contact electrodes. The method is employed mainly for welding the longitudinal seams of thin-walled tubes and other hollow sections. A third process widely employed in present-day tube manufacture is induction welding (Fig. 4), in which the joint is heated by induction produced by a medium-frequency (10 kilocycles/sec.) alternating current. The inductor is in the form of a coil which encloses the tube or is a linear inductor placed on the joint.

In the process known as thermit pressure welding, the heat is generated by a chemical reaction between powdered aluminum and iron oxide that develops a temperature of 3000° C. The aluminum powder is converted into aluminum oxide, and molten iron is formed by the reduction of the iron oxide. A recently developed method is friction welding (Fig. 5). The parts to be joined are mounted, with the joint faces in contact with each other, in a device somewhat like a lathe. One of the parts is then set in rotation. The friction generates heat, and when the requisite welding temperature has been reached, the rotating part is stopped and the two parts are then forced together with increased pressure so that they unite. Explosion welding (Fig. 6) is a form of cold pressure welding in which the pressure is produced by the shock wave from the detonation of an explosive. It is sometimes used for the joining of thin overlapping plates. The plates are gripped together and are covered with a "buffer layer" of rubber sheet and a layer of a special explosive, which is detonated electrically. Another new method is ultrasonic welding (Fig. 7), which is effected by high energy concentrations developed at the joint by ultrasonic vibrations, in combination with pressure.

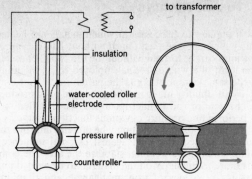

FIG. 3 TUBE MANUFACTURE BY RESISTANCE WELDING

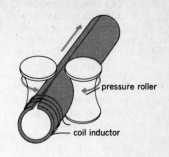

FIG. 4 INDUCTION WELDING

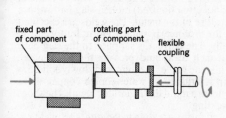

FIG. 5 FRICTION WELDING

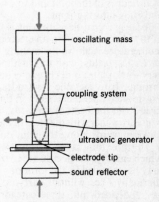

FIG. 7 ULTRASONIC WELDING

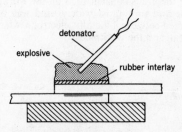

FIG. 6 EXPLOSION WELDING

The term "fusion welding" is applied to processes in which metals are heated to the temperature at which they melt and are then joined without hammering or the application of pressure. The joint can be formed without the use of a filler metal, but usually a filler metal in the form of a wire or rod is employed to fill the joint. Normally the filler metal has the same composition as the parent metal, but may contain alloying metals to improve its fluidity in the molten condition or to produce a fine-grained weld structure. The wire or rod of filler metal may be sheathed in a special coating. Such coatings perform one or more of various functions: serve as a flux, remove oxides or other disturbing substances that may be present, improve the wettability of the material surface, protect the weld against external influences, prevent excessively rapid cooling, and (in arc welding) stabilize the arc. The composition of the coating depends more particularly on the material to be welded and on the welding method. Mixtures of oxides of iron, manganese and titanium, alkaline-earth carbonates, fluorite, and organic compounds are used for coatings. Sources of heat employed in fusion welding are gas, electricity, chemical reactions, etc. Gas welding (Fig. 1) uses a flame produced by the burning in oxygen of acetylene (oxyacetylene welding) or sometimes another fuel gas (e.g., propane, butane, hydrogen) to heat and liquefy the metal at the joint to be welded. This is a very widely employed method of welding iron, steel, cast iron, and copper. The flame is applied to the edges of the joint and to a wire of the appropriate filler metal, which is melted and runs into the joint.

A fairly recent development is the electroslag process (Fig. 2), in which the metal at the joint is melted in an electrically conducting (ionized) molten-slag bath whose temperature is above the melting temperature of the metal. The welds are executed as vertical welds; with this method it is, for instance, possible to form butt welds in very thick plates quickly and economically. The current is supplied to the slag bath through bare metallic electrodes, which melt away and provide the filler metal. The molten filler metal sinks in the slag, fills the gap of the joint and slowly solidifies in it, from the bottom upwards. The gap is bridged by water-cooled copper shoes which, together with the faces of the joint, form a mold for the molten metal. The shoes move upwards along the joint during welding.

The most important and most widely used fusion-welding technique is arc welding, which employs an electric arc to melt the parent metal and the filler metal. The latter may be provided in the form of an electrode which melts away or it may be melted thermally—i.e., without carrying the welding current. The general technique can be subdivided into three categories: open-arc welding, covered-arc welding, and gas-shielded-arc welding. Open-arc welding by Benardos' method (Fig. 3a) employs direct current, the arc being formed between the parent metal and a carbon electrode. In Zerener's method (Fig. 3b) the arc is formed between two carbon electrodes; the heat of the arc is concentrated on the workpiece by the action of a magnetic coil. The method now most widely used was originated by Slavjanov (Fig. 3c): the arc is formed between a metallic electrode, which gradually melts away to supply the filler metal, and the workpiece.

(more)

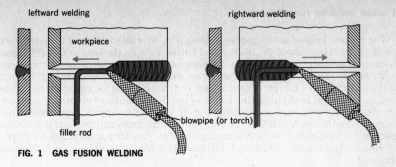

leftward welding   rightward welding

workpiece

blowpipe (or torch)

filler rod

FIG. 1 GAS FUSION WELDING

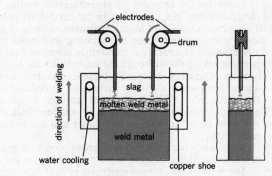

electrodes

drum

direction of welding

slag

molten weld metal

weld metal

water cooling

copper shoe

FIG. 2 ELECTROSLAG WELDING

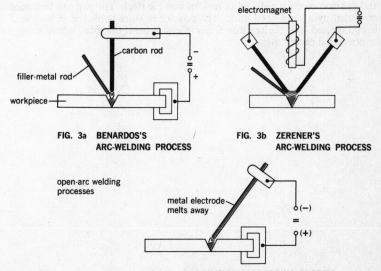

carbon rod

filler-metal rod

workpiece

FIG. 3a BENARDOS'S ARC-WELDING PROCESS

electromagnet

FIG. 3b ZERENER'S ARC-WELDING PROCESS

open-arc welding processes

metal electrode melts away

FIG. 3c SLAVJANOV'S ARC-WELDING PROCESS

The process known as firecracker welding (Fig. 4) is an example of a covered-arc method. A heavily coated electrode is laid horizontally on the joint to be welded and is covered with an insulating layer of paper and a covering bar of copper or some other metal. The workpiece is connected to one pole and the electrode is connected to the other pole of a current source. An arc is struck between the end of the electrode and the joint, and burns along the length of the electrode. Another form of covered-arc welding is submerged-arc welding (Fig. 5). The flux is supplied separately in the form of powder which blankets the arc. The powder melts and protects the molten filler metal from atmospheric contamination. Any powder not melted is recovered by suction and reused. When cool, the fused powder forms a slag, which peels off the weld.

Shielded-arc welding is based on the principle of protecting the molten filler metal by an envelope of chemically inert gas, which may be helium (heliarc process), argon (argonarc process) or carbon dioxide. In atomic-hydrogen welding (Fig. 6a) the heat liberated by monatomic hydrogen when recombining into molecules is used to fuse the metal. An alternating-current arc is maintained across two tungsten electrodes. A stream of hydrogen gas is passed through the arc, in which the hydrogen molecules are split up into atoms. Outside the actual arc these atoms recombine into molecules. This produces great heat, which melts the parts to be welded and unites them, with or without the addition of a filler metal. The inert-gas tungsten-arc process (Fig. 6b) and the inert-gas metal-arc process (Fig. 6c) are two shielded-arc welding processes that are used both for manual techniques and for automatic welding by mechanized equipment.

Thermit welding (Fig. 7) has already been referred to in connection with pressure welding. It is also used as a fusion-welding process, more particularly for iron and steel castings and forgings. The source of heat is not electricity or gas but a chemical reaction that produces intense heat ($3000°$ C): the combustion of a mixture of aluminum powder and iron oxide by which the aluminum is converted into aluminum oxide and the iron oxide is reduced to molten iron (or steel). The parts to be joined are surrounded by a sand-lined mold. The powder mixture is packed in a conical crucible and ignited. The molten iron flows in and around the joint, where it fuses with the preheated parent metal.

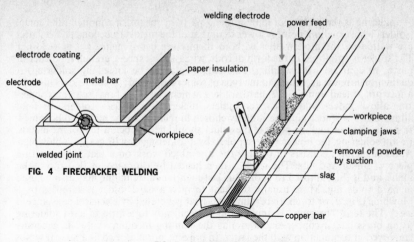

FIG. 4 FIRECRACKER WELDING

FIG. 5 SUBMERGED-ARC WELDING

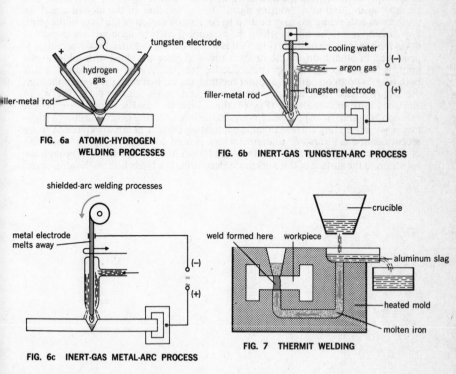

FIG. 6a ATOMIC-HYDROGEN
WELDING PROCESSES

FIG. 6b INERT-GAS TUNGSTEN-ARC PROCESS

FIG. 6c INERT-GAS METAL-ARC PROCESS

FIG. 7 THERMIT WELDING

SOLDERING

Soldering is the process of joining metal parts by means of a molten filler metal (solder) whose melting point is lower than that of the metals to be joined. The latter are wetted by the molten filler without themselves being melted (as in welding). The solder is employed in the form of rods, wires, strips, sheets, granules, powder or paste. In contrast with welding, different metals can be joined by soldering. A distinction is made according to the type of solder employed : (1) soft solders (usually a mixture of lead and tin); (2) hard solders, which comprise brass solders (copper-zinc alloys), silver solders, copper solders, nickel-silver solders, solders for light alloys, etc. The solder must be suitably chosen in relation to the metals to be joined. In particular, the melting point of the solder should be well below that of the metals. In soft soldering, the heat may be supplied by a soldering iron (Fig. 1) or a blowpipe. Another method consists in placing the assembled work on a plate along with a piece of solder and flux. The work is then heated in a furnace, so that the solder melts, and is then allowed to cool. This last-mentioned technique may also be used in hard soldering. More usually the heat to melt a hard solder is supplied by a blowpipe (Fig. 2) or torch. In certain cases heat generated by electrical resistance is used. The term "brazing" is applied more specifically to a form of hard soldering using brass (i.e., a copper-zinc alloy) as the jointing medium. A flux is generally employed in conjunction with the latter. In brazing, borax is used as a flux ; it serves as a means of preventing the formation of an oxide coating on the joint faces, as a cleaning agent, and as a "wetting agent" to aid the flow of the molten metal. In some cases a shielding gas may be used to prevent oxidation of the faces of the joint. Dip brazing is a technique in which the assembled parts to be joined are immersed in the molten jointing medium (Fig. 3). It is widely used in industrial mass-production processes. In other industrial techniques the workpiece, provided with solder at the joint, is heated to the appropriate soldering temperature by immersion in a salt bath (Fig. 5) or an oil bath. In another method, the molten solder is poured through the highly heated joint until the metal cools and unites the two parts (Fig. 4). In electrical-resistance soldering (Fig. 6), the solder, flux and workpiece are heated between tungsten or copper electrodes. Induction soldering (Fig. 7) utilizes a high-frequency alternating current to induce a heating current in the workpiece. A more recent method is ultrasonic soldering, which is used, for example, for the soldering of aluminum. The ultrasonic vibrations are transmitted by a nickel rod through the solder on to the surface of the workpiece, destroying the oxide film on the aluminum.

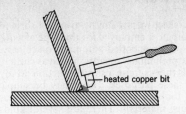

FIG. 1 SOFT SOLDERING WITH A SOLDERING IRON (OR SOLDERING BIT)

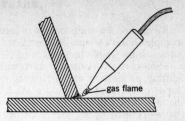

FIG. 2 HARD SOLDERING WITH A BLOWPIPE

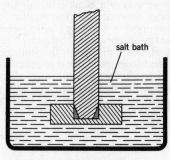

FIG. 3 DIP BRAZING

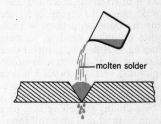

FIG. 4 SOLDERING WITH MOLTEN METAL

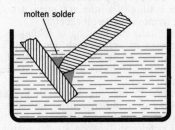

FIG. 5 SALT-BATH SOLDERING

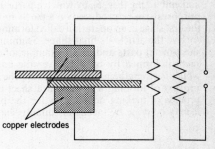

FIG. 6 RESISTANCE SOLDERING

FIG. 7 INDUCTION SOLDERING

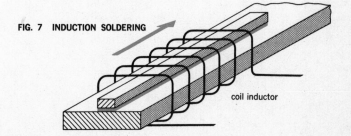

By means of a wide variety of techniques, sheet metals can be given all sorts of complex hollow shapes and sections. The equipment employed for this type of work ranges from simple hand tools to elaborate power-operated automatic machinery.

Sheet-metal work usually starts with a basic preliminary operation such as cutting, slitting, perforating, etc., performed with tools that exercise a shearing action. Many kinds of shearing devices are available for these purposes. These may be simple hand shears, which are scissorslike cutting tools of various shapes and sizes, or hand-operated bench shears for heavier-gauge materials. There are several kinds of power-driven shearing machines. In general, such a machine comprises a fixed blade and a movable blade. The term "punching" refers to operations performed with the aid of a punching machine (or press) and comprising blanking (the shearing of a metal blank from the sheet), piercing (the cutting of a hole in a metal article with tools fitted in the machine), and clipping (removal of surplus metal).

A wide range of shaping operations coming under the heading of folding and bending are done on presses and similar machines, as well as such operations as stamping, crimping, beading, seaming, grooving, etc. These processes are used for the shaping or stiffening of metal sheets, the forming of tubular sections of circular or other shape, and numerous other purposes.

Angles and sections of all kinds can be formed by bending (Fig. 1) or folding (Fig. 2). Coiling (Fig. 3) is the process of coiling over the edge of a sheet-metal component to increase the strength or to provide a suitable edge finish. It can be done by a coiling or rolling tool on a press. Cylindrically shaped articles such as tubes can be produced on a roll-forming machine (Fig. 4). Press forming operations include cupping (Fig. 5), which denotes the conversion of a blank into cup form, and embossing (Fig. 5), by which a particular design (for decorative or strengthening purposes) can be produced on a partly finished component. Cupping is often merely the first stage in an operation called forming, in which an appropriate tool is employed to give the article its final shape. Seaming (Fig. 6) is used mainly for the joining of sheet-metal parts and is, for example, often applied to joints in metal roofing. The seam is formed by bending over the adjacent edges to produce an interlocking, which is strengthened by hammering the seam to flatten it. Flanging (Fig. 7) is the process of forming a flange on a sheet-metal component. Beading (Fig. 8) is the process of making depressions for the purpose of stiffening, embellishment, etc.; it may be done by means of suitably shaped rollers.

(more)

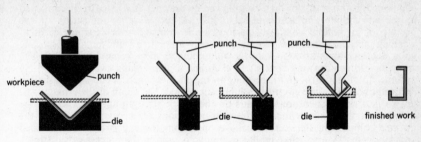

FIG. 1 BENDING

FIG. 2 FOLDING

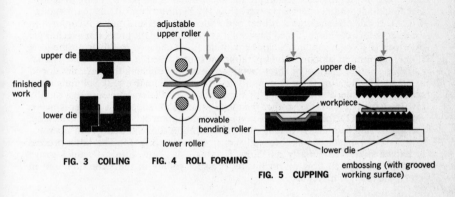

FIG. 3 COILING

FIG. 4 ROLL FORMING

FIG. 5 CUPPING

embossing (with grooved working surface)

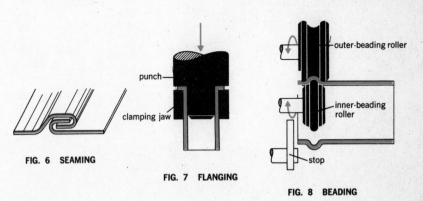

FIG. 6 SEAMING

FIG. 7 FLANGING

FIG. 8 BEADING

Deep drawing (Fig. 9) is the forming of sheet or strip metal into a cup-shaped component by a process of shaping followed by drawing. It is extensively employed in the automobile and aeronautical industries and in other industries for the production of hollow objects of various kinds, ranging from cartridge cases to washing-machine tubs. The process is usually associated with metals possessing good ductility, such as copper, brass, aluminum, cupronickel and mild steel. The principle of deep drawing consists in clamping a metal blank over a die opening and then forcing it through the opening by means of a punch. A cup-shaped shell of reduced thickness is thus formed. A lubricant is used. The depth of shell that can be drawn in a single operation depends on the tensile strength and thickness of the metal. As a rule, one or more redrawing operations are necessary to obtain the finished component. Between successive operations the cup is annealed, cleaned and lubricated. A somewhat similar technique is used in the process called stretch forming, in which the blank is strained beyond its elastic limit over a form tool or stretch block by the application of a tensile load, so that plastic deformation takes place (Fig. 10). In another technique the sheet metal is shaped by being distended in a mold under radial internal pressure (Fig. 10). In the process called "marforming" (Fig. 11) the displacement of a rubber pad under high pressure makes it conform to the contour of a die block placed between it and the table of the press. It is a technique by which shallow components can be manufactured cheaply. In "hydroforming" a rubber diaphragm supported by hydraulic pressure is used instead of a solid rubber pad. Yet another method of sheet-metal forming is by utilizing the force developed by an explosion acting through a liquid or gaseous medium (Fig. 12). Spinning is the process of forming a hollow shape by the application of lateral pressure to a rapidly revolving blank on a lathe, so that the metal assumes the shape of a former which is rotating with it. Deformation is effected by a combination of bending and stretching, pressure being exerted by means of a steel forming tool worked by an operator (Fig. 13) or by means of a mechanically controlled roller (Fig. 14). This latter process is known more particularly as "flow forming" or "flow turning": thick-gauge material is made to flow plastically by pressure rolling in the same direction in which the roller is traveling.

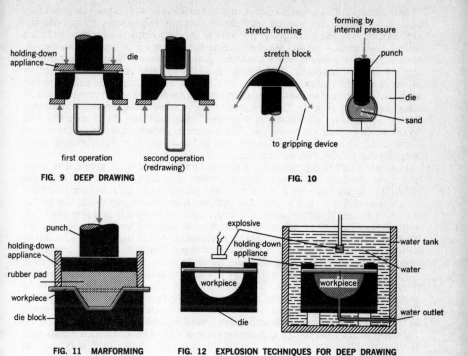

stretch forming

forming by internal pressure

holding-down appliance

die

stretch block

punch

die

sand

to gripping device

first operation

second operation (redrawing)

FIG. 9 DEEP DRAWING

FIG. 10

punch

holding-down appliance

rubber pad

workpiece

die block

explosive

holding-down appliance

workpiece

die

water tank

water

workpiece

water outlet

FIG. 11 MARFORMING

FIG. 12 EXPLOSION TECHNIQUES FOR DEEP DRAWING

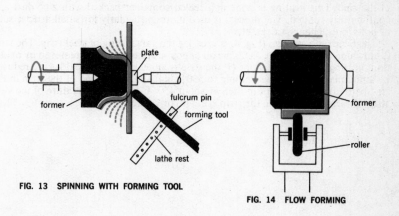

plate

former

fulcrum pin

forming tool

lathe rest

former

roller

FIG. 13 SPINNING WITH FORMING TOOL

FIG. 14 FLOW FORMING

GALVANIZING

Among the various metallic coatings applied to iron and steel to provide protection against corrosion, zinc plays a very important part. The process of applying the zinc coating is called galvanizing. It is very extensively employed for products such as bar, tube, strip, wire and sheet as well as for all manner of articles and utensils such as buckets, watering cans, washtubs, garbage cans, etc. The most commonly applied process is hot-dip galvanizing, in which the zinc coating is obtained by immersion of the materials or articles in a bath of molten zinc. The zinc combines with the iron, so that iron-and-zinc-alloy crystals are formed which provide a firmly adhering coating (Fig. 1). The characteristic crystalline surface patterns presented by hot-dip coatings are known as "spangles"; their size and shape are influenced by the surface condition of the steel, the impurities present in the bath, the rate of cooling, etc. Less widely used zinc-coating processes are electrogalvanizing, metal spraying (see p. 146), and sherardizing.

For successful hot-dip galvanizing, the steel must be free of oil, grease, dirt, scale, and corrosion products. Preparatory treatment may comprise some or all of the following: degreasing with a suitable solvent, pickling with acid, rinsing, treatment with a flux, and drying. The object of pickling is to remove any oxide film by the action of hydrochloric or sulphuric acid. Castings to which molding sand still adheres may have to be subjected to mechanical cleaning treatments such as grit blasting or tumbling, the latter being an operation in which small articles are mixed with an abrasive and rotated in a cylindrical drum. The flux, usually a mixture of zinc chloride and ammonium chloride, serves to remove any remaining traces of impurities and increases the wettability of the steel surface. In "wet galvanizing" (Fig. 3) the flux is deposited in molten form on the zinc bath, and the metal to be galvanized is introduced into the bath through the layer of flux. In "dry galvanizing" (Fig. 2) the metal components are first dipped in a solution of flux and are then dried, so that they become precoated with a thin film of flux, which melts in the zinc bath. The molten-zinc bath is kept at a temperature of 450°–470° C. Certain metals such as tin and aluminum may be added to the bath; they promote fluidity, and tin imparts brightness to the coated metal.

Sherardizing is a process for forming intermetallic compounds of iron and zinc on a steel surface by heating it in the presence of zinc dust below the melting point of the zinc. This heating is done in a sealed container packed with zinc dust and continuously rotated. The process is used more particularly for small articles such as bolts, nuts, chains, valves, etc.

The Sendzimir process (Fig. 4) is used for the galvanizing of steel strip. The strip is unwound from a coil (1). At (2) the oil or grease adhering to it is removed by oxidation (heating). In the next stage (3) the strip is annealed and the oxides are reduced by ammonia. Then follows cooling to 500° C (4) and immersion of the strip in the zinc bath (5), which is kept molten at about 450° C by the temperature of the steel strip. On leaving the bath, the strip is cut (6) or coiled (7).

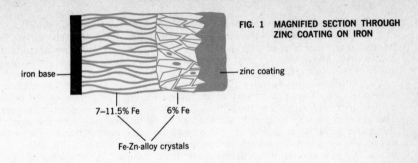

FIG. 1 MAGNIFIED SECTION THROUGH ZINC COATING ON IRON

iron base

zinc coating

7–11.5% Fe   6% Fe

Fe-Zn-alloy crystals

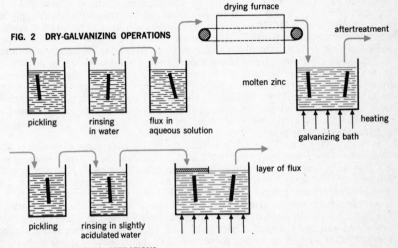

drying furnace

aftertreatment

FIG. 2 DRY-GALVANIZING OPERATIONS

molten zinc

pickling   rinsing in water   flux in aqueous solution

galvanizing bath   heating

pickling   rinsing in slightly acidulated water   layer of flux

FIG. 3 WET-GALVANIZING OPERATIONS

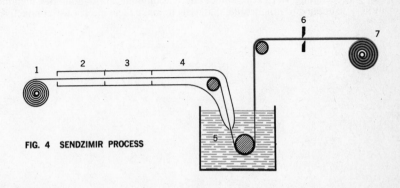

FIG. 4 SENDZIMIR PROCESS

METAL SPRAYING

Metal spraying, or metallizing, is a process for applying protective coatings to iron and steel. It consists in spraying particles of molten metal on to the surface to be treated and can be used with most of the common metals, including aluminum, copper, lead, nickel, tin, zinc, and various alloys. The coating metal, in the form of a wire, is fed into a spray gun in which it is melted by the combustion of a fuel gas—e.g., a mixture of oxygen and acetylene. A spray gun of this kind (Fig. 1) comprises two rollers which are powered by an air turbine (driven by compressed air) or an electric drive and which feed the wire through the central part of a special nozzle. The gas at the nozzle is ignited; the wire is melted on emerging from the nozzle; the molten metal is "atomized" by compressed air and is projected at high velocity against the surface to be coated. In another type of spray gun (not illustrated) the wire is melted in a combustion chamber in the head of the gun. Although the particles of molten metal are cooled instantly, the impact causes them to adhere firmly to the steel surface, provided that it has been cleaned and roughened thoroughly, as by machining or by sandblasting. Special measures may have to be taken in certain cases to ensure good bonding of the sprayed metal to the steel surface. A special adhesion-promoting intermediate layer may be provided, the steel component may be preheated, or a subsequent heat treatment (after spraying) may be applied. A process known as fuse bond may be employed, in which the surface is roughened by low-voltage electric arcs. In a somewhat different technique, the metal to be sprayed is fed to the spray gun not in the form of wire, but as powder supplied from a container by suction or blowing (Fig. 2). This method is generally used in cases where the metal to be sprayed cannot be drawn into a wire—e.g., hard alloys—or has a high melting point.

Metal spraying is a very versatile technique, offering advantages of flexibility and portability. Thus it can be applied in the field to steel bridges, storage tanks, pylons, etc. For such structures, zinc spraying is most extensively used. Spraying is used not only for the application of thin coatings as protection against corrosion, but also for building up surfaces—e.g., for reconditioning worn or damaged parts, for filling holes and cavities, and for the application of friction surfaces to bearings. The metallized coating can be built up to any reasonable thickness and can be filed, turned, ground and polished. Coatings of lead, aluminum, silver or stainless steel are sometimes used for providing protection against corrosion in special apparatus employed in the chemical and foodstuff industries. Steel or hard-alloy coatings are used as wearing surfaces: for instance, light-alloy pistons can be surfaced with a sprayed steel coating. In the electronics and telecommunication industries, metallic coatings are applied to nonmetallic materials to make them electrically conductive.

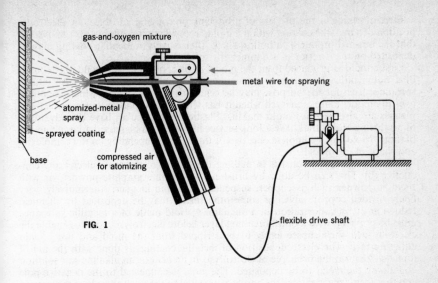

gas-and-oxygen mixture

metal wire for spraying

atomized-metal spray

sprayed coating

base

compressed air for atomizing

FIG. 1

flexible drive shaft

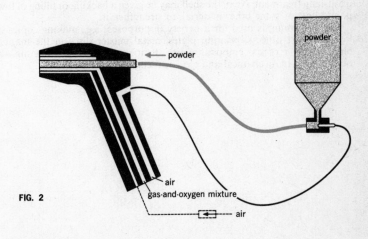

powder

powder

air

gas-and-oxygen mixture

air

FIG. 2

ELECTROFORMING

Electroforming is the process of producing or copying an object by electrode-position of a metallic coating within a mold, from which it is afterwards removed so that the finished product is a hollow shell. Alternatively, a metallic coating may be deposited on the exterior of a nonmetallic solid object.

First a mold is produced from the model to be copied (a). The mold may consist of a nonmetallic substance or sometimes a low-melting-point alloy. A suitable substance used for the purpose may be celluloid, wax or stearin, which is poured over the model and is removed when it has set. Gypsum plaster, gutta-percha and plastics are also used for mold making. Plastics, in particular, have the advantage of producing molds that have a long service life—i.e., can be reused a large number of times. Molds may comprise one, two or three parts, depending on the complexity and shape of the model.

The next operation consists in making the surface of the mold electrically conductive (b). This can be done by brushing it with fine graphite powder, or with metallic powder such as copper, suspended in a thin lacquer. Alternatively, very finely divided copper, silver or some other metal may be deposited by chemical reduction or by vaporization in a vacuum. A mold made of a metallic substance must be provided with a bond-breaking layer before electrolysis, so as to enable the electroformed shell subsequently to be stripped from the mold and not remain adhering to it. The electrodeposition of metallic coatings is done with the aid of direct current on the principle of electrolysis in an acid or an alkaline salt solution containing the metal to be deposited. The mold is connected to the negative pole and thus forms the cathode; the anode, connected to the positive pole of the current source, usually consists of a plate of the metal to be deposited and is gradually consumed (c). Various auxiliary techniques are applied—such as the use of internal anodes, masking, etc.—to ensure that a uniform and smooth metallic coating is formed. By the addition of special substances it is possible to enhance the smoothness, fineness and luster of the coating. When a coating of the desired thickness has been attained, the shell is rinsed, removed from the mold (d) and, if necessary, given a finishing treatment. Next, the shell may be given a backing or filling of low-melting-point alloy, or some other material, to strengthen it.

Electroforming is used for a variety of purposes: e.g., making copies of archeo-logical or art objects, printing plates, metal master discs in the manufacture of phonograph records, embossing dies, templates, molds for casting, and many objects used in mechanical and electrical engineering.

object to be copied

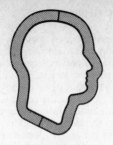

forming the mold

a

b

making the mold electrically conductive by brushing it with graphite powder

c making the electroformed copy

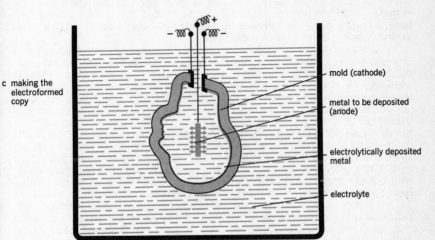

mold (cathode)

metal to be deposited (anode)

electrolytically deposited metal

electrolyte

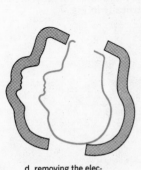

d removing the electroformed copy from the mold

e filling the copy

finished copy

ELECTROPLATING

Electroplating is the process of producing a metallic coating on a surface by electrodeposition—i.e., by the action of an electric current. Such coatings may perform a mainly protective function, to prevent corrosion of the metal on which they are deposited: e.g., plating with zinc (electrogalvanizing) or with tin; or a decorative function: e.g., gold or silver plating; or both functions: e.g., chromium plating. The principle of electroplating is that the coating metal is deposited from an electrolyte—an aqueous acid or alkaline solution—on to the base: i.e., the metal to be coated (Fig. 2). The latter forms the cathode (negative electrode), while a plate of the metal to be deposited serves as the anode (positive electrode). A low-voltage direct current is used; the anode is gradually consumed. Various substances (addition agents) are added to the electroplating bath to obtain a smooth and bright metal deposit. These are principally organic compounds, usually colloidal. Sometimes the objects to be plated are coated with two or more layers of different metals; for example, chromium plating cannot suitably be applied directly to a zinc-sprayed base; a coating of copper followed by a coating of nickel must be applied intermediately before the chromium is deposited.

To obtain a good and firmly adhering coating it is necessary to subject the objects or components to a thorough cleaning. This may be achieved by mechanical treatment—e.g., sandblasting, grinding, wire brushing, scraping, etc.; or by physical methods such as degreasing with organic solvents; or by chemical methods such as pickling with acid, or degreasing by the action of alkalies (saponification); or by electrocleaning, which is a method of cleaning by electrolytic action (more particularly the scrubbing action exercised by the evolution of gas at the surface of the metal). Wetting agents or emulsifiers may be added. The vats for electroplating baths differ greatly in size, shape and lining material (glass, lead, etc.), depending on the size and shape of the components to be plated and on the chemical character of the bath. Electroplating is normally done with direct current. However, particularly with cyanide copper baths, improved smoothness and uniformity of the coating can be obtained by means of the so-called periodic-reverse process, in which the polarity is periodically reversed, so that the metal is alternatively plated and deplated.

Steel strip is plated with zinc or with tin by continuous and largely automated high-speed processes. The electrolytic tin-plating process illustrated schematically in Fig. 3 comprises the following operations: electrolytic cleaning in dilute sulphuric acid, pickling, electrodeposition of tin, melting of the coating to give it a brilliant surface, chemical dipping in chromate solutions, oiling, shearing. The steel strip travels through the installation at a speed of about 25 m/min. (80 ft./min.). A continuous zinc-plating installation is illustrated schematically in Fig. 4.

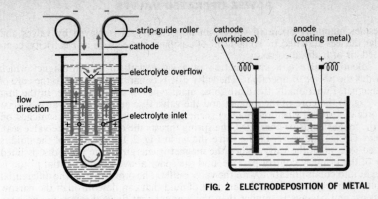

FIG. 1 ELECTROGALVANIZING CELL

FIG. 2 ELECTRODEPOSITION OF METAL

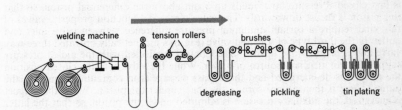

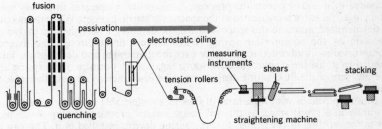

FIG. 3 ELECTROLYTIC TIN-PLATING PLANT
FOR STEEL STRIP

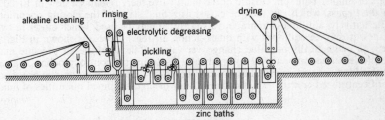

FIG. 4 ELECTROGALVANIZING PLANT FOR STEEL STRIP

Besides manually operated valves, an important part is played by valves and similar devices actuated by some form of auxiliary power such as electricity, compressed air or hydraulic pressure.

A solenoid valve is a combination of a valve with an electromagnet which provides the power to operate it. The valve disc is connected by a rod to the core of the magnet. Functionally there are three main types of solenoid valve. In the first type (Fig. 1) the core of the magnet and the valve disc are pulled upwards against the force of a spring when the magnet is energized. When the current is switched off and the magnet thus de-energized, the spring thrusts the disc against the valve seat, thereby closing the valve. In the valve shown in Fig. 2, the pressure of the fluid is utilized to control the valve. When the magnet is energized, the valve disc is lifted clear of the primary control passage, and the space above the differential piston is brought into communication with the valve outlet. The pressure over the differential piston is thus reduced. Since the amount of fluid that can flow through the narrow compensating passage is smaller than the amount that flows through the primary control passage, a difference in pressure is developed, causing the differential piston to be lifted off the valve seat. On removal of the pressure, the primary control passage is first closed. Pressure now builds up again above the differential piston, so that this piston is thrust downwards. The valve closes. To function properly, valves of this kind require a certain minimum pressure difference between valve inlet and outlet. In the third type of solenoid valve (Fig. 3) a magnetically operated three-way valve and a piston valve form a unit. Control is effected with the aid of pressure supplied by an auxiliary source of power. While the valve is in the closed position the magnet is de-energized and the bypass passage is in communication with the outlet. There is then no pressure in the space under the piston. When the magnet is energized, the auxiliary pressure is admitted under the piston, so that the latter rises, causing the valve to open (left-hand diagram in Fig. 3).

A diaphragm valve (Fig. 4) is controlled by the action of a diaphragm which is actuated by liquid or pneumatic pressure. A magnetically operated three-way valve or changeover valve is connected to the space over the diaphragm. While the magnet is de-energized, access to this space is closed, the diaphragm being held in the closed position by the compression spring. When current flows through the magnet, compressed air is admitted to the space over the diaphragm and develops the force needed to thrust the diaphragm downwards (against the pressure of the spring and the pressure of the fluid acting against the valve disc), thereby causing the valve to open.

In long pipelines with high rates of flow it is undesirable to effect valve closure abruptly in a single stage, as this will cause sudden pressure buildup which may harm the pipeline or the fittings and measuring devices installed in it. This can be avoided by employing two valves, one in the pipeline itself and the other in a bypass pipe of smaller bore. On closure of the main valve there remains a flow of 10–20% in the bypass, which can then be closed as a final stage of the closing operation. Alternatively, a double-diaphragm valve (Fig. 5) may be used, which closes in two stages. It may be equipped with quick-action air-relief valves (as shown in Fig. 5) or with magnetically operated changeover valves. Release of the air pressure may be effected either manually or automatically by closing the control pipelines. Valves of this kind are widely used in industrial processes, especially automatic processes with centralized control: e.g., for the delivery of predetermined quantities of fluid.

(more)

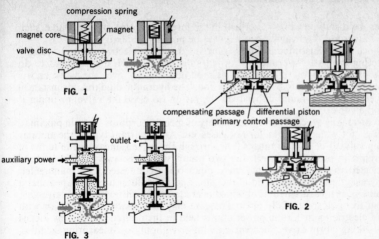

FIG. 1

compression spring

magnet core — magnet

valve disc

compensating passage — differential piston
primary control passage

FIG. 2

auxiliary power → outlet →

FIG. 3

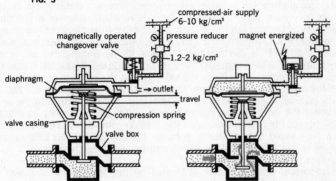

compressed-air supply 6–10 kg/cm²

magnetically operated changeover valve — pressure reducer — magnet energized

—1.2–2 kg/cm²

diaphragm

→ outlet ↓
travel

compression spring

valve casing

valve box

FIG. 4 DIAPHRAGM VALVE WITH MAGNETICALLY OPERATED CHANGEOVER VALVE

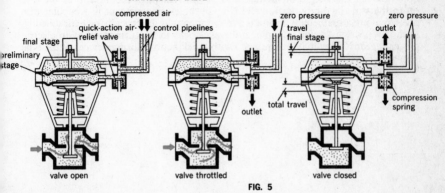

compressed air

quick-action air-relief valve — control pipelines

final stage

preliminary stage

zero pressure

zero pressure

travel
final stage

outlet

outlet

total travel

compression spring

valve open

valve throttled

valve closed

FIG. 5

153

Besides the disc-type valves described in the foregoing, other types of valve, such as sluice valves and flap valves, can likewise be power-operated.

The sluice valve illustrated in Fig. 1 is provided with an electric motor for raising and lowering the gate. Alternatively, a hydraulic drive system may be used to do this. The hydraulic actuating cylinder is mounted directly over the valve and is connected to the valve gate by means of a rod. The hydraulic fluid (oil, for instance) is admitted into the cylinder either over the piston (to close the valve) or under it (to open the valve).

There exist many types of valve for a variety of specific purposes—e.g., in pipelines, in refineries, for water turbines, for hot gases, etc. An important type is the annular measuring valve (Fig. 2). The rate of flow through the valve is measured in terms of the difference in pressure between the two points where the two small side pipes are connected to the valve casing, these pipes being connected to a manometer. The inlet casing, the upstream end of the gate, and the diffuser are so shaped that a reliable flow-rate measurement is obtained over almost the entire range of valve-gate movement from closed to fully open. The gate may be actuated manually or with the aid of electric or hydraulic power and is held in the desired position by means of a self-locking worm drive. Such valves, with or without flow-measuring facilities, are used in water engineering as valves for pipelines, pumping stations, etc. In such circumstances they often have to perform the additional function of a nonreturn valve in the event of pump failure or as a check valve to protect the pump itself from reverse flow. For this purpose a special quick-action closing mechanism may be employed (Fig. 3). It is electrically connected to the pump-drive motor and functions as follows. The electromagnet is normally energized, the clutch is engaged, and the self-locking worm drive holds the valve gate in the position to which it has been set. In the event of a power failure, the magnet becomes de-energized and the clutch is automatically disconnected by spring action. The connection to the worm drive is broken, the drop weight descends and—through the agency of a crank mechanism—moves the gate to the closed position. To slow down the final stage of closure and thus avoid too sudden a pressure buildup, an oil braking cylinder (dashpot) is provided, whose braking action can be controlled by means of valves in the oil-bypass pipes.

Another important class of valves is formed by the flap valves, of which the automatically acting check valve (Fig. 4) is a particular type. It comprises a disc, or gate, which is pivoted at its upper end and is held open by the flow; but if the flow reverses, the weight of the disc and the movement of the fluid force the disc on to the seat. In the butterfly valve (Fig. 5) the closing element is a circular disc pivoted along a diameter. Closure is effected manually, electrically or hydraulically. Such valves are used, for example, in the penstocks to the turbines of hydroelectric power stations. They may be equipped with quick-action closing devices of the type shown in Fig. 3.

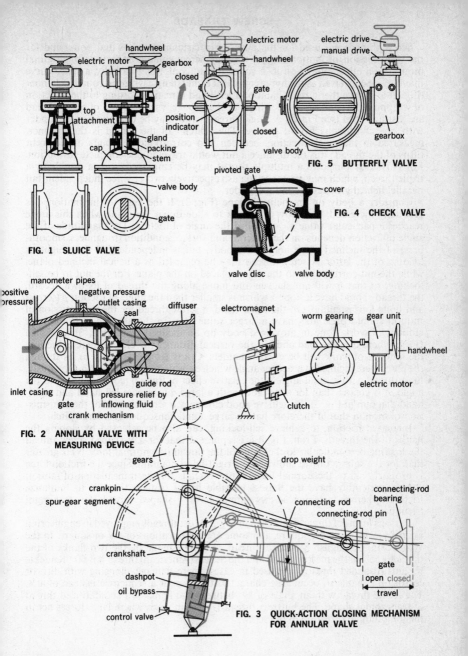

FIG. 1 SLUICE VALVE

handwheel
electric motor · gearbox
closed
position indicator
closed
gate
valve body

FIG. 5 BUTTERFLY VALVE

electric drive
manual drive
gearbox

top attachment
gland
packing
cap
stem
valve body
gate

pivoted gate
cover
FIG. 4 CHECK VALVE
valve disc · valve body

manometer pipes
positive pressure
negative pressure
outlet casing
seal
diffuser
inlet casing
gate
guide rod
pressure relief by inflowing fluid
crank mechanism

FIG. 2 ANNULAR VALVE WITH MEASURING DEVICE

electromagnet
worm gearing
gear unit
handwheel
electric motor
clutch

gears
drop weight
crankpin
spur-gear segment
connecting rod
connecting-rod pin
connecting-rod bearing
crankshaft
gate
dashpot
oil bypass
open closed
travel
control valve

FIG. 3 QUICK-ACTION CLOSING MECHANISM FOR ANNULAR VALVE

SCREW THREADS

Screw threads are used for the purpose of fastening (screws and bolts) and for the transmission of motion: e.g., a rotating screw spindle imparts a longitudinal motion to a nut mounted on it. A screw thread conforms to a helix, a space curve that may be conceived as the hypotenuse of a right triangle wrapped round a cylinder (Fig. 1) to form either a right-hand or a left-hand thread, according to the direction of wrapping. Depending on the shape of the groove, various types of screw thread are distinguished (see Fig. 5). A single-thread screw may be conceived as a cylinder with one continuous helical groove. The "pitch" of the thread is the distance, measured in the axial direction, between two corresponding points on adjacent turns of the thread; it is the distance a nut would travel in one complete revolution. For particular purposes a multiple thread may be employed—e.g., a double or a triple thread, which may be conceived as respectively two or three independent but parallel helical grooves around a cylinder.

Consider a body on an inclined plane (Fig. 2). If the angle of inclination α is gradually increased, the body will begin to slide down the plane when this angle reaches a particular value $\alpha = \rho$, called the angle of friction. The magnitude of the angle of friction depends on the material and surface condition (roughness, smoothness) of the inclined plane and of the body, but is independent of the weight or loading of the latter. A screw thread may be regarded as a helical inclined plane, while the nut corresponds to the object placed on the plane. For the nut to be self-locking, so that it will not slacken and move along the thread of its own accord, the thread should have a slope α which is smaller than the angle of friction ρ (screws and bolts for fastening). On the other hand, a screw thread for the transmission of motion should preferably have a larger value of α. In the absence of friction the amount of work done in moving a body up an inclined plane (Fig. 3) would be $G \times h$, where G is the load and h is the vertical distance traveled. If there is friction, the amount of work will be greater, namely, $G \times H$ (H greater than h), as though the plane were inclined at a steeper angle which exceeds α by an amount ρ. Obviously, for a thread with a small angle α the relative effect of friction (corresponding to the angle ρ) is greater than for a threader with a large α (compare left-hand and right-hand diagram in Fig. 3). To achieve good efficiency a screw thread for the transmission of motion should therefore have a large α and consequently a large pitch.

Increased friction, to achieve self-locking, can also be achieved by sloping the flanks of the threads. From Fig. 4 it is apparent that for a V-shaped screw thread the loading perpendicular to the plane of the thread (the normal force N) is greater than for the square thread shown in the right-hand diagram. Since the friction force is proportional to the normal force, screw threads for the transmission of motion should preferably have the flattest possible flanks (square threads) to minimize friction, whereas fastening screws should have V-shaped threads for maximum friction.

Various forms of thread are shown in Fig. 5. Most threads employed in engineering are of V form, some are square, and some are modifications of a V or square. In the Whitworth, or English Standard, thread, the angle made by the two flanks of the thread is 55°. In the metric thread and the Sellers (American) thread it is 60°. Knuckle threads (rounded threads) are used in cases where damage, clogging with dirt, or corrosion is liable to occur; the characteristic shape is a semicircle instead of a V. The Acme thread with an angle of 29° between the flanks is a standardized thread of trapezoidal shape. The buttress thread is used in cases where large forces act in the longitudinal direction of the screw.

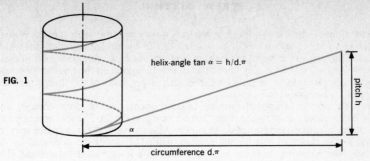

FIG. 1

helix-angle tan $\alpha = h/d.\pi$

pitch h

circumference d.π

α

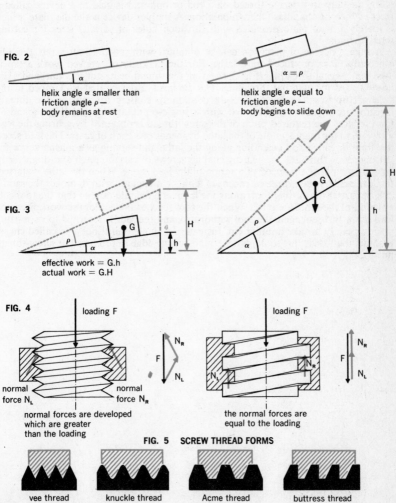

FIG. 2

helix angle α smaller than friction angle ρ — body remains at rest

$\alpha = \rho$

helix angle α equal to friction angle ρ — body begins to slide down

FIG. 3

G

H

h

effective work = G.h
actual work = G.H

ρ α

G

H

h

ρ α

FIG. 4

loading F

N_R
F
N_L

normal force N_L

normal force N_R

normal forces are developed which are greater than the loading

loading F

N_R
F
N_L

N_R

N_L

the normal forces are equal to the loading

FIG. 5   SCREW THREAD FORMS

vee thread

knuckle thread

Acme thread

buttress thread

Screw threads can be produced in various ways: by hand with the aid of such devices as a screw tap, a die plate or a screw die, or by mechanical methods which comprise turning (on a lathe), milling, rolling, pressing or casting. Which of these methods is most suitable in any particular case will depend on the number of screw-threaded components to be produced, the desired precision of the thread, and the quality of surface finish to be attained.

Screw cutting for a limited number of components can be done cheaply and simply by hand with a screw tap (Fig. 1) or a screw die (Fig. 2). The tool is rotated by hand, the workpiece being gripped in a vise. Alternatively, the workpiece may be gripped in the chuck of a lathe and rotated, while the tap or die is guided by the tailstock. The screw tap cuts an internal thread in a hole drilled beforehand. The screw die cuts an external thread on a rod or bolt and is held in a device called a stock, which has handles for manipulation. A simpler device is the die plate, which is merely a steel plate provided with threaded holes of various sizes for cutting screw threads.

A higher degree of precision can be obtained with machine-cut screw threads, especially those produced on a lathe. For the majority of screwed work a tap is used for internal threading (Fig. 3, showing the thread being cut in a nut) and a die head is used for external threading. The die head, a device that is clamped to the lathe, comprises a cylindrical body containing chasers for cutting the thread. Greater precision can be attained with a single-point cutting tool (Figs. 4 and 5). On the lathe the required pitch of the screw thread is obtained by gearing the lead screw up to the main spindle of the lathe by means of a train of gears. The lead screw is a long threaded rod extending along the lathe and serving as a master screw for cutting screw threads. The gearing enables screws of varying pitch and diameter to be cut, by varying the speed of rotation of the lead screw. When the lathe is started, the lead screw rotates and, by means of a nut that engages with it, moves the saddle (which carries the cutting tool) along the lathe at a definite rate, so that the tool cuts a thread of the requisite pitch. When the tool reaches the end of the workpiece, the lead screw is disengaged, the tool withdrawn, and the saddle returned to its starting position ready to take another cut. Instead of a single-point tool a so-called chaser is sometimes used; this tool has a serrated cutting edge to produce the screw-thread profile.

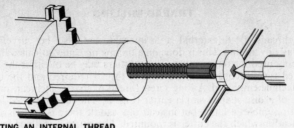

**FIG. 1 CUTTING AN INTERNAL THREAD
WITH A SCREW TAP**

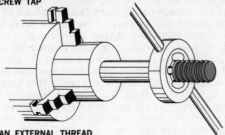

**FIG. 2 CUTTING AN EXTERNAL THREAD
WITH A SCREW DIE**

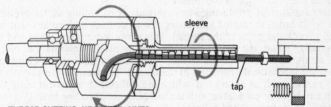

sleeve

tap

FIG. 3 THREAD-CUTTING HEAD FOR NUTS

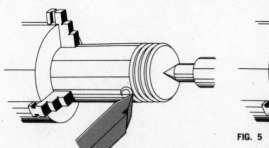

**FIG. 4 CUTTING AN EXTERNAL THREAD
WITH A SINGLE-POINT TOOL**

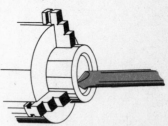

**FIG. 5 CUTTING AN INTERNAL THREAD
WITH A SINGLE-POINT TOOL**

THREAD MILLING

Screw threads, both external and internal, can also be cut efficiently and economically by milling. This is done on a milling machine, so-called thread-milling cutters being used for the purpose. These cutters may be of either the single or the multiribbed type, according to the kind of thread required and the design of the milling machine employed. A long screw thread of coarse pitch can suitably be cut by means of a disc-shaped single cutter (Fig. 1). The machine used for this work somewhat resembles a lathe, but instead of a saddle there is a carriage supporting a cutter head in which the cutter is mounted. The cutter is inclined to produce the correct helix angle of the thread. The feed (longitudinal motion) of the carriage and the rotation of the workpiece are interlinked by means of a lead screw and gearing so as to obtain the correct pitch of the thread being milled. The "long-thread milling" technique with a disc-shaped cutter can also be used for internal threading (Fig. 2) if the hole is of sufficiently large diameter to admit the cutter. In "short-thread milling" the tool is a multiribbed cutter (or multiple cutter) of the ring or shell type (Figs. 3 and 4) or of the taper-shank type (not illustrated). The ribs on the cutter have the shape of the screw-thread profile they have to cut. The cutter is usually as long as, or longer than, the required length of threading on the workpiece; it is fed radially toward the workpiece, and either the cutter or the workpiece is moved axially in synchronization with the slow rotary motion of the workpiece. Generally the latter performs only a little more than one revolution, the axial motion being only little more than one pitch. The axis of the cutter is usually parallel to the axis of the workpiece, but in some cases tilting may be necessary to produce the helix angle of the thread.

Another method of producing screw threads is by rolling under pressure. The part to be threaded is rotated and brought into contact with rollers that have the required profile and pitch of the thread. This is not a cutting method and involves no removal of metal, the thread being formed by plastic deformation. Not only is there a saving in metal, but the rolling operation causes cold working and thus improves the mechanical properties of the thread. The method is especially suitable for soft metals, such as aluminum, which are difficult to screw-cut with a smooth finish. It can be done on a lathe with the aid of a thread-rolling head (Fig. 6), which is equipped with three ribbed rollers that can move in and out radially. The rollers open out automatically at the end of the operation. A variant of this process consists in rolling the thread between grooved flat plates on a thread-rolling machine (Fig. 5).

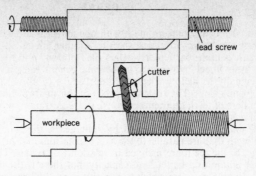

FIG. 1   LONG-THREAD MILLING (EXTERNAL)

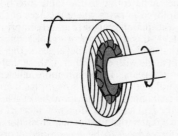

FIG. 2   LONG-THREAD MILLING (INTERNAL)

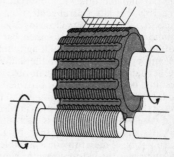

FIG. 3   SHORT-THREAD MILLING (EXTERNAL)

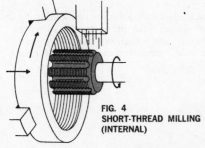

FIG. 4
SHORT-THREAD MILLING
(INTERNAL)

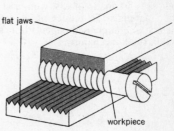

FIG. 5

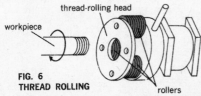

FIG. 6
THREAD ROLLING

GEARS

Toothed wheels whose teeth mesh with one another and which serve to transmit rotary motion or power from one shaft to another are called "gears." The smaller of a gear pair is more particularly called the "pinion" and the larger is the "gear." The term "gear wheel" is also used in the same general sense as "gear" or sometimes more specifically to denote a wheel comprising a toothed rim mounted on arms radiating from a boss. The rotation speeds of the shafts are inversely proportional to the numbers of teeth on their respective gears. A train of gears denotes a number of gears in mesh with one another.

Parallel shafts may be connected by spur gears, the simplest and commonest type, with teeth that are parallel to the axis of rotation (Fig. 2a). In the case of helical gears (Fig. 2b) the teeth are inclined in relation to the axis. This ensures smoother and quieter action and better load capacity, but there is the disadvantage that the teeth set up a side pressure which causes thrust on the bearings. To overcome this, double helical gears (or herringbone gears) may be used; these have teeth of V formation, so that lateral thrust is compensated. When two meshing gears both have external teeth, their shafts rotate in opposite directions (Fig. 1a); when one gear has external and the other has internal teeth, the shafts rotate in the same direction (Fig. 1b). This latter arrangement may be employed in cases where the axes of the respective shafts are very close together. Rack-and-pinion gearing (Fig. 1c) comprises a pinion with a straight toothed bar (the rack); this arrangement enables rotary motion to be converted into longitudinal motion, and vice versa. Bevel gears (Fig. 3) are employed in cases where the shafts form an angle with each other. If these shafts are at right angles, such gears are called miter gears. To obtain more efficient meshing, bevel gears may be provided with spiral teeth instead of straight teeth and are then known as spiral bevel gears. Hypoid gears are similar to spiral bevel gears, except that the axis of the pinion is offset from the gear axis—i.e., does not intersect it. In a case where shafts cross each other at a greater distance apart, crossed-axis helical gears may be used (Fig. 5). If large forces have to be transmitted, worm gearing (Fig. 4) may provide a suitable solution; one of the gears is called the worm and the other the worm wheel (or simply the "gear"). The worm is of helical shape, forming a continuous "tooth"; the worm wheel has independent teeth and is driven by the rotation of the worm with which it meshes.

Gears are usually made of metal and are formed by machining or molding. Spur gears may be made of cast iron, cast steel, forged steel, brass, special alloys, and other materials. To reduce noise, gear pinions are sometimes made of layers of rawhide or compressed paper. Plastics are widely employed as a material for gears used in small mechanisms.

(more)

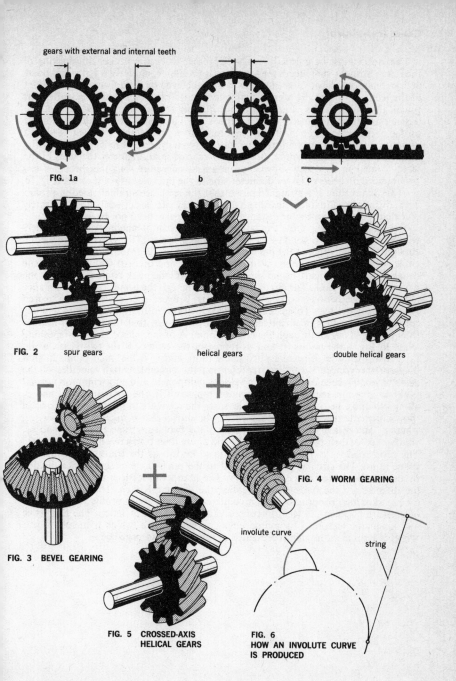

gears with external and internal teeth

FIG. 1a   b   c

FIG. 2   spur gears   helical gears   double helical gears

FIG. 3   BEVEL GEARING

FIG. 4   WORM GEARING

FIG. 5   CROSSED-AXIS HELICAL GEARS

FIG. 6 HOW AN INVOLUTE CURVE IS PRODUCED

involute curve

string

Gear teeth must be correctly meshed together to ensure proper interlocking of the teeth. Straight-sided teeth (of trapezoidal profile) would have a jerky and noisy action. Smooth transmission of power can be achieved by suitably curving the sides of the teeth. Theoretically a great many gear-tooth profiles would be possible, but practical considerations of gear-cutting technique require shapes that are not too difficult and therefore uneconomical to produce. The shape usually adopted for the profile of each side of the tooth is an involute curve, this being the curve traced by a point on a taut string when it is unwrapped from a cylinder (Fig. 6, p. 163). Such a tooth is known as an involute tooth. On a gear of small diameter the sides of the teeth formed in this way show a pronounced curvature, but this becomes less pronounced on gears of larger diameter, and in the limiting case of the rack (Fig. 1c, p. 163) the involute is a straight line, so that the teeth are straight-sided in profile. The advantages of involute gear teeth are that they can be cut easily and accurately and that the center-to-center distance of meshing gears need not be exact.

Two gears that are in mesh can be conceived as a pair of imaginary friction rollers whose diameters correspond to the so-called pitch circles of the gears. The two pitch circles touch each other at the pitch point. The part of a gear tooth outside the pitch circle is called the addendum (h_k in Fig. 2, p. 165), and the part within is called the dedendum (h_1). The part of the side profile outside the pitch circle is the face; the part within is the flank. The circular pitch (t in Fig. 2) is the center-to-center distance (or the distance between corresponding points) between successive teeth, measured along the pitch circle. To ensure that the velocity ratio of a pair of gears remains constant, the profiles of their teeth must be so shaped that the common normal (NN in Fig. 1) passes through the contact point A of the sides of a pair of meshing teeth. In Fig. 1 the points 0_1 and 0_2 represent the centers of the two gears, while r_{01} and r_{02} are the respective pitch-circle radii, and C is the pitch point. In that diagram the vectors v_1 and v_2 representing the circumferential velocities of the gears have each been resolved into a velocity component along the common normal NN and a component along the common tangent TT. The components c_1 and c_2 along NN must be of equal magnitude, since the teeth are in mesh. On the basis of these kinematic relationships it can readily be shown that $n_1/n_2 = r_{02}/r_{01} = z_2/z_1$, where n_1 and n_2 are the rotational speeds of the two gears respectively, r_{01} and r_{02} are the radii of their pitch circles, and z_1 and z_2 are their respective numbers of teeth. The ratio n_1/n_2 is called the gear ratio (also known as the transmission ratio or speed ratio). The circular pitch t is equal to the pitch-circle circumference divided by the number of teeth—i.e., $t = 2\pi r_0/z$. The number of teeth in a gear divided by the diameter of the pitch circle is called the diametral pitch. The line of action (or pressure line) is normal to the involute curve of the tooth profile; the direction of pressure between the teeth in contact is along the line of action. The geometric and kinematic features of a pinion engaging with a rack (which is merely a gear wheel of infinite radius) are represented in Fig. 3. See also page 206.

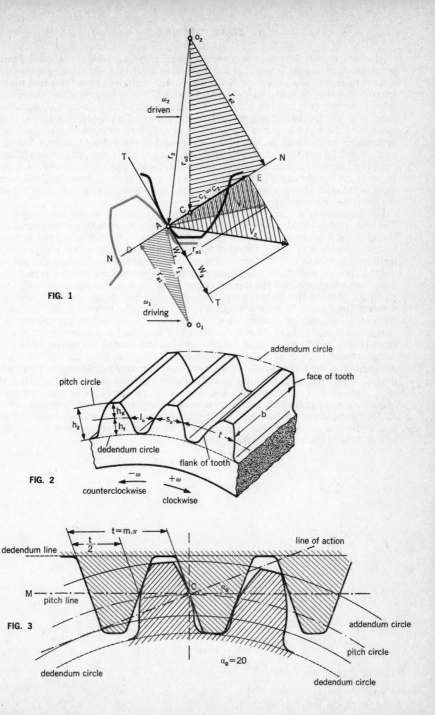

FIG. 1

ω_2 driven

ω_1 driving

FIG. 2

addendum circle

face of tooth

pitch circle

h_K

l_o — s_o

h_z

h_f

b

t

dedendum circle

flank of tooth

$-\omega$ counterclockwise

$+\omega$ clockwise

FIG. 3

$t = m \cdot \pi$

$\dfrac{t}{2}$

dedendum line

line of action

M

pitch line

α_o

C

addendum circle

pitch circle

dedendum circle

$\alpha_o = 20$

dedendum circle

GEAR CUTTING

Most metal gears for precision work are produced by machining—various cutting processes involving the removal of metal. A widely used method of cutting spur-gear teeth is by means of a rotary cutter whose shape corresponds to the required tooth space (a so-called formed cutter, Fig. 1). Mounted on a milling machine, the cutter takes successive cuts from the gear blank, which is rotated the correct distance after each cut to give the desired spacing of the teeth. This precisely controlled step-by-step rotation of the blank is called "indexing" and is performed by an indexing device, or dividing head, which is part of the equipment of the milling machine. Similar gear-cutting operations can also be performed on a suitably equipped lathe. Helical gears can likewise be produced by the formed-cutter technique: the gear blank must be set at the correct angle to the cutter, and the work table of the milling machine must be swiveled through the helix angle of the teeth being cut, so that the plane of the cutter becomes tangent to the helix.

For the quantity production of gears, the hobbing process (Fig. 2) is advantageous. The tool used is called a hob, a screw-shaped rotary cutter with gashes extending across the screw threads so as to form cutting edges. The hob is rotated and its teeth cut the blank. The tooth form of the hob is similar to that of the straight-sided teeth on a rack. Any involute gear will mesh with such teeth, so that one hob can be used to generate gears having any number of teeth which will mesh with one another. This is an advantage over the formed cutter, which in theory can produce only one particular tooth shape, so that a whole range of such cutters are needed for cutting different-sized gears. In spur-gear cutting, the teeth of the hob mesh with the gear teeth in the manner of a worm wheel. The motions performed by gear blank and hob are indicated in Figs. 2a and 2b. The blank is rotated at uniform speed and is geared to the spindle that rotates the hob, which is traversed slowly across the face of the gear blank in its axial direction (downwards in Fig. 2a). Since the hob teeth are disposed on a helix, the spindle of the hob is inclined at the helix angle of the hob thread, so that the tangent to this thread is parallel to the axis of the blank. Helical teeth also can be cut by hobbing (Fig. 3). In this case the indexing and feed gear system are so designed that the hob is appropriately advanced or retarded in order to obtain the correct helix angle of the gear. Alternatively, a so-called differential mechanism may be employed to increase or decrease the relative rotation of the hob and the blank to produce the helix angle.

(more)

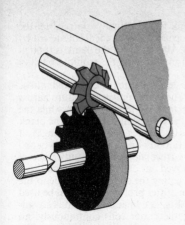

**FIG. 1 GEAR CUTTING WITH
A FORMED CUTTER**

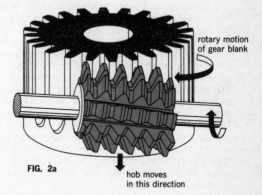

rotary motion
of gear blank

FIG. 2a

hob moves
in this direction

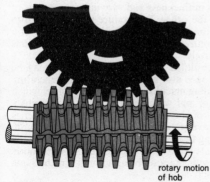

rotary motion
of hob

FIG. 2b

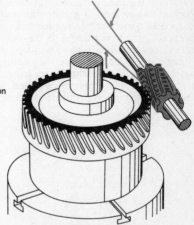

**FIG. 3 HOBBING APPLIED TO
A HELICAL-GEAR CUTTING**

Another method of cutting gear teeth in a blank is by slotting performed on the rotating table of a slotting machine (Fig. 1). The spaces between the teeth are cut by the action of a tool with an up-and-down motion. After each cut the blank is rotated an appropriate distance by the indexing mechanism, to obtain the required pitch of the teeth. This method of gear cutting is relatively seldom employed.

In the rack-cutter generating process, a rack-shaped cutter is employed whose teeth are provided with hardened cutting edges. The cutter moves to and fro across the blank, parallel to its axis, and is at the same time fed longitudinally, while the blank is rotated at appropriate speed (Fig. 2a). The teeth being formed on the blank are rolling in mesh with those on the cutter. As the cutter is of limited length, it cannot cut all the teeth of the gear in one continuous operation; a "step-back" motion is used to restart the cutting operation farther along the circumference of the blank when the teeth cut on the latter have moved out of engagement with those on the cutter. A helical gear can be produced by setting the cutter at an angle to the axis of the blank. Similar in principle is the pinion-cutter process, which can be used both for external (Fig. 2b) and for internal gears (Fig. 2c). In addition to this advantage, a favorable feature of this process is that the cutter rolls continuously, no "step-back" motion being necessary. There is, however, the disadvantage that the pinion cutter is more difficult to make than the rack cutter. The cutter performs its reciprocating cutting motion parallel to the axis of the blank (just as the rack cutter does); the gear teeth are generated by slow rotation of the cutter and the blank in synchronization, just as they would rotate if their pitch circles rolled together without slip.

Bevel gears can be produced with great precision on a bevel-gear planing machine (Fig. 4). The teeth of a bevel gear diminish in thickness and vary in the curvature of their sides from one end of the tooth to the other. To produce the correct tooth shape the machine employs two planing tools, each of which shapes one side of a tooth, while the machine imparts synchronized reciprocating and rotary motions to the tools and to the gear blank respectively.

To improve their precision and surface finish, the sides of gear teeth may be subjected to a grinding treatment. Grinding is also employed as a gear-generating process in its own right. A gear-tooth-grinding machine uses a grinding wheel which is shaped to correspond to the tooth space and which, while rotating, performs a reciprocating motion parallel to the axis of the gear being ground. The latter is indexed one tooth after each to-and-fro traverse of the wheel. In another and more frequently employed technique, the grinding tool consists of a pair of saucer-shaped grinding wheels (Fig. 3), whose faces correspond to the straight sides of a rack tooth. The "rack" is rolled in synchronization with the gear being ground, on the same principle as the motion performed by a rack cutter.

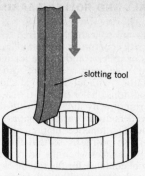

FIG. 1   GEAR CUTTING ON A SLOTTING MACHINE

slotting tool

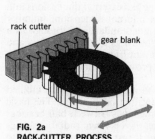

rack cutter

gear blank

FIG. 2a RACK-CUTTER PROCESS

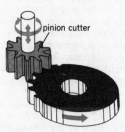

pinion cutter

FIG. 2b PINION-CUTTER PROCESS APPLIED TO EXTERNAL TEETH

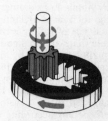

FIG. 2c PINION-CUTTER PROCESS APPLIED TO INTERNAL TEETH

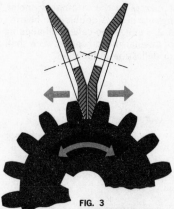

FIG. 3 GEAR GRINDING WITH SAUCER-SHAPED GRINDING WHEELS

FIG. 4   BEVEL-GEAR PLANING

BALL AND ROLLER BEARINGS

Ball and roller bearings are collectively referred to as "antifriction bearings." A bearing of this type normally comprises two annular components known as races, the rolling elements (balls or rollers), and a cage for retaining the rolling elements in position. The races and other components are so designed as to achieve as far as possible pure rolling motion without additional sliding motion associated with friction. Under certain conditions one or both of the races may be omitted, in which case the rolling elements run directly in contact with the shaft and/or the housing in which it is mounted. Rollers are of various shapes: cylindrical, tapered, barrel-shaped, needle-shaped, etc. The function of the cage is to maintain the balls or rollers in their correct relative positions, so that they do not touch one another, and to hold them in one of the races when the bearing is dismantled. The cage also provides a certain amount of guidance for the rolling elements.

Antifriction bearings are characterized by low frictional losses. In particular, the starting friction is low, so that savings in driving power for machinery can be effected. Other features of these bearings are long service life, low lubricant consumption, hardly any wear, possibility of rapid fitting and dismantling, high degree of interchangeability. Antifriction bearings are normally supplied as units ready for installing. For commercially available types of bearing the external dimensions and the tolerances are internationally standardized. The internal dimensions (number and size of rolling elements, race diameter, etc.) are not standardized, though as a rule there are no significant differences between the bearings of a certain class and quality supplied by different manufacturers. Most antifriction bearings have a cylindrical bore, but in some cases a tapered bore is provided to enable the bearing to be fitted directly on to a tapered shaft.

Types of bearing: Differences in the functions and requirements to be fulfilled have led to the development of different kinds of antifriction bearings. The basic distinction, according to the shape of the rolling elements, is between ball bearings and roller bearings. Another important distinction is between radial bearings (designed to resist only or mainly radial loads) and thrust bearings (designed to resist thrust—i.e., loads acting in the axial direction of the shaft). Certain types of bearing can resist radial as well as axial loading. A single-thrust bearing resists thrust in one direction only; a double-thrust bearing resists thrust in both directions. To obtain greater mechanical strength without increasing the external diameter, double-row bearings instead of single-row bearings are used. A double-row bearing has two rows of balls or rollers side by side (Figs. 2, 3 and 6). So-called self-aligning bearings (Figs. 3 and 6) allow a certain amount of angular movement between shaft and housing, thus correcting any misalignment or deflection of the shaft.

(more)

FIG. 1 SINGLE-ROW GROOVED BALL BEARING

FIG. 2 DOUBLE-ROW ANGULAR CONTACT BEARING

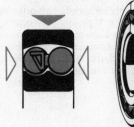

FIG. 3 SELF-ALIGNING BALL BEARING   **FIG. 4 CYLINDRICAL ROLLER BEARING**

FIG. 5 NEEDLE ROLLER BEARING

Radial bearings: Single-row grooved ball bearings (Fig. 1) are most widely employed. They can take some thrust as well as considerable radial loading, even at fairly high speeds. Double-row bearings of this type are also used. Angular contact bearings (double-row type illustrated in Fig. 2) can resist larger amounts of thrust than grooved ball bearings. The term "magneto bearing" is applied to a single-thrust ball bearing with separately detachable outer and inner races, making for greater convenience of assembly. The self-aligning ball bearing (Fig. 3) has a spherical track in the outer race which enables it to compensate for misalignment. Such bearings have only a fairly limited load-carrying capacity. Cylindrical roller bearings (Fig. 4) have a much higher radial load-carrying capacity than ball bearings of equivalent size, but are generally not able to resist any considerable amount of thrust. In the bearing illustrated, the locating lips, or shoulders, are formed on the outer race. Needle roller bearings (Fig. 5) have relatively long rollers of small diameter. Their one advantage over normal cylindrical roller bearings is the radial saving in space; in other respects they are much less efficient. Needle roller bearings are used in low-speed heavily loaded positions. A double-row self-aligning roller bearing is illustrated in Fig. 6. This type of bearing can compensate for shaft misalignment and can resist thrust in both directions as well as considerable radial load. The taper roller bearing (Fig. 7) is suitable for any combination of thrust and radial load. The single-row type, as illustrated, can take thrust in one direction only, and as with angular contact ball bearings it is necessary to fit two bearings on each shaft if there is any possibility of two-way thrust. Barrel-shaped rollers are suitable for bearings subjected to heavy impact-type loading; such bearings are moreover self-aligning, but the thrust capacity is fairly low.

Thrust bearings: Ball thrust bearings (Fig. 8) can take thrust loads only and are used chiefly where a shaft revolving at not too high a speed requires rigid axial support. Angular contact thrust bearings are suitable for higher speeds. Self-aligning roller thrust bearings (Fig. 9) differ in performance from other types of thrust bearing in being able to resist radial loads as well as thrust, besides allowing angular movement between shaft and housing. Other forms of thrust bearing are the cylindrical roller thrust bearing, the needle roller thrust bearing, and the taper roller thrust bearing. These types of bearing are suitable only for fairly low speeds.

(more)

FIG. 6   DOUBLE-ROW SELF-ALIGNING ROLLER BEARING

FIG. 7   TAPER ROLLER BEARING

FIG. 8   BALL THRUST BEARING

FIG. 9   SELF-ALIGNING ROLLER THRUST BEARING

Materials: High local stresses of varying magnitude occur in the rolling elements of antifriction bearings under the action of loading. For this reason very-high-grade steels of uniform texture and high resistance to wear are used in the manufacture of such bearings. In addition, they must be suitably machinable and hardenable. In Europe, low-alloy fully hardenable chromium steels are widely employed (chromium content ranging from 0.4 to 1.65%, depending on the purpose and size of the bearings). Case-hardening steels have hitherto not been much used there—in contrast with American practice, which makes fairly extensive use of these steels for antifriction bearings. Furthermore, for special applications other types of steel are employed, such as stainless steels, nonmagnetic steels and manganese-silicon steels. The cage does not participate in the transmission of forces, but may in certain circumstances be severely stressed by inertial forces, vibrations and impact effects. The choice of material for the cage is determined by considerations of strength and also of machinability and sliding capacity. The materials mainly used for the purpose are steel and brass. Sintered metals, cast iron, light alloys or plastics are also sometimes employed.

Manufacture (Fig. 10): The initial material for the manufacture of the ball or roller races are tubes, rods, rolled material and forged material. Small and medium-sized races are machined on single-spindle and multiple-spindle automatic lathes or on turret lathes, large races on turning and boring mills. After the turning operation comes heat treatment, in which the races are heated by gas or electricity to above 800° C in continuous furnaces or chamber furnaces, quenched in an oil or salt bath, and then tempered. The hardness attained in this way is in the range of 60 to 66 on the Rockwell scale. Then follows grinding of the sides and the internal and external faces, which are finished by polishing and honing.

Manufacture of the rolling elements (balls or rollers) starts from rod or wire. In a typical production process, pieces of wire are cut off in a press (a) and are then formed into balls or rollers by upsetting between dies (b). Large rollers are produced by machining (turning). The flash (fin of surplus material) formed in the pressing process is removed between rotating file discs (c). By means of grinding (d) and tumbling (e) the diameter is reduced, the specified roundness attained, and the surface finish improved. (Tumbling consists in rotation of the elements with an abrasive in a horizontal cylindrical container.) After hardening and tempering (f) come further polishing operations; then the rolling elements are given a high polish by further tumbling with a suitable polishing agent, and finally the elements are graded according to diameter (g).

Cages for antifriction bearings are made from deep-drawing steel strip or brass sheet or from tube material. Manufacture involves a series of press-tool operations (h) or, alternatively, various machining processes (i). The components are finally assembled into complete bearings (k). At all stages of manufacture appropriate checks for precision are maintained. The finally assembled bearings are tested for noiseless running (l), precision of bore, external dimension, clearance and play (m).

(more)

FIG. 10

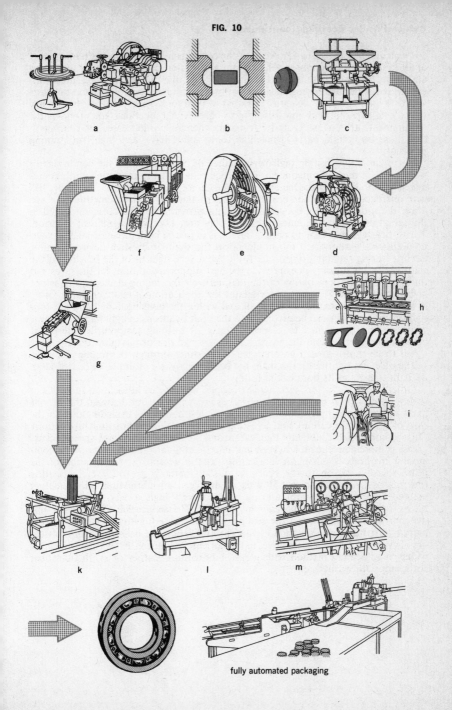

a

b

c

f

e

d

h

g

i

k

l

m

fully automated packaging

Applications (Figs. 11 to 13): Most of the bearings installed in modern machinery of all kinds, including motor vehicles, are antifriction bearings. These range from miniature bearings with bore diameters of less than 1 mm for fine precision-engineered apparatus to huge bearings with external diameters of 2 m and upwards and with load-carrying capacities in excess of 1000 tons. Speeds ranging from a few rpm to 300,000 rpm and upwards have to be provided for. For some purposes the requirements are not particularly stringent, whereas in other cases the bearings— e.g., those on certain parts of machine tools—must have very high true-running precision.

For reliable and lasting high performance of bearings under the conditions of use for which they are intended, certain requirements as to mounting and maintenance have to be fulfilled. One of these is the choice of the correct fit. The term "fit" with reference to antifriction bearings relates particularly to the diametral clearance —i.e., the total clearance between the rolling elements and the races, measured in the radial direction. The amount of end play may also be a significant criterion. Thrust bearings that have to resist axial loads in one or both directions will have to be additionally supported by shoulders on the shaft or by such devices as nuts, circlips (spring retaining rings), etc. Although a very tight fit of the bearing on the shaft and in the housing does provide the best support conditions for the races and the most efficient distribution of the loading over the rolling elements, this is not always practicable. A tight fit necessitates the application of relatively large forces in assembling the bearing on the shaft and within the housing. Small and not-too-tight-fitting bearings can be driven on to the shaft by means of light hammer blows, which should be applied to a special sleeve or bushing temporarily interposed between the hammer and the bearing. Also, it is bad practice to allow the assembly forces to be transmitted from one race to the other through the rolling elements. In the case of a tight fit on the shaft, the bearing may, for example, be heated in an oil bath to expand its bore before fitting.

Other important factors affecting bearing performance are lubrication and efficient sealing. Lubrication serves several purposes: reducing friction between the various parts of the bearing, providing protection against corrosion, excluding dirt (in the case of grease lubrication), and assisting the dissipation of heat (oil lubrication). Oil is a more reliable lubricant than grease and is essential where high speeds occur; grease is more convenient, however, and offers certain advantages. One of the commonest methods is oil-splash lubrication (e.g., in gearboxes: the rotation of the gears splashes oil on to the bearings). Other methods are oil-level lubrication (bearing partly immersed in oil), wick or drip-feed oil lubrication, oil circulating system (force feed by pump), and oil mist (for very high speeds: compressed air blows a fine spray of oil through the bearing). The service life of a bearing also depends on the efficiency of the sealing system, whose function is to prevent the entry of dirt, moisture, etc., and to retain the lubricant in the bearing. Sealing devices are of various kinds: grease grooves in the bore of the housing, felt washers (the most widely used sealing method with grease lubrication), leather or synthetic-rubber seals, labyrinth washers, etc.

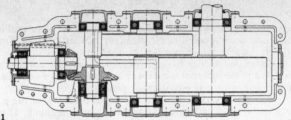

FIG. 11
BEVEL-AND-SPUR GEAR SYSTEM
WITH DOUBLE-ROW ANGULAR CONTACT
BEARINGS AND SINGLE-ROW GROOVED
BALL BEARINGS

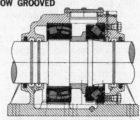

FIG. 12
HORIZONTAL WATER-TURBINE SHAFT
(HIGH AXIAL LOADING) WITH SELF-ALIGNING
ROLLER THRUST BEARING AND DOUBLE-ROW
SELF-ALIGNING ROLLER BEARING

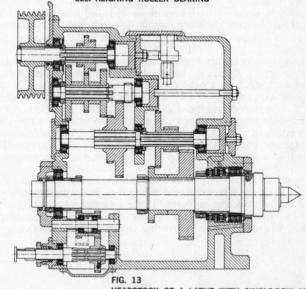

FIG. 13
HEADSTOCK OF A LATHE WITH SINGLE-ROW AND
DOUBLE-ROW CYLINDRICAL
ROLLER BEARINGS AND BALL THRUST BEARINGS

LATHE

A lathe is a machine for revolving a piece of material so as to enable a cutting tool to shape it into a component of circular cross section or to perform a screw-cutting operation. Lathes, which are among the most important machine tools, vary widely in design. What they have in common is that the workpiece is given a rotational motion and the material is cut away by a tool that is given an appropriate combination of linear (axial and radial) movements.

The most widely used type is the center lathe, also known as the engine lathe (Fig. 1), in which the work is held between centers or in a chuck. The rotational movement is imparted to the workpiece by the work spindle mounted in the head-stock (at the left-hand end of the lathe in Fig. 1). To enable long bars to be accommodated, the work spindle may be of hollow construction. The end of this spindle is threaded to take various chucks (gripping devices for holding the work), as required. Alternatively, instead of a chuck, a center—a pointed steel attachment for mounting the work—may be fitted into a taper socket at the end of the work spindle. The required speed of rotation of the spindle, and therefore the cutting speed of the tool, is controlled by suitable selection of the transmission ratio of the main gear-box. Mounted on the guideways of the lathe bed is the saddle, or carriage, which carries the cutting tool and is constructed as a compound slide (Fig. 1): the saddle itself moves in the longitudinal direction of the lathe, whereas the cross slide can be moved only in the transverse direction. Mounted on the cross slide is the top slide, which in turn carries the tool post in which the tool is held. In its normal position the top slide can be moved longitudinally; it can, however, also be swiveled about a vertical axis and clamped in any position, so that conically tapered surfaces can be machined. The feed (advancing) and adjustment movements of the slides can be performed by means of crank handles on the saddle. Automatic control of the feed motion may be provided by means of the so-called feed shaft, which receives its rotational motion from the work spindle. The feed shaft is provided with a worm which rotates with this shaft, but can slide longitudinally in relation to it. When the longitudinal feed motion is started (position LF in Fig. 3), the worm rotates the worm wheel, which in turn rotates the gear Z_1 (mounted on the same shaft). Z_1 drives the gears Z_2, Z_3 and Z_4, which engages with the rack that moves the saddle longitudinally. When the cross-feed motion is engaged (position CF in Fig. 3), the gear Z_2—which can be swiveled—is brought into mesh with the gear Z_5 instead of with Z_3 and thus drives the shaft for moving the cross slide. The carriage can also be moved longitudinally by means of the lead screw, which is a long bar extending along the lathe and provided with a square screw thread. Two half nuts can be brought into engagement with the lead screw (Fig. 2), so that the rotation of the latter imparts a longitudinal motion to the saddle. The lead screw is intended essentially for screw cutting.

At the opposite end of the lathe bed from the headstock is the tailstock, which can move along the guideways and clamped in any desired position. The center sleeve in the tailstock can be moved in the longitudinal direction of the lathe by means of a handwheel and screw spindle and can thus be brought toward the workpiece. The sleeve is provided with a taper socket to take a center or a boring or reaming tool.

(more)

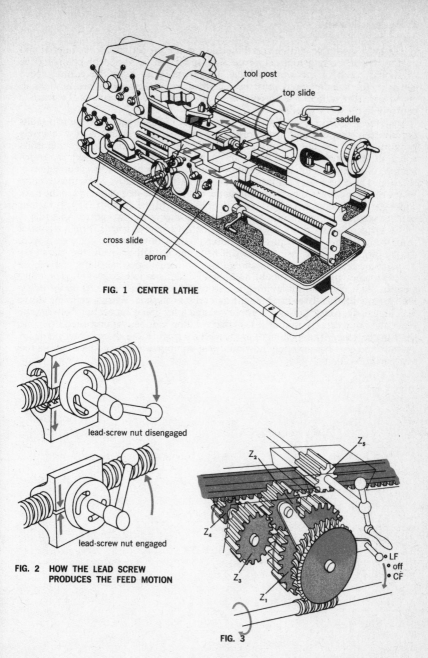

FIG. 1 CENTER LATHE

tool post

top slide

saddle

cross slide

apron

lead-screw nut disengaged

lead-screw nut engaged

FIG. 2 HOW THE LEAD SCREW
PRODUCES THE FEED MOTION

Z_2

Z_5

Z_4

Z_3

Z_1

LF
off
CF

FIG. 3

The most favorable cutting speed depends on the properties of the material and on the type of cutting tool employed. The speed can be varied by changing the rotational speed of the work spindle. Furthermore, to keep the cutting speed constant, the spindle speed must be increased when the tool moves inwards to cut at a smaller radius. The speed change can be effected by means of the gearbox controlling the work spindle.

With the belt drive (Fig. 1) and the gear drive (Fig. 2) the speeds of the two shafts are inversely proportional to the respective pulley or gear diameters (or inversely proportional to the number of gear teeth). The shafts of a pair of meshing gears rotate in opposite directions. When an intermediate gear is introduced, the transmission ratio remains unchanged, but now the two shafts have the same direction of rotation. The cone-pulley transmission system illustrated in Fig. 3 permits selection of any of four different speeds by shifting the belt from one pair of pulleys to the next. This type of transmission is now virtually obsolete. In the gearbox (Fig. 4) various pairs of gears are brought into mesh by shifting one of the two gears. As a rule, a whole range of gears is employed to provide a variety of speeds. For instance, if three gears, each with two possible combinations with other gears, are employed, a total of $2 \times 2 \times 2 = 8$ basic speeds will be available. Alternatively, the gears may be permanently in mesh and mounted loose on their respective shafts, but can be locked to the latter by means of multiple-disc clutches (see page 194). The outer element of the clutch forms an integral part of the gear and is provided on the inside with longitudinal grooves in which the outer discs can slide. When a coupling sleeve is actuated, the pack of discs is compressed and a frictional connection between the outer and inner discs is established, so that rotation motion is transmitted from one shaft to the other. These clutches are used for engaging and disengaging the main gearbox and also for engaging individual gear combinations to produce the desired transmission ratios.

(more)

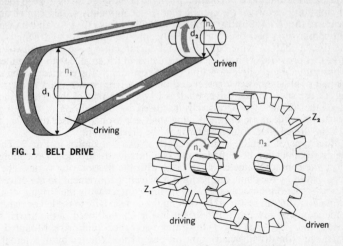

FIG. 1 BELT DRIVE

FIG. 2 GEAR DRIVE

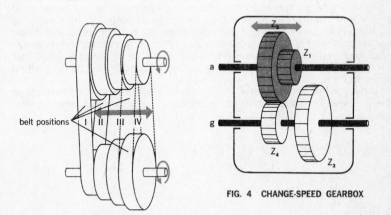

belt positions

FIG. 3 CONE-PULLEY SYSTEM

FIG. 4 CHANGE-SPEED GEARBOX

Adjustment of the speed of rotation to any desired value, so as to keep the cutting speed constant even when the cutting diameter continually varies, can be achieved by means of the PIV ("positive infinitely variable") drive (Fig. 1). This transmission system comprises a pair of radial-toothed conical pulleys on the driving shaft and on the driven shaft. A wide belt of special construction connects the two pairs of pulleys; it is provided with projecting elements which engage with the pulley teeth and thus provide a positive nonslip drive. Speed control is effected by the shifting of one pair of pulleys closer together and the other pair farther apart, and vice versa. In Fig. 1a the speed of the driven shaft is lower than that of the driving shaft; the speed of the former is progressively increased by the shifting of its two pulleys farther apart, while those on the driving shaft are brought closer together; thus in Fig. 1b the driven shaft rotates faster than the driving shaft.

Another type of variable-speed mechanical drive is shown in Fig. 2. The cone on the driven shaft can be shifted in the axial (arrowed) direction. Depending on its position, the contact diameter of the cone with the friction ring varies. The latter, which is driven by the cone, can perform a swiveling movement on the driven shaft, so that the friction surface is always in full contact with the driving cone. The transmission ratio can be varied by shifting the cone axially to the left or right. When the cone is moved inwards (to the right in Fig. 2), the speed of the driven shaft is increased, and vice versa. Fig. 3 illustrates a so-called fluid drive. Mounted on the left-hand shaft (the driving shaft) is the oil-pump rotor, rotated by an external power source (usually an electric motor). It rotates eccentrically in a movable casing. The space between the rotor and the casing is subdivided into compartments which increase and decrease in size in consequence of the rotation of the rotor, so that oil is alternately sucked into them and then discharged into the inlet of the oil motor, whose rotor is mounted on the right-hand shaft (the driven shaft). The motor is very similar in construction to the pump. The oil delivered by the pump causes the rotor of the motor to rotate. The greater the eccentricity of the pump rotor in relation to its casing, the higher the rate of delivery of the oil and the higher the rotational speed of the motor. When the rotor of the pump is shifted to the central position within its casing, delivery of oil ceases, so that the motor stops. When the pump rotor is shifted farther to the left (in Fig. 3), the direction of flow of the oil is reversed, and the motor therefore also reverses its direction of rotation. The tank merely serves as a container for a reserve supply of oil. Speed control is therefore effected by shifting the position of the pump rotor.

Infinitely variable speed control by electrical—as distinct from mechanical or hydraulic—means is usually effected by means of variable-speed direct-current motors.

(more)

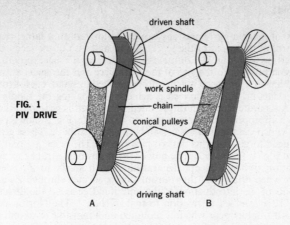

FIG. 1
PIV DRIVE

driven shaft

work spindle

chain

conical pulleys

driving shaft

A   B

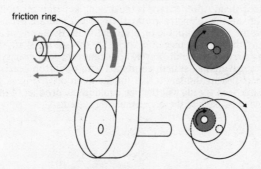

friction ring

FIG. 2   CONE-AND-FRICTION-RING
VARIABLE-SPEED DRIVE

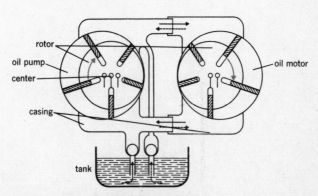

rotor

oil pump

center

casing

oil motor

tank

FIG. 3   FLUID-DRIVE OPERATING PRINCIPLE

Lathe (continued)

A wide variety of machining operations can be performed on a lathe, requiring appropriate control of the speed of the work spindle and also of the feed—i.e., the advance of the cutting tool (Fig. 1). For maximum efficiency it is necessary to adjust the feed correctly in relation to the size of the workpiece and the speed at which it rotates. For finishing cuts the speed is generally controlled by hand, but for roughing cuts and medium cuts an automatic feed is achieved by locking the saddle to the lead screw or to a separate feed shaft (if provided). This shaft is geared to the work spindle to give the appropriate traverse motion to the saddle. Fig. 2 shows a system of gears through which the feed shaft is driven by the work spindle. These gears can readily be exchanged to give various desired feed speeds. The reversing gear unit is shown in detail in Fig. 4. The lever can be moved up or down to engage the required direction of rotation. When it is up, the transmission from Z_1 to Z_4 is effected through Z_3 only; when it is down, the transmission is effected through Z_2 and Z_3, so that (as a result of the introduction of the additional gear Z_2) the direction of rotation of shaft II is reversed in relation to that of shaft I. The Norton gearbox (Fig. 3) comprises a tumbler gear which is mounted on a movable lever and can be brought into mesh with any one of a number of other gears, permitting rapid change in the speed of the feed shaft. Another type is the driving-key transmission, in which a number of gears of varying diameter are fixed to a shaft and are permanently in mesh with gears that are freely rotatable on a second shaft. By means of a movable driving key any one of the latter gears can be locked to this second shaft, so that this gear and its meshing partner transmit power. Sometimes the feed gear system on a lathe will comprise a driving-key system combined with a Norton gearbox and an additional sliding gear system similar to that for the main drive (Fig. 4, page 181). The number of possible feed speeds will then be equal to the product of the number of possibilities provided by each of these transmission systems.

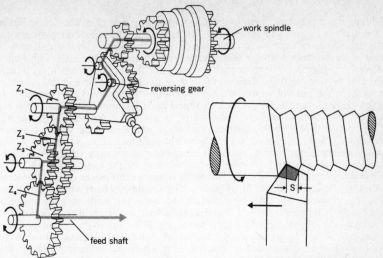

work spindle

reversing gear

Z_1

Z_2

Z_3

Z_4

feed shaft

FIG. 2   GEAR SYSTEM WITH INTERCHANGEABLE GEARS

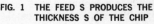

S

FIG. 1   THE FEED S PRODUCES THE THICKNESS S OF THE CHIP

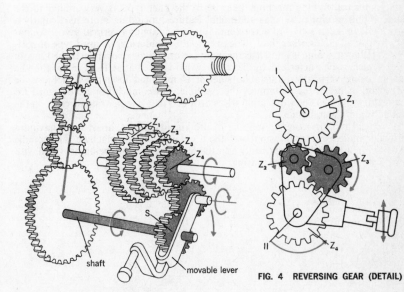

Z_1 Z_2 Z_3 Z_4

S

shaft

movable lever

FIG. 3   NORTON GEARBOX

I

Z_1

Z_2 Z_3

II Z_4

FIG. 4   REVERSING GEAR (DETAIL)

MILLING MACHINE

Milling is a machining operation in which a workpiece is given the desired shape by the action of a rotating cutter while the workpiece performs linear movements. In its simplest form the milling cutter is a circular disc whose rim is provided with specially shaped teeth (cutting edges). Cutters are of many different kinds and shapes. The work is fed against the teeth of the cutter, while the feed motion is longitudinal, transverse or vertical, depending on the type of milling machine and the nature of the work. Milling machines are of the horizontal or the vertical type. A commonly employed horizontal machine is the knee type (Fig. 1). It comprises a massive column which contains the gearbox and spindle-drive motor and is provided with bearings for the spindle. The spindle speed can be varied by means of the gearbox, shown schematically in Fig. 2. Projecting from the front of the column is the knee, whose top surface carries the saddle. The latter in turn carries the work table, which slides in guideways. The motor providing the feed motions for the knee, saddle and work table is accommodated in the knee. The whole knee assembly can be raised or lowered by means of a crank handle or by power. The saddle can be traversed across the knee, and the work table can be moved to and fro in the guideways, by means of handwheels or by power drive. On some machines the work table can perform an automatic cycle of predetermined movements: e.g., a fast run to the cutting position, a change to slow feed motion during the actual cutting, and a quick return to the initial position on completion of the cut, after which the cycle is repeated. The milling cutter is mounted on a shaft called an arbor, whose extremity fits into a tapered socket in the driving spindle. The outer end of the arbor is supported on bearings mounted in the overarm. When the machine is taking a cut, the saddle is clamped to the knee, and the latter is clamped to the column. The knee-type horizontal milling machine illustrated in Fig. 1 is a so-called plain milling machine.

The universal milling machine, likewise of the knee type, is very similar to the plain milling machine, but has additional features, including more particularly a work table that can swivel in a horizontal plane (i.e., about a vertical pivot), so that it can move at angles other than 90 degrees to the spindle axis. A third type of horizontal milling machine is known as the manufacturing type (particularly the Lincoln type), which is characterized by having a work table that is fixed in height, the spindle being vertically adjustable, since it is mounted in a head that can be moved up or down the column of the machine. These machines are designed for heavy-duty milling. The work table slides on a bed that is supported directly on the foundation of the machine.

A vertical milling machine may be of the knee type and, apart from having a vertical spindle, is generally similar to the horizontal milling machine. The spindle is carried in a head that is vertically adjustable on the column, being provided with down feed by means of worm gearing.

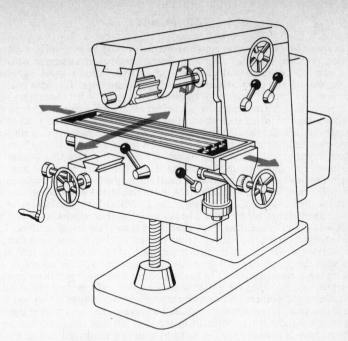

FIG. 1   KNEE-TYPE HORIZONTAL MILLING MACHINE

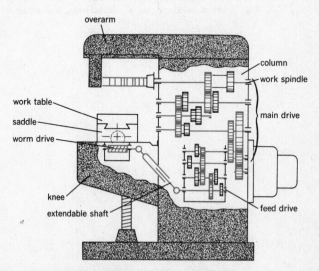

FIG. 2   MAIN PARTS OF KNEE-TYPE HORIZONTAL
MILLING MACHINE

COUPLINGS

A coupling is a device that connects two shafts end to end, while a clutch is a coupling provided with some form of sliding or other arrangement whereby the shafts can be connected and disconnected at will. Broadly speaking, couplings may be divided into rigid couplings and flexible couplings. The rigid type is used where accurate lineal alignment of the shafts is ensured. Where accurate alignment is not possible, a flexible coupling is used; it allows for a certain amount of mis-alignment, besides acting as a shock absorber for vibrations and jerks in torque transmission. The flanged coupling (Fig. 1), one of the simplest types, comprises two halves, each consisting of a flange mounted on the end of a shaft. The boss of each flange is keyed to its shaft, and the flanges are bolted together, thus connecting the two shafts. The split-type muff coupling (Fig. 2) is easier to install and remove because the two halves can be fitted around the aligned shaft ends and clamped by bolting. The muff is keyed to the shafts. A more elaborate form of construction is the serrated coupling (Fig. 3), comprising contact surfaces with interlocking teeth that are held meshed together by bolts. The jaw coupling (Fig. 4) consists of two flanged bosses with "jaws" projecting from their inner faces. This coupling allows longitudinal movement of the two shafts in relation to each other, so that it can compensate for thermal expansion or inaccuracies in assembly. If one of the halves is so mounted on its shaft that it can be slid into or out of engagement with the other half, the jaw coupling can serve as a clutch. The floating-center coupling (Fig. 5) may be used in a case where the two shafts are not accurately aligned and have a slight parallel shift in relation to each other. This coupling comprises two flanged halves and a center piece with lugs which engage with slots in the flanges. The lugs are set at right angles to each other and have a sliding fit in the slots, so that compensation for slight axial movements is provided. The toothed coupling illustrated in Fig. 6 allows a certain amount of parallel, angular and axial displacement between the two shafts. The two coupling bosses fitted on the shaft ends are provided with teeth which mesh with internal teeth in a coupling sleeve. The teeth have a "crowned" (convex) shape, thereby permitting some movement in all directions, including angular movement.

(more)

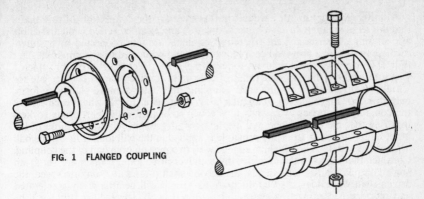

FIG. 1 FLANGED COUPLING

FIG. 2 SPLIT-TYPE MUFF COUPLING

FIG. 3 SERRATED COUPLING

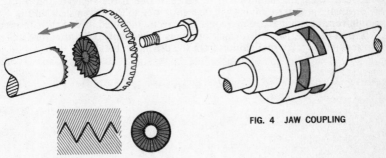

FIG. 4 JAW COUPLING

floating center

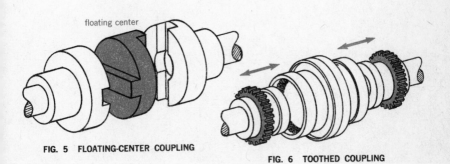

FIG. 5 FLOATING-CENTER COUPLING

FIG. 6 TOOTHED COUPLING

A universal coupling (or cardan joint) is used for the connection of two shafts that are set at an angle to each other and whose angle can be varied while the shafts are rotating. An arrangement whereby two shafts are interconnected by an intermediate shaft with a universal coupling at each end is referred to as a universal shaft (Fig. 1). This principle is employed, for example, in propeller shafts of motor vehicles. The intermediate shaft is sometimes of telescopic construction to compensate for variations in length. The universal coupling may take a more elaborate form, permitting greater amounts of angular movement, as in Fig. 2, where each half of the coupling comprises two swivel pins which so engage with appropriate sockets in a ring that the pins of one half are set at 90 degrees in relation to those of the other half. Essentially the same principle is applied in the ball joint (Fig. 3): the ball is provided with four holes which engage with two pins on each half of the coupling. Consider two shafts, interconnected by a universal coupling, which are set at an angle β in relation to each other. If the driving shaft rotates at a uniform speed, the driven shaft will undergo speed fluctuations—i.e., it will be alternately accelerated and retarded according to a sinusoidal pattern (Fig. 4). These fluctuations will be accordingly greater as the angle β is larger. Such fluctuations can be eliminated by the interposition of an intermediate shaft and two universal couplings. If the two angles α (Fig. 5) are equal, so that the driving and the driven shaft are parallel to each other, these fluctuations will be canceled out. If the angles α and β (Fig. 6) are unequal, speed fluctuations will be transmitted to the driven shaft. In Fig. 6 the two universal couplings are moreover incorrectly mounted in that their respective pivot pins are not parallel to each other, as they ought to be to ensure uniform transmission of rotational motion from the driving to the driven shaft.

(more)

FIG. 1 UNIVERSAL SHAFT

FIG. 2 UNIVERSAL COUPLING

FIG. 3 BALL JOINT

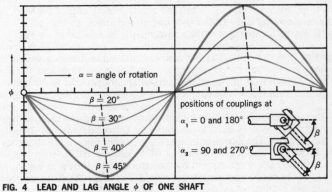

α = angle of rotation

ϕ

$\beta \doteq 20°$

$\beta \doteq 30°$

$\beta \doteq 40°$

$\beta \doteq 45°$

positions of couplings at

$\alpha_1 = 0$ and $180°$

$\alpha_2 = 90$ and $270°$

β

β

FIG. 4 LEAD AND LAG ANGLE ϕ OF ONE SHAFT IN RELATION TO THE OTHER

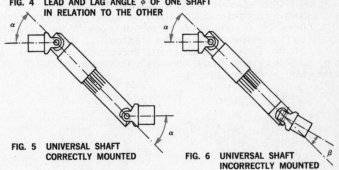

α

α

α

β

FIG. 5 UNIVERSAL SHAFT CORRECTLY MOUNTED

FIG. 6 UNIVERSAL SHAFT INCORRECTLY MOUNTED

In a *flexible coupling* the connection between the two halves is formed by a "yielding" intermediate element which may consist of rubber, leather, steel springs, or some other flexible material. This element allows small amounts of parallel and/or angular movement of the shafts in relation to each other, besides absorbing impact ("shock") due to irregularities in the motion of the driving shaft. Shock absorption may be achieved by storage of energy or by conversion of energy or both. Thus the coil-spring coupling (Fig. 1) stores the impact energy in its coil springs when one flange of the coupling undergoes rotation in relation to the other in consequence of a sudden variation in speed or torque. When the springs subsequently return to their original length, they transmit the temporarily stored-up impact energy to the driven shaft. Every resilient mechanism forms an oscillating system whose natural frequency of oscillation will depend on the spring characteristic and the oscillating masses. The extension or the shortening of the springs in the clutch, and thus the angle of relative rotation of the two clutch halves, is proportional to the magnitude of the torque applied (Fig. 3a). This oscillating system has a particular natural frequency, and if the coupling is rotated with an impact frequency corresponding to this natural frequency, the phenomenon known as resonance will occur, causing objectionable oscillations of large amplitude. As against a linear characteristic of the type represented in Fig. 3a, the steel-band coupling (Fig. 2) has a progressively curved characteristic (Fig. 3b): in this the angle of relative rotation is not proportional to the torque to be transmitted by the coupling. When the effective lever arm of the steel bands changes, the natural frequency of the coupling also changes, so that no resonance will occur. The area located under the characteristic line or curve in Fig. 3 (abc) represents a certain amount of energy, namely, the impact energy which is absorbed by the "yielding" system of the flexible coupling and subsequently given off by it. If the coupling, in addition to presenting a curved characteristic, also develops a so-called damping action (Fig. 3c), its recovery characteristic (the lower of the two curves) will differ from that of the characteristic for initial deformation. In general, the energy given off is less than the energy absorbed by the "yielding" system. The difference between these two energy amounts corresponds to the area between the curves; this "lost" energy may, for example, be converted into heat by internal friction in the coupling. The flexible coupling illustrated in Fig. 4 has a characteristic of this type because of the rubber bushings which enclose the bolts in one of the two halves. A similar effect is achieved by the arrangement shown in Fig. 5, where the two halves of the coupling are interconnected by a rubber "tire" which provides flexibility and shock-absorbing capacity. In the disc-type flexible coupling (Fig. 6), each of the two shafts is provided with a boss (called a "spider") having three radial arms set at 120 degrees in relation to one another. Between the two spiders is a flexible disc made of rubber and canvas bonded together. This disc has six equally spaced holes for bolting. The spider arms on each shaft are bolted to the disc, but at different positions from those on the other shaft. Thus "give" or "yield" of the disc will occur when power is transmitted.

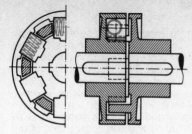

FIG. 1 COIL-SPRING COUPLING

FIG. 2 STEEL-BAND COUPLING

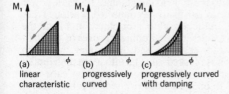

FIG. 3 COUPLING CHARACTERISTICS

(a) linear characteristic
(b) progressively curved
(c) progressively curved with damping

FIG. 4 FLEXIBLE BOLTED COUPLING

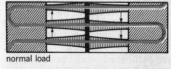

normal load

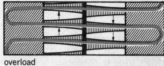

overload

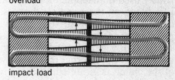

impact load

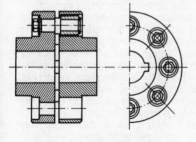

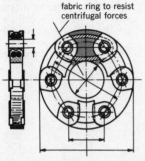

fabric ring to resist centrifugal forces

FIG. 6 DISC-TYPE FLEXIBLE COUPLING

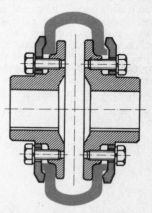

FIG. 5 PERIFLEX COUPLING

CLUTCHES

As p. 188 explained, a *clutch* is any coupling that enables shafts or other rotating parts to be connected or disconnected at will—i.e., without the removal or refitting of any components. In the claw clutch (Fig. 1) one half of the clutch can slide on its shaft, so that the claws can be engaged or disengaged. This type of clutch can be engaged only when the shafts are stationary or rotating at low speed. The geared clutch (Fig. 2) is widely employed in machine tools and motor vehicles. The two clutch bosses are each provided with external teeth which can mesh with a sleeve that has corresponding internal teeth and can be slid over both bosses so as to establish a positive connection between the shafts. To permit engagement of the clutch while the shafts are rotating, the sleeve and the shaft end to be coupled are respectively provided with friction surfaces which are brought into contact with each other and thus equalize the speed of the rotating parts before the teeth on the shaft and inside the sleeve are brought into mesh. Friction clutches transmit power through contact friction surfaces on the two halves to be connected. Various types of friction clutch are illustrated in Fig. 3. In the disc clutch (or plate clutch) and the cone clutch, the boss of the movable part slides in longitudinal grooves in the shaft on which it is mounted. The movements for engaging and disengaging the clutch are performed by the action of a lever whose forked ends fit into a circumferential recess in the boss of the clutch plate or cone. The internal-expanding shoe-type clutch comprises an outer shell attached to one shaft and two semicircular "shoes" which are mounted on arms attached to a sliding sleeve on the other shaft and which can be brought internally into contact with the shell. The forks of the clutch-operating lever engage with a recess on the sliding sleeve. A type of friction clutch widely used in machine tools and motor vehicles is the multiple-disc clutch (Fig. 4). It is based on the principle that a series of discs or plates alternately connected to the driving and the driven shaft will increase the power-transmitting capacity in proportion to the number of pairs of contact surfaces. In the form of clutch illustrated in Fig. 4 the boss mounted on the driving shaft is provided with external teeth with which the internal teeth on a series of thin steel plates engage. The outer shell of the clutch is mounted on the driven shaft and has internal teeth with which the external teeth of a second series of plates (alternating with those of the first series) likewise engage. When the clutch lever shifts a collar to the left, the plates are pressed together and thus transmit power by friction. Multiple-disc clutches in machine tools usually operate immersed in oil; those in motor vehicles are usually of the dry type. A magnetic clutch is a friction-disc clutch that is engaged by the energizing of a magnet coil, which attracts a set of steel friction discs and thus establishes the connection. A double-acting clutch based on this principle is illustrated in Fig. 5. When the coil "a" is energized, the discs "b" are compressed together by magnetic attraction, thereby connecting the gear "c" to the shaft "d." When the coil "e" is energized, the discs "f" are pressed together, so that now power is transmitted from the shaft "d" to the gear "g."

(more)

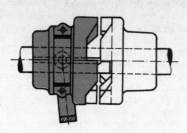

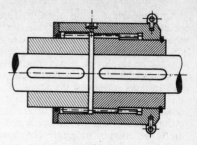

FIG. 1   CLAW CLUTCH

FIG. 2   GEARED CLUTCH

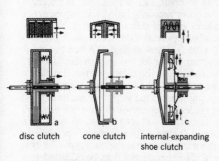

disc clutch   cone clutch   internal-expanding
                                          shoe clutch

FIG. 3   FRICTION CLUTCHES

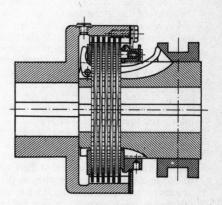

FIG. 4   MULTIPLE-DISC CLUTCH

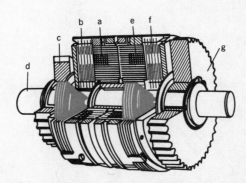

FIG. 5   MAGNETICALLY OPERATED MULTIPLE-DISC CLUTCH

An *automatic clutch* is often installed between the driving shaft of a motor and the machinery it drives. It enables the shaft to reach a predetermined speed before engagement is effected and is especially useful in a case where the driven machinery requires a high starting torque. For such purposes a centrifugal clutch (Fig. 1) may suitably be employed. It comprises two or more "shoes" which, when the driving shaft on which they are mounted has reached a certain speed, overcome the pressure of restraining springs by the action of centrifugal force and move outwards to press against the inner surface of the rim mounted on the driven shaft. In this way the transmission of power to the driven shaft is gradually and automatically increased, so that smooth engagement is effected. The speed at which engagement takes place can be increased by fitting the clutch with more power-restraining springs, and vice versa. When the shafts are not rotating, the shoes are retracted and not in contact with the rim. Various other types of automatic clutch are likewise based on the centrifugal principle.

Freewheeling clutches drive in one direction only and permit free movement when the speed of the driven shaft exceeds that of the driving shaft. In the grip-roller type of freewheeling clutch (Fig. 2) each roller is gripped, i.e., jammed, in the wedge-shaped space as soon as the movement of the outer race in relation to the inner race causes the roller to move into the "shallower" part of this space. With clockwise rotation this occurs when the outer race tends to overtake the inner race; the two shafts then become locked together: i.e., the clutch is now engaged. When the outer race slows down and tends to lag behind the inner, the roller moves into the "deeper" part of the space in which it is housed. This disengages the clutch. In the slip clutch (Fig. 3), springs produce the contact pressure between the two clutch bosses and the interposed (longitudinally movable) friction plate provided with friction linings on both its faces. The friction developed at these faces will depend on the contact pressure exerted by the springs. If the pressure is low, the friction will also be low, so that slip in the clutch will occur at a low value of the torque. By means of a screw it is possible to increase the spring pressure and therefore the friction, so that the clutch will be able to transmit a greater torque without slipping. The torque can thus be adjusted to a predetermined value, and the clutch can serve as a safety device against overloading of the driven machinery. A simpler safety device for this purpose is the shear-bolt coupling (Fig. 4). It comprises two flanges connected by bolts that are designed to fail in shear (i.e., to "break off") when the torque exceeds a predetermined value.

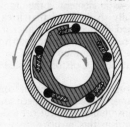

spring   centrifugal weight

FIG. 1 CENTRIFUGAL CLUTCH

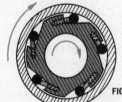

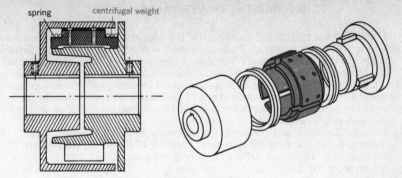

	when rotated	
	clockwise	counterclockwise
clutch is engaged	when outer race tends to overtake inner race	when inner race tends to overtake outer race
clutch is disengaged	when inner race overtakes outer race	when outer race overtakes inner race

FIG. 2 GRIP-ROLLER FREEWHEELING CLUTCH

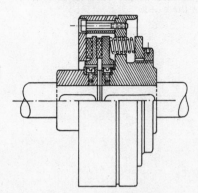

FIG. 3 SLIP CLUTCH

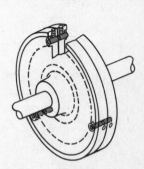

FIG. 4 SHEAR-BOLT COUPLING

Devices for the mechanical transmission of power, or "mechanisms," constitute the basic units from which all kinds of machinery are built up. They are devices whereby certain actions can be exercised when certain motions are performed. Every mechanism consists of individual elements whose movements in relation to one another are "positive": i.e., the motion of one element produces an accurately determinable and definable motion of every individual point of the other elements of that mechanism. Numerous combinations and modifications are possible, but six basic types of mechanism are to be distinguished:

1. Screw mechanism (Fig. 1): When the screw spindle is rotated, the element attached to the nut will move in the longitudinal direction of the screw. (Examples: vise, cross slide of a lathe, work table of a milling machine). Conversely, if the nut is rotatably mounted in the frame of the mechanism and driven, the screw spindle will move longitudinally.

2. Linkage or crank mechanism (Fig. 2): The characteristic element is the crank, which is rotatably mounted on the frame and is usually so designed that it can perform complete revolutions. Its motion is transmitted through the coupler (or connecting rod) to the lever (or rocker arm), likewise rotatably mounted, but not performing complete revolutions. Alternatively, instead of being connected to a lever, the coupler may be attached to a sliding element—e.g., a piston in a steam engine or internal-combustion engine.

3. Gear mechanism (Fig. 3): This type of mechanism transmits rotary motion from one shaft to another, usually in conjunction with a change in rotational speed and torque. In a gear mechanism of the usual type the transmission is effected by the meshing of gear teeth, but in the friction-gear mechanism this positive drive is replaced by frictional contact of wheels or rollers.

4. Pulley mechanism (Fig. 4): Connection between the pulleys on their respective shafts is effected by flexible elements (belts, ropes, etc.).

5. Cam mechanism (Fig. 5): A cam plate mounted on a frame is driven and thus moves a lever or slider which thus performs a desired predetermined motion depending on the shape of the cam (example: valve control mechanism in an internal-combustion engine).

6. Ratchet mechanism (Fig. 6): This serves to arrest a motion or to produce an intermittent rotation in the driven element. The pawl allows the ratchet wheel to rotate in one direction only, preventing rotation in the opposite direction by engaging with the specially shaped teeth on the wheel.

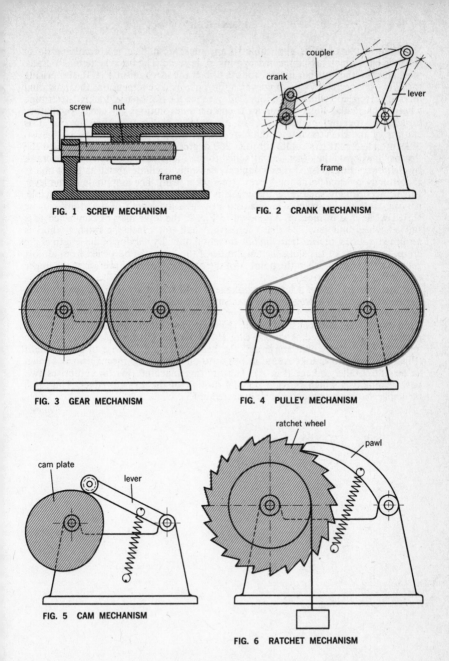

FIG. 1 SCREW MECHANISM

FIG. 2 CRANK MECHANISM

FIG. 3 GEAR MECHANISM

FIG. 4 PULLEY MECHANISM

FIG. 5 CAM MECHANISM

FIG. 6 RATCHET MECHANISM

LINKAGES

The term "linkage" is applicable to any mechanism that is a combination of links or bars which are connected by pins, sliders, etc. The basic system of a crank mechanism is the four-bar linkage (or quadric-crank mechanism, Fig. 1), comprising four links connected by pin joints which form pivots. The dimensions the individual links are given, and which of the four links is made the stationary "frame," determine whether particular links will perform complete revolutions or merely oscillatory (to-and-fro rocking) movements. For example, if the bottom link in Fig. 1 is stationary and thus constitutes the frame of the mechanism, the shorter of the two links attached to it can rotate through 360 degrees (and is accordingly termed the "crank"), whereas the other link attached to the frame (and connected to the crank by the fourth link, termed the "coupler") can only oscillate about its pivot and is accordingly referred to as the "lever" (or rocker arm). The amplitude of the lever will be accordingly smaller as the crank is shorter. On the basis of this principle, it is possible to construct a mechanism in which the length of the crank can be varied while it is in motion. As shown in Fig. 2, the lever may be connected to a ratchet wheel and pawl, so that the driven shaft (on which the ratchet wheel is mounted) rotates intermittently in one direction only. By varying of the length of the crank (by means of the slot), the amplitude of the lever can be varied from almost zero to a maximum, when the point A of the coupler is at the lower or at the upper end of the slot respectively.

If the shortest link of a four-bar linkage is held stationary (Fig. 3), the resulting mechanism is called a drag-link mechanism. Here both of the links (crank and lever) attached to the stationary frame can perform complete revolutions. When the left-hand crank rotates at constant speed, the right-hand crank (originally the "lever") rotates at varying speed. A special case is the parallel linkage, in which the frame and coupler are of equal length and the two cranks are likewise of equal length (Fig. 4). If the two cranks rotate in opposite directions, the mechanism is known as an antiparallel linkage (Fig. 5). The drafting machine (Fig. 6) comprises two parallel linkages which provide parallel motion of the straight edges. The same principle is applied to the toolbox illustrated in Fig. 7.

(more)

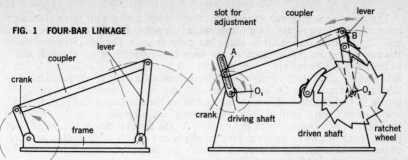

FIG. 1 FOUR-BAR LINKAGE

FIG. 2 COMBINED LEVER-AND-RATCHET MECHANISM
WITH INFINITELY VARIABLE DRIVE

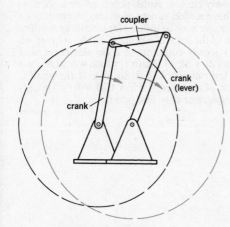

FIG. 3 DRAG-LINK MECHANISM

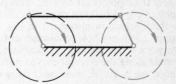

FIG. 4 PARALLEL LINKAGE

FIG. 5 ANTIPARALLEL LINKAGE

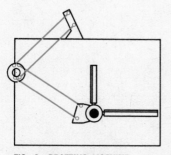

FIG. 6 DRAFTING MACHINE

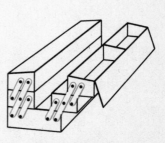

FIG. 7 TOOL BOX

The *pantograph* (Fig. 8a) utilizes a parallel linkage for the proportional enlargement or reduction in scale of a given drawing. The points B and C of this mechanism trace figures that are similar in shape but differ in scale. The linear dimensions of the two figures are proportional to the respective distances of the points B and C to the pivot A. For example, if the distance CA is four times the distance BA, and if B is made to trace the outlines of a drawing, then a pencil at C will reproduce this drawing at four times the original size, and vice versa. A multiple pantograph is shown in Fig. 8b. Here all the points located on the horizontal line—i.e., B, C, C_1 and C_2—will trace similar figures of varying size, depending on the respective distances from those points to the pivot at A. The familiar device known as "lazy tongs," used for such purposes as lamp or telephone supports, consists of an assembly of parallel linkages. Fig. 9 shows a typical application of a four-bar linkage for producing a to-and-fro swinging motion of a fan.

An interesting feature of the beam-and-crank mechanism (Fig. 10) is that whereas all points of the crank and lever trace only circular paths, points on or associated with the coupler trace paths that may have a wide variety of shapes, depending on where these points are located. This principle may be utilized for obtaining motions conforming to paths of particular shape. If the trace of a point C on the coupler AB in the mechanism illustrated in Fig. 11 locally conforms to a circular curve with radius r, it is possible to connect at C a link CM, of length r, which will produce a temporary standstill of the oscillating lever attached to it. So long as the point C travels along the circular curved portion of the trace, the link CM will rotate only about the point M, without causing displacement of this point.

(more)

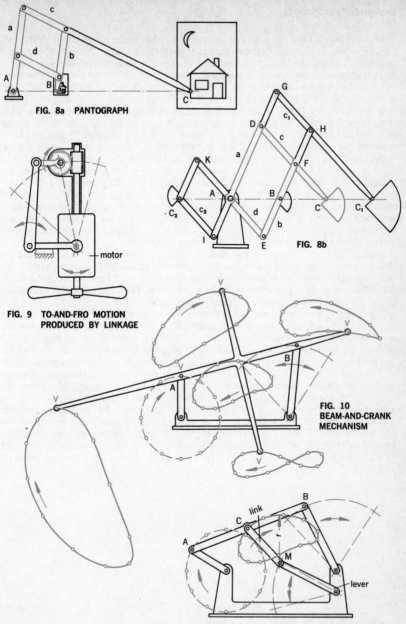

FIG. 8a PANTOGRAPH

FIG. 8b

FIG. 9 TO-AND-FRO MOTION PRODUCED BY LINKAGE

FIG. 10 BEAM-AND-CRANK MECHANISM

FIG. 11 BEAM-AND-CRANK MECHANISM FOR PRODUCING TEMPORARY STANDSTILL OF OSCILLATING LEVER

One or more of the pin joints (pivots) of a four-bar linkage may be replaced by a "slider"—i.e., a pivoted element that can perform a guided linear motion in a slideway. The mechanism thus obtained is called a slider-crank linkage (Fig. 1). In this form it is used for the crankshaft, connecting rod, piston and cylinder of an internal-combustion engine or air compressor. The stroke performed by the slider is equal to twice the length of the crank if the center line of the slideway passes through the crank pivot. In the eccentric crank mechanism (Fig. 2) the stroke is less than twice the crank length. The farthest points reached by the slider in the course of its to-and-fro motion are called the "dead centers." The crank angles α_1 and α_2 between the two dead-center positions in the eccentric crank mechanism are not equal; the return movement of the slider is therefore faster than the forward movement. In the case of the oscillating crank mechanism (Fig. 3) the crank end is pivotably attached to a slider which moves up and down in a slot in an oscillating bar which in turn moves a horizontal rod to and fro. For the direction of crank rotation indicated in Fig. 3 the movement of the slotted bar to the left is associated with a larger crank angle than the movement to the right. Thus the horizontally moving rod (which may, for example, actuate the work table of a machine tool) will perform a slow forward movement and a rapid return movement. If the length of the crank is greater than the distance designated as "frame length" in Fig. 3, the slotted bar will perform a rotational instead of an oscillatory movement. A combination of a rotating and an oscillating slotted bar (Fig. 4) is often used for obtaining the work-table motion on machine tools in order to ensure as far as possible a constant low speed for the forward movement (feed) and a rapid return.

Fig. 5 shows a cross-shaped bar assembly with slot-and-crank drive. The distance traveled by the cross, measured from its middle position, is proportional to the sine of the angle α or the cosine of the angle β. This characteristic of the mechanism is utilized for certain purposes, such as the production of sinusoidal cam drives. The device known as the elliptic trammel (Fig. 6), which is used for drawing ellipses, comprises two slideways at right angles to each other. A crank is pivotably connected to a slider in each slideway. All points on the crank will trace elliptical curves; only the center point on the crank (midway between A and B) traces a circle. When an ellipse is traced by a point C located beyond B, the length BC will be equal to the semiminor axis and AB will be equal to the difference between the semimajor and semiminor axes of the ellipse. If the point that traces the ellipse (e.g., the point D) is located between A and B, the length AD will be equal to the one semiaxis (in this case the semiminor axis), and AB will be equal to the sum of the two semiaxes of the ellipse.

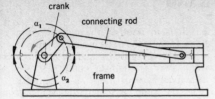

FIG. 1 SLIDER-CRANK MECHANISM

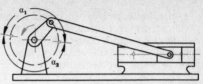

FIG. 2 ECCENTRIC SLIDER-CRANK MECHANISM

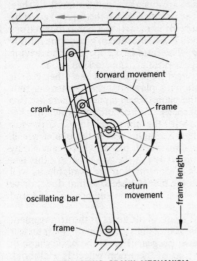

FIG. 3 OSCILLATING CRANK MECHANISM

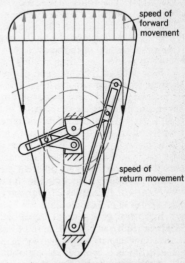

FIG. 4 WORK-TABLE DRIVE MECHANISM
(COMBINATION OF ROTATING
AND OSCILLATING SLOTTED BAR)

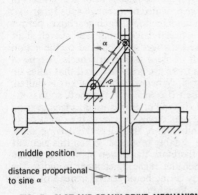

FIG. 5 SLOT-AND-CRANK-DRIVE MECHANISM

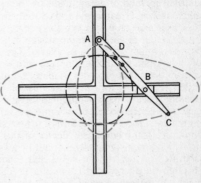

FIG. 6 ELLIPTIC TRAMMEL

GEAR MECHANISM

Gears transmit rotary motion from one shaft to another through meshing perimetral teeth on wheels mounted on the shafts. Usually the transmission of motion is associated with a change in torque. It is effected without slip and with a high degree of mechanical efficiency. The simplest gear mechanism comprises two meshing gears (toothed wheels) mounted in a fixed cage or frame (Fig. 1). Since the two gears intermesh, they must rotate with the same circumferential velocity. This means that the smaller gear (usually referred to as the "pinion") has to rotate at a higher angular velocity, i.e., has to perform more revolutions per minute, than the larger. The rotational speeds of the gears are inversely proportional to their respective diameters or their respective numbers of teeth. Since the force exerted by two teeth in contact must be of equal magnitude for both (action is equal to reaction : $F = F_R$), the torque acting on the larger gear ($= F \times R$) must be of greater magnitude than that acting on the smaller gear ($= F \times r$). The ratio of the speed of the driving shaft to that of the driven shaft is called the gear ratio or transmission ratio.

If, instead of the cage, one of the gears is held stationary and the other gear is driven, the cage will perform a rotary motion; a planetary gear system is thus obtained (Fig. 2), comprising the sun wheel, the planet wheel (as a rule there are two or more planet wheels), the cage, and the internally toothed annulus which meshes with the panel wheel, while the latter meshes with the sun wheel. In Fig. 2 the sun wheel is conceived as being stationary, while the annulus drives the planet wheel. Alternatively, any one of the three elements—sun wheel, planet wheel, cage—may be assigned the role of the stationary element and any of the two others of this trio may be driven. Thus six different possibilities of transmission are available, as well as the possibility of locking the sun wheel and the planet wheels, so that direct drive is obtained. Examples of planetary gear systems are given on pp. 462–463 and pp. 494–495 of Volume I.

In general, gear systems are named after the shape of the gears or the arrangement of their teeth (see page 162 et seq.). Shafts whose center lines cross but do not intersect are connected by spiral gears (Fig. 3). In the worm gear (Fig. 4) the basic shape of the worm is a cylinder or a globoid, while that of the worm wheel is a globoid. A globoid is a body of revolution that is generated by the rotation of a circular arc about any axis.

Change-speed gears permit the selection of various transmission ratios. The simplest method of doing this is by the removal and replacement of gear wheels of different sizes. Greater ease and convenience are provided by a change-speed gearbox, such as is used on motor vehicles and certain machine tools. Such gearboxes permit the selection of various transmission ratios between driving shaft and driven shaft by appropriate operation of a lever. The speed changes are obtained by the action of sliding pinions that are moved into or out of mesh with gear wheels. A type of transmission that provides as many ratios as there are gear pairs and that takes up little space is the draw-key transmission (Fig. 5). Whenever the draw key is shifted, one of the gears that are mounted loose on shaft II is locked to the shaft, so that the gear pair concerned then transmits power. This type of gear is used as feed gearing on machine tools (see page 184). A reversing gear serves to reverse the direction of rotation. In the bevel-gear device of this kind (Fig. 6) shaft I is the driving shaft, on which an engaging element with claws is slidably mounted. When this element is shifted to the left, the bevel gear C and therefore the driven shaft II is driven through the agency of the bevel gear A; when the engaging element is slid to the right, the drive is effected through the agency of the bevel gear B instead of A, so that now the shaft II rotates in the opposite direction.

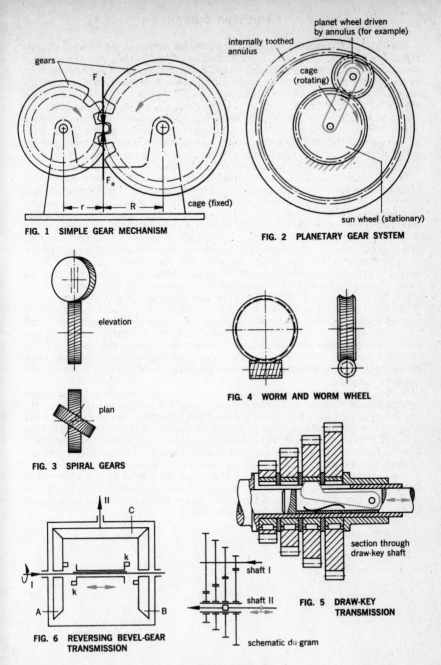

FIG. 1 SIMPLE GEAR MECHANISM

FIG. 2 PLANETARY GEAR SYSTEM

gears

F

F$_R$

r

R

cage (fixed)

planet wheel driven by annulus (for example)

internally toothed annulus

cage (rotating)

sun wheel (stationary)

elevation

plan

FIG. 3 SPIRAL GEARS

FIG. 4 WORM AND WORM WHEEL

section through draw-key shaft

FIG. 5 DRAW-KEY TRANSMISSION

II

C

I

k

k

A

B

FIG. 6 REVERSING BEVEL-GEAR TRANSMISSION

shaft I

shaft II

schematic diagram

207

FRICTION DRIVES

In a friction-drive mechanism the friction elements are usually cylindrical, conically tapered or globoid wheels which are pressed together so firmly that the frictional force developed at the point or line of contact transmits power. The power that can be transmitted in this way will depend on the magnitude of the contact pressure and the coefficient of friction of the surfaces in contact. In every friction drive a small amount of slip develops between the friction elements: i.e., at the point of contact the driven wheel always has a slightly lower circumferential velocity than the driving wheel. The drive is therefore not a fully positive one. Such forms of drive have the advantage that the effective radius can be varied quite simply by a shifting of the point of contact of the friction wheels toward or away from the axis of rotation, so that "infinitely variable" control over the transmission ratio is obtained.

In the simplest form of friction-wheel drive, cylindrically shaped wheels roll against each other. The resilience of the (often rubber-covered) wheels, spring pressure or weights produces the necessary contact pressure and thus develops the friction through which power is transmitted. Just as with toothed gears, the transmission ratio is determined by the ratio of the respective radii, while the rotational speeds of the two wheels (rpm) are inversely proportional to their radii. (Examples: record-player drive, tape-recorder drive.) Every powered traction wheel of a road vehicle or rail-mounted vehicle is in effect a friction drive. The contact pressure due to the weight of the vehicle enables the wheel to develop friction and thus "get a grip" on the road or rail surface. This combination is comparable to a rack and pinion, the wheel being the "pinion" and the road or rail the "rack." Fig. 1 shows a friction drive that is similar in construction to a planetary gear (see page 206) and is of the type used for the tuning mechanism on some radio receivers. The balls correspond to the planet wheels; they roll on the inner shaft, which corresponds to the sun wheel. When the fine-adjustment knob mounted on the inner shaft is rotated, the "cage" formed by the hollow shaft (on which the large coarse-adjustment knob giving direct drive is mounted) will rotate at a lower speed and thus make possible precision tuning.

A friction drive for infinitely variable speed control comprising two conically tapered friction elements and an intermediate ring is illustrated in Fig. 2. When the ring is in the right-hand position (shown in black), the upper shaft will rotate faster than the lower shaft, which is the driving shaft, because in this position the driving radius R is larger than the driven radius r. When the ring is shifted to the left-hand position (shown in red), the driving radius r will be smaller than the driven radius R, so that now the upper shaft will rotate more slowly than the lower shaft. Fig. 3 shows a friction drive comprising a large flat wheel and a small friction wheel which can be slid to different positions on the driving shaft on which it is mounted. The small wheel can thus engage with the large wheel at any desired radius of the latter, so that the speed of the driven shaft can be varied.

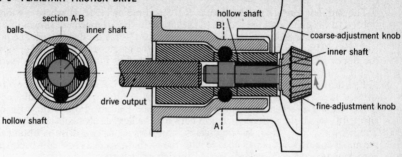

FIG. 1 PLANETARY FRICTION DRIVE

section A-B

balls   inner shaft

hollow shaft
B

coarse-adjustment knob

inner shaft

hollow shaft

drive output

fine-adjustment knob

A

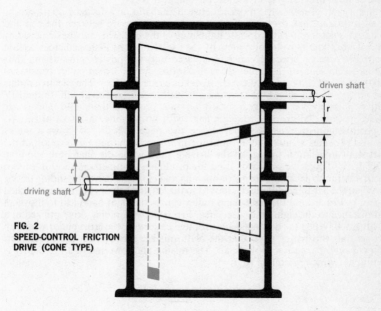

driven shaft

r

R

R

r

driving shaft

FIG. 2
SPEED-CONTROL FRICTION
DRIVE (CONE TYPE)

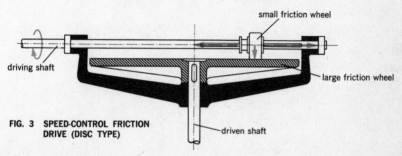

small friction wheel

driving shaft

large friction wheel

FIG. 3 SPEED-CONTROL FRICTION
DRIVE (DISC TYPE)

driven shaft

PULLEY DRIVES

The simplest form of pulley drive is the "fixed pulley" (Fig. 1), where the force Z needed to raise the load Q is equal to the latter. In the arrangement shown in Fig. 2 the load is suspended from the pulley; the (upward) force needed to raise the pulley with the load suspended from it is now only half the magnitude of the load (assuming the pulley itself to be weightless). Pulleys employed as lifting tackle are described on pp. 230-231 of Vol. I.

The flexible connecting elements used in pulley drive systems are ropes, belts, wires, chains, etc., made of a variety of materials (rubber, leather, textile fabrics, metal, etc.). These elements can take only tensile loading. Power transmission is effected through the action of friction between the flexible elements and the pulleys around which they pass. In some cases, however, a positive drive is obtained by means of chains whose links engage with the teeth on special toothed wheels called sprockets (example: bicycle or motorcycle chain drive).

Transmission of rotational motion between shafts at any distance apart can be effected through a belt drive. The necessary contact pressure between the belt and the pulleys is ensured by appropriate tightening of the belt. This may be achieved by means of a tensioning roller or pulley. By the use of a belt of V-shaped cross section (V-belt) the contact pressure can be increased and the power transmission thus made more efficient. In the pulley drive illustrated in Fig. 3 each pulley consists of two halves which can be moved farther apart or closer together. The effective radius of the pulleys and therefore the transmission ratio can be varied at will. In the PIV (positive infinitely variable) gear based on this principle the pulley halves are provided with radial grooves which engage with projections on a special belt, so that positive nonslip drive is achieved (see also page 182). Fig. 4 shows a pulley drive that provides a simple solution of the problem of shifting the position of the driven shaft in relation to that of the driving shaft while the shafts are rotating. In Fig. 5 a pulley drive system is used for producing symmetrical motion of two parts in relation to each other (e.g., for opening and closing of curtains, sliding doors, etc.). When the left-hand door leaf is moved to the left, the belt or rope that is attached to it and passes around the right-hand pulley causes the right-hand leaf to move an equal distance to the right, and vice versa. Fig. 6 shows a pulley drive utilized in a high-lift truck. When the driving shaft is rotated clockwise, the large pulley winds up the rope, thus shortening it, so that the platforms are raised. They are lowered by counterclockwise rotation of the shaft. The raising and lowering of a fire ladder is based on the same principle (Fig. 7).

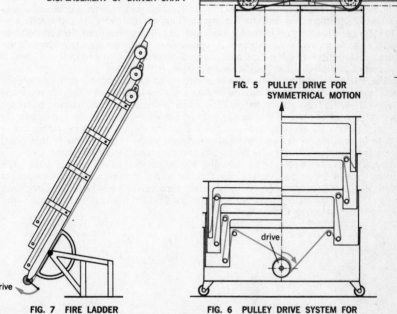

FIG. 1 SIMPLE PULLEY DRIVE (FIXED PULLEY)

pulley drives

FIG. 2 LOOSE PULLEY

FIG. 3 PULLEY DRIVE FOR INFINITELY VARIABLE SPEED CONTROL

FIG. 4 PULLEY DRIVE ALLOWING DISPLACEMENT OF DRIVEN SHAFT

FIG. 5 PULLEY DRIVE FOR SYMMETRICAL MOTION

FIG. 7 FIRE LADDER

FIG. 6 PULLEY DRIVE SYSTEM FOR HIGH-LIFT TRUCK

CAM MECHANISM

A cam is a specially shaped component that serves to guide the motion of a component called a follower. The cam may have a linear or a rotary motion. Like the crank mechanism, the cam mechanism serves to convert a given input motion (usually a uniform motion) into a desired output motion of particular form. With a crank mechanism it is not always possible to produce a motion whose path is of the desired shape, whereas a suitably designed cam mechanism will make possible practically any shape or pattern of motion. A great advantage of the cam principle is that it is quite conveniently possible to introduce pauses of any desired length into the motion. This advantage is widely utilized in machinery of all kinds, such as packaging machines and many others. With cams it is possible to perform simple oscillatory or sliding movements as well as precisely controlled movements of elaborate shape (e.g., guiding a milling cutter along a curved outline of any desired shape).

Every cam mechanism (Figs. 1, 2 and 3) essentially comprises three parts: the frame or base (a), on which the cam (b) is mounted, and the follower (c) whose motion is controlled by the cam, which is given a linear (Fig. 1) or a rotary (Figs. 2 and 3) motion. As a rule, the follower is held in contact with the cam by a spring or a guiding groove or some other appropriate device.

A typical example of a cam mechanism is the valve gear of an internal-combustion engine (Fig. 4). The rotating cam has an approximately pear-shaped profile comprising two circular curves joined by two straight lines which are tangential to those curves. The follower consists of a roller tappet which is moved up and down by the cam and imparts this motion to the rod that controls the opening and closing of the valve. The center of the roller traces a curve of similar shape to the cam profile (equidistant curve). In Fig. 5 the stroke, the speed and the acceleration of the roller tappet have been plotted against the angle of rotation of a cam of the type shown in Fig. 4. At a certain angle of rotation the acceleration undergoes a sudden change in value, which imparts a jerk to the tappet. This occurs every time the radius of curvature of the cam profile changes abruptly (e.g., transition from circular curve to tangent, and vice versa). Conversely, it is possible to start from a certain acceleration curve that comprises no abrupt changes (red curve in Fig. 5) and design a cam so shaped as to produce this "gentle" acceleration, free of jerks.

Fig. 6 shows a cam mechanism whose two paths can so move a milling cutter in two mutually perpendicular directions that a cut conforming to a specific predetermined shape (in this case the letters "HB") can be produced. The two cam paths are determined as follows: The trace is subdivided into a number of approximately equal portions (22 in the present example). Then the distances that the cutter has to move in the horizontal and vertical directions, respectively, to reach this point from the initial position 0, are plotted in two diagrams for all the points (0 to 22). Then the circumference of the cam disc is also divided into 22 equal parts. The distances from the respective cam paths for horizontal and vertical cutter movements to a certain reference radius on the disc are then marked out in radial directions. From 19 to 22 the cutter must moreover be lifted off the work, since this constitutes the return motion to the starting point.

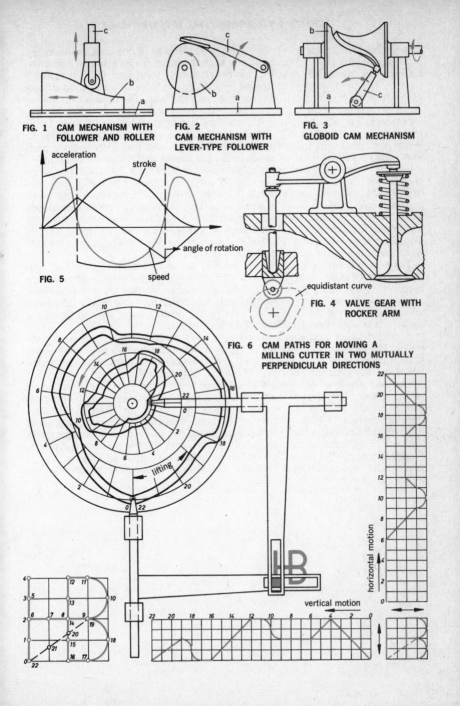

FIG. 1 CAM MECHANISM WITH FOLLOWER AND ROLLER

FIG. 2 CAM MECHANISM WITH LEVER-TYPE FOLLOWER

FIG. 3 GLOBOID CAM MECHANISM

FIG. 5

acceleration

stroke

angle of rotation

speed

equidistant curve

FIG. 4 VALVE GEAR WITH ROCKER ARM

FIG. 6 CAM PATHS FOR MOVING A MILLING CUTTER IN TWO MUTUALLY PERPENDICULAR DIRECTIONS

lifting

horizontal motion

vertical motion

A ratchet mechanism is usually employed as a means of arresting a motion and producing intermittent action of a force so that it develops its action at particular instants. The ratchet bar A in Fig. 1 is provided with teeth with which the pawl B engages. The pawl, which is controlled by a spring, is pivotably mounted in a frame C in which the ratchet bar is also mounted in a slideway. The pawl is thus able to arrest the motion of the ratchet bar when the latter is thrust to the right and can, for example, periodically release it, so that the bar moves in successive jerks. In the grip-roller locking device (Fig. 2a) the bar A can move in relation to the cage C. When A moves to the right, the roller (or ball) jams and thus locks the two parts immovably together. When A moves to the left, the roller is released. Fig. 2b shows a locking or clamping device for a belt or strap working a roller blind or some such device. In the friction brake (Fig. 3), counterclockwise rotation of the wheel causes friction to build up as a result of the thrust exerted by the brake block on its pivoted angle lever, so that the wheel is braked; no braking action is developed when the wheel rotates in the clockwise direction. An ordinary lock is also essentially a device of this general type. In the catch lock (Fig. 4) the sliding element A is slid forward in the guide C and is locked by means of the catch B. The latter is released when it is lifted by the key, so that A can then be slid back.

The grip brake (Fig. 5) that acts as a safety device for elevators (passenger lifts) in the event of cable fracture is a locking mechanism of this class. When the rope exerts an upward pull, the gripping jaws are released, so that the grip brake as a whole can be moved in relation to the fixed guide rod. If the cable breaks, the powerful spring will push the slide down, causing the pivoted angle lever to thrust the brake blocks with considerable force against the rod.

A ratchet-wheel mechanism is used to produce intermittent motion, as already stated. The mechanism illustrated in Fig. 6 is equipped with two pawls. The upper pawl is operated by an eccentric drive, and its successive thrusts cause the ratchet wheel to rotate counterclockwise in an intermittent motion. The lower pawl prevents clockwise rotation of the ratchet wheel while the driving pawl is performing its return motion. The Maltese cross mechanism used, for instance, in motion-picture equipment also belongs to this class of devices (see Vol. I, pp. 190–191). The star-wheel mechanism (Fig. 7) operates on the same principle. The large driving wheel drives the small driven wheel only as long as the drive pins on the former engage with the teeth of the latter. As there are (in this particular case) eight pins and eight teeth, the small wheel will always perform one complete revolution as long as there is engagement. A mechanism of this type is used in most counting devices: e.g., the mileage counter in a motor vehicle.

(more)

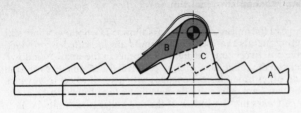

FIG. 1 RATCHET BAR AND PAWL

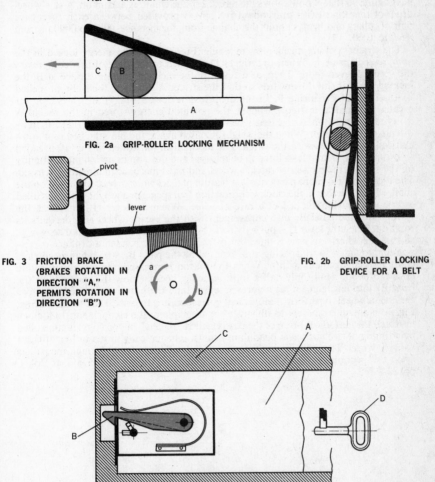

FIG. 2a GRIP-ROLLER LOCKING MECHANISM

FIG. 3 FRICTION BRAKE
(BRAKES ROTATION IN
DIRECTION "A,"
PERMITS ROTATION IN
DIRECTION "B")

pivot

lever

a

b

O

FIG. 2b GRIP-ROLLER LOCKING
DEVICE FOR A BELT

FIG. 4 CATCH LOCK

The operating principle of the mileage counter is as follows: The hollow cylindrical roller on which the ten numerals representing the units are indicated is provided internally with one drive pin. Every time the figure 9 appears in the small "window" on the instrument panel and the roller has therefore performed a complete revolution (but only then and at no other time) this pin engages with a small wheel to advance the roller on which the "tens" are marked to the next figure (e.g., from 0 to 1); at the same time, the figure 0 appears in the window of the first roller (the "units" roller). When the 9 again appears on this roller, the "tens" roller is again rotated on to the next figure, so that it now shows the figure 2 in the window, and so on. A star-wheel drive of this kind (with only one drive pin) is provided between each successive pair of rollers and brings about the change from, for instance, 0089 to 0090 or from 0999 to 1000.

The striker-cocking mechanism in a rifle (Fig. 8) keeps the energy locked in the compressed spring F. When the trigger D is pulled to the right, it first compresses the spring F even more. Then, as a result of the inclined guiding surface at B, the locking pin B is slid downwards so that the striker A can fly to the right, propelled by the force of the spring F. In the drop-weight device (Fig. 9), which is used for certain mechanical testing purposes, the potential energy stored in the weight A in its raised position is released when the lever B is swung to the right, thereby disengaging the catch. When the weight is raised again, the pin attached to it automatically re-engages with the catch. The mechanism illustrated in Fig. 10 is called an escapement. It enables a force to be released and develop its action intermittently on the same principle as the ratchet-wheel-and-pawl mechanism already discussed. The escapement is more particularly a feature of clockwork drives. Mounted on the shaft, which is driven in the clockwise direction by a spring or a weight, is the so-called escape wheel, which is in fact a ratchet wheel. When the lever H is released, the spring pulls the pawl B_1 into engagement with the escape wheel and prevents its rotation. When the lever H is pushed down (by pressing the button), B_1 is disengaged, so that the wheel can now rotate; but by the time it has rotated a distance corresponding to half a tooth spacing, it is arrested by the pawl B_2, which has meanwhile come into engagement with it. When the button is released, the wheel can rotate once more, but again only half a tooth spacing, because now the spring pulls the pawl B_1 into engagement, as before. So every time the button is pressed, it allows the escape wheel to perform a movement corresponding to twice a half tooth spacing. Fig. 10 is intended merely to illustrate the principle of the escapement. In a clock or watch the periodic motion of the pawls (called "pallets" in horology) is controlled by a timing device such as a pendulum (shown schematically in red in Fig. 10) or a balance wheel. The reaction impulse of the escape wheel acting through the pallets upon the pendulum or the balance wheel maintains the motion. See also Vol. I, pp. 214–215.

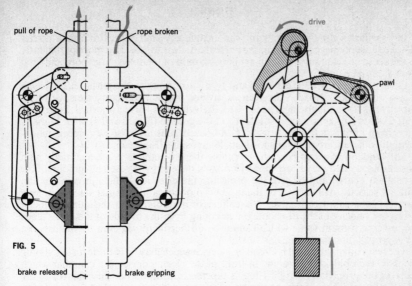

pull of rope

rope broken

brake released

brake gripping

FIG. 5

drive

pawl

FIG. 6   RATCHET-WHEEL MECHANISM

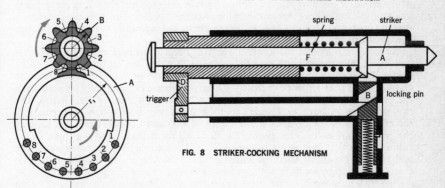

5 4 B
6   3
7   2
8   1
         A

1
8   2
7 6 5 4 3

FIG. 7   STAR-WHEEL MECHANISM

spring   striker

F   A

D

trigger   B   locking pin

FIG. 8   STRIKER-COCKING MECHANISM

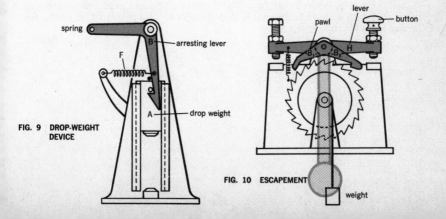

spring

B   arresting lever

F

A   drop weight

FIG. 9   DROP-WEIGHT
DEVICE

pawl   lever   button

B   H

FIG. 10   ESCAPEMENT

weight

In its widest sense the term "gun" can include any kind of firearm from a pocket pistol to a heavy siege gun, but in the present article it will be used more particularly in the sense of "cannon"—i.e., an artillery weapon of relatively large bore and fired from a carriage or a fixed mount.

Guns are of many different sizes and types, according to the purposes for which they are intended. Broadly, they can be divided into flat-trajectory guns with long barrels (Fig. 1) and high-angle guns with relatively short barrels (mortars, howitzers) (Fig. 2). Guns for action against targets on the ground (Figs. 3 and 4) differ in many ways from antiaircraft guns (Fig. 5). Land-based guns are in some respects fundamentally different from naval guns installed on ships (Figs. 6 and 7).

The function of a gun barrel is to enable the projectile to reach a suitably high initial velocity in a very short time by utilizing the energy released by ignition of the propellant charge and to give the projectile the direction which, in combination with its velocity, will carry it to the target. To obtain a high initial velocity, a long barrel is necessary. Irrespective of the caliber, or bore, of the gun, the muzzle velocity —i.e., the velocity of the projectile on emerging from the muzzle of the gun—tends toward a maximum value which cannot be further increased even by the use of a large propellant charge (Fig. 8).

The rear end of the barrel is closed by a device which may be in the form of a sliding block (breech block) or a screw. In front of the closing device is a chamber which receives the propellant charge. There is tapered transition from the chamber to the barrel proper. The interior of the latter is provided with spiral grooves (rifling), which impart a spinning motion to the projectile. The pitch of the rifling is expressed as a multiple of the bore; for instance, the pitch may be forty times the bore, meaning that the projectile performs one revolution about its longitudinal axis over a distance equal to forty times the bore of the barrel.

When a gun is loaded, the soft-metal driving band at the back end of the projectile (see page 226) is pressed into the rifling grooves and thus centers the projectile in the barrel. When the gun is fired, the firing pin strikes the primer and this ignites the charge, which may either be enclosed in a cartridge case or be entirely separate from the projectile (separate-loading ammunition, usually for guns of large bore). The explosive powder burns extremely rapidly and develops a very high gas pressure in the chamber—of the order of 45,000 lb./in.^{2} (3000 atm.) and upwards. As soon as the gas pressure exceeds the pressure with which the driving band is gripped in the rifling, the projectile is set in motion, so that the space behind increases in volume and the gas expands. After the initial pressure buildup in the chamber there is therefore a drop in pressure as the projectile makes its way along the barrel (Fig. 9). At the instant when it leaves the muzzle, the gas still has a high pressure, which causes the report when it escapes into the atmosphere. The maximum pressure developed by the propellant can be measured by means of a device called a crusher cylinder (Fig. 10). With the aid of a quartz crystal utilizing the piezoelectric effect, in combination with an oscillograph, it is possible to plot the variation of the gas pressure as a function of time.

(more)

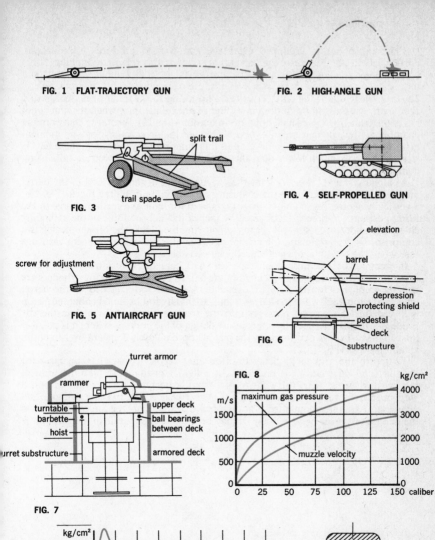

FIG. 1 FLAT-TRAJECTORY GUN

FIG. 2 HIGH-ANGLE GUN

split trail

trail spade

FIG. 3

FIG. 4 SELF-PROPELLED GUN

screw for adjustment

FIG. 5 ANTIAIRCRAFT GUN

elevation

barrel

depression
protecting shield
pedestal
deck
substructure

FIG. 6

turret armor

rammer

turntable
barbette
hoist
turret substructure

upper deck
ball bearings
between deck
armored deck

FIG. 7

FIG. 8

m/s

maximum gas pressure

1500

muzzle velocity

1000

500

0   25   50   75   100   125   150 caliber

kg/cm²
4000

3000

2000

1000

0

kg/cm²

3000

2000

gas pressure curve

1000

0

FIG. 9

distance traveled by
projectile

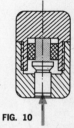

FIG. 10

219

The gas pressure developed when the gun is fired produces high stresses, particularly in the interior of the barrel (Fig. 11). For this reason, monobloc ("one-piece") barrels are now used only for guns of fairly small caliber. Medium- and large-caliber guns have barrels of composite construction, built up from several layers. In heavy guns the outer layers are shrunk on to the inner ones. As a result, the innermost layer of the tube is in a state of precompression when the gun is not in action, while the layers around it are in tension (Fig. 12). When the charge in the chamber is fired, the gas pressure first overcomes the precompression and momentarily causes tensile stresses to develop in this innermost layer, though these tensile stresses are now much lower than they would be if there had been no initial precompression.

As already stated, in heavy guns the closing device at the rear end of the barrel may take the form of a screw mechanism or a horizontally sliding block (Fig. 13); a vertically sliding breech block (Fig. 14) is used for light and medium guns. In the firing position the breech block must be locked against the force of the exploding charge. In addition to suitable means for moving the block, the breech mechanism comprises the firing pin and the extracting system for removing the spent cartridge case when the breech is opened after the gun has been fired.

In every gun of the conventional type dealt with here the expulsion of the projectile is accompanied by a recoil movement of the barrel. Hydraulic braking cylinders are usually employed to arrest this movement. A piston rod attached to the barrel pulls the piston back when the barrel recoils, so that hydraulic fluid behind the piston is forced through narrow passages into the space in front of the piston, thereby producing a braking effect. By appropriate design of the braking system it is possible to keep the resistance developed by the piston approximately constant over its entire length of travel (Fig. 15).

To return the gun to its firing position after recoil, a special recuperator (or counterrecoil) mechanism is provided. In a light gun this may take the form of recuperator springs (Fig. 16), while medium and heavy guns are generally equipped with pneumatic recuperators (Fig. 17). To save weight, the braking cylinder and recuperator are often combined. Another device for reducing the forces acting upon the gun mount is the muzzle brake. It may be fitted to the muzzle of the gun barrel,

(more)

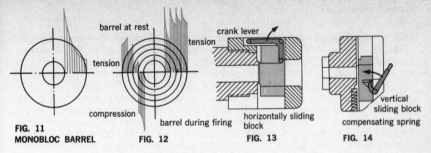

FIG. 11
MONOBLOC BARREL

barrel at rest

tension

tension

compression

FIG. 12

barrel during firing

crank lever

horizontally sliding block

FIG. 13

vertical sliding block

compensating spring

FIG. 14

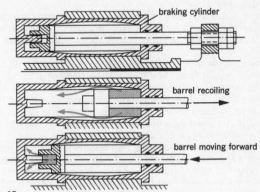

braking cylinder

barrel recoiling

barrel moving forward

FIG. 15

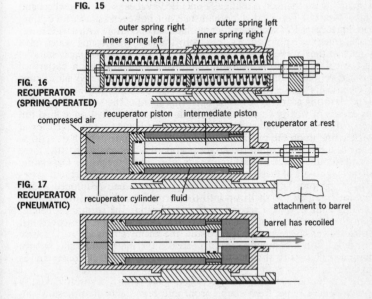

outer spring right

inner spring left

outer spring left

inner spring right

FIG. 16
RECUPERATOR
(SPRING-OPERATED)

recuperator piston

intermediate piston

compressed air

recuperator at rest

FIG. 17
RECUPERATOR
(PNEUMATIC)

recuperator cylinder   fluid

attachment to barrel

barrel has recoiled

and its function is to deflect part of the propellant gases sideways or to the rear (Fig. 18). A properly designed muzzle brake can absorb 50–60% of the recoil, permitting the use of a lighter gun mount. On the other hand, the gun barrel is more severely stressed at the muzzle in consequence of this brake and there is more lateral air-pressure buildup (Fig. 19), which may be objectionable to the gun's crew.

As a rule, the gun mount has two axes of rotation, for aiming the barrel (Fig. 20). The latter can be swung about a vertical axis (this movement being called traversing or training) and also about a horizontal axis (this movement is called elevating). Mounts for ships' guns which have to be aimed at high angles of elevation for anti-aircraft defense may have a third axis of rotation to compensate for the rolling motion of the ship (Fig. 21).

The gun proper is mounted in a cradle attached to the top carriage of the mount. The top carriage rotates about a vertical axis for traversing the gun, while the cradle has trunnions which form the horizontal axis about which the gun can be rotated in elevation.

One of the main requirements with regard to the mount is that it provide adequate stability when the gun is fired. Stability can be improved by varying the recoil length. With low angles of elevation a long recoil with a small braking force is employed, whereas with high angles a shorter recoil with a larger braking force is more suitable. The variable recoil of the gun barrel is obtained by appropriate varying of the cross-sectional area of the flow passages for the hydraulic fluid in the braking cylinder with the angle of elevation.

The range of traversing and elevating movements that can be performed by the aiming mechanism depends on the type and purpose of the gun concerned. Thus a field gun generally has only a limited range of traversing movement, whereas an antiaircraft gun can be swung around in all directions and be elevated to high angles. The simplest aiming mechanism is manually operated by means of a handwheel and gearing (Figs. 22 and 23). This kind of mechanism is not, however, suitable for dealing with fast-moving targets. For such purposes mechanized operation, with a wide range of control adjustment, is employed: e.g., electrohydraulic drive systems, which have the advantage that they are relatively simple to control and are particularly suitable for remote control of guns from a fire-direction center. In this way a battery of several guns can be jointly aimed and fired—e.g., at fast aircraft. Heavy naval guns may have electrically powered traversing gear (Ward-Leonard control system) and hydraulic elevating gear (Fig. 24). The operating speed of the hydraulic cylinder is varied by means of valves which control the flow rate of the hydraulic fluid (oil or a mixture of glycerine and water). The two piston areas F and f are so dimensioned that the force acting upon the piston rod is of equal magnitude in both directions. Only the flow of fluid acting upon the larger piston area F is controlled; the force thus developed is counteracted by the force on the other side of the piston. Depending on the relative magnitude of these forces, the piston moves in the cylinder and varies the elevation of the gun. Turret-mounted naval guns are usually equipped with a hoist system for bringing the shells up into the turret from the shell room. With separate-loading ammunition, a second system for hoisting the cordite propellant charges from the magazine is provided. A modern warship's guns are controlled from a control tower where information about the enemy's range, speed, course and heading is fed into a computer, which processes the information and in turn passes aiming instructions to the gun turrets. Each turret rotates on rollers and is operated hydraulically.

For aiming the gun at a directly visible target, some form of sighting device is required. Such devices range from simple front and rear sights, like those on an ordinary rifle, to complex "fire-control systems" for large guns. The sighting device may move with the gun barrel (Fig. 25) or may be separate from the gun-elevating

(more)

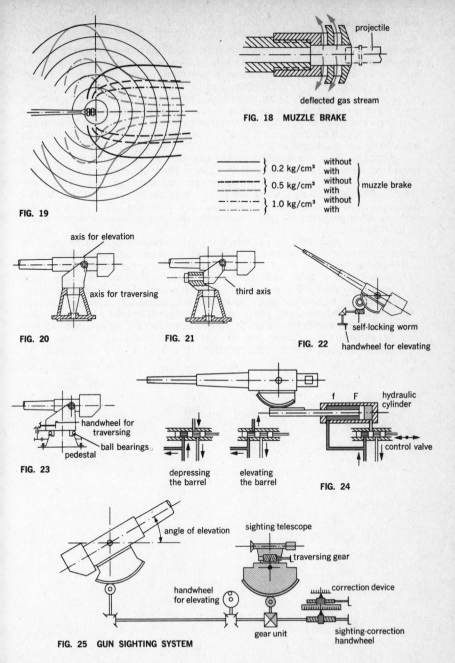

projectile

deflected gas stream

FIG. 18   MUZZLE BRAKE

FIG. 19

——————	0.2 kg/cm²	without / with
– – – – –	0.5 kg/cm²	without / with
–·–·–·–	1.0 kg/cm²	without / with

muzzle brake

axis for elevation

axis for traversing

FIG. 20

third axis

FIG. 21

self-locking worm

handwheel for elevating

FIG. 22

handwheel for traversing

ball bearings

pedestal

FIG. 23

depressing the barrel

elevating the barrel

f   F   hydraulic cylinder

control valve

FIG. 24

angle of elevation

sighting telescope

traversing gear

handwheel for elevating

correction device

gear unit

sighting-correction handwheel

FIG. 25   GUN SIGHTING SYSTEM

mechanism (Fig. 26). Precise prediction of the details of the trajectory of the projectile is based on ballistic calculations. The results of such calculations are embodied in a firing table, summarizing the trajectory information needed for correctly aiming the gun to hit the target. In the simplest case, the table will comprise a list of angles of elevation corresponding to various ranges. The elevation data for aiming the gun may be plotted in graph form, as in Fig. 27. This graph gives the so-called tangent-elevation values for various ranges. The tangent elevation is the angle between the direction in which the gun is actually aimed and the line of sight—i.e., the straight line joining gun and target.

A projectile fired from a gun is subject to a number of different forces. In the first place, there is the force exerted by gravity. There are also forces due to the rotation of the earth, including more particularly the centrifugal force. Then there are the aerodynamic forces, which are produced by the resistance of the air. The most important of these is called drag, which acts along the axis of the projectile. Stability of the projectile in flight is ensured by the spinning motion imparted to it by the rifling.

To obtain a practically serviceable firing table, various corrections must be applied to the standard trajectories. Corrections must be made for meteorological conditions (temperature, wind, etc.; see also page 226), the decrease in muzzle velocity as the barrel wears, and other factors. The firing table may be incorporated into a machine called a director, into which observations of the target position are fed and which computes the projectile-release conditions. A radar sighting device, a director and a power-aimed gun (or a battery of guns) provide a fully automatic system for antiaircraft defense (Fig. 28). Manual adjustment of the gun's aim against such fast-moving targets as modern aircraft is impracticable, and for this reason the high-speed computerized control provided by a director system is essential.

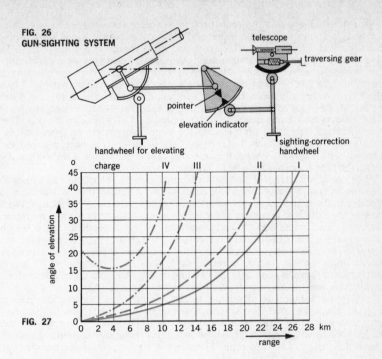

FIG. 26
GUN-SIGHTING SYSTEM

telescope

traversing gear

pointer

elevation indicator

handwheel for elevating

sighting-correction
handwheel

FIG. 27

charge   IV   III   II   I

angle of elevation

range

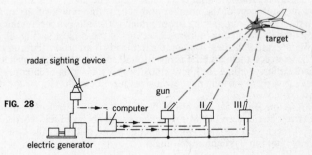

FIG. 28

radar sighting device

target

gun

II   III

computer

electric generator

PROJECTILES

The external shape of a projectile fired from a gun is so designed as to enable it to achieve a favorable ballistic trajectory. A typical projectile has a tapered point which is called the ogive and which is joined to a cylindrical portion (Fig. 1). The ogive usually contains a fuse for detonating the bursting charge of the shell (in certain types of shell, however, the fuse is located at the base instead of at the nose). At the transition from the ogive to the cylindrical part is an accurately machined band called the bourrelet. It serves to center and guide the projectile in the gun barrel. The driving band (or rotating band) is a ring of softer metal which engages with the rifling grooves and forms a seal to the gas pressure developed by the propellant. With fixed and semifixed ammunition the propellant charge is contained in a cartridge case which is attached to the base of the shell—e.g., by being crimped so as to engage with a groove. Within the cartridge case is the primer, a metal tube containing the primary explosive which detonates the main propellant.

When the gun is fired, the driving band engages with the rifling and causes the projectile to spin about its longitudinal axis. This spinning motion stabilizes the projectile in flight and keeps it aligned—i.e., with its nose always pointing forward. The projectile in flight is subject to the action of gravity and to the air resistance it encounters. The magnitude of this resistance is largely dependent on the velocity of the projectile. The density and temperature of the air and the velocity of the wind (if any) are important factors which affect the trajectory. A 28-cm (11-in.) shell fired with an initial velocity of 900 m/sec. (3000 ft./sec.) at an elevation of 45 degrees would attain a range of about 80 km (50 miles) in a vacuum, but under actual atmospheric conditions the range is less than half this distance (Fig. 2).

Artillery ammunition is of various kinds, depending on the purpose (type of target) at which it is fired. The high-explosive shell (Fig. 3), with a sensitive fuse in the nose, is used mainly against unarmored targets. The shell wall is relatively thin, enclosing a large bursting charge, whose weight corresponds to between 7 and 10% of the total weight of the projectile. The effect of such a shell is due mainly to the blast and the splinters from the shattered shell wall. The effect of a shell can be determined by measurement of the number and penetrating power of the splinters formed. This is done on a special testing site provided with walls of a certain thickness arranged at various radial distances from the center of the explosion (Fig. 4). After the shell has been exploded, these walls are examined to ascertain the number of splinters that have struck them and the proportion of splinters that have penetrated them. The usual explosive filling for high-explosive shells has long been either TNT or amatol. In recent years various new explosives have been developed for use in artillery ammunition.

Tracer ammunition (Figs. 5a and 5b) is fired at a rapid rate from automatic weapons. It permits the course of the projectiles to be observed and corrected as necessary. The "tracer" is a cartridge of pyrotechnic composition at the base of the shell and is visible both in daylight and at night.

(more)

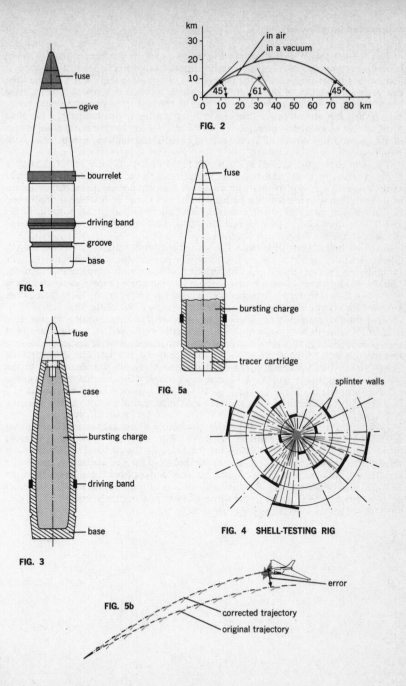

fuse

ogive

bourrelet

driving band

groove

base

FIG. 1

km

in air

in a vacuum

45°   61°   45°

0 10 20 30 40 50 60 70 80 km

FIG. 2

fuse

bursting charge

tracer cartridge

FIG. 5a

fuse

case

bursting charge

driving band

base

FIG. 3

splinter walls

FIG. 4   SHELL-TESTING RIG

error

corrected trajectory

original trajectory

FIG. 5b

227

The high-explosive shell shown in Fig. 6 has its fuse at the base instead of at the nose and is used mainly against lightly armored targets. The wall of the shell is somewhat thicker than that of the ordinary high-explosive shell, while the bursting charge is about 6–7% of the weight of the shell. The fuse is screwed into the base plug, which in turn is screwed into the base of the shell. The penetrating power of such a shell depends on the caliber, the velocity and the angle of impact; it is not so great as that of an armor-piercing shell. In Fig. 7 the curves present a comparison of the penetrating power of armor-piercing shells and high-explosive shells with base fuses.

The armor-piercing shell (Fig. 8) is employed against heavily armored targets. It has a relatively thick wall and a small bursting charge (about 2% of the weight of the projectile). It is provided with an armor-piercing nose and sometimes a thin-walled windshield to improve the ballistic properties, as in Fig. 8, where the hardened nose moreover has a cap of softer metal which flattens out on impact with the target and thus forms a kind of guiding pad for the hard armor-piercing nose. In this type of shell the fuse is always located at the base.

Artillery ammunition of various kinds used for producing light for illuminating enemy positions or for signaling is collectively referred to as "pyrotechnics." The illuminating shell (Figs. 9a and 9b) contains an illuminant (sometimes called the "candle"), which may consist of magnesium or aluminum powder in combination with an oxidizing material. Attached to the candle is a parachute which opens when an expelling charge, detonated by a time fuse, forces it out of the shell.

Most mortar ammunition differs from other artillery ammunition in that it is fired from smooth-bore weapons. The projectile employed is fin-stabilized, not spin-stabilized. Fig. 10 shows a typical teardrop-shaped mortar projectile, with a fuse in the nose and metal fins in the tail. It is dropped, tail first, into the muzzle of the mortar; when the ignition charge strikes the firing pin, the propellant in the tail of the projectile is ignited. To increase its range, the projectile may contain a booster propellant charge (as in Fig. 10) which provides "rocket propulsion." In general, projectiles fired from mortars are characterized by low muzzle velocity, high trajectory and short range.

To obtain very high muzzle velocities, projectiles whose caliber is less than that of the gun barrel may be employed (Fig. 11). For firing, the projectile is inserted into a special propelling base which fits the barrel of the gun and which drops off shortly after the projectile emerges from the muzzle of the gun. Because of its smaller size, the projectile receives a more powerful acceleration from the gas pressure than would a shell of normal size. By this means it is possible to attain muzzle velocities of 1600 m/sec. (5300 ft./sec.) and upwards, whereas approximately 1200 m/sec. (4000 ft./sec.) is the maximum for normal projectiles.

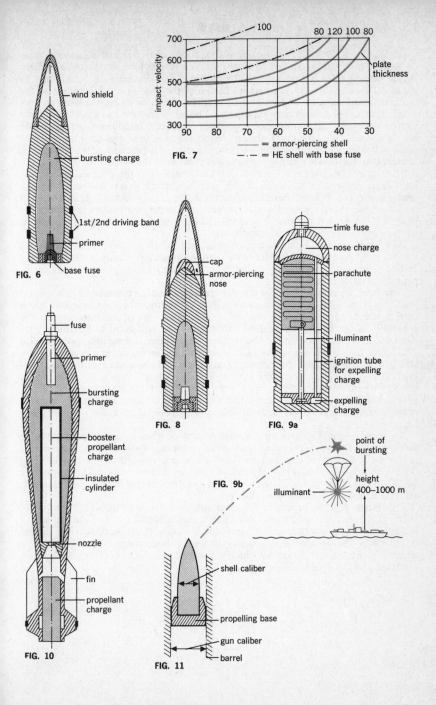

FIG. 6
- wind shield
- bursting charge
- 1st/2nd driving band
- primer
- base fuse

FIG. 7
impact velocity
700 600 500 400 300
90 80 70 60 50 40 30
100 — 80 120 100 80
plate thickness
— = armor-piercing shell
-·- = HE shell with base fuse

FIG. 8
- cap
- armor-piercing nose

FIG. 9a
- time fuse
- nose charge
- parachute
- illuminant
- ignition tube for expelling charge
- expelling charge

FIG. 9b
- point of bursting
- height 400–1000 m
- illuminant

FIG. 10
- fuse
- primer
- bursting charge
- booster propellant charge
- insulated cylinder
- nozzle
- fin
- propellant charge

FIG. 11
- shell caliber
- propelling base
- gun caliber
- barrel

229

FUSES

A "fuse" is a device for detonating the explosive charge in a shell, missile, mine or bomb. This article is concerned more particularly with fuses employed in artillery shells. According to the position of the fuse in the projectile, a distinction can be made between the nose fuse (or point fuse) and the base fuse. With regard to the mode of functioning, a fuse may be an impact fuse (with or without delayed action), a time fuse or a proximity fuse. Fuses should be safe to handle and store and be safe against accidental detonation due to jolting or shaking. In addition, a fuse should be "bore-safe"—i.e., it should not be able to function until the shell has traveled some considerable distance from the gun muzzle. This precaution is necessary to protect the gun's crew against premature explosion of the shell.

The impact fuse, without delay action and installed in the nose of the shell, is used mainly against unarmored targets such as aircraft, boats and small vessels, which present sufficient resistance to actuate the highly sensitive fuse. Delayed-action impact fuses are used against targets into which the shells are required to penetrate before they explode. In general, the time lag is only a few hundredths of a second. Fig. 1 shows a typical impact fuse.

Base fuses are used in shells employed against lightly armored targets and also in armor-piercing shells, the latter being used for attacking heavily armored targets. In both cases the fuses function with a delayed action, so as to give the shell time to penetrate before exploding.

Time fuses are of two types, the powder-train type and the mechanical (clockwork) type. They are used in cases in which it is required that the shell explode after a certain precise length of time—i.e., at a particular point of the trajectory in the proximity of the target; for example, an antiaircraft shell is set to burst at a predetermined altitude. Setting the fuse to the required time is done in a special device just before the shell is loaded into the breech of the gun for firing. Certain types of shell are fitted with a fuse that detonates on impact and also embodies a self-destroying time fuse. The latter causes the shell to explode while still in the air if it has missed the target; this arrangement obviates the danger that the unexploded shell will plunge back to earth, where it could harm one's own personnel. When the functions of an impact fuse and a time fuse are embodied in one device, it is called a combination fuse.

A proximity fuse (Fig. 2) causes the shell to explode when it passes within a specific distance of the target. This result is achieved by means of electronic devices carried in the shell. A well-known example is the VT fuse developed in World War II. It contains a miniature radio transmitter and also a miniature receiver. The transmitter sends out a continuous signal, and when this signal is reflected back by a solid object, the fuse detonates the explosive charge.

The impact fuse illustrated in Fig. 1 comprises a device to make it "bore-safe." When the projectile is fired from a rifled gun, a number of rotations of the shell about its longitudinal axis are needed to "arm" the fuse—i.e., to release the striker so that it can pierce the detonating cap when the shell hits the target. This "arming" is effected by centrifugal action by which a number of blades or other elements are retracted and the striker thus released.

(more)

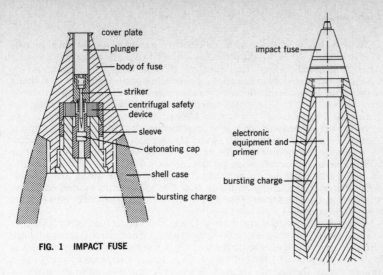

FIG. 1   IMPACT FUSE

cover plate
plunger
body of fuse
striker
centrifugal safety device
sleeve
detonating cap
shell case
bursting charge

impact fuse

electronic equipment and primer

bursting charge

FIG. 2   PROXIMITY FUSE

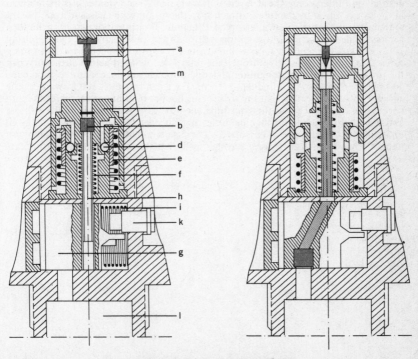

a
m
c
b
d
e
f
h
i
k
g
l

FIG. 3   IMPACT FUSE FOR SMOOTH-BORE GUN,
BEFORE AND AFTER HITTING TARGET

A highly sensitive impact fuse with self-destroying time action for rifled guns is illustrated in Figs. 4a–4d. When the projectile is fired, it acquires such a high spinning speed that the steel balls a are forced outwards by centrifugal action and cause the sliding element c to move forward toward the nose of the shell, against the restraining force of the spring b. As a result, the two pins d are released and are then able to move outwards against the pressure of the annular spring e. When this happens, the sliding element f is released, which now likewise moves radially outwards and thereby brings the detonating cap h into alignment with the striker g. The fuse is now "armed"—i.e., ready to function; when the impact occurs, the striker pierces the cap, which ignites the priming charge, and this in turn detonates the bursting charge. Self-destruction occurs when the spin of the projectile has slowed down to such an extent that the counteracting force developed by the spring b is stronger than the centrifugal force acting on the balls a. The balls are thus forced back into the sliding element c. The action of the spring b now thrusts the striker into the cap, so that the shell explodes.

Fig. 3 illustrates the functioning of a highly sensitive impact fuse for smooth-bore guns which is convertible from instant to delayed action. The fixed striker a is attached to the front cover of the fuse. The detonating cap b is located in a piston c, which is held in its lower position by the balls d, which in turn are retained by the collar e. The spring f forces the piston c, braked by the air cushion m, on to the striker a. In the bottom part of the fuse is the device for interrupting the ignition passage, if desired, and interposing the delayed-action element. The cylinder g is maintained in the safe position by the tube h attached to the piston c; the passage from the cap to the priming charge is now interrupted. When this tube is withdrawn, the cylinder g rotates (under the action of the spring i) through the angle permitted by the setting knob k. Thus the ignition flash issuing from the cap is conducted to the priming either direct or, alternatively, through the interposed delayed-action element. When the projectile is fired, the longitudinal acceleration causes the collar e to slide back and release the balls d. Consequently, the piston c can be slowly slid forward by the spring f to such an extent that the striker touches but does not as yet pierce the thin cover plate over the cap. At the same time, the tube h is withdrawn from the cylinder and the latter is rotated through its preset angle (as determined by the setting given to the knob k before the shell is fired). The fuse is now armed for functioning on impact, when the striker is forced through the thin plate and into the detonating cap.

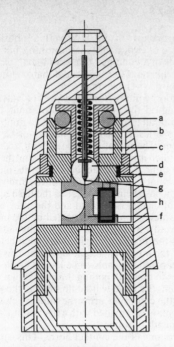

FIG. 4a FUSE SAFE

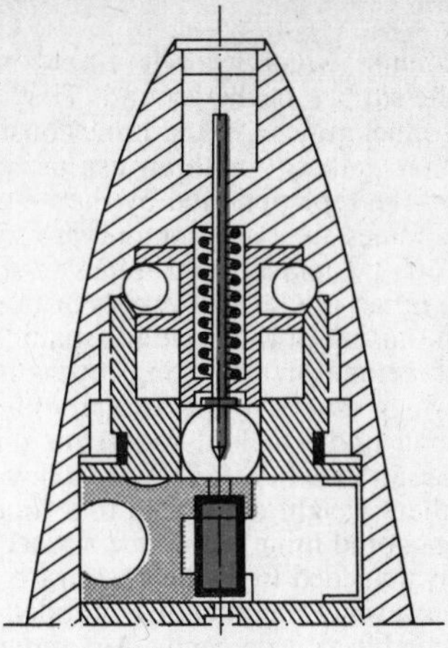

a
b
c
d
e
g
h
f

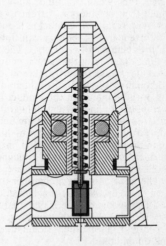

FIG. 4b FUSE ARMED

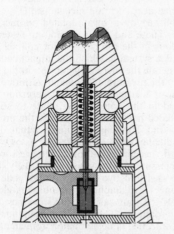

FIG. 4c STRIKER PIERCES CAP
ON IMPACT

FIG. 4d SELF-DESTROYING ACTION

MINES

Land mines, which were developed particularly in World War II, are buried just below the surface of the ground. They are of two main types: antitank mines and antipersonnel mines. A land mine consists of a container (made of metal, plastic or some other material) with an explosive charge, a fuse and a detonator. Fuses are mostly of the type operated by the weight of a vehicle or a man.

Naval mines are of two main types: moored mines (used in depths of water up to about 1300 ft.) and ground mines (resting on the bottom down to about 130 ft.). Moored mines are laid by vessels or by aircraft; Fig. 1 shows a contact mine of this type. The casing of the mine is usually spherical and contains the explosive charge with the detonator and firing mechanism. The mine is provided with an arming device which makes it operational only after it has been deposited from the mine-laying craft and which also disarms the mine if it breaks loose from its mooring. The mine, attached to its anchor, is thrown into the sea. At first this assembly floats. An auxiliary weight connected to a rope whose length is equal to the desired depth of the anchored mine below the surface of the sea is released and the mine is automatically detached from the anchor (to which it remains connected by the mooring cable, however). The anchor is flooded with water, so that it sinks, unwinding the mooring cable as it descends. As soon as the auxiliary weight, suspended below the anchor, touches the bottom, the rope of the auxiliary weight slackens. This causes the unwinding of the mooring cable to stop and the cable to be locked at the length it has then attained. The anchor continues to descend, pulling the mine down with it. Thus when the anchor reaches the bottom, the mine will be floating at the desired depth, attached to its cable. The pull on the cable now arms the mine by closing an electrical contact. Alternatively, the mine may be provided with a hydrostatic arming device which reacts to a predetermined depth of water.

The mine explodes when a vessel strikes one of the contact horns. This causes an electric current to actuate the firing mechanism, which in turn sends a strong current from a battery through the detonator. One type of contact horn contains an acid-filled glass tube which fractures, allowing the acid to enter a zinc-carbon electric cell which then produces a current to energize the firing mechanism (Fig. 2a). In another type (Fig. 2b), contact is established by a spring-loaded contact piece, so that current from a battery can flow. These and other detonating systems are illustrated schematically in Fig. 3. The four in the top row are for contact mines. In the third diagram of that row is shown a system in which the impact causes sea water to enter an electric cell and thus generate the energizing current. In the hydrostatic detonating system (fourth diagram) the pressure of sea water admitted to the interior of the mine when the contact horn is struck causes the energizing circuit to be completed. The systems illustrated in the bottom row relate to so-called influence mines—i.e., mines that are actuated not by contact but by the proximity of a ship. Ground mines are always of this kind. The magnetic mine is actuated by the change in the earth's magnetic field, producing, when a steel vessel passes within a certain distance, a deflection of a magnetic needle. Actuation of the pressure mine is brought about by the change in water pressure under a vessel in relatively shallow water. The mine contains a chamber divided into two parts, with one side of the chamber open to the sea; the deflection of the diaphragm establishes electrical contact. The acoustic mine is actuated by the sound of the ship's engines or propellers, which is picked up by sensitive microphones. The last diagram in the bottom row illustrates a detonating system for a controlled mine detonated from an observation station on land.

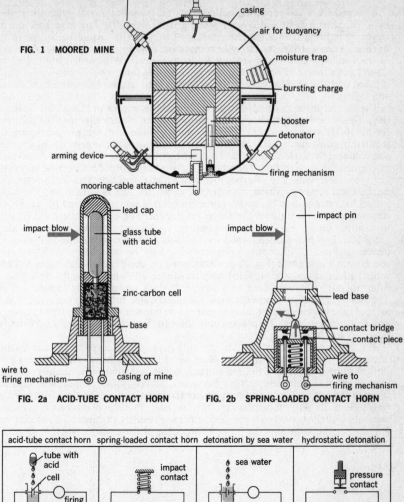

FIG. 1 MOORED MINE

- contact horns
- casing
- air for buoyancy
- moisture trap
- bursting charge
- booster
- detonator
- arming device
- firing mechanism
- mooring-cable attachment

impact blow

- lead cap
- glass tube with acid
- zinc-carbon cell
- base

wire to firing mechanism
casing of mine

FIG. 2a ACID-TUBE CONTACT HORN

impact blow

- impact pin
- lead base
- contact bridge
- contact piece

wire to firing mechanism

FIG. 2b SPRING-LOADED CONTACT HORN

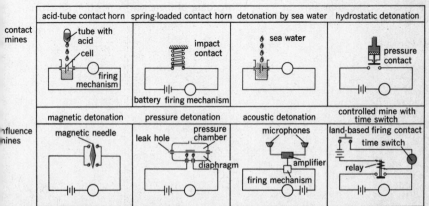

contact mines

acid-tube contact horn
- tube with acid
- cell
- firing mechanism

spring-loaded contact horn
- impact contact
- battery firing mechanism

detonation by sea water
- sea water

hydrostatic detonation
- pressure contact

influence mines

magnetic detonation
- magnetic needle

pressure detonation
- leak hole
- pressure chamber
- diaphragm

acoustic detonation
- microphones
- amplifier
- firing mechanism

controlled mine with time switch
- land-based firing contact
- time switch
- relay

FIG. 3 MINE-DETONATING SYSTEMS

TORPEDOES

A torpedo is a self-propelled underwater missile with its own guidance system. In some cases it is provided with homing equipment enabling it to seek out its target. It is fitted with an exploder which detonates the explosive charge in the warhead when it strikes the target or comes close to it. Fig. 1 shows a typical torpedo of the ordinary kind employed against surface craft. With a diameter of 53 cm (21 in.) and a length of 7 m (23 ft.), it weighs about $1\frac{1}{2}$ tons, roughly one-fifth of which is taken up by the explosive charge. Torpedoes used against submarines are shorter and lighter.

The typical torpedo comprises the main sections shown in Fig. 1: the warhead, the air-flask section (or battery compartment in an electrically powered torpedo), the afterbody (comprising the "engine room" and the compartment containing the regulating equipment), and the tail. The warhead carries the exploder mechanism and contains the explosive charge. The homing mechanism, if any, is accommodated in a detachable nose section. The air-flask section of the torpedo contains compressed air, water and fuel. The afterbody contains the gyroscope, the depth-regulating mechanism, the combustion chamber (in which the fuel is burned and the water turned into steam), and the turbine or reciprocating engine (powered by the air and steam mixture) which drives the propellers (mounted on coaxial shafts and rotating in opposite directions). The tail section also contains the tail blades and rudders.

Surface ships may launch torpedoes from tubes which can be aimed in the desired direction (Fig. 3). Alternatively, the torpedo may be launched sideways from a special launching frame (Fig. 2), a method used on small craft. Submarines and also some surface vessels are provided with launching tubes built into the hull. Above-water torpedo tubes are fired by a charge of black powder or by compressed air. The latter propellant is always used for underwater tubes. Alternatively, a sub-merged tube is sometimes of the type that can be flooded, so that the torpedo emerges under its own power. Torpedoes may also be launched from aircraft flying at relatively low altitudes.

A fairly recent development is the rocket-propelled torpedo for use against submarines. This may take the form of an Asroc (antisubmarine rocket) for use from surface ships (Fig. 4) or a Subroc (submarine rocket) fired from the torpedo tube of a submarine; it emerges from the water, travels some considerable distance through the atmosphere, and then reenters the water on approaching its submerged target (Fig. 5). With these devices, the torpedo propulsion mechanism is auto-matically switched on when the torpedo enters, or reenters, the water. This mechanism is preset before launching: i.e., it is fed the data of the enemy's position (distance, course, speed).

(more)

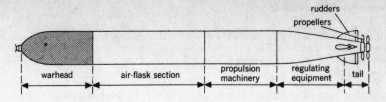

FIG. 1   MAIN PARTS OF A TORPEDO

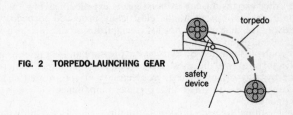

FIG. 2   TORPEDO-LAUNCHING GEAR

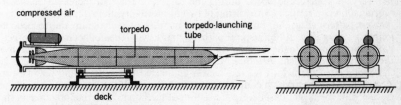

FIG. 3   TRIPLE LAUNCHING TUBES

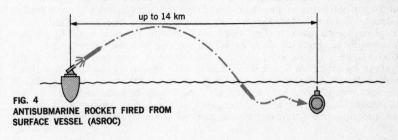

FIG. 4
ANTISUBMARINE ROCKET FIRED FROM
SURFACE VESSEL (ASROC)

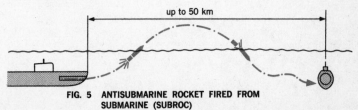

FIG. 5   ANTISUBMARINE ROCKET FIRED FROM
SUBMARINE (SUBROC)

Torpedoes are driven by multicylinder reciprocating engines, turbines or battery-powered electric motors. The propulsive agent may be compressed air, a mixture of compressed air and steam, electricity, etc. Fuels may be oil, alcohol, hydrogen peroxide, etc.

With compressed air and steam drive (Fig. 6a), air is supplied from a flask (a) through a pressure-reducing valve which is preset to enable the torpedo to develop its appropriate speed. Oil, atomized by the compressed air, is burned in the combustion chamber (b), into which water is sprayed so that steam is generated. The resulting mixture of air and steam is fed to the four-cylinder radial-type engine (Fig. 6b). After expansion, the exhaust gas is expelled, leaving a wake of bubbles. The telltale wake is absent in the electrically propelled torpedo, which has the further advantage that it does not lose weight during the run (as a steam torpedo does, since it burns up its fuel).

Depth control is effected by hydrostatic pressure and is preset to give a depth at which the torpedo will be most effective against its target. Lateral guidance is controlled by a gyroscope (driven by electricity or compressed air), by which the torpedo can be made to travel in a predetermined linear course or a curved path (Figs. 7a and 7b).

The conventional torpedo fired in a straight line is effective only if the target does not vary its course and speed during the torpedo's run. An increased chance of scoring a hit is obtained by the firing of two or more torpedoes simultaneously in a fanwise pattern (Figs. 8a, 8b, 8c). Another method is to program the torpedo to follow a zigzag (Fig. 10) or a spiral path, so that it will repeatedly cross the target's path. For this purpose the torpedo is equipped with a cam mechanism which actuates the steering mechanism. Another means of increasing the accuracy is provided by the wire-guided torpedo (Fig. 9), which is connected to the attacking vessel by an electric wire through which control signals are fed to keep the torpedo on a collision course with the target despite evasive action of the latter. A further advance in this direction is the homing torpedo, which is provided with a special device, mounted in the nose, which may consist of a sensitive acoustic receiver that picks up sounds emitted by the enemy vessel and controls the torpedo's steering equipment accordingly (passive acoustic torpedo). Alternatively, the device itself may generate a sound signal and home on the echo reflected from the target (active acoustic torpedo). Modern antisubmarine torpedoes are almost invariably of the homing kind.

Exploder mechanisms are of various types. Detonation may be caused by physical impact with the target, by acoustic influence (noise of enemy vessel's engines or propellers), by magnetic influence (change of magnetic field in the vicinity of the target), or by optical influence (the shadow of the enemy vessel when the torpedo passes under it). After launching, a torpedo must run for some distance before it becomes "armed"—i.e., ready to explode. Also, it must be fitted with a device that will automatically sink it when it completes its run without having hit the target. These precautions are necessary to minimize risk to the attacker's own ships.

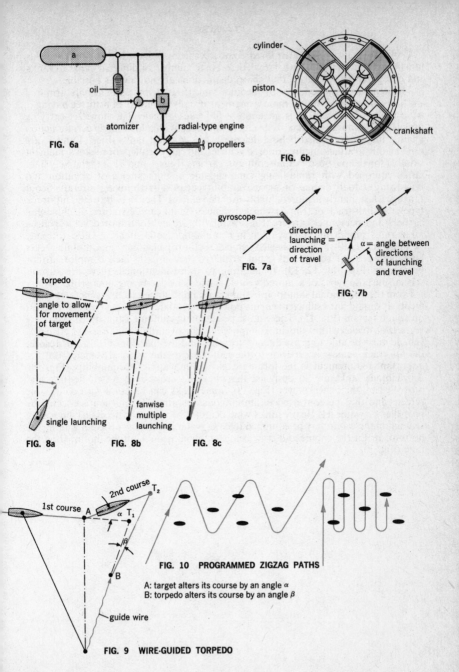

oil

atomizer

radial-type engine

propellers

FIG. 6a

a

b

cylinder

piston

crankshaft

FIG. 6b

gyroscope

direction of launching = direction of travel

FIG. 7a

α = angle between directions of launching and travel

FIG. 7b

torpedo

angle to allow for movement of target

single launching

FIG. 8a

fanwise multiple launching

FIG. 8b

FIG. 8c

1st course A

2nd course T₂

α T₁

β

B

guide wire

FIG. 9 WIRE-GUIDED TORPEDO

FIG. 10 PROGRAMMED ZIGZAG PATHS

A: target alters its course by an angle α
B: torpedo alters its course by an angle β

TANKS

Tanks are self-propelled armored combat vehicles designed to perform various functions in modern land warfare. There are a number of different types of tanks and comparable vehicles, so that sharp distinctions are not always possible.

The general-purpose heavy or medium tank (Figs. 1 and 2) is heavily armored and heavily armed, with its main armament usually mounted in a turret having a wide range of traverse. It is generally a full-track vehicle. The howitzer-carrying tank (Fig. 3) is a more specialized vehicle, equipped with a high-angle gun for action against targets concealed behind defense works. Assault tanks (Figs. 4 and 5) are used mainly to attack enemy armored vehicles and carry artillery or rocket armament suitable for the purpose. Antiaircraft gun carriers (Figs. 6 and 7) are relatively light tanks equipped with rapid-firing guns capable of high angles of elevation and possessing a high degree of maneuverability against fast-moving aircraft. Scout tanks are fast, lightly armored, highly mobile vehicles. They may be of the full-track type or be half-track or (Fig. 8) wheeled vehicles (scout cars). Armored machine-gun carriers and personnel carriers (Figs. 9 and 10) are lightly armored vehicles used for various supporting duties, often in combination with infantry. These too may be full-track, half-track or wheeled vehicles. Self-propelled guns and mobile rocket launchers are not really tanks, but artillery; they usually lack complete armor protection (Figs. 11, 12, 13). In addition to those mentioned, there are various special-purpose types of armored vehicle, such as bridge-laying tanks (Fig. 14).

From the operational standpoint, a tank must fulfill some exacting general requirements. It should have sufficient engine power to negotiate slopes of at least 30 degrees on rough terrain (Fig. 15). There must be adequate climbing capacity to overcome vertical obstacles within the limits imposed by the overall design (Fig. 16). Also, the vehicle must be able to cross ditches, trenches and the like (Fig. 17). In this respect the full-track vehicle is superior to the half-track or the wheeled vehicle. Another important requirement is the fording capacity, or water-crossing ability (Fig. 18). Amphibious tanks are so designed that they can cross deep water, floating and propelled by their own power (Fig. 19). Some tanks can be fitted with collapsible screens and thus be converted to amphibious operation. Propulsion is effected by propeller or water jet. Heavy tanks which cannot be given the necessary buoyancy to float on the water may be equipped to cross water obstacles by traveling along the bottom, air for the engine and crew being supplied through a tube mounted on the tank (Fig. 20).

(more)

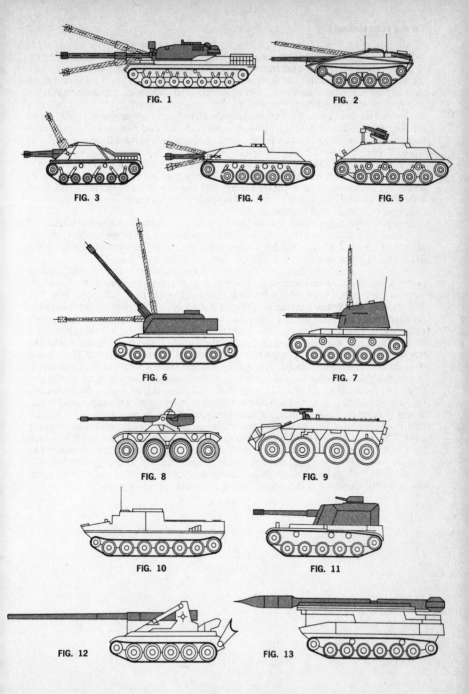

FIG. 1

FIG. 2

FIG. 3

FIG. 4

FIG. 5

FIG. 6

FIG. 7

FIG. 8

FIG. 9

FIG. 10

FIG. 11

FIG. 12

FIG. 13

The hull (Fig. 21) constitutes the basic structural element, the chassis, of the tank. It carries the power unit and transmission, the running gear and suspension, and the armament. It also has the function of protecting the crew. It is constructed of steel armor plates and cast steel and is so designed in shape and armoring as to provide the best possible protection against projectiles fired against it. Thus, the surfaces of the hull are sloped to increase the likelihood that projectiles will glance off. The whole silhouette of the tank should be squat and low so as to present a target that is as difficult as possible to hit. Especially the frontal armor must be strong and sloped well back. Complete protection of the crew by heavy armor is not, however, compatible with adequate mobility and maneuverability of the tank. The designer will always have to effect a compromise between weight and mechanical performance. In general, the weight of the armor corresponds to at least a third of the total weight. The hulls of amphibious tanks and fording tanks must of course be of watertight construction.

The power unit is almost invariably an internal-combustion engine, usually of the diesel type. Multifuel engines which can run on other fuels besides diesel oil are sometimes used, offering obvious advantages from the standpoint of supply in the field. The engine is normally installed in the rear of the hull (Fig. 22).

Wheeled armored vehicles are steered in the same way as ordinary road vehicles (front-wheel or all-wheel steering). The steering of tracked vehicles is effected by stopping the track on one side or by controlled variation of the track speeds (Fig. 23). In the clutch-and-brake steering gear (Fig. 24), the engine power output is transmitted through the shaft a and the bevel gears b to the clutch shaft c, on which the primary components of the friction clutches d are mounted. Each of the two driven shafts e carries the fixed secondary component of the clutch d, the track brake f and the track drive wheel g, which drives the crawler tracks. When the vehicle is traveling straight ahead, both clutches d are engaged, and the shafts e rotate at the same speed. When the driver wishes to make a left turn, for example, the left-hand clutch d is gradually disengaged, so that the speed of the left-hand driven shaft e decreases and the left-hand track is therefore slowed down in relation to the right-hand track. When the clutch d is completely disengaged, the shaft e stops rotating; the vehicle then swivels about the stationary track and thus describes its minimum turning circle. The effect of the clutches can be intensified by the two track brakes f, by which the shafts e can be additionally slowed down or indeed completely locked. The driving shafts for the change-speed gearbox and steering gear pass through the sides of the hull and are connected to side gear units (Fig. 25). The latter transmit the power to the track drive wheels through planetary gear systems.

(more)

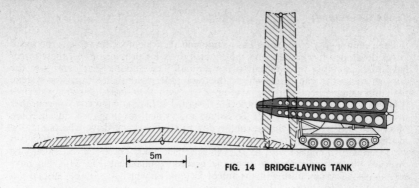

FIG. 14 BRIDGE-LAYING TANK

5m

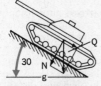

FIG. 15 CLIMBING

30
Q
N
g

H
H

FIG. 16 SURMOUNTING OBSTACLES

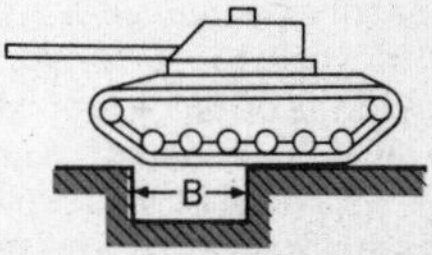

B

FIG. 17 CROSSING DITCHES

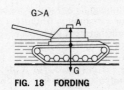

G>A
A
G

FIG. 18 FORDING

G=A
A
drive
G

FIG. 19 FLOATING

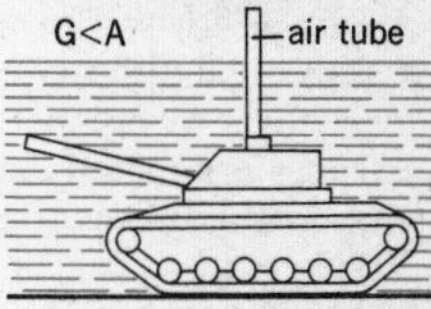

G<A
air tube

FIG. 20 SUBMERGENCE

turret   exhaust slits
idler   drive wheel
rollers

FIG. 21 LONGITUDINAL SECTION
THROUGH HULL OF TANK

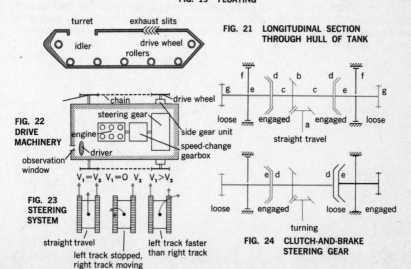

FIG. 22
DRIVE
MACHINERY

chain   drive wheel
steering gear
engine   side gear unit
observation   driver   speed-change
window   gearbox

$V_1 = V_2$ $V_1 = 0$ V_2 $V_1 > V_2$

FIG. 23
STEERING
SYSTEM

straight travel
left track stopped,
right track moving
left track faster
than right track

f   d b d   f
g e c c e g
loose engaged engaged loose
a
straight travel

e d   d e
loose engaged loose engaged
turning

FIG. 24 CLUTCH-AND-BRAKE
STEERING GEAR

The running gear of the tank has to transmit the weight of the vehicle, the recoil forces developed by the gun, and the engine power for propulsion. All these forces have to be suitably transmitted to the ground. The running gear comprises the crawler tracks, the track drive wheels, the track rollers with their suspension system, the supporting rollers, and the take-up idlers with their track-tensioning devices (Fig. 26). The individual links composing the crawler tracks are joined to one another by pivoted connections (Fig. 27). To reduce noise when traveling on roads, they are often provided with detachable rubber coverings. The track drive wheels may be located either at the front or at the rear (Figs. 26 and 29). Each drive wheel is provided with teeth which engage with recesses in the track links. The track rollers may be disposed one behind the other or in a staggered arrangement (Fig. 29). They have sprung suspension mountings in the hull of the tank, for which torsion-bar suspension is now particularly favored (Fig. 30).

The main armament installed in tanks varies according to the purpose for which it is intended. Action against enemy tanks requires long-barreled guns firing projectiles of high muzzle velocity and great penetrating power. As the range at which the enemy is engaged is usually not very great, the guns need not be elevated to high angles. Instead of conventional guns, recoilless guns may be fitted. A weapon of this kind is often equipped with a retractable mount; this can be raised for firing the gun, which is loaded, aimed and fired from within the tank (Fig. 31). Concealed targets have to be attacked with relatively short-barreled high-angle guns (howitzers) firing projectiles of low muzzle velocity. The guns constituting the main armament of tanks usually have calibers between 7 cm and 12 cm, but calibers up to 20.3 cm are employed for self-propelled guns and howitzers. In addition to the main armament, tanks usually carry light secondary armament (machine guns or small-caliber artillery) for use against aircraft, unarmored vehicles or personnel. Antiaircraft gun carriers are usually equipped with turret-mounted multibarreled guns capable of a rapid rate of fire; the turrets may be entirely closed or be open on top (Figs. 6 and 7). These guns can be traversed and elevated smoothly and quickly.

The main gun of a tank may be mounted in a revolving turret (Fig. 32) or it may be mounted in the hull itself, in which case only a limited traverse is possible (Figs. 33 and 34). Sometimes no independent traverse motion of the gun is possible at all, or indeed the gun may be rigidly fixed to the hull and thus incapable of independent traversing or elevation. Traversing must then be effected by aiming the whole tank at the target, while elevation is effected by means of the hydropneumatic suspension system.

(more)

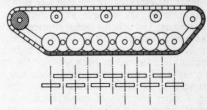

FIG. 26 CRAWLER TRACK WITH DRIVE WHEEL
AT REAR AND SUPPORTING ROLLERS

FIG. 28 CRAWLER TRACK WITH DRIVE WHEEL
AT REAR, NO SUPPORTING ROLLERS

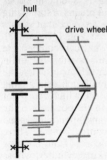

FIG. 25 SIDE GEAR UNIT

FIG. 29 CRAWLER TRACK WITH DRIVE WHEEL
IN FRONT AND STAGGERED ARRANGEMENT
OF TRACK ROLLERS

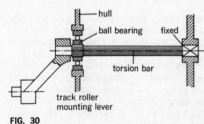

FIG. 30

FIG. 31 TANK WITH RECOILLESS GUN
ON RETRACTABLE MOUNT

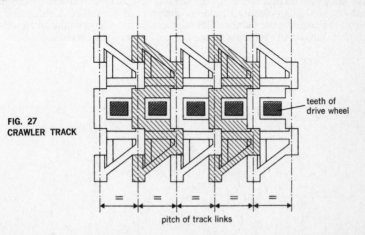

FIG. 27
CRAWLER TRACK

When traveling on rough ground, a tank performs irregular angular movements (rolling and pitching) about its longitudinal and its transverse axis. Such conditions would make it impossible to use the main armament effectively. A means of stabilizing the gun and maintaining its aim is provided by the gyroscope, a device that strives to maintain the position of its axis of rotation unchanged (see Vol. I, page 458). Employed in combination with a mechanical sighting device that can compensate sufficiently quickly for changes in the vehicle's position, it is thus possible to obtain a fire-control system that will enable the gun to be used effectively even when the going is rough (Fig. 35). Fig. 36 is a simplified schematic diagram of such a stabilizing system for the elevation of the gun barrel. Attached to the barrel, which can pivot on a horizontal axis, is a gyroscope that responds to variations in position in the vertical plane. The control signal emitted by the gyroscope is amplified and passed to the electromagnet coils of the mechanically interlinked control valves. These valves continuously control the flow of hydraulic fluid, which comes from the gear pump and is admitted to both sides of the piston of the working cylinder, whose piston movements elevate or depress the gun barrel. The flow is controlled in such a manner that the barrel is constantly maintained in the same position in the vertical plane. Greater accuracy of stabilization is achieved by the use of gyroscopes that measure and respond with corrective action to the angular velocity of the pitching motion of the tank. Lateral stabilization, for rolling motion, is achieved on the same principle, except that for this purpose a rotary machine (hydraulic or electric motor) is used instead of a hydraulic cylinder.

A modern tank is fitted with a variety of auxiliary equipment. To give the crew a good field of vision, various windows and observation slits are provided, together with optical devices such as telescopes, periscopes, etc. For dusk and nighttime operation, searchlights and infrared steering and sighting devices may be provided. These devices are of the kind that either detects infrared rays emitted by the target (e.g., heat from the engine of an enemy vehicle) or picks up the reflection of infrared rays directed on to the target by an infrared projector mounted on the tank itself. Steering devices based on infrared rays provide a range of vision of some 200 or 300 yards, while infrared gunsights may range up to about 2000 yards. The actual effective range of these devices is considerably affected by weather conditions. Because of disturbing reflections from the ground and natural obstacles, radar is generally unsuitable for use in tanks. Communication between one tank and another and between tanks and their base command post is provided by radio.

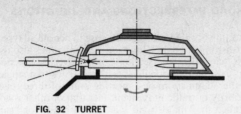

FIG. 32 TURRET

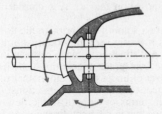

FIG. 33 GUN MOUNTED FOR
TRAVERSING AND ELEVATION

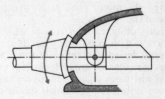

FIG. 34 GUN MOUNTED FOR
ELEVATION ONLY

FIG. 35 TANK WITH STABILIZING SYSTEM FOR GUN

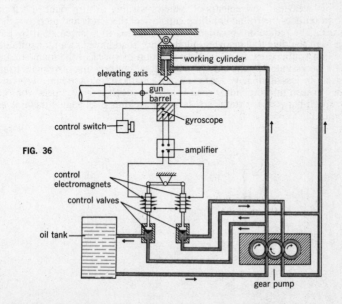

FIG. 36

working cylinder

elevating axis

gun
barrel

control switch

gyroscope

amplifier

control
electromagnets

control valves

oil tank

gear pump

ROAD INTERSECTIONS AND JUNCTIONS

The term "intersection" is used in a general sense to denote all types of connection between similar or dissimilar roads. "Junction" is sometimes also used in a general sense—for example, "cloverleaf intersection" and "cloverleaf junction" are synonymous—but is often used more particularly with reference to a single connection of one road with another (T junction) or a meeting point of three roads (Y junction). A road that has priority of traffic movement over that of other roads at a junction or intersection is called the "major road"; a "minor road" (or "subsidiary road") has a lesser traffic value than a major road. These are relative terms in the sense that even the minor road at a particular intersection may carry a considerable volume of traffic.

Properly designed intersections which ensure a smooth and brisk traffic flow in conjunction with good safety conditions are obviously very important features in the overall design of an efficient road system. A distinction is made between intersections of roads that are all located at the same level (grade intersections) and intersections where one or more roads are routed over or under one or more other roads by means of bridges or tunnels (grade-separated, or multilevel, intersections). The latter category obviously offers advantages in terms of smooth traffic flow and safety because of the segregation of vehicles traveling in different directions and the reduction or elimination of the number of points where different traffic streams have to cross one another. On the other hand, the bridges (overpasses, or flyovers) and tunnels (underpasses) make grade-separated intersections much more expensive, so that they are usually employed only on important roads for fast traffic (in particular, expressways).

Figs. 1 and 2 show the basic traffic streams at a four-way intersection and a T junction respectively. The design of important intersections is based on traffic surveys which provide information on the anticipated traffic flows. The slowing down and obstruction of the general flow of traffic on the intersecting roads in consequence of "weaving" movements of vehicles wishing to turn right or left at the intersection reduces the traffic-handling capacity of the roads and increases the accident hazard. The reduction in capacity is determined to a large extent by the delay arising from the time that vehicles on the minor road have to wait for suitable gaps in the main traffic stream and by the reduction of speed of the main-stream vehicles that is necessitated by the entry ("weaving in") or crossing of vehicles from the minor road. The loss of capacity due to vehicles leaving the main stream ("weaving out") in order to turn into the minor road is less serious. It largely depends on the difference in speed between the traffic streams that are thus disengaging themselves from each other.

(more)

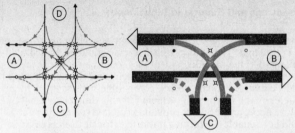

FIG. 1 → = TRAFFIC GOING STRAIGHT AHEAD
        ⬦ = CROSSING POINT
        → = LEFT-TURNING TRAFFIC
        ⤏ = RIGHT-TURNING TRAFFIC

FIG. 2 ● = WEAVING-IN POINT
        ○ = WEAVING-OUT POINT

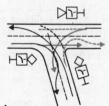

FIG. 3a

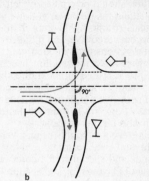

b

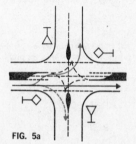

FIG. 4a

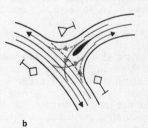

b

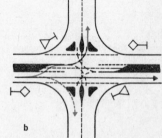

FIG. 5a   b

249

Some general principles governing the design and layout of road intersections will now be considered. The main traffic streams should be routed along the most direct possible paths; the subsidiary streams should be routed as far as possible at right angles to the main streams. The use of signs and signals that stop the flow of traffic should be avoided to the degree compatible with safety. The most serious disturbing factor is the left-turning vehicle.* If this drawback cannot be obviated altogether by an appropriate layout providing suitable segregation of traffic streams, the aim should be to enable left-turning traffic to extricate itself as easily and smoothly as possible from the main stream by the provision of a special traffic lane and an appropriate layout of the intersection. The main stream of traffic traveling in the opposite direction should be crossed by left-turning traffic as nearly at right angles as possible. In situations where large flows of traffic intersect, it becomes necessary to provide signals (traffic lights, etc.). If the total delay times for traffic at such intersections become excessive, and if linked control of the traffic lights at a number of adjacent intersections (coordinated control system) fails to solve the problem, then the grade-separated intersection becomes a necessity. In such solutions the left-turning traffic is routed either indirectly through a 270-degree loop or semi-indirectly via individual ramps, so that the movements of this traffic are thus reduced to basic weaving-in (merging) and weaving-out (disengaging) movements. Grade-separated intersections are a standard feature of modern expressways and other important highways for motor traffic.

The disturbing effect of right-turning vehicles on the general flow of traffic is relatively slight and can be minimized by appropriate routing with the aid of triangular islands and special traffic lanes (acceleration and deceleration lanes).

Some examples of layouts for intersections will now be briefly discussed.

Fig. 3a shows a simple intersection of two roads not at right angles to each other. Though commonly employed, this is an unsatisfactory solution in cases where appreciable flows of traffic occur, inasmuch as the points of encounter (potential points of collision) between vehicles moving in different directions are scattered all over the area of the intersection, the curbs fail to provide sufficient guidance, and the roads cross each other obliquely. The efficiency is poor, since only low vehicular speeds can be tolerated. This type of intersection is at best acceptable for residential roads. In Fig. 3b an improvement has been achieved in that the main stream of traffic (i.e., on the major road) has been given preferential treatment by suitable constructional arrangements. The turning movements have been more precisely defined by means of teardrop-shaped islands in the minor road, right-angled intersection of major and minor road, large-radius curb alignments, and traffic lane markings by which points of encounter are more definitely located. This layout is suitable for medium traffic flows and may be used in conjunction with control signals.

(more)

---

*The text and illustrations are based on the assumption that traffic normally travels on the right-hand side of the road. In countries where left-hand driving is the rule, as in Great Britain, the problem of course arises from right-turning vehicles.

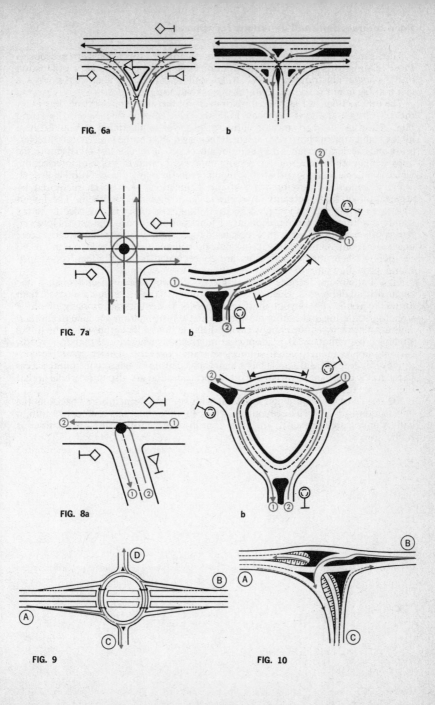

FIG. 6a   b

FIG. 7a   b

FIG. 8a   b

FIG. 9   FIG. 10

The simple T junction in Fig. 4a is unsatisfactory for much the same reasons as those relating to Fig. 3a. It is sometimes possible to improve this type of junction into the layout illustrated in Fig. 4b, which is more particularly characterized by the fact that the minor road now joins the major road at right angles.

The intersections in Fig. 5 are developments of the simple intersection. In Fig. 5a the left-turning traffic is detached from the main stream well before the actual intersection. A more advanced solution of this type is illustrated in Fig. 5b: here all the traffic streams have been assigned their own lanes at the intersection; acceleration and deceleration lanes may additionally be provided. Control signals for traffic approaching the intersection are usually unnecessary with this type of intersection, which is suitable for major urban thoroughfares and heavily trafficked rural highways.

Fig. 6 represents developments of the T junction, likewise characterized by segregation of traffic streams on separate lanes provided for them. The layout illustrated in Fig. 6a, embodying a relatively large triangular island, used to be the standard solution for certain secondary T junctions on German highways. However, it is now recognized that it incorporates danger spots where traffic streams cross one another obliquely and has accordingly been abandoned. In the layout in Fig. 6b the points where the streams cross are concentrated before the teardrop-shaped island, providing more favorable conditions of intersection.

Fig. 7a shows an obsolete form of intersection which is in effect a small traffic circle, or roundabout, in which the central "rotary island" is more an obstacle than an aid to traffic. The layout illustrated in Fig. 7b reduces all crossing of traffic streams to entering and exiting movements and transition from one lane to another. This arrangement requires long weaving distances and a layout in which the traffic streams cross obliquely. It is adopted at intersections where more than four roads, carrying approximately equal volumes of traffic, converge. Under favorable conditions this type of traffic circle has a high traffic capacity. Signals controlling access to the traffic circle are at variance with the principle and are also undesirable in that they reduce the capacity of the circle.

Fig. 8a is an obsolete junction layout with the same general drawbacks as the intersection in Fig. 7a. The layout shown in Fig. 8b represents a modern solution with a high traffic capacity. The large central island ensures smooth, continuous traffic flow.

(more)

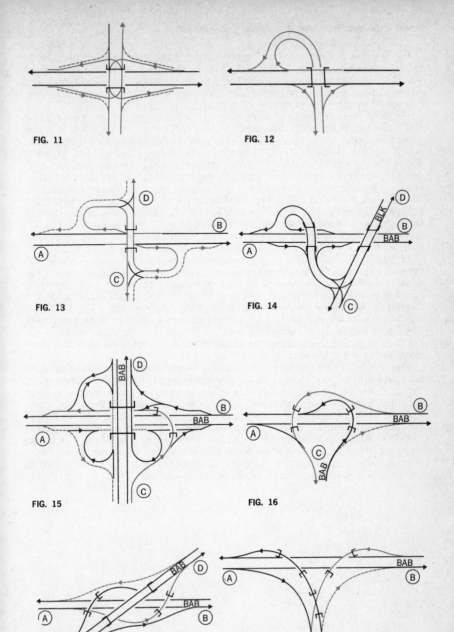

FIG. 11

FIG. 12

FIG. 13

FIG. 14

FIG. 15

FIG. 16

FIG. 17

FIG. 18

Fig. 9 shows a familiar form of grade-separated (or multilevel) intersection. In this case the traffic streams A–B and B–A are routed under or over the minor road, which is connected to the major road by means of a traffic circle and ramps (slip roads). This is a good solution for an interchange where access between a street (the minor road) and an expressway in a built-up area has to be provided.

Fig. 10 is a grade-separated junction in which the large volumes of left-turning traffic have been routed at different levels. Fig. 12 is a somewhat different grade-separated layout for a junction. A type of intersection that originated and is still widely used in Holland (and elsewhere) is shown schematically in Fig. 11, while Fig. 13 is a so-called semicloverleaf intersection in which the main streams A–D and B–C are kept entirely free of grade intersections. In this latter layout the major road (A–B) is usually an expressway and routed under the minor road (C–D). The solution illustrated in Fig. 14 is a different type of grade-separated intersection between a major road (A–B) and a minor road (C–D): there is only one junction of the link road with the minor road, as compared with two such junctions in Fig. 13; against this, the layout in Fig. 14 involves the construction of two overpasses and moreover has a smaller traffic capacity than the semicloverleaf intersection.

Fig. 15 illustrates what could be termed a three-quarter cloverleaf intersection between two roads of equal importance—e.g., two highways. All traffic turning out of the main stream is routed on to parallel service lanes. Fig. 16 is a grade-separated junction between two highways. The through road (A–B) is routed over or under the link roads. An oblique intersection of two highways is illustrated in Fig. 17. This is an efficient solution, but relatively expensive because it necessitates three overpasses. Fig. 18 is another grade-separated junction. It achieves the same result as the junction in Fig. 16, but presents smoother traffic-flow conditions; it is also more expensive in that it has three overpasses instead of two.

Figs. 19 and 20 are perspective drawings of intersections of the types illustrated schematically in Figs. 13 and 18. In the cloverleaf type of intersection, turning traffic has to make considerable detours. The space occupied by a cloverleaf is so great that the adoption of this form of intersection is seldom practicable in built-up areas. A traffic circle, though minimizing detours, imposes speed restraint on the flow of both direct and turning traffic. A grade-separated layout of the kind illustrated in Fig. 19, which is suitable for built-up areas, permits free flow of through traffic, while exercising no more restraint on turning traffic than would be imposed by a normal traffic circle. Generally speaking, the layouts in Figs. 11–18 are typical of main roads in open country or relatively sparsely populated areas.

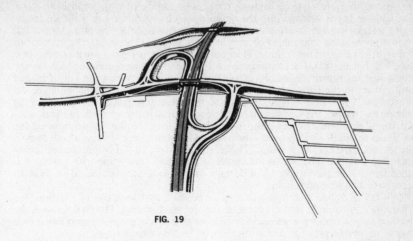

FIG. 19

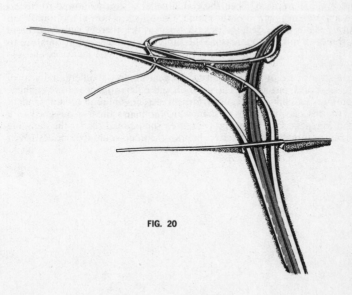

FIG. 20

Before the engine of a standard-shift car is started, the gear lever should be shifted to the neutral position (Fig. 2b) to disconnect the engine from the drive shaft. Now the ignition key is inserted into the ignition switch. When the key is turned to the right (clockwise), the steering-wheel lock (an antitheft device fitted to some cars) is first released, so that the wheel can be freely rotated. Turning the key farther to the right switches on the ignition. The ignition coil and distributor are now energized (see Vol. I, page 482). The charging control light and oil-pressure control light in the combined instrument assembly should light up (Fig. 1). When the key is turned still farther to the right, against the spring pressure that is then encountered, the starter (see Vol. I, page 476) is energized, so that the engine is rotated. As soon as the engine has started, the key should be released, so that it springs back to the normal "driving" position. The key must not be held in the "starting" position while the engine is running, as this would soon damage the starter. For starting a cold engine, the choke, if the car has a manual choke, should be pulled out before the ignition is switched on. The function of the choke is to supply the engine with a very rich gasoline-and-air mixture (see Vol. I, page 480) to facilitate starting. Most modern cars have an automatic choke.

When the engine is running and has reached a certain speed of rotation, the charging control light and oil-pressure control light go out. The choke should be pushed in as soon as possible, as the rich starting mixture contains too much fuel, which washes away the film of lubricating oil on the cylinder walls.

Depressing of the clutch pedal (Fig. 1) causes the mechanical connection between engine and gearbox to be interrupted (see Vol. I, page 486). To enable the vehicle to move off, the gear lever, which at first was in neutral, is shifted to "first gear" and the hand brake is released. Then the clutch pedal is slowly allowed to rise so as to achieve smooth engagement of the clutch. The vehicle is now in motion. When the accelerator pedal is depressed, the throttle opens wider, thereby supplying more air to the carburetor and thus increasing the supply of the fuel-and-air mixture to the engine (see Vol. I, page 478). As a result, the engine speed increases and the vehicle goes faster.

The drive shaft (or cardan shaft) connects the gearbox to the differential gear of the rear axle (see Vol. I, page 500). The gearbox and differential gear have resilient rubber mountings, so that they can perform a certain amount of independent movement. For this reason the drive shaft which connects them is designed as a universal shaft (see page 190). A rigidly connected shaft would render the damping action of the rubber mountings ineffective; besides, the shaft would soon be destroyed

(more)

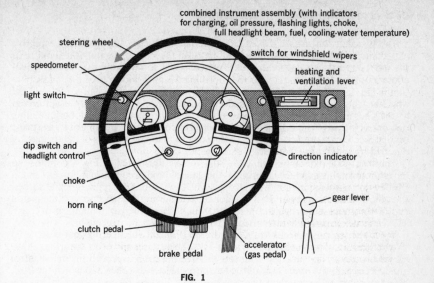

combined instrument assembly (with indicators
for charging, oil pressure, flashing lights, choke,
full headlight beam, fuel, cooling-water temperature)

steering wheel

switch for windshield wipers

speedometer

heating and
ventilation lever

light switch

dip switch and
headlight control

direction indicator

choke

gear lever

horn ring

clutch pedal

brake pedal

accelerator
(gas pedal)

FIG. 1

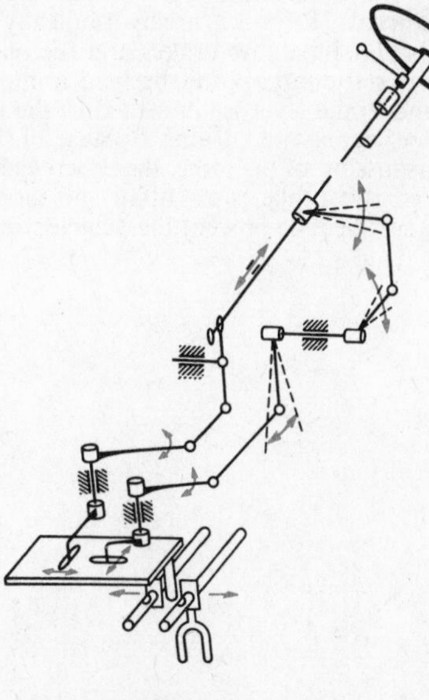

**FIG. 2a DIAGRAM OF
STEERING SYSTEM**

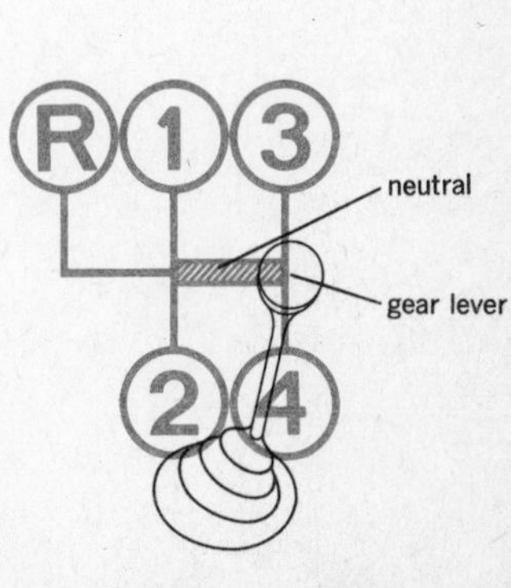

R 1 3

neutral

gear lever

2 4

FIG. 2b GEAR-LEVER POSITIONS

The two rear wheels are driven by the differential gear through their respective half-shafts, which are likewise universal shafts. In the suspension illustrated in Fig. 1, the independently sprung rear wheels are attached to swinging arms which pivot about a point adjacent to the rear-axle housing. Another widely used type of suspension is the rigid axle. In this system the differential gear, the two half-shafts, and the wheel bearings are all accommodated in one sheet-steel casing, which is connected to the chassis through two springs (usually leaf springs disposed longitudinally), one on each side of the vehicle. When a wheel encounters an irregularity, the spring on that side is compressed and the whole axis tilts. The oscillations of the springs are damped by shock absorbers (see Vol. I, pp. 502–508).

The front-axle assembly (Fig. 2a) of the vehicle illustrated in Fig. 1 comprises a lower control arm, which transmits the lateral and longitudinal forces, and a McPherson suspension unit, which consists of a two-tube telescopic shock absorber, the outer tube of which carries a seat for the coil spring. At its upper end the piston rod of the shock absorber is attached to the body of the car. This form of front-axle suspension is now widely used in passenger cars.

When the steering wheel is turned to the left (Fig. 2a), the steering gear is actuated by the steering column (see Vol. I, page 510). This causes the drop arm (Fig. 2b) to swivel inwards. The track rod connects the steering drop arm with the axle-control arm. As a result, the outer tube of the telescopic shock absorber—to which the front wheel is attached—rotates in the direction indicated by the red arrow in Fig. 2a. The vehicle accordingly makes a left turn. When the steering wheel is turned to the right, the two front wheels are correspondingly swiveled to the right.

The motorist signifies his intention of turning left or right by means of the self-canceling flashing direction indicators (see Vol. I, page 520). Braking of the vehicle is normally done by means of the foot brake, the braking action varying with the force exercised by the foot on the brake pedal. The brakes are hydraulically actuated (see Vol. I, page 512). The front wheels may have disc brakes and the rear wheels have drum brakes. When the vehicle is stationary, it can be held immovable by application of the hand brake. The hand-brake lever is connected to the two rear-wheel brakes and presses the brake shoes against the drums. Because of the mechanical—as opposed to hydraulic—transmission of the force, the effectiveness of the hand brake is relatively poor. However, this brake is essentially no more than a parking brake and its action should be sufficient to prevent the vehicle from rolling down a slope.

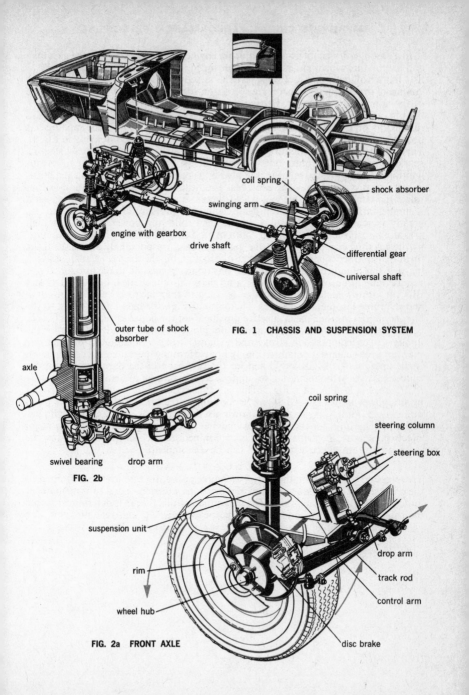

coil spring

shock absorber

swinging arm

engine with gearbox

drive shaft

differential gear

universal shaft

FIG. 1 CHASSIS AND SUSPENSION SYSTEM

outer tube of shock absorber

axle

swivel bearing

drop arm

FIG. 2b

coil spring

steering column

steering box

suspension unit

rim

wheel hub

drop arm

track rod

control arm

disc brake

FIG. 2a FRONT AXLE

In a gasoline-driven internal-combustion engine (see Vol. I, page 466) the gasoline-and-air mixture which enters the cylinder during the suction (or intake) stroke at first fills a volume equal to $V_Z + V_B$, where V_Z represents the piston-swept working volume of the cylinder (= piston area × piston stroke) and V_B represents the volume of the combustion chamber. During the compression stroke which then follows, the mixture is forced into a smaller space by the rising piston. When the latter has reached its highest point, referred to as "top dead center" (TDC), the mixture has been compressed into the relatively small volume V_B (right-hand drawing in Fig. 1). This compression of the mixture is associated with a rise in temperature which, in a spark-ignition engine, must not, however, be so high as to cause spontaneous ignition. The compressed mixture is ignited by a spark from the spark plug and burns in a fraction of a second—i.e., explodes. The heat released on combustion further raises the temperature of the gas, which therefore strives to expand. As the piston is momentarily stationary at top dead center, the pressure of the imprisoned gas greatly increases (firing pressure). In the then following power stroke, this high pressure thrusts the piston down to "bottom dead center" (BDC). The space occupied by the gas thus increases, so that the gas pressure correspondingly decreases. If the pressure is plotted against the piston stroke in a graph, the diagram shown in Fig. 2 (page 263) is obtained. The line AB is so positioned in relation to the horizontal line representing atmospheric pressure (1 atm) that, with the vertical lines at TDC and BDC, it forms a rectangle whose area is equal to that enclosed between the curve and the three last-mentioned lines. The distance from the base line at 1 atm to the line AB represents the "mean effective pressure" for the power stroke (p_{m3} in Fig. 2).

If a similar diagram representing pressure plotted against stroke is drawn for the intake stroke, compression stroke and exhaust stroke, a continuous diagram for the four-stroke engine is obtained (Fig. 3). For performing the intake and compression of the mixture (strokes 1 and 2) and for the expulsion of the burned gas from the cylinder (stroke 4) the engine has to supply a certain amount of energy corresponding to the areas shaded in red in Fig. 3. This energy is deducted from the energy delivered by the power stroke (stroke 3), corresponding to the area shaded in black. In its more usual form this diagram—known as the indicator diagram (or cylinder-pressure diagram)—for the four-stroke engine is represented in Fig. 4. If the mean effective pressure, as defined above, is determined for each of the four strokes, the "indicated mean effective pressure" can be determined:

$$p_{mi} = p_{m3} - (p_{m1} + p_{m2} + p_{m4})$$

The pressure p_{mi} acting on the piston area produces a force which is transmitted to the crankshaft by the connecting rod and develops a torque—i.e., a turning moment about the axis of the crankshaft. The magnitude of this torque is determined by the indicated mean effective pressure p_{mi}, the piston area F_k, and the crank radius r (which is equal to half the stroke: i.e., $r = \frac{1}{2} s$; see Fig. 1). The total torque M_d

(more)

FIG. 1 WORKING VOLUME AND COMBUSTION CHAMBER OF A CYLINDER

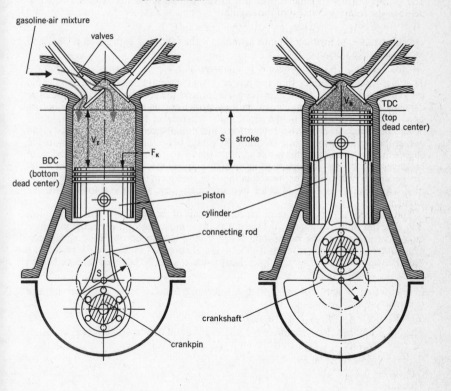

developed at the crankshaft is of course also dependent on the number of cylinders i. Hence:

(1) $\quad M_d = k_1 \cdot p_{mi} \cdot F_k \cdot s \cdot i$

where k_1 denotes a constant. The product of F_k and s represents the swept volume of the individual cylinder; multiplied by i (the number of cylinders) it gives the total cubic capacity (or cylinder capacity) V_H of the engine. The torque can thus be expressed in terms of the indicated mean effective pressure and the total cubic capacity:

(2) $\quad M_d = k_1 \cdot p_{mi} \cdot V_H$

The power output and the torque are linked by the following relation, where n denotes the rotational speed of the engine:

(3) $\quad N = k_2 \cdot M_d \cdot n$ (where k_2 is another constant)

On substitution of formula (2) into formula (3) the following expression is obtained for the effective output:

(4) $\quad N = K \cdot p_{mi} \cdot V_H \cdot n$ (where K is a constant)

This last formula signifies that the power output depends on the speed, the cubic capacity, the indicated mean effective pressure, and the constant K. The value of K depends on the type of engine. In the two-stroke engine the power stroke occurs once in every revolution of the crankshaft, whereas in the four-stroke engine it occurs only once in every two revolutions. For equal speed and cubic capacity, the two-stroke engine would thus theoretically attain twice the output of a four-stroke engine. In actual practice this is not so, since the two-stroke engine cannot—because of scavenging losses—be operated at as high a mean effective pressure as the four-stroke engine. When V_H is expressed in liters, p_{mi} in kg/cm^2, and n in rpm, the values of K are as follows: $K = 1/450$ for two-stroke engines, $K = 1/900$ for four-stroke engines.

The theoretical output determined on the basis of the cylinder-pressure diagram (Fig. 3) is not fully available at the output shaft. Some loss of power occurs in consequence of friction in the piston and bearings and also in driving pumps, fan, etc. The ratio of the effective output N_e to the theoretical output N_i is called the "mechanical efficiency" η_m; its numerical value is usually between 0.75 and 0.85:

(5) $\quad \eta_m = N_e/N_i$

In combination with formula (4) the following expression is obtained for the effective output:

(6) $\quad N_e = K \cdot p_{mi} \cdot \eta_m \cdot V_H \cdot n$ (in horsepower)

This formula can be simplified by introduction of the so-called "brake mean effective pressure" p_{me}, which is the indicated mean effective pressure p_{mi} multiplied by the mechanical efficiency; i.e.:

(7) $\quad p_{me} = p_{mi} \cdot \eta_m$

Therefore:

(8) $\quad N_e = K \cdot p_{me} \cdot V_H \cdot n$ (in horsepower)

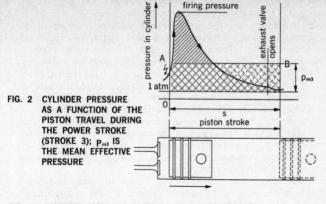

FIG. 2 CYLINDER PRESSURE AS A FUNCTION OF THE PISTON TRAVEL DURING THE POWER STROKE (STROKE 3); p_{m3} IS THE MEAN EFFECTIVE PRESSURE

FIG. 3 CYLINDER-PRESSURE DIAGRAM OF A FOUR-STROKE ENGINE

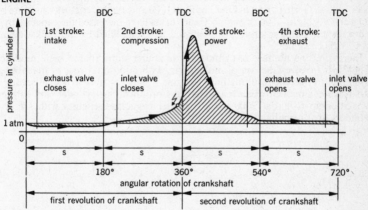

FIG. 4 CYLINDER-PRESSURE DIAGRAM (THE MORE USUAL FORM, DERIVED FROM FIG. 3)

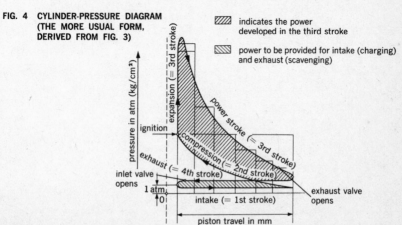

indicates the power developed in the third stroke

power to be provided for intake (charging) and exhaust (scavenging)

INCREASING THE ENGINE'S CUBIC CAPACITY

Increasing the cubic capacity—i.e., the piston-swept working volume of the cylinders—is the most obvious means of increasing the power output of an internal-combustion engine. According to formula (1) on page 262, the cubic capacity increases when the piston area F_k, the piston stroke s, and the number of cylinders i are increased. Any or all of these factors can be increased to obtain the desired effect. In an existing engine the cylinders can be bored out to a larger internal diameter so as to increase the piston area, though obviously there is a limit to what can be achieved in this way without unduly weakening the cylinder walls. See Fig. 2.

The piston stroke s is determined by the crank radius r, this being the distance from the center of the crankshaft to the center of the crankpin (to which the connecting rod is attached). When the radius is increased from r_1 to r_2 (Fig. 1), the piston stroke is correspondingly increased from s_1 to s_2. To make this modification it is of course necessary to fit a different crankshaft.

The measures indicated above are relatively inexpensive to apply and are a popular means of increasing the output of an engine. Thus a number of automobile manufacturers offer more powerful versions of standard engines, the improvement in performance having been achieved by an increase in the cubic capacity. For example, the German firm of Opel supplies 1700 and 1900 cc engines which are bored-out versions of the 1500 cc engine. Other firms have increased the capacity by using a crankshaft giving a larger piston stroke, as exemplified by the Volkswagen 1200 and 1300 cc engines.

Increasing the number of cylinders is of course more than a mere modification (Fig. 3). It involves the use of more parts, with additional cost of machining and assembly, so that the resulting bigger engine is of necessity also a significantly more expensive engine. European manufacturers produce four-cylinder cars with engine capacities up to about 2000 cc. For higher capacities, engines with six or eight cylinders are normally used.

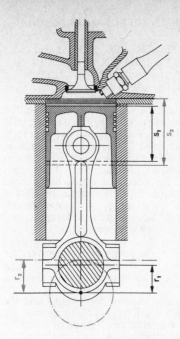

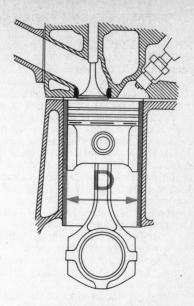

FIG. 2   INCREASING THE CYLINDER
DIAMETER (BORE)

FIG. 1   INCREASING THE STROKE

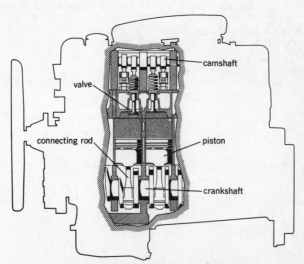

FIG. 3   INCREASING THE NUMBER OF CYLINDERS (TWO
EXTRA CYLINDERS ADDED TO A FOUR-CYLINDER
ENGINE)

INCREASING ENGINE EFFICIENCY: MEAN EFFECTIVE PRESSURE

The mean effective pressure is not constant, but varies with the rotational speed developed by the engine. With increasing speed, the mean effective pressure attains a maximum and then decreases (Fig. 1, curve A, representing more particularly the brake mean effective pressure p_{me}). According to formula (2) on page 262, the torque M_d is dependent on p_{mi} and therefore on p_{me}, while the two other factors in that formula are constant for an engine of given cubic capacity. Hence the curve that represents the torque as a function of engine speed (curve B) has the same general shape as curve A. Thus the torque and the brake mean effective pressure have their respective maximum values at the same engine speed (2500 rpm, for example, in Fig. 1). This is not, however, the speed at which the engine develops its maximum power output. The output is represented by curve C; its maximum at 5000 rpm (in this particular example) corresponds to 73 hp, associated with a brake mean effective pressure p_{me} of only 7.5 kg/cm^2 (as compared with the maximum value of 9 kg/cm^2 that p_{me} attains at 2500 rpm). In general, the value of p_{me} at which maximum power output is attained is between 7 and 8 kg/cm^2 for ordinary cars; for sports cars it is higher, up to about 10 kg/cm^2, while for racing cars it is between 12 and 13 kg/cm^2. The general object is to obtain the flattest possible shape for the p_{me} curve, so as to have a wide range of engine speeds within which the torque is nowhere very far below its maximum value. A relatively constant torque over a wide speed range ensures "flexibility" of the engine in the sense that there is no need for frequent gear changing.

Since p_{me} is the product of the indicated mean effective pressure p_{mi} and the mechanical efficiency η_m (formula 7 on page 262), it will be possible to increase p_{me} by increasing p_{mi} and also by increasing η_m, the latter result being achieved by reduction of the power losses due to friction and other causes.

One method of increasing p_{mi} is by increasing the *compression ratio*. At the beginning of the compression stroke the piston is at bottom dead center. As already explained, the gas then occupies the volume $V_Z + V_B$ (see page 260). When the piston moves to top dead center, the cylinder contents are compressed into the combustion chamber (volume V_B). The ratio between the total volume at the beginning and at the end of the compression stroke is termed the "compression ratio" (ε):

$$(9) \quad \varepsilon = (V_Z + V_B)/V_B$$

An increase in compression ratio results in an increase in effective pressure and therefore in engine power output. Also, the fuel is more efficiently utilized, so that the thermal efficiency of the engine is improved.

(more)

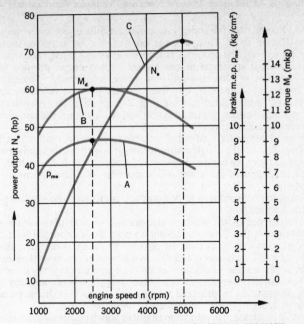

FIG. 1 ENGINE PERFORMANCE CURVES FOR AN ORDINARY CAR ENGINE

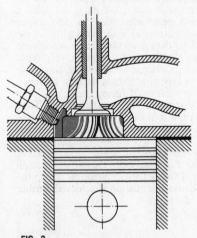

FIG. 2
DETONATION DUE TO EXCESSIVELY HIGH COMPRESSION: THE CLASH OF PRESSURE WAVES CAUSES KNOCKING

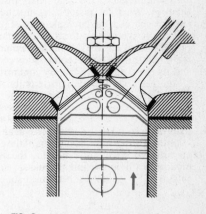

FIG. 3
HIGH-PERFORMANCE COMBUSTION CHAMBER (AS USED FOR RACING ENGINES)

According to formula (9), the compression ratio can be increased by an increase in V_Z and a reduction in V_B. However, the ratio cannot be increased indefinitely. If it were made too high, the powerful compression would cause excessive heating of the mixture, so that spontaneous combustion would occur, giving rise to the phenomenon known as "detonation"—or more specifically, "knocking" or "pinging" of the engine—caused by interference of the pressure waves set up by the self-ignited and the spark-ignited mixture (Fig. 2, page 267). This may damage the valves or the pistons, besides imposing severe sudden loads on the connecting-rod bearings and crankshaft bearings.

A badly designed combustion chamber is liable to cause detonation at relatively low compression. To achieve a high compression ratio without these adverse effects it is therefore necessary to use an antiknock gasoline (which has a high octane number), and the design of the combustion chamber should conform to certain principles, namely:

(a) The combustion chamber should be compact, presenting the smallest possible surface area for heat loss.

(b) Intensive turbulence of the mixture is essential, so that the ignition spreads rapidly throughout the chamber, leaving no pockets of unburned mixture.

(c) The spark plug should be as close as possible to the exhaust valve, this being the hottest spot in the chamber, and also in a central position so that the flame-travel path is of the same length (and as short as possible) in every direction.

(d) The combustion chamber should be located over the piston, and any recesses or cavities necessitated by practical reasons of construction should be as small as possible in order to minimize pockets of unburned gas.

Fig. 3 (page 267) illustrates a combustion chamber which comes closest to satisfying the above conditions. It is approximately hemispherical in shape and permits large valves to be installed. The central location of the spark plug in the compact hemispherical space ensures that the flame-travel distances are the same in every direction. These distances can be further reduced by the use of two spark plugs. To achieve high compression, the chamfered piston head penetrates into the combustion chamber. When the piston rises, it squeezes the mixture at high velocity out of the annular space at the circumference of its head, so that good turbulence is achieved and gas pockets liable to cause detonation are eliminated. In addition, very rapid combustion is ensured. With a premium-grade gasoline a 10:1 compression ratio is attainable. This type of combustion chamber is expensive to manufacture, however, and its use is therefore mainly confined to sports cars and racing cars.

Another favorably designed combustion chamber is illustrated in Fig. 4. Here the inclined valves are located off center in relation to the cylinder. The combustion chamber is in the shape of two intersecting hemispheres differing in size. Excellent turbulence is ensured, especially because the base of the chamber is smaller in area than the piston head, so that here again the gas is squeezed at high velocity out of a confined space into the interior of the chamber. Fig. 5 indicates the swirling motion imparted to the mixture entering the cylinder. During the compression stroke the mixture is still in swirling turbulence, so that good combustion is achieved, further enhanced by the above-mentioned squeezing effect.

(more)

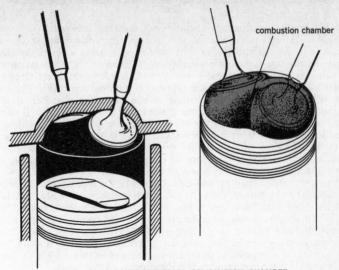

combustion chamber

FIG. 4 DOUBLE-HEMISPHERICAL COMBUSTION CHAMBER
USED IN BMW ENGINES

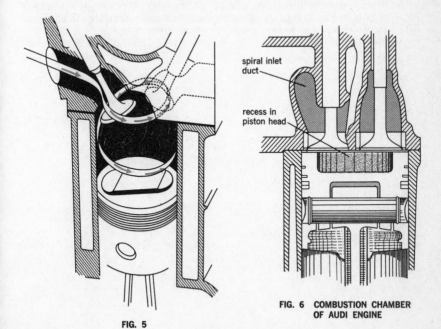

spiral inlet
duct

recess in
piston head

FIG. 5

FIG. 6 COMBUSTION CHAMBER
OF AUDI ENGINE

The highest compression ratio ever obtained in an ordinary passenger car—a ratio of 11.2:1—has been achieved with the form of construction illustrated in Fig. 6 (page 269). In this design the inlet duct terminates in a spiral before the inlet valve, so that the mixture has a swirling motion on entering the combustion chamber (Fig. 7a). This is formed by a recess in the piston head, which causes acceleration of the swirling motion when the piston approaches top dead center (Fig. 7b). The rising piston moreover squeezes the swirling mixture into the chamber. In this way thorough mixing and turbulence are obtained, so that combustion is very rapid. There is a drawback, however. The shape of the inlet duct and the swirling motion it produces present a fairly high resistance to the flow of the mixture. Because of this, proper charging of the cylinder with the mixture cannot be achieved if the suction stroke is too rapid. This form of chamber is therefore not suitable for high-speed engines that have to develop high peak outputs.

With a combustion chamber as illustrated in Fig. 8 it is possible to fit large valves in a staggered arrangement. Only the exhaust valve (shown dotted in red) is located in the chamber. The spark plug is positioned directly beside the hottest part. The face of the inlet valve is flush with the cylinder top. Between this and the head of the piston at top dead center a narrow gap (clearance) remains from which the mixture is squeezed into the combustion chamber, where it is ignited by a spark. With this form of chamber it is possible to obtain a compression ratio of 9:1 for carburetor engines and 9.3:1 for engines with fuel injection.

Finally, the wedge-shaped combustion chamber in Fig. 9 calls for mention. It is comparatively cheap to manufacture and has been used in numerous variants. The valves are in a parallel disposition, one behind the other; they cannot be made as large as those in Fig. 8. Here again squeezing action and turbulence are achieved. This type of chamber can be used for compression ratios up to 9:1.

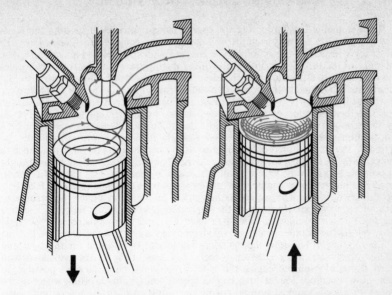

FIG. 7a MIXTURE IS SWIRLING AS IT ENTERS THE CYLINDER

FIG. 7b THE SWIRLING MOTION OF THE MIXTURE IS INTENSIFIED

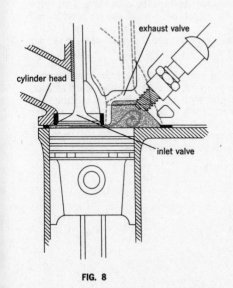

exhaust valve

cylinder head

inlet valve

FIG. 8

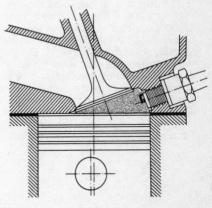

FIG. 9 WEDGE-SHAPED COMBUSTION CHAMBER

IMPROVING VOLUMETRIC EFFICIENCY

Another method of improving the engine performance by increasing the mean effective pressure consists in improving the so-called *volumetric efficiency*—i.e., the efficiency with which the cylinders are charged with the fuel-and-air mixture. The quantity of mixture drawn into the cylinder during the suction stroke determines the mean effective pressure and therefore the power output. This quantity should theoretically be equal to the working volume of the cylinder (= piston area × stroke). In reality the quantity of mixture drawn into the cylinder is less. The ratio of the actual to the theoretical quantity is known as the "volumetric efficiency." It depends on the size and shape of the inlet and exhaust ducts and ports, the shape of the combustion chamber, and the method whereby the fuel is introduced into the cylinder. The ducts should be so designed as to offer the least possible resistance to the gas flowing through them at high velocity. They should therefore be as straight as possible and with the fewest possible changes in diameter. Fig. 1 shows a section through a four-cylinder engine. The inlet pipe has been designed to ensure favorable gas-flow conditions. The water-heating jacket at the bend under the carburetor serves to prevent condensation of the vaporized gasoline in the mixture. The pipe functions as an "oscillation tube" (see page 274) and is, for this reason, relatively long. Within the cylinder head itself the duct is very short, to avoid excessive heating of the mixture drawn into the cylinder. The diameter adopted for the inlet and exhaust ducts depends on the cross-sectional areas of the valves associated with them. For a high-output engine the inlet valve should be as large as possible, to keep flow resistance low at high engine speeds (minimum throttling effect at the valve). In a well-designed normal engine operating at its maximum output, a gas velocity of about 300 ft./sec. (90 m/sec.) should occur in the opening between the valve head and its seat when the valve is fully open (Fig. 2). The exhaust valve may have a 15% lower discharge capacity than the inlet valve.

In many engines the inlet duct is made to taper toward the inlet valve. This helps to keep the flow velocity and therefore the resistance low in the carburetor and inlet manifold, while providing a suitably high velocity at the actual inlet valve. The exhaust duct also is given a divergent shape toward its outlet end for similar reasons (Fig. 3). Sudden changes in direction and cross section of the exhaust duct must likewise be avoided.

The combustion mixture is normally produced in the carburetor, which is connected to the inlet manifold that delivers this mixture to the inlet duct of each cylinder. Quite often the manifold is integral with the cylinder-head casting. The manifold and ducts system inevitably comprises bends which have an adverse effect on gas-flow conditions. Engines designed for very high outputs have an independent intake and mixture-producing system for each individual cylinder. Twin and multiple carburetors are used in such engines (Fig. 4).

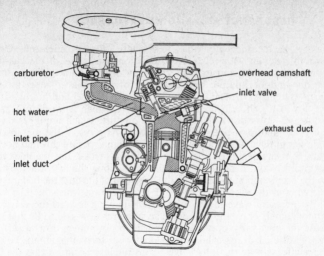

carburetor
overhead camshaft
inlet valve
hot water
exhaust duct
inlet pipe
inlet duct

FIG. 1   FOUR-CYLINDER GAS ENGINE

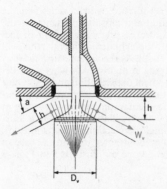

a = angle of valve seating

h = valve lift

W_v = gas velocity in valve-flow section

D_v = valve diameter

FIG. 2   CROSS SECTION THROUGH VALVE

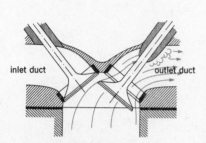

inlet duct
outlet duct

**FIG. 3   INLET AND OUTLET DUCTS
WITH FAVORABLE FLOW**

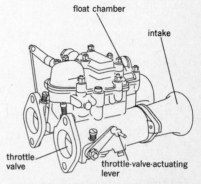

float chamber
intake
throttle valve
throttle-valve-actuating lever

FIG. 4   TWIN CROSS-DRAFT CARBURETORS

RESONANCE AND ENGINE EFFICIENCY

In addition to the measures described in the foregoing, an inlet-and-exhaust system of favorable design with regard to oscillation conditions of the gas does much to improve the intake and exhaust efficiency of the engine. The pulsating intake of the fuel-and-air mixture and discharge of the exhaust gases initiates oscillation in the system. At the end of the suction stroke the fuel-and-air mixture in the inlet duct flows at high velocity to the inlet valve, which is in the process of closing. This closure slows down the rush of the mixture, which impinges on the valve and causes a buildup of pressure in front of it. As a result, the fuel mixture continues to flow into the cylinder even after the piston has passed through bottom dead center and is rising again to start the compression stroke. But the inlet valve now closes completely and deflects the mixture back along the inlet duct. This is what initiates the oscillation in the duct. At the open end of the duct the fuel mixture is again deflected, and the cycle is repeated. If, at the instant when the inlet valve opens again, the pressure wave in the inlet duct is moving toward the valve, the mixture will immediately enter the cylinder. The system is now said to be in a state of *resonance*. As the inlet valve opens wider and the piston moves downwards, the pressure in the inlet duct drops, while the velocity rises to its maximum. Toward the end of the suction stroke the inlet valve begins to close again, so that the flow is again retarded, the pressure builds up, and the oscillation phenomena are repeated.

Optimum charging of the cylinder with the fuel mixture is achieved when the frequency of oscillation coincides with the opening and closing frequency of the valve so as to produce resonance, as envisaged above—when the inrush of fuel mixture finds the valve just opening to let it into the cylinder. Evidently this will occur only at one particular engine speed. At other speeds the volumetric efficiency will be lower. Long inlet ducts provide good charging at high speeds, whereas short ones are better at relatively low speeds. The development of oscillation and resonance is counteracted by the flow resistance in the inlet duct, the constriction and turbulence at the throttle valve (in the carburetor), and the damping effect occurring at the open intake end of the duct. Figs. 1 and 2 show the oscillation system in a four-cylinder engine: the inlet pipes all emerge from a single connection at the carburetor, so that damping occurs there. The charging is therefore poorer than in the case of the inlet system of a fuel-injection spark-ignition engine as illustrated in Figs. 3 and 4 (further described on page 278).

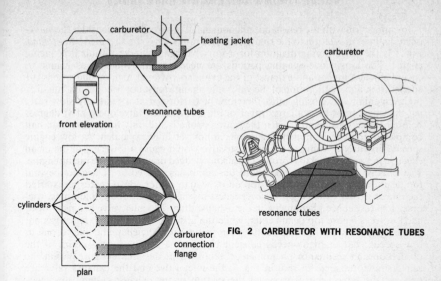

carburetor

heating jacket

carburetor

front elevation

resonance tubes

resonance tubes

cylinders

carburetor
connection
flange

FIG. 2   CARBURETOR WITH RESONANCE TUBES

plan

**FIG. 1   CARBURETOR ENGINE
WITH RESONANCE TUBES (SCHEMATIC)**

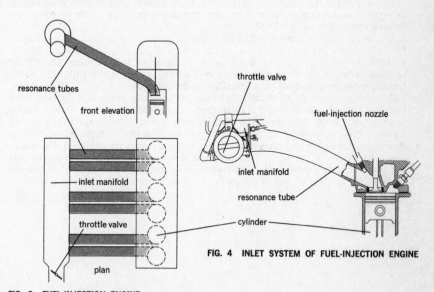

resonance tubes

front elevation

throttle valve

fuel-injection nozzle

inlet manifold

inlet manifold

throttle valve

resonance tube

cylinder

plan

FIG. 4   INLET SYSTEM OF FUEL-INJECTION ENGINE

**FIG. 3   FUEL-INJECTION ENGINE
WITH RESONANCE TUBES (SCHEMATIC)**

In connection with the resonance phenomena, it is necessary to consider the valve-opening times in relation to the rotation of the crankshaft—the so-called *valve timing*. Fig. 1 is the valve-timing diagram for a Porsche eight-cylinder Grand Prix racing engine. The long valve-opening periods are necessary in a high-speed engine to ensure efficient intake and exhaust of the cylinders. Also, at high engine speeds the acceleration and deceleration of the valve movements must not become inadmissibly high. Opening and closing must therefore be performed at a relatively slow rate: i.e., the valves must have a large travel, or lift. The exhaust valve opens at 81 degrees before bottom dead center, when the power stroke is still only little more than half completed and the combustion gases have not yet fully expanded. At low engine speeds this early opening of the exhaust valve would cause a lowering of the mean effective pressure and of the torque—an acknowledged drawback of the racing engine. The exhaust valve remains open until the crank has rotated to 51 degrees beyond top dead center. Although the piston on its way to bottom dead center has started the intake stroke, exhaust gas is nevertheless discharged from the cylinder in consequence of resonance phenomena in the exhaust duct. The inlet valve begins to open at 81 degrees before top dead center, while the piston is forcing the exhaust gas out of the cylinder. This likewise reduces the volumetric efficiency and the torque at low speeds; but at high speeds efficient charging is achieved on account of the oscillation and resonance phenomena established in the inlet system. Within the range shown hatched (in red) in Fig. 1, the inlet valve and the exhaust valve are open at the same time. Because of the suction in the exhaust system, this overlapping of the valve-opening periods promotes the development of a low pressure (partial vacuum) in the cylinder and thus assists the intake of mixture and improves volumetric efficiency. The inlet valve closes at 71 degrees after bottom dead center, during the compression stroke. Thus the charging action due to the inertia of the flowing gas is utilized.

This kind of valve timing, while appropriate to a racing engine, is not suitable for an ordinary car engine because of the low torque at low and medium speeds, so that the engine would be deficient in flexibility of performance. The timing approximately suited for ordinary engines is also indicated in Fig. 1 (points 1 to 4): the valve-opening periods now overlap much less (points 3 and 1), the exhaust valve does not open so far in advance of bottom dead center (2), and the inlet valve does not close so late (4). Fig. 2 is a diagram showing the valve lift (or travel) plotted against the angular rotation of the crankshaft. It is seen that for equal valve timing it is nevertheless possible to have different amounts of lift and different cross-sectional flow areas through the opened valves (black and red curves respectively). The intake and exhaust can be improved by an increase in the valve lift. The valve movements are controlled by cams on the camshaft (Figs. 3 and 4), which rotates at half the speed of the crankshaft and is driven from the latter by a chain drive or gearing.

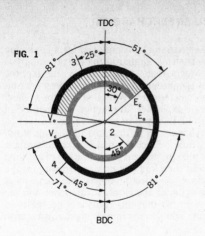

FIG. 1

Valve timing for a racing engine

V_o = inlet valve opens

V_c = inlet valve closes

E_o = exhaust valve opens

E_c = exhaust valve closes

TDC = top dead center

BDC = bottom dead center

The angles relate to the rotation of the crankshaft

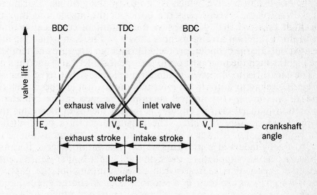

FIG. 2
GRAPH OF VALUE PATH PLOTTED AGAINST CRANKSHAFT ROTATION

FIG. 3 CAMSHAFT

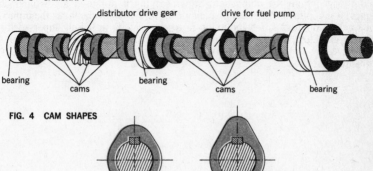

FIG. 4 CAM SHAPES

Instead of a carburetor, a *fuel-injection* system may be employed for introducing fuel (gasoline, petrol) into the cylinder. It is basically similar to the system employed in a diesel engine (see Vol. I, page 470), except that with gasoline as the fuel the ignition is initiated by an electric spark. A somewhat higher effective pressure and better output can be achieved by injection as compared with a carburetor system. Against this, the injection equipment is more expensive. In practice, this method of introducing the fuel is therefore confined to high-output or racing engines.

With injection, the inlet pipe for each cylinder can be designed to give optimum performance as an individual "oscillation tube." Since the fuel is injected straight into the cylinder, the need to heat the inlet pipe (to prevent condensation of gasoline vapor) is obviated. Consequently, cooler and therefore denser air is drawn into the cylinder, thus improving the volumetric efficiency. Injection of gasoline commences during the suction stroke. On entering the cylinder, the gasoline vaporizes, and the heat for evaporation is extracted from the air, so that this cools and decreases in volume, thus causing more air to be drawn in and thereby improving the volumetric efficiency.

Fig. 1 shows an arrangement in which the injection nozzle (colored red) is aimed at the hot exhaust valve, which is cooled by the gasoline. During the compression stroke the piston sweeps past the outlet of the nozzle and thus protects it from the high pressure that develops at the instant of combustion (initiated by spark ignition). A different arrangement is shown in Fig. 2, in which the injection nozzle is located outside the cylinder, protected from high pressure and temperature. It injects the fuel through the inlet port on to the opened inlet valve and thus into the cylinder.

The measures to improve volumetric efficiency that have been described in the foregoing relate to four-stroke internal-combustion engines which draw in the fuel-and-air mixture by the natural suction developed in the cylinder (self-aspirating, or suction-induced-charge, engine). A further means of increasing the power output is provided by *supercharging*. The supercharger is a compressor (axial-flow or centrifugal type) or blower which supplies air, or a combustion mixture of fuel and air, to the cylinders at a pressure greater than atmospheric. Because of this higher pressure, the air supplied to the cylinders has a higher density and absorbs more gasoline vapor. This increases the power output, but the gas consumption per horsepower is higher than in a suction-induced-charge engine, and wear and tear becomes more severe. Fig. 3 is a partial section through an American V8 engine equipped with a Roots supercharger with three-lobed rotors. The supercharger is usually driven from the crankshaft. Supercharging is not used for ordinary car engines; it is confined to aircraft or racing cars, and even for the latter the improvements achieved with suction induction in recent years have largely superseded the supercharger. Large diesel engines are often supercharged, the centrifugal compressor being driven by a small gas turbine motivated by the exhaust gases (exhaust-driven turbosupercharger).

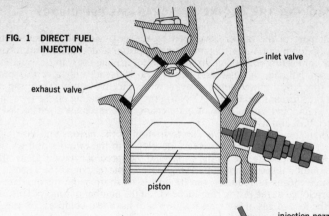

FIG. 1   DIRECT FUEL
         INJECTION

inlet valve

exhaust valve

piston

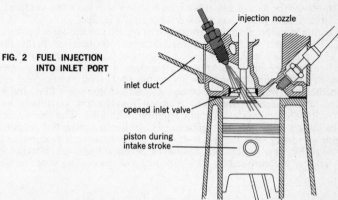

FIG. 2   FUEL INJECTION
         INTO INLET PORT

injection nozzle

inlet duct

opened inlet valve

piston during
intake stroke

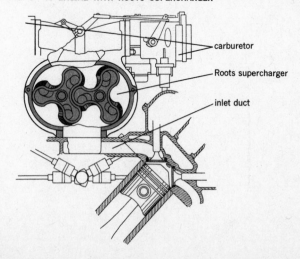

FIG. 3   V8 ENGINE WITH ROOTS SUPERCHARGER

carburetor

Roots supercharger

inlet duct

IMPROVING THE ENGINE'S MECHANICAL EFFICIENCY

A significant part of the power developed by the expansion of the gas in the cylinders is used for overcoming friction (between piston and cylinder and in the bearings of the connecting rod and crankshaft) and for driving the water-circulation pump, oil pump, dynamo, camshaft and valves (Fig. 1). Hence only a certain proportion of the theoretical power output is available as effective output. This proportion is termed the "mechanical efficiency" of the engine. Depending on the type and design of the engine and on its state of maintenance, the mechanical efficiency is usually between 0.75 and 0.85.

More than half the loss of power is due to friction of the pistons and bearings. The piston friction depends on the pressure developed in the cylinder and on the piston speed, which is determined by the stroke and the speed of rotation. Generally speaking, the rotational speed should be as high as possible (cf. page 266). Therefore the only possible means of reducing the friction is to shorten the piston stroke. The friction developed at the piston rings depends on the number of rings per piston. To minimize the loss of gas, it is necessary always to have two compression rings; in addition, each piston has an oil-scraper ring (Fig. 2).

Friction in the crankshaft bearings can be reduced by the use of lighter connecting rods. This also reduces the lubricant requirement of the bearings, so that the oil-pump power input is lessened. A crankshaft rotating at high speed causes frictional losses due to turbulence and foaming of oil in the sump. For this reason high-speed engines have dry-sump lubrication (Fig. 3). In this system, oil entering the crankcase is immediately extracted by suction and is returned through a filter and a cooler to the oil tank. A second pump delivers the oil from the tank to the bearings.

A water-cooled engine is usually equipped with a fan. The fan is necessary only when the cooling-water temperature is high. For a substantial proportion of the engine's running time the fan is absorbing power without performing any useful function. For this reason fans have been developed that are switched on and off automatically, controlled by the temperature of the cooling water or air.

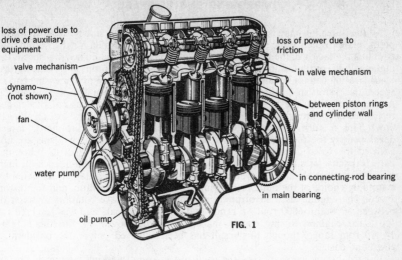

loss of power due to drive of auxiliary equipment

valve mechanism

dynamo (not shown)

fan

water pump

oil pump

loss of power due to friction

in valve mechanism

between piston rings and cylinder wall

in connecting-rod bearing

in main bearing

FIG. 1

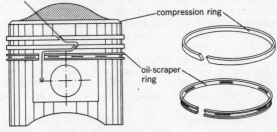

compression ring

oil-scraper ring

FIG. 2 PISTON RINGS

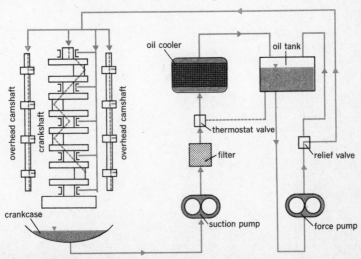

overhead camshaft

crankshaft

overhead camshaft

crankcase

oil cooler

oil tank

thermostat valve

filter

relief valve

suction pump

force pump

FIG. 3 DRY-SUMP LUBRICATION

All measures for increasing the power output that have been described up to this point relate to the torque (see page 260). However, the formula for the output also contains the factor n, the rotational speed of the engine. The higher this speed, the higher will the output of an engine generally be. Unfortunately, this purely theoretical consideration cannot be fully translated into practical terms. With increasing engine speed, the piston speed increases and the frictional losses likewise become higher. At the same time, the mean effective pressure diminishes because of the higher resistance encountered in the inlet-and-exhaust system (throttling effect on the gas flow). This in turn reduces the volumetric efficiency. Besides, the inertia forces developed by the reciprocating parts of the crank and valve mechanisms are not allowed to exceed certain values, otherwise damage is liable to occur. When the cubic capacity for a new engine design has been determined, the influences of high speed that adversely affect power output and engine life can be largely obviated by a suitable choice of the number of cylinders, the stroke-bore ratio, and the piston speed. The engines used in ordinary present-day cars have rotational speeds of between 5000 and 6000 rpm—a range that only a few years ago was reserved for sports-car engines. Racing engines have meanwhile moved up into the 11,000–14,000 rpm range, though this result has been achieved only with considerable effort and cost.

The total cubic capacity (i.e., the total piston-swept working volume V_H) of an engine should be divided over the largest possible number of cylinders, to ensure that the reciprocating masses of the individual pistons and connecting rods will be small. The lighter these components are, the easier and less power-consuming will be their acceleration and deceleration at the ends of the piston stroke. For a given cubic capacity, the capacity of the individual cylinder is reduced, the bore and stroke are likewise reduced, and the piston speed is lower. However, an increase in the number of cylinders also has its drawbacks. For one thing, there are now more bearings in which friction occurs. In addition, the cost of manufacture goes up because of the more numerous components that have to be made, machined and assembled. For reasons of economy, the cubic capacity of a cylinder of an ordinary car engine is normally between 250 and 500 cc. A racing engine usually has many relatively small cylinders ranging from, for example, 62 cc (Honda) to about 200 cc.

(more)

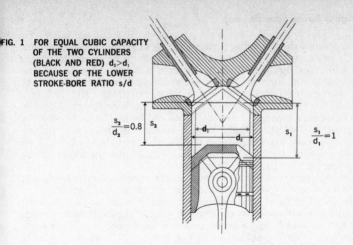

FIG. 1 FOR EQUAL CUBIC CAPACITY OF THE TWO CYLINDERS (BLACK AND RED) $d_2 > d_1$ BECAUSE OF THE LOWER STROKE-BORE RATIO s/d

$\frac{s_2}{d_2} = 0.8$

$\frac{s_1}{d_1} = 1$

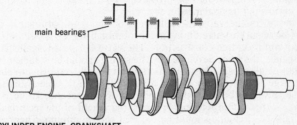

main bearings

FIG. 2 FOUR-CYLINDER-ENGINE CRANKSHAFT WITH FIVE MAIN BEARINGS

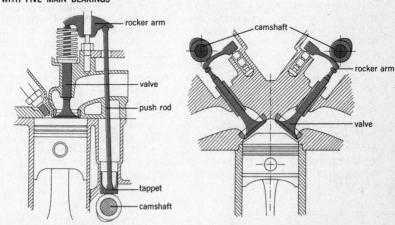

rocker arm

valve

push rod

tappet

camshaft

FIG. 3 ENGINE WITH LOW CAMSHAFT AND PUSH ROD

camshaft

rocker arm

valve

FIG. 4 ENGINE WITH INCLINED VALVES AND TWO OVERHEAD CAMSHAFTS

In addition to dividing the total cubic capacity among a large number of cylinders, each of relatively small capacity, other measures to reduce the reciprocating masses of the pistons and crank mechanism consist in the use of light-alloy pistons and connecting rods made from titanium, a metal not unlike steel, but lighter. When the capacity of the individual cylinder has been determined, the stroke s and the bore d can be determined from the stroke-bore ratio (s/d) that has been chosen. As a rule this ratio is somewhere between 0.7 and 1.0. It should be as low as possible for high-speed engines, so that the cylinder bore is larger than the stroke: i.e., the cylinder is relatively wide, making possible the use of large valves (see Fig. 1 on page 283). Besides, the piston speed is then also lower, so that the frictional and throttling losses during the suction stroke are less.

At high speeds the crankshaft functions under severe stress conditions because at each power stroke it is subjected to sudden impactlike torsional loading. The crankshaft must therefore be of very rigid construction; it must not deflect. Better resistance to deflection is obtained by closer positioning of the crankshaft bearings (usually called the main bearings). For this reason a high-speed crankshaft has a bearing on each side of each crankpin, as in Fig. 2 (page 283), which shows the crankshaft of a four-cylinder engine with five main bearings. (In a cheaper engine only three main bearings would be provided.) To achieve better balancing of the masses, the crankshaft has balanced webs.

Efficient design of the valve mechanism is of major importance in high-speed engines because accurate valve timing at all rotational speeds is essential. This calls for rigid and vibration-free construction. The valve is opened against the closing action of a spring; the force developed by the spring should be sufficiently powerful to ensure that at all speeds the valve motion accurately conforms to the shape of the cam (see page 277). At high speeds there is only very little time available in which closure of the valve can be effected, a mere fraction of a second. To keep the spring force needed for this within reasonable limits, the weight of the reciprocating valve parts in a high-speed engine should be reduced to a minimum. There are various methods of achieving this. Dividing the total cubic capacity among a large number of cylinders permits the use of correspondingly smaller and lighter valves. The high speeds of present-day engines have been attained partly as a result of using overhead camshafts, thereby eliminating transmission elements which make the valve mechanism slower and more cumbersome (compare Figs. 3 and 4, page 283). In the arrangements illustrated in Figs. 1a and 1b (page 285), the camshaft is located above the valve. Interposed between the cam and the valve is a rocker arm (Fig. 1a) or a cup-type hollow tappet (Fig. 1b); these intermediate elements protect the valve stem from friction forces exerted by the cam. For high-speed engines the arrangement in Fig. 1a is preferable to that in Fig. 1b because the moving masses in the former are smaller. Sports-car and racing-car engines have hemispherical combustion chambers, so that the valves have to be inclined. For this reason each row of valves is provided with its own camshaft (Fig. 4 on page 283). This solution is too expensive for the engines of ordinary cars. Alternatively, two rows of inclined valves can be actuated by one camshaft (Fig. 2), though in this arrangement the rockers constitute a larger moving mass. Fig. 3 shows a different overhead camshaft arrangement embodying a tappet.

(more)

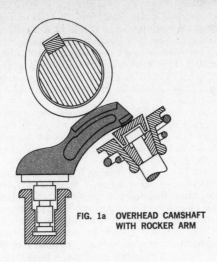

FIG. 1a OVERHEAD CAMSHAFT
WITH ROCKER ARM

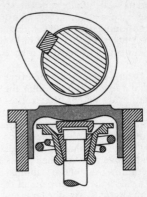

FIG. 1b OVERHEAD CAMSHAFT
WITH CUP TAPPET

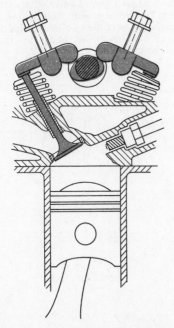

FIG. 2 ENGINE WITH INCLINED VALVES
AND ONE OVERHEAD CAMSHAFT

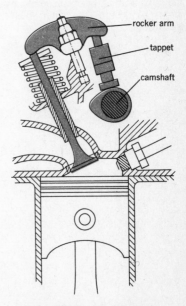

rocker arm

tappet

camshaft

FIG. 3 ENGINE WITH OVERHEAD
CAMSHAFT

In ordinary car engines, the overhead camshaft is usually driven by a chain from the crankshaft and at half the speed of the latter. To avoid objectionable noise arising from wear and thermal expansion, the chain is kept under uniform tension by a tensioning device. In some instances a silent valve drive in the form of a toothed plastic belt (reinforced with steel wire) is used instead of a metal chain. The camshaft drive systems illustrated in Figs. 1 and 2 are suitable for engine speeds up to 7000 rpm.

In racing engines, which operate at considerably higher speeds, the overhead camshafts are driven through the agency of gear systems or bevel-geared shafts (the latter are shown schematically in Fig. 3; actually these shafts are vertical). Such systems are preferable to chain drives because they are free of vibration and backlash effects. They are, of course, also more expensive.

Another means of reducing the weight of the valves consists in using valves with hollow stems. To improve the heat conduction and cooling of the exhaust valves, which become very hot, their stems are partly filled with sodium. At the high working temperatures the sodium is molten and its movements help to conduct heat from the valve head to the cooler parts of the stem, thus cooling the head (Fig. 4).

As an alternative to one large and heavy valve it is possible to employ two smaller, lighter valves. Thus the cylinders of some racing engines are each provided with two inlet valves and one exhaust valve. This is a very expensive form of construction and therefore unsuitable for ordinary engines. Various types of valve embodying "positive" actuation, as distinct from the spring-controlled reciprocating action of the usual poppet (or mushroom) valve envisaged here, have been devised, including more particularly the rotary valve, but have never achieved much practical significance.

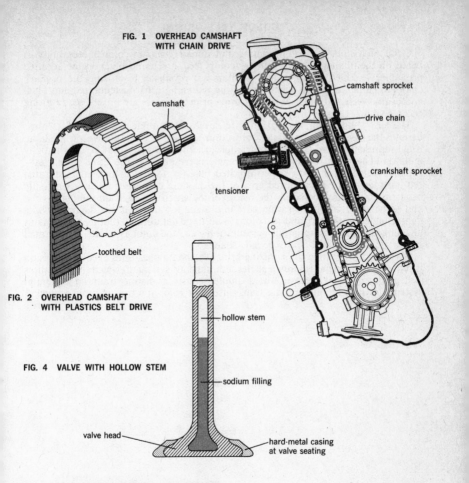

FIG. 1 OVERHEAD CAMSHAFT
WITH CHAIN DRIVE

camshaft

camshaft sprocket

drive chain

crankshaft sprocket

tensioner

toothed belt

FIG. 2 OVERHEAD CAMSHAFT
WITH PLASTICS BELT DRIVE

hollow stem

FIG. 4 VALVE WITH HOLLOW STEM

sodium filling

valve head

hard-metal casing
at valve seating

FIG. 3 DRIVE OF THE FOUR OVERHEAD CAMSHAFTS
OF A PORSCHE OPPOSED CYLINDER ENGINE

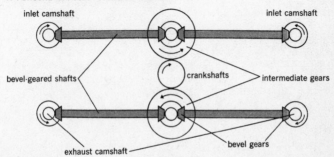

inlet camshaft

inlet camshaft

bevel-geared shafts

crankshafts

intermediate gears

exhaust camshaft

bevel gears

Color television, like color printing and some systems of color photography, is based on the mixing of three primary colors (red, green and blue) on the additive principle. So-called complementary colors are produced by mixing the primary colors in pairs: yellow or ocher (by mixing red and green), magenta (red and blue) and cyan (green and blue). When all three primary colors are mixed, the resulting color is white.

The terms "hue" and "saturation" are employed with reference to colors: "hue" denotes the essential color, while "saturation" signifies its dilution with white light. Fig. 1 represents the so-called color triangle in a form known as the CIE chromaticity diagram. The three corners of the triangle correspond to the three primaries—supersaturated green (top), supersaturated blue (bottom left), and supersaturated red (bottom right). The colored area of the diagram comprises all colors actually occurring in nature, including the colors of the spectrum. At R all the light is red and is therefore said to be saturated. On moving toward W in the diagram, white light is added to the red, and its color is said to be desaturated (but its hue remains unchanged); at W there is full desaturation—i.e., the color is white. Desaturated colors are called pastel colors or pale shades.

Applying the additive color-mixing principle, the simplest form of color television is represented by what is known as the simultaneous system, in which three pictures (red, green and blue) are transmitted simultaneously, as distinct from the sequential system, in which the pictures are transmitted one color at a time in rapid sequence.

(more)

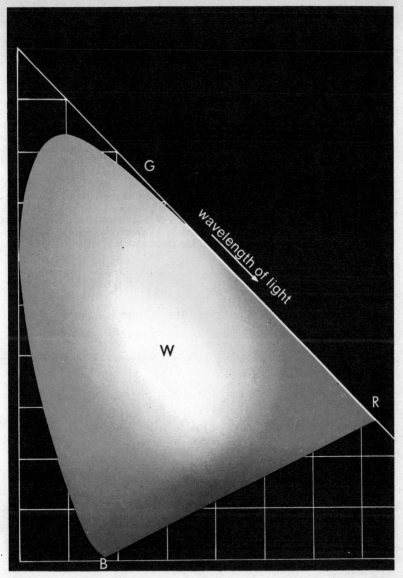

FIG. 1

In the simultaneous arrangement shown schematically in Fig. 2 the camera comprises three tubes (e.g., image orthicon tubes; see Vol. I, page 124). By means of a beam-splitting device (a system of mirrors and color filters), the first tube forms a red image (R), the second forms a green image (G), and the third forms a blue image (B). The three camera tubes have identical scanning patterns, so that the picture signals they produce are identical, except that they differ in color. In the receiver the primary color signals E_R, E_G and E_B are simultaneously fed to three color picture tubes and thus converted back into three separate color images (red, green and blue). These are superimposed optically by means of mirrors, so that the viewer sees them as one composite picture in which the correct colors are obtained by addition of the three primary colors. The great drawback of this system is the large band-width required for the simultaneous transmission of the three color signals.

The three main color television systems in present-day regular use are the NTSC system (developed in the U.S.A. by the National Television System Committee and first demonstrated in 1953), the SECAM system (developed in France, 1958), and the PAL system (a German development of the NTSC system, 1961, now used also in Britain). The basic principles are shared by all three systems, the only major difference being in the manner in which the chrominance signals are transmitted.

(more)

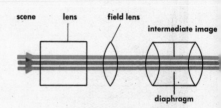

FIG. 2
SIMULTANEOUS SYSTEM OF
COLOR TELEVISION

In the NTSC system the primary color signals E_R, E_G and E_B are converted by a device called a color coder into a "luminance" (i.e., brightness or brilliance) signal E_Y and two "chrominance" signals. Chrominance comprises hue and saturation as its two component characteristics. The luminance signal gives no information about the color of the picture; it is a monochrome signal which can be received by an ordinary monochrome (black and white) receiver. It is this separation of the color information from the luminance information that is essential to a so-called "compatible" color television system, i.e., one which transmits signals that can also be satisfactorily received by existing monochrome receivers.

The luminance signal is subtracted from the primary color signals, and the color-difference signals thus obtained are further combined, in the transmitter, to produce two signals I and Q which serve to modulate the chrominance subcarrier signal. This signal is amplitude-modulated in accordance with the saturation values and phase-modulated in accordance with the hues. The luminance and chrominance components are combined to form the overall color picture signal, which is transmitted. The picture signal wave is a composite wave in which the chrominance wave is superimposed upon part of the luminance wave.

(more)

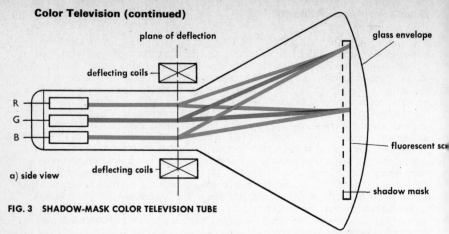

FIG. 3   SHADOW-MASK COLOR TELEVISION TUBE

The primary color signals, which have been converted into luminance and chrominance components at the transmitter, have to be reconverted into primary color signals before they can be applied to the color picture tube. Instead of using three separate picture tubes, as envisaged in the simultaneous system, the NTSC and the other two modern systems use only one tube, called a shadow-mask tube or tricolor tube. This contains three electron guns; these guns produce three separate electron beams, which move simultaneously in the scanning pattern over the viewing screen and produce a red, a green and a blue image respectively. The screen is covered with three separate sets of uniformly distributed tiny phosphor dots. The dots of each of these three sets glow in a different color when struck by an electron beam. Electrons discharged by the "red" gun, i.e., the gun controlled by the red primary color signal, impinge only on the red-glowing phosphor dots and are prevented from impinging on the green- and blue-glowing dots by the shadow mask, which is a metal sheet containing a large number of tiny holes, each of which is accurately aligned with the different colored phosphor dots on the screen. Similarly, the electrons from the other two guns fall only on the blue and the green dots respectively. In this way three separate color images are formed simultaneously on the screen. The dots producing the three colors are so small and so close together that the eye does not see them as separate points of light, but forms an overall impression of continuous color. See Fig. 3.

The color receiver contains a tuner and intermediate frequency amplifiers Color reproduction is divided into luminance and chrominance functions. A video detector applies the luminance component, after amplification, to all three electron guns of the picture tube. The inverse operations of the addition and subtraction circuits at the transmitter are performed by appropriate circuits in the receiver. In this way three color-difference signals are obtained (the difference between the luminance signal and the primary color signals), which are applied to the respective electron guns in addition to the luminance signal. The net control signal applied to each gun conforms to the primary color signal coming from the corresponding tube in the camera.

The modern shadow-mask tube in the color television receiver has the phosphor dots applied directly to the curved glass front of the tube (in the earlier tubes the dots were on a flat glass plate inside the tube, as in Fig. 3). The shadow mask of a 25-inch tube has about 440,000 holes, and three times that number of dots on the screen. The principle of the tube is simple, but its actual operation is complex. Various adjustments are necessary to compensate for inevitable inaccuracies and distortions, particularly convergence adjustments and purity adjustments.

(more)

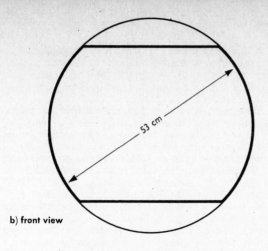

b) front view

phosphor dots
on screen

blue beam

red beam

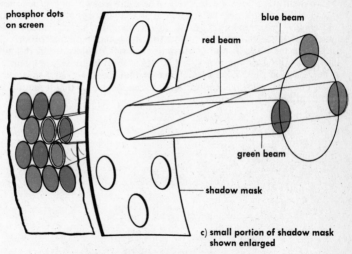

green beam

shadow mask

c) small portion of shadow mask
shown enlarged

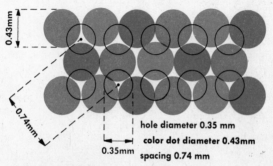

hole diameter 0.35 mm

color dot diameter 0.43mm

spacing 0.74 mm

d) small portion of screen
shown enlarged

The type of color television camera at present in extensive use has three image orthicon tubes, the scene being split up by means of special mirrors (dichroic mirrors) into red, green and blue components (see Fig. 2). Other three-tube cameras have so-called plumbicon tubes; vidicon tubes are sometimes also used for the purpose (see Vol. I, page 124).

As already mentioned, the three main modern color systems differ in the manner in which the two chrominance signals are transmitted on a single subcarrier (i.e., a carrier wave which is applied as a modulating wave to another carrier). In the NTSC system the two signals are transmitted on the subcarrier by using a special modulating system (quadrature modulation). In the PAL system the same principle is applied, except that one of the chrominance signals is reversed on alternate lines (hence the name "phase alternation, line"). On the other hand, in the SECAM system the chrominance signals are transmitted sequentially, one at a time.

The subcarrier in the NTSC system is modulated by the I and Q chrominance signals, these being respectively situated 33 degrees ahead of the E_R-E_Y axis and the E_B-E_Y axis in the vector diagram. In the PAL system this modulation is performed by the so-called V and U signals, whose vectors correspond in direction to the two above-mentioned axes respectively. These are shown in Fig. 4, where the colors of the color triangle are represented in a circular diagram, together with subcarrier vectors for various colors; "burst" denotes the burst signal, which is the transmitted synchronizing signal which controls the phase and frequency of the color oscillator in the receiver. The V and U chrominance signals are used to modulate two subcarriers of the same frequency, but 90 degrees out of phase, i.e., quadrature modulation, just as in the NTSC system, except that the V signal is reversed on alternate lines. The basic reason for adopting this complication is that it reduces certain errors and improves the performance.

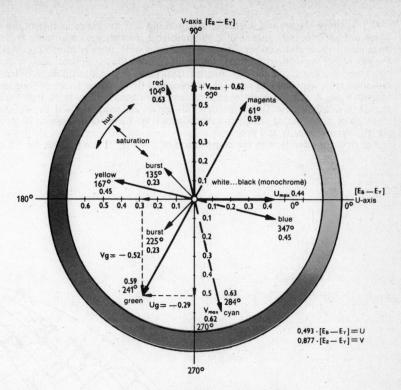

FIG. 4   COLOR VECTORS FOR PAL SYSTEM

Cartography, the art and science of map making, had its origins in ancient times. Maps are an important means of communicating information concerning angles, distances, areas and directions. They are used for a wide range of purposes: as aids for sea or air navigation (charts); for military, scientific, technical, agricultural, educational, commercial and administrative purposes (e.g., cadastral maps of land-holdings, etc.); for local and regional planning and development; for motoring and tourism; etc. A very important requirement, besides accuracy, is that the map should be clearly legible. To help achieve this, colors are generally used. Present-day methods of cartographic design and reproduction enable modern maps to be given a high degree of clarity and precision.

The first consideration in the compilation of any map is the purpose for which

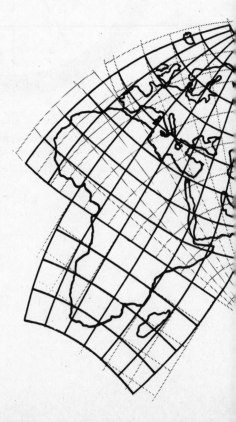

Europe, Africa, Asia and Australia represented on a so-called interrupted projection (a compromise projection)

it is to be used. A suitable map projection must be chosen, depending on whether the map will have to show only a small part of the earth's surface, a relatively large area such as a country or indeed a continent, or the whole surface of the earth. It will to some extent depend also on whether the map must fulfill geographical, navigational or traffic engineering requirements. Many different projections and variants thereof have been developed. The basic problem arises from the need for mapping the features of a spherically curved surface (the terrestrial sphere) upon a plane (a flat map). It is impossible to make this transformation without modifying the geometric relationships in some manner. However, there are a great many possibilities of transformation that can retain one or more of the spherical relationships. The actual process of transformation is called "projection."

(more)

When the projection has been decided, the meridians and parallels of latitude are constructed (this system of lines is called the map graticule). A small-scale map has to be compiled from data given on large-scale survey maps of the region concerned. These basic data are coastlines, rivers, lakes and political boundaries, which have to be transferred to the map which is being compiled. When these data have been drawn in, the cartographer has, as it were, a skeleton on which to hang the other

Contour lines of a mountainous region

details. Clear representation of relief features (mountains, valleys, etc.) is important on certain maps. The most important method is by contouring, i.e., by drawing contour lines (or contours), which each have a constant elevation above sea level. If the vertical contour interval is small enough, so that the contours are sufficiently close together, they will express the relief character of the terrain (slopes, mountain peaks, plateaus, plains, etc.). When sufficient contours and information concerning the terrain features are available, the technique known as shading may be applied. This gives a kind of "plastic" (three-dimensional) effect, so that the map gives the

The same region as above, with shading

impression of a topographic model of the landscape. Oblique shading of hills and mountains is usually based on the assumption that the incident light is inclined at 45 degrees and comes from the northwesterly direction. Features of the sea or ocean bed can similarly be represented by depth contour lines.

Depth-contour lines of the ocean bed

Another commonly employed method of relief representation is by means of hypsometric coloring (also known as altitude tinting or layer tinting): a range of different colors is used to represent different zones of elevation, which can thus be easily visualized. Conventionally, low land is colored green, followed by shades of buff or yellow, and ranging to dark brown (and sometimes red or purple) for high altitudes.

Before a map can be reproduced for publication, it must be redrawn in finished form. Freehand lettering has now largely been superseded by other techniques, especially stick-up (or preprinted) lettering: the names are printed, cut out, and stuck in position on the map. Another and more recent method which is coming into wider use is photolettering. In multicolor map reproduction the problem of register (precise fit of individual drawings which have to be superimposed) has largely been solved by the introduction of synthetic plastics which are free from shrinkage effects and generally retain their shape better than does paper.

Most maps are printed by either letterpress or lithography (see Vol. I, pp. 414 and 418), the latter process being used more particularly for large maps. The basic steps are: photographing the original drawing; processing the negative; making the plate; presswork. Printing plates for lithography and letterpress printing are made of various metals or of plastics. Presswork involves placing the completed plate on the press, inking, and feeding the paper through the press. The reproduction of maps in color differs from black reproduction only in that different colored inks are used. Each separate ink requires a separate printing plate, so that all the steps in the process have to be repeated each time, thus greatly increasing the cost. This is more particularly true of the "flat color" reproduction process, which is most often employed. Alternatively, "process color" may be used in which the map is reproduced from a single color drawing by the four-color process in essentially the same way as color photographs and painted art work are reproduced in popular magazines (see Vol. I, page 164). On the other hand, the flat color process requires a number of separate drawings: a number of line copies of the map are made. On these copies

(more)

the cartographer makes separate finished drawings in black for each solid printing color to be used. A color guide for background tints is provided by him for the printer's information. A wide variety of tints can be produced in the plate-making and printing processes. The accompanying illustrations show the eight successive

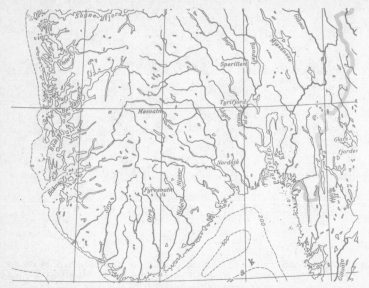

Coastline, rivers and lakes

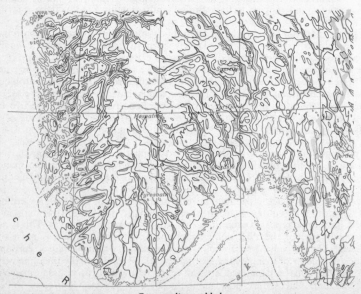

Contour lines added

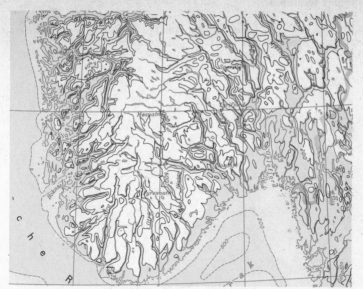

Light-blue tints are printed first

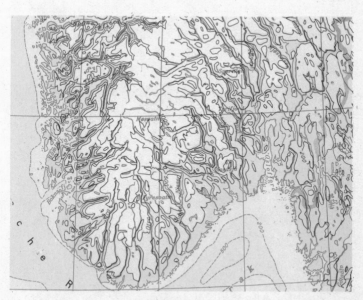

Next, yellow (where superimposed on blue it produces green)

printing stages in the reproduction of a modern map on the scale of 1:5,000,000. Each stage (a separate printing plate) registers accurately on the preceding one, the finished map being produced in one continuous operation on a multiple-color press.

On small-scale maps much topographical and physical detail has to be omitted.

(more)

Those details which are retained must be suitably generalized. For instance, cities and towns are represented by artificial symbols (conventional signs, e.g., dots, circles, etc.). Different styles and sizes of lettering used for the names of towns may serve

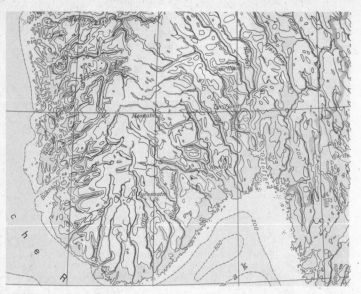

Brown altitude tinting is obtained by superimposing light red on yellow

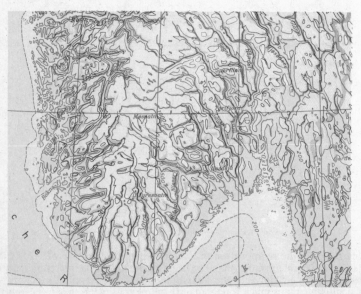

Shading of mountains added

as indications of size (number of inhabitants). An explanation of the symbols is usually given in the margins of the map or in the so-called map legend, together with the title and indication of scale.

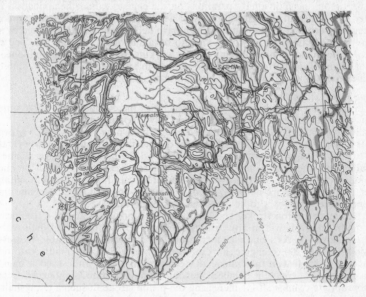

Railways and roads added in dark red

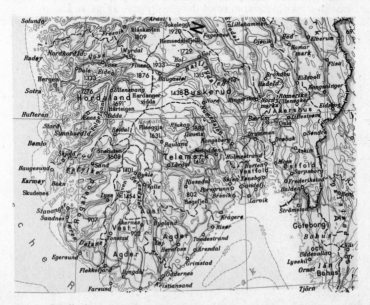

Finally, lettering and certain other details are printed in black

ENGINE TUNING

The term "tuning" applied to automobile engines refers to a number of measures all aimed at getting better performance, particularly in the case of standard production engines. Because of mass-production requirements, some of these measures requiring individual operations on the engines cannot be applied in the factory. Other measures consist in partly or wholly removing certain limitations on the engine speed and power output which have deliberately been embodied in the engine by the manufacturer with a view to obtaining longer service life. The enhanced performance is attended with severer working conditions of the components, so that engine life is shortened. However, it is possible partly to offset this by suitable modifications to the cooling system and lubrication.

Increasing the cubic capacity:
Within limits, an increase in the cubic capacity of the cylinders is a fairly simple technical matter (cf. page 264). There are two ways of doing this:
1. Increasing the cylinder bore. This procedure, known as "overboring," is practicable only if pistons of appropriately larger diameter are obtainable. If the cylinder wall is thick enough to permit overboring beyond the maximum piston diameter available from the engine manufacturer, it will be necessary to obtain or make pistons specially. Many automobile manufacturers can, however, supply suitably larger pistons, which are normally used in a higher-powered type of engine than the one whose capacity is to be increased by overboring.
2. Increasing the piston stroke. This, on the other hand, is not always a technically easy modification. It will in any case be necessary to fit a new crankshaft with a larger crank radius. As a rule such a crankshaft is not readily obtainable. However, the Volkswagen 1200 cc engine, for example, can be fairly easily modified to near 1300 cc by fitting it with the crankshaft normally employed in the 1500 cc engine. For some types of engine it is indeed possible to obtain from components dealers the appropriate crankshaft for increasing the piston stroke. If such a crankshaft is not obtainable, however, it is still possible to achieve the desired result by obtaining a crankshaft forging and eccentrically grinding the crankpins, bearing in mind that the increase in stroke is equal to twice the eccentricity e (Figs. 1a and 1b).

(more)

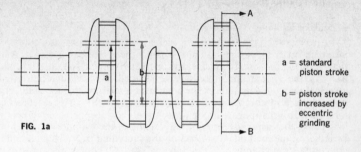

FIG. 1a

a = standard
piston stroke

b = piston stroke
increased by
eccentric
grinding

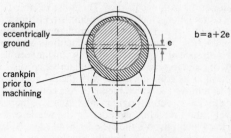

crankpin
eccentrically
ground

crankpin
prior to
machining

$b = a + 2e$

FIG. 1b   section A–B

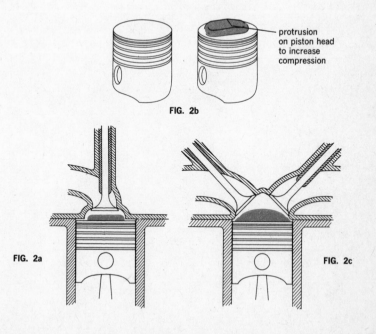

protrusion
on piston head
to increase
compression

FIG. 2b

FIG. 2a

FIG. 2c

Increasing the mean effective pressure:

1. Increasing the compression. An increase in the compression ratio of an engine results in better fuel utilization and a higher mean effective pressure. A drawback is that the firing pressure is considerably increased as well, so that all the main moving parts of the engine are more severely loaded. Besides, detonation phenomena, associated with self-ignition of the combustion mixture, impose limits on the attainable increase in compression ratio, as was explained on page 266. According to formula (9) on that page, the compression ratio can be increased in two ways. In the first place, the piston-swept working volume of the cylinder (V_H) can be increased by either of the methods described on the preceding page. Secondly, the volume of the combustion chamber (V_B) may be reduced. Of course, these two measures can be applied in combination with each other. The most reliable method of reducing the volume of the combustion chamber is to employ pistons of special design of which the head protrudes into the combustion chamber at top dead center (see Figs. 2a and 2c, page 305). Many engine manufacturers and components dealers can supply pistons with specially shaped heads (Fig. 2b) for various makes of engines. Another possibility of reducing the combustion chamber consists in removing metal from the cylinder head at its contact surface with the cylinder block. However, care must be taken not to remove too much metal, as this could result in piercing of the water jacket, undue loss of structural rigidity, and excessive increase in compression (Fig. 1). Alternatively, the individual cylinder or the whole cylinder block can be reduced in height, for which purpose the cylinder head must be suitably grooved to receive the top of the cylinder (Fig. 2). Finally, another way to reduce combustion-chamber volume is by means of buildup welding on the inside.

(more)

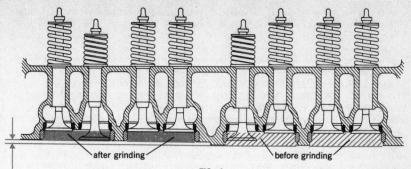

after grinding   before grinding

FIG. 1

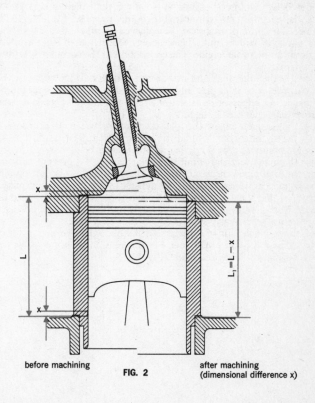

before machining

FIG. 2

after machining
(dimensional difference x)

With all the measures so far described it is essential to determine the actual volume of the combustion chamber with the aid of kerosene and a graduated pipette (Fig. 1). The piston should be at top dead center and the inlet and exhaust valves closed when this is done.

2. Improving the gas flow. The fuel-and-air mixture for the cylinders of a multicylinder engine is usually supplied by only one carburetor. A relatively narrow, long and often bent or curved inlet pipe connects the carburetor to each cylinder. This shape is unfavorable, from the point of view of gas flow, as it causes throttling effects which adversely affect the volumetric efficiency. Quite often the shape of the inlet pipe is utilized by the engine manufacturer as a means of throttling down the gas flow and thus limiting the engine performance. Changing the inlet pipes and employing more than one carburetor often constitute the easiest means of tuning. It does not involve dismantling the engine. Multiple-carburetor assemblies are available for many engines from components dealers, so that in most cases the owner of the vehicle is spared the effort and cost of developing such a system for his engine. The greatest gain in performance is achieved when each cylinder is provided with its own individual carburetor. However, if the inlet duct is cast integrally with the cylinder head, it is not possible to do more than merely enlarge the opening to which the carburetor is connected. For this purpose, the cylinder head must be removed. The inlet opening can be enlarged as much as the wall thickness of the water passages will permit. Fig. 2 shows the increased inlet cross section in the cylinder head of a Fiat engine. The connection to the carburetor is formed by means of an adapter.

Fitting of the inlet pipes and carburetors should be done with care. The transitions from carburetor to inlet pipe and from the latter to the cylinder must not comprise any projecting edges or constrictions (e.g., the edges of gaskets) that will reduce the cross-sectional area of flow. .

In a Volkswagen engine the inlet ducts of the two rows of cylinders are located far apart, with the carburetor midway between them. The gas-flow passages comprise bends and are quite long (Figs. 3a and 3b), besides being narrow (in order to throttle down the engine and thus limit its performance). To prevent condensation of gas vapor, the inlet pipe is heated (Fig. 3b). This heating lowers the volumetric efficiency of the engine, however, because the density of the combustion mixture is thereby reduced, so that the actual quantity (in terms of weight) drawn into the cylinders is reduced. Measures to improve the gas flow to the cylinders of a Volkswagen engine are therefore particularly rewarding in that they result in a marked increase in performance. Each pair of cylinders shares one inlet opening in the cylinder head, giving access to a forked inlet duct. For this reason it is not possible to fit more than two carburetors to such an engine (Fig. 4).

(more)

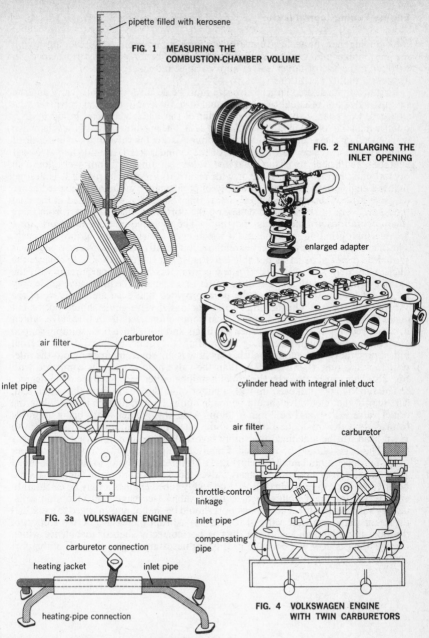

pipette filled with kerosene

FIG. 1 MEASURING THE COMBUSTION-CHAMBER VOLUME

FIG. 2 ENLARGING THE INLET OPENING

enlarged adapter

cylinder head with integral inlet duct

air filter

carburetor

inlet pipe

FIG. 3a VOLKSWAGEN ENGINE

air filter

carburetor

throttle-control linkage

inlet pipe

compensating pipe

FIG. 4 VOLKSWAGEN ENGINE WITH TWIN CARBURETORS

carburetor connection

heating jacket

inlet pipe

heating-pipe connection

FIG. 3b INLET MANIFOLD OF VOLKSWAGEN ENGINE WITH INLET-PIPE HEATING SYSTEM

Modifying the cylinder head constitutes a further important measure for improving the gas intake and exhaust and thus obtaining better engine performance. In particular, the use of larger valves and seatings and enlargement of the ports are important tuning operations.

First of all, the desired larger valve size must be decided. If the inlet valve and the exhaust valve are of equal or nearly equal size, then only the inlet valve need be enlarged. For equal valve lift, the diameter of the outlet valve can be up to 15% smaller than that of the inlet valve (cf. page 272). The maximum size of the inlet valve will depend on the available space for accommodating the valve in the cylinder head (Figs. 1a and 1b). The wall thickness of the cylinder head around the valve seat should be sufficient to ensure that the valve-seat rings will not work loose in operation. A further limiting factor with regard to valve size arises if it is desired that the engine have adequate flexibility of performance in the middle speed range (cf. page 276). When new larger valve-seat rings have been shrink-fitted at the inlet ports, the work of shaping and enlarging the ports and ducts can start from there (these operations are known as "porting"). The tools required for this work are a flexible-shaft grinding machine and a selection of rotary files and grinding wheels. The inlet ports must be given very careful treatment. All constrictions and projections should be removed as far as possible (see Fig. 1a). It is often possible to enlarge the ducts accurately for a distance of a few centimeters from the carburetor with the aid of a drill fitted with a face-milling cutter or a reamer. The inlet-valve guide may be cut off level with the wall of the port, provided that accurate centering of the valve is still ensured. On the other hand, the outlet-valve guide should not be cut down (Fig. 1b), as that would impair heat conduction away from this valve, which becomes red-hot in operation. When the ducts and ports have been suitably shaped and enlarged by rough machining or grinding, they must be given a fine internal finish, preferably to a shining polish. Gas flow is improved by opening out the inlet port to a size only slightly smaller than the valve head (Fig. 2a as compared with Fig. 2b). This can be done by the use of a milling cutter with an angle of 75 degrees to enlarge the port, followed up with a cutter of 45 degrees to form the actual seat. The seat of the inlet valve should be made about 1.5 mm (0.06 in.) wide, while the outlet-valve seat should be 2 mm (0.08 in.) wide, this greater width being desirable to provide better dissipation of heat. Quite often the transition from valve stem to valve head can be machined to a more favorable shape from the viewpoint of gas flow. The edges should be rounded. Finally, the head should be polished in order to obviate any harmful notch effects (Fig. 3). This treatment is not to be recommended for exhaust valves, however; instead, valves faced with hard metal (stellite) should be fitted on the exhaust side (Fig. 4). For highly tuned engines it may be essential to employ sodium-filled hollow-stem exhaust valves (see page 286) to obtain satisfactory valve service life. The exhaust port should be shaped and enlarged to a gradual divergent shape from the valve seat (Fig. 5a). Abrupt increases in cross section of the duct must be avoided. If the duct is not correctly shaped, turbulence which impairs the gas flow will occur in it. It is not necessary to give the wall of the port and duct a very fine finish, however.

(more)

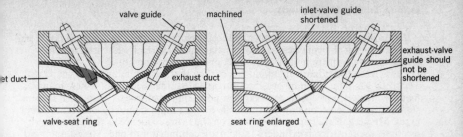

FIG. 1a CYLINDER HEAD BEFORE MACHINING

FIG. 1b CYLINDER HEAD AFTER ENLARGEMENT OF THE PORTS

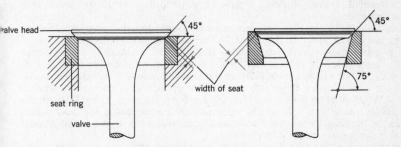

FIG. 2a VALVE SEAT IN A STANDARD ENGINE

FIG. 2b VALVE SEAT MODIFIED BY MACHINING

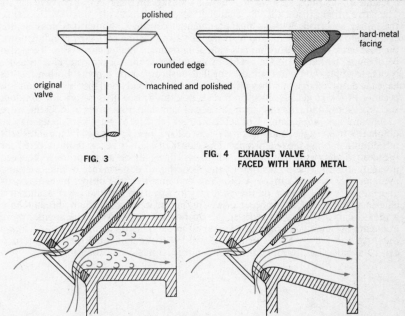

FIG. 3

FIG. 4 EXHAUST VALVE FACED WITH HARD METAL

FIG. 5a TURBULENCE IN EXHAUST DUCT

FIG. 5b EXHAUST DUCT ENLARGED TO GIVE GOOD FLOW CONDITIONS

The pipe connections to the inlet and to the exhaust side of the engine should be modified to suit the enlarged gas-flow passages. It is specially essential to achieve a clean, smooth transition at such connections (Figs. 1a and 1b).

Quite often the exhaust manifold is a casting that has an unfavorable shape with regard to gas flow. To improve the situation, it is advisable to fit a new manifold made from steel pipes, with each exhaust port connected to its own pipe. If there is sufficient space to accommodate them, the exhaust pipes should be left as individual pipes. Merging of the pipes will depend on the firing sequence of the cylinders: for example, the pipes from cylinders 1 and 4 are combined into one exhaust manifold, and those from cylinders 2 and 3 are combined into a second exhaust manifold. The two manifolds are continued separately to the exhaust muffler. For various cars the components industry supplies complete high-performance exhaust systems with special manifolds.

In addition to the measures already referred to, changing the valve lift and valve timing (see page 277) are further important means of improving the gas flow.

In engines whose valves are actuated through push rods, the lift can, within limits, be increased by modification of the rocker-arm transmission (Fig. 2). It is necessary to fit new rocker arms. A higher transmission ratio is obtained, for instance, when the Volkswagen 1200 cc engine is fitted with the rocker arms of the 1500 cc engine. When different rocker arms are used, it is important to check that there is still sufficient clearance between the valves and the piston head at top dead center. Increasing the valve lift is a worthwhile modification in a case where the lift is equal to less than a quarter of the inlet diameter.

An increase in the valve lift results in an increased cross-sectional area of the opening, so that the corresponding curve in the diagram is higher (red curve in Fig. 3). The valve timing is not thereby changed, however. It is hardly possible to change the timing effectively without fitting a different camshaft, though it is possible, for example, to increase the lift and increase the valve-opening period by grinding down the base circle (Fig. 4). Without any change in timing, a higher curve (Fig. 3) can also be obtained by flattening the profile of the tappet or the rocker arm. This measure is easy to apply, but is not suitable for flat mushroom-type tappets. Both measures have the disadvantage that they impose an additional load upon the valve mechanism because of the larger acceleration forces. For this reason it is better not to adopt such measures and, instead, to fit a special camshaft that gives a different valve timing and has larger cams producing a larger lift (Fig. 5). Such camshafts may be obtainable from engine manufacturers, who often supply sports and racing camshafts in addition to standard camshafts. The camshaft should be carefully selected for the desired purpose. The choice will always have to be a compromise between maximum obtainable power output and flexibility of performance in the lower and medium engine-speed ranges. A camshaft that improves the output at high speeds of rotation will, because of the greater valve overlap, give a lower volumetric efficiency and therefore a poorer power output at low and medium speeds than a standard camshaft. It thus reduces the operational flexibility of the engine; more frequent gear changing becomes necessary. On the other hand, measures to increase the output in the medium speed range will lower the maximum output. Extremes can be avoided by appropriate adjustment of the inlet-and-exhaust system.

(more)

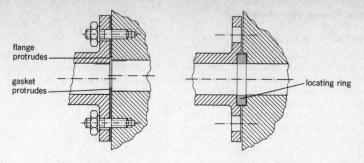

FIG. 1a POOR TRANSITION   FIG. 1b GOOD TRANSITION

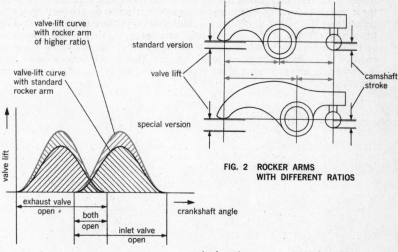

FIG. 2 ROCKER ARMS
WITH DIFFERENT RATIOS

FIG. 3 VALVE-LIFT CURVE

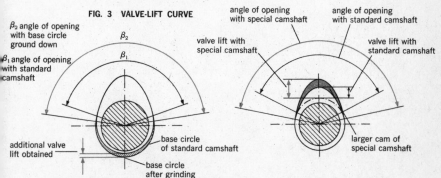

FIG. 4 CAMSHAFT MODIFIED TO INCREASE
VALVE-OPENING PERIOD

FIG. 5 SPECIAL CAMSHAFT

Increasing the speed of rotation:

In the internal-combustion engine, all reciprocating parts have to be accelerated from a momentarily stationary condition to maximum speed and then at once to be decelerated to a standstill again. The forces associated with this acceleration and deceleration exert quite severe loads upon the piston, the connecting rod, the camshaft and the bearings concerned and may cause damage to these parts.

1. Work on the valve mechanism. Acceleration of the reciprocating parts of the valve mechanism results from the rotational speed of the engine, the shape of the valve-lift curve (determined by the shape of the cams), and the magnitude of the valve lift (see page 313). In a new engine model with push-rod-operated valves, the mass of the valve mechanism can be reduced by suitable choice of materials for the moving parts concerned and by appropriate design. This can be achieved by the use of lighter materials (e.g., titanium, Fig. 1). On the other hand, reducing the weight of these severely stressed components in an existing engine by grinding or machining them down to smaller dimensions is inadmissible because of the increased risk of fracture and failure. It is possible to fit an existing engine with an overhead camshaft and thus dispense with tappets and push rods, so that the reciprocating masses are reduced (Fig. 2b as compared with Fig. 2a). This is a laborious and expensive modification, however, since it involves making a new cylinder head of appropriate design.

(more)

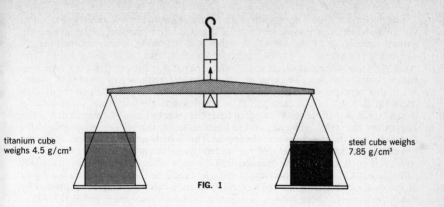

titanium cube weighs 4.5 g/cm³

steel cube weighs 7.85 g/cm³

FIG. 1

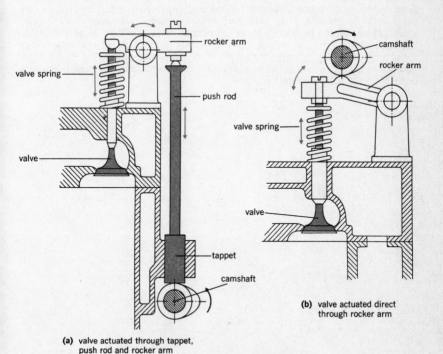

valve spring

rocker arm

push rod

valve

tappet

camshaft

(a) valve actuated through tappet, push rod and rocker arm

camshaft

rocker arm

valve spring

valve

(b) valve actuated direct through rocker arm

FIG. 2   VALVE AND CAMSHAFT ARRANGEMENTS

The maximum speed of rotation of an engine is limited by the throttling effect of the inlet-and-exhaust system (see page 272 et seq.) and by valve flutter due to the natural oscillation of the valves. A valve and its spring together constitute an oscillating system with a particular frequency of its own (its so-called natural frequency). When the engine is run up to high speeds, it may happen that the reciprocating movements of the valve come within the natural frequency range of this system. When this happens, the valve and spring tend to move independently of the motion imparted to them by the camshaft. The opening and closure of the valve now no longer conform to the correct timing, so that the power output of the engine decreases. This can be counteracted by steps taken to increase the natural frequency of the valve system beyond the speed range of the engine. This is fairly simple to achieve. A washer of suitable thickness can be inserted under the valve spring so as to increase its stiffness (Fig. 1). It should be checked that the coils of the spring are not in contact with one another when the maximum valve lift is attained. Another method consists in fitting a stiffer spring; or the natural frequency behavior may be improved by use of two or even three springs of different stiffness telescoped together (Fig. 2).
2. Work on the crank mechanism. The acceleration developed in the crank mechanism depends on the engine speed and the piston stroke. The total reciprocating mass is determined by the weight of the pistons, gudgeon pins and connecting rods. In order to reduce the forces that they exert upon the crankpins, it is necessary to reduce the weight of these reciprocating parts as far as possible. This must not be done by removing metal from the pistons and connecting rods, however, as this could result in fracture of the parts and serious damage to the engine. Very-high-speed engines should be fitted with forged pistons because of their greater strength. Replacement of steel connecting rods by lighter ones made of titanium brings about a significant reduction of the weight of the crank mechanism (Fig. 3). A further saving in weight can be effected by removal of surplus material (by machining on a lathe) from the gudgeon pin (Fig. 4).

(more)

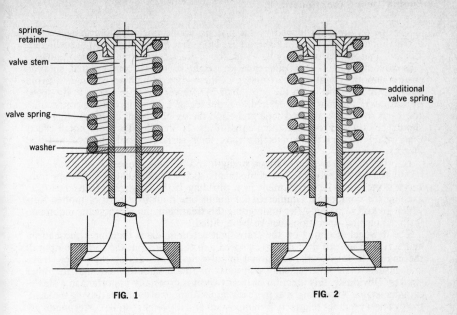

spring
retainer

valve stem

valve spring

washer

additional
valve spring

FIG. 1

FIG. 2

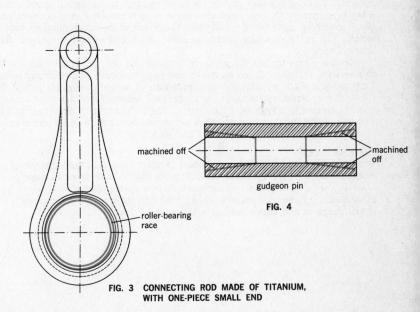

machined off

machined off

gudgeon pin

FIG. 4

roller-bearing
race

FIG. 3 CONNECTING ROD MADE OF TITANIUM,
WITH ONE-PIECE SMALL END

The inertia forces produced by the pistons, connecting rods and gudgeon pins and acting upon the crankshaft should be of the same magnitude for all the cylinders. For this reason it is necessary to weigh the pistons, connecting rods and gudgeon pins carefully (Fig. 1). Any differences in weight should be compensated, so as to ensure smooth, vibration-free running of the engine at high speeds and to avoid uneven stress conditions in the crankshaft. Where very large increases in the speed of rotation of the engine are desired, the rotating and the reciprocating connecting-rod parts should be weighed separately, and the weights of all corresponding parts should be adjusted to make them equal (Figs. 2a and 2b). To avoid notch effects that would weaken the connecting rods, their surfaces should be polished and inspected for cracks.

In order to improve their fatigue strength and running properties, it is advisable to subject all severely stressed components such as crankshaft, connecting rods, rockers, etc., to a heat treatment in a nitriding bath. This treatment consists in keeping the components immersed for about one hour in a bath of molten salts which give off nitrogen. After undergoing this treatment, the components require no further finishing operations before being fitted.

The flywheel mounted on the crankshaft accommodates the clutch mechanism and serves to absorb fluctuations in speed and thus even out the torque output of the engine. It receives the rotational impulse developed during the power stroke and ensures that the crankshaft continues to rotate smoothly through the other strokes. Obviously, this function of the flywheel is especially important in a single-cylinder engine. Accordingly, as there is a larger number of cylinders driving the shaft, the flywheel becomes relatively less important and its weight can be correspondingly reduced. A heavy flywheel prevents a rapid increase in rotational speed when the throttle is open, as its large mass also has to be accelerated. Faster acceleration can be obtained by removal of metal from the flywheel (by machining) to reduce its weight (Fig. 3). There will, however, be some sacrifice of smooth running in the low speed range.

After the foregoing operations have been carried out, the crankshaft—with the flywheel and clutch mounted on it—should be dynamically and statically balanced. Correction is made by addition or subtraction of weight at suitable points. When the shaft runs steadily, it is said to be in dynamic balance; it is then automatically in static balance also. Dynamic balancing is done with the aid of electronic machines. The engine can then be reassembled. Care should be taken to check and ensure that all moving parts rotate or slide easily and smoothly.

Other measures:
With a highly tuned engine, it is necessary to adapt the clutch to the increased power output. The thrust springs of the pressure plate should be replaced by more powerful ones (Fig. 4). If obtainable, it is advisable to fit a spring plate clutch, which has the advantage over the conventional clutch with coil springs that it develops a higher thrust pressure and embodies less rotating mass (Fig. 5).

(more)

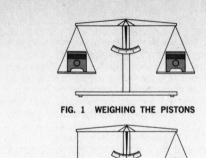

FIG. 1 WEIGHING THE PISTONS

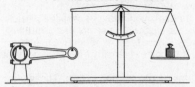

FIG. 2a WEIGHING THE ROTATING PART
OF THE CONNECTING RODS

FIG. 2b WEIGHING THE RECIPROCATING PART
OF THE CONNECTING RODS

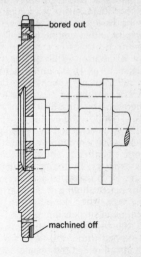

bored out

machined off

FIG. 3 REDUCING THE WEIGHT
OF THE FLYWHEEL

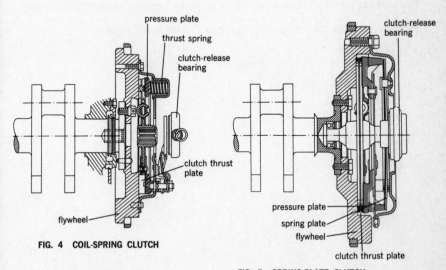

pressure plate

thrust spring

clutch-release
bearing

clutch thrust
plate

flywheel

FIG. 4 COIL-SPRING CLUTCH

clutch-release
bearing

pressure plate

spring plate

flywheel

clutch thrust plate

FIG. 5 SPRING-PLATE CLUTCH

Since the object of tuning is to improve the gas flow in the engine, it results in a higher fuel consumption and correspondingly greater evolution of heat, which has to be dissipated so as to prevent overheating. The existing radiator of a water-cooled rear-mounted engine is liable to be inadequate for the tuned engine. If there is no room to fit a large radiator, the effective cooling surface will have to be increased by installation of an auxiliary radiator, which will have to be mounted under the vehicle (Fig. 1). In some cases it is possible to install a radiator of sufficient capacity at the front of the vehicle instead.

The cooling-air flow rate for an air-cooled engine can, within certain limits, be increased by modification of the V-belt drive. It must be ensured that sufficient air can flow into the engine compartment. With a rear-mounted engine this can nearly always be achieved by opening the cover of the engine compartment (Fig. 1).

In addition, attention must be paid to the cooling of the lubricating oil. If the engine is already equipped with an oil cooler, it will usually be possible to increase its capacity. By fixing cooling fins to the sump and increasing its oil capacity it is likewise possible to achieve a substantial lowering of the oil temperature (Fig. 2). For better cooling of the bearings, the oil flow rate should be increased by increasing the capacity of the oil pump (Fig. 3). Contaminants consisting of abraded particles remain in suspension in the oil and thus enter the engine. To prevent premature wear of the bearings, the oil must be continually filtered. Most modern engines are equipped with a full-flow or a bypass filter. If no filter is already fitted, it is possible to install one in the bypass oil flow (Fig. 4).

Finally, there remains some work to be done on the ignition system. The ignition wire may be of the resistance type having a graphite core, which is liable to fracture easily. It is advisable to fit a copper-core wire instead, in which case it is necessary also to fit a radio-shielded connector for the spark plug. To obtain a stronger ignition spark at higher speeds of rotation, the standard ignition coil should be replaced by a high-output coil. It is also necessary to fit spark plugs with a higher thermal value.

Of course, the brakes, spring suspension, shock absorbers and tires must be properly suited to the enhanced performance of the tuned engine.

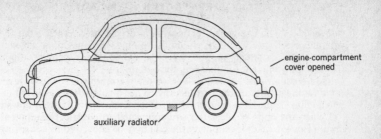

engine-compartment cover opened

auxiliary radiator

FIG. 1

FIG. 2 CAST-ALUMINUM SUMP WITH COOLING FINS

connection for oil-temperature measurement

cover plate

inlet-pipe extension to oil pump

cooling fin

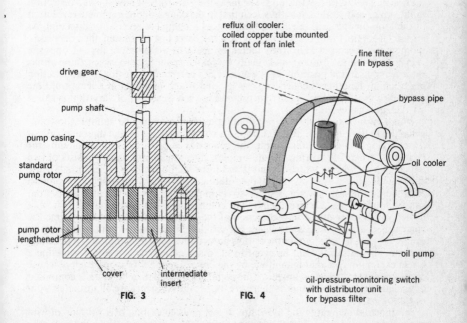

drive gear

pump shaft

pump casing

standard pump rotor

pump rotor lengthened

cover

intermediate insert

FIG. 3

reflux oil cooler: coiled copper tube mounted in front of fan inlet

fine filter in bypass

bypass pipe

oil cooler

oil pump

oil-pressure-monitoring switch with distributor unit for bypass filter

FIG. 4

GENERATOR (DYNAMO)

Principles: The car generator is usually driven by a V-belt from the engine. It generates electricity when the engine is running. Its capacity is sufficient to supply current to all consumer equipment such as the ignition system, headlights, etc., and also to supply current for charging the battery.

The functioning of a dynamo—or, to use the more general term, a generator—is based on the phenomenon of induction: When a coil of wire is moved in a magnetic field so that it intersects magnetic lines of force, an electric current flows through the coil when the ends of the latter are connected to electrical equipment to form a circuit. For the sake of simplicity the coil may be conceived as consisting of a single loop of wire (Fig. 1) which is rotated in a magnetic field between the poles of a permanent magnet. Each of the two ends of the loop is connected to a slip ring. A carbon brush is in contact with each ring, and the two brushes are connected to a sensitive current-measuring instrument. When the loop of wire, which is assumed to be initially in the horizontal position, is rotated clockwise (Fig. 2), the loop will progressively—as its angle of rotation increases—intersect more and more lines of force. As a result, the current generated in the circuit increases in intensity, so that the needle of the measuring instrument shows a correspondingly larger deflection. The current increases until the loop reaches the vertical position—i.e., when its plane is parallel to the magnetic lines of force. With further clockwise rotation of the loop, the current flowing through the circuit decreases until it becomes zero when the rotating loop reaches the horizontal position—i.e., when its plane is perpendicular to the lines of force. According to the so-called "right-hand rule" (Fig. 3), the current flows in the direction indicated by the red arrows in Fig. 1. When the clockwise rotation of the loop is continued through the horizontal position, the current once again increases, but now flows in the opposite direction to that in the first half revolution of the loop. The measuring instrument now shows a deflection in the opposite direction. It is thus seen that in the course of one complete revolution of the loop (or the coil) the current undergoes a complete reversal of direction. As appears from the lower diagram in Fig. 2, it increases to a maximum, then decreases, momentarily becomes zero, then increases again to a maximum (but of opposite sign to the first maximum), and finally decreases again to zero, when the coil has returned to its initial position. This sequence of a current flowing first in one direction and then in the other is called the "cycle" and is accomplished in a length of time and called the "period." A current presenting such a pattern of variation is called an alternating current.

Direct-current generator: An alternating current may not be suitable, however, since the battery needs a direct current—a current flowing in one direction only—for charging. To obtain this result, the generator is provided with a commutator instead of a pair of slip rings. In its simplest form the commutator consists of two halves of a metal ring which are insulated from each other (Fig. 4), each half being connected to one of the two ends of the rotating loop. In the course of one revolution each half of the commutator is successively in contact with the top and the bottom carbon brush (C and D) respectively. The commutator thus functions as a switching device which reverses the flow direction of the current in the circuit every time the current becomes zero. As a result, the current does indeed vary from zero to a maximum and back to zero, but it now always flows in the same direction: it is a pulsating current. The pulsations can be evened out to something more nearly resembling a direct current of constant magnitude by the use of a number of loops, each of whose ends is connected to one segment of the commutator, which in this case is composed of as many pairs of segments as there are loops; the segments are all insulated from one another (Fig. 5).

In an actual generator the loops are in fact coils consisting of a winding of wire around an iron core. The whole rotating assembly comprising the windings, core
(more)

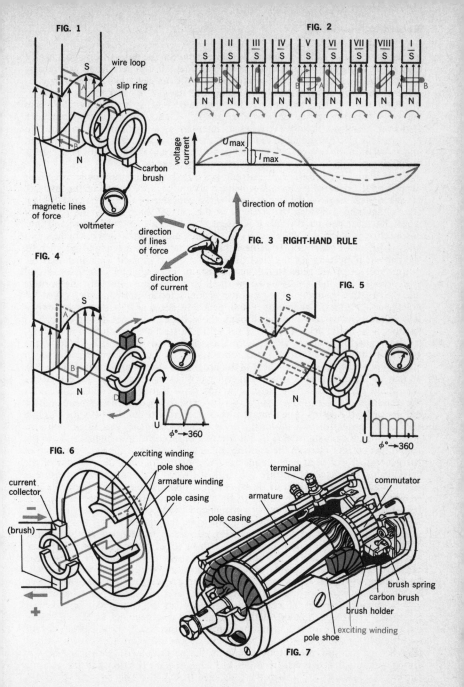

FIG. 1

wire loop
slip ring
S
A
B
N
carbon brush
magnetic lines of force
voltmeter

FIG. 2

I S   II S   III S   IV S   V S   VI S   VII S   VIII S   I S
A   B   N   N   N   B   N   A   N   A   B

voltage current
U_{max} I_{max}

FIG. 3   RIGHT-HAND RULE

direction of motion
direction of lines of force
direction of current

FIG. 4

S
A
B
C
N
D
U
$\phi° \rightarrow 360$

FIG. 5

S
N
U
$\phi° \rightarrow 360$

FIG. 6

exciting winding
pole shoe
armature winding
pole casing
current collector
(brush)
−
+

FIG. 7

terminal
commutator
armature
pole casing
brush spring
carbon brush
brush holder
exciting winding
pole shoe

and commutator is called armature. It rotates between the poles of a magnet, which may be a permanent magnet made of special steel (only in very small generators), but which, for an automobile generator, is normally an electromagnet energized by some of the current generated by the generator itself (Fig. 6). When the armature begins to rotate, a weak current is at first generated because the magnetic field is as yet very weak, being due only to the residual magnetism of the electromagnet. However, this field rapidly increases in intensity as a result of the current that energizes the electromagnet, so that in turn the generated current becomes stronger, and so on until the magnetic field reaches saturation and the generator is operating at full power (Fig. 7). This process is known as self-excitation.

Regulator: The voltage and current generated by an automobile generator depend on the speed of rotation at which the generator is driven and on the intensity of the magnetic field. As the engine speed varies considerably during normal driving, the voltage and therefore the strength of the current supplied by the generator would fluctuate continually. Yet it is essential to have a constant voltage and thus ensure effective charging of the battery at all times, despite variations in engine speed and in load on the generator. To achieve this, a regulating device is employed whereby the current that energizes the electromagnet is varied in such a manner that the output voltage of the generator is kept constant. The regulator comprises several switch contacts which are actuated by small electromagnets. Between the positive and the negative terminal of the generator is connected an electromagnet, called the voltage coil. When the generator is running at low speed and thus generates only a low voltage, the current flowing through the voltage coil is not strong enough to produce a magnetic field of sufficient strength to attract the armature of the spring-loaded regulator contact. The current then flows from the positive pole of the generator through the spring-loaded contact and through the exciting coil—which helps to produce the magnetic field of the generator—to the negative pole (Fig. 8a), so that the field magnet is strongly energized. When, as a result of this excitation, the voltage rises, the stronger current that now flows through the voltage coil causes the armature of the regulator contact to be attracted against the restraining force developed by the spring. The contact is thus opened. The current from the positive pole of the generator now no longer flows only through the exciting coil but also goes through the resistance *B* (Fig. 8b). Because of this resistance interposed into the circuit, the current through the exciting coil is weakened, and the voltage output of the generator decreases in consequence. This in turn reduces the strength of the current that energizes the voltage coil, with the result that the contact armature is released and springs back to its initial position, thus closing the contact again and enabling the current to flow direct to the exciting coil. The sequence of events is repeated, with the regulator contact opening and closing at a rate of between 50 and 200 times per second. With increasing engine speed the contact closing times become progressively shorter, so that the exciting current is correspondingly weakened. When the speed of the engine (and thus the drive speed of the generator) increases, the magnetic field produced by the strong current flowing through the voltage coil causes the armature of the regulator contact to oscillate between the contact positions for high and low speed respectively. The exciting current is then constantly weakened by the resistance *B*. With a further increase in engine speed, the output voltage of the generator and therefore the current flowing through the voltage coil are likewise increased. The magnetic field of the voltage coil now becomes so strong that it draws the armature of the regulator contact to the "high speed" position; the exciting coil of the generator is short-circuited and therefore receives no current (Fig. 8c). As a result, the output voltage of the generator goes down, so that the spring pulls back the armature from the "high speed" position and the exciting current can again flow (through the resistance *B*, Fig. 8b). This sequence of events is repeated in rapid succession.

(more)

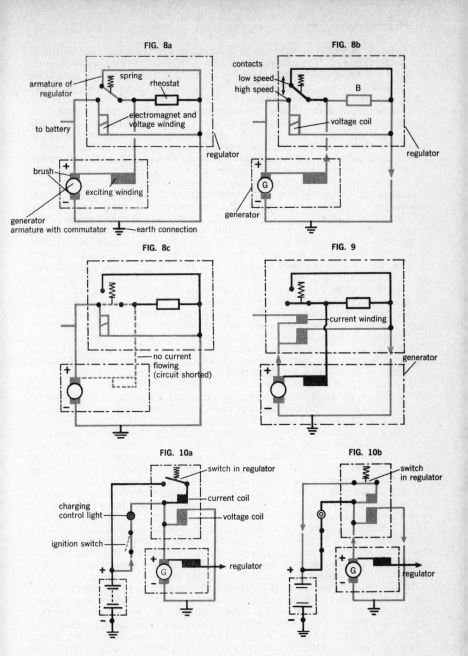

FIG. 8a

armature of regulator — spring — rheostat

to battery — electromagnet and voltage winding

regulator

brush

exciting winding

generator armature with commutator — earth connection

FIG. 8b

contacts

low speed — high speed — B

voltage coil

regulator

generator

FIG. 8c

no current flowing (circuit shorted)

FIG. 9

current winding

generator

FIG. 10a

switch in regulator

current coil

voltage coil

charging control light

ignition switch

regulator

FIG. 10b

switch in regulator

regulator

The power output of the generator is measured in terms of the product of voltage and current strength. The voltage is regulated in the manner described in the foregoing. It is still necessary to limit the maximum current. The iron core of the voltage coil is provided with a second winding (the current coil) through which the whole current output of the generator flows (Fig. 9). The magnetic field produced by this coil intensifies the field produced by the voltage coil. When a high current flows through the current coil, this too brings about a sequence of events, as already described, whereby the energizing current of the exciting coil is weakened. The output voltage of the generator, and therefore also the current it yields, are thus reduced.

When the generator is not rotating, or is being driven only at very low speed, there exists a difference in voltage between the battery and the generator. If the engine is not then switched off, the battery will discharge itself through the generator. To prevent this, the regulator is provided with an additional magnetic switch, whose iron core is likewise provided with a voltage coil and a current coil (Figs. 10a and 10b). The voltage coil is connected to the positive and the negative pole of the generator. It is so adjusted that the armature of the switch contact, attracted by the magnetic field produced by the voltage coil, closes the contact to the battery only when the voltage of the generator is higher than that of the battery. The current flowing to the battery energizes the current coil, whose magnetic field augments that of the voltage coil (Fig. 10b). If the battery voltage is higher than the generator voltage, the current flows in the opposite direction through the current coil; its magnetic field weakens that of the voltage coil, so that the spring-loaded switch contact automatically opens (Fig. 10a). The charging control light on the instrument panel is connected between the positive pole of the battery and that of the generator. If—when the ignition is switched on—the battery voltage is higher than the generator voltage, current will flow through the control light when the switch contact of the regulator is open. The light glows and thus indicates that the ignition has been switched on. When the switch contact of the regulator closes, the control light is short-circuited and is extinguished, thus indicating that the generator is working and the consumer-equipment circuits are connected. Figs. 11a to 11d show the electrical connections and functioning of a widely used type of two-element regulator (contact 1 for low speeds, contact 2 for high speeds).

Three-phase generator: In principle, every generator produces an alternating current. In the case of the direct-current generator this is "rectified" by the commutator, which reverses the direction of the current just as it is about to change its direction naturally, so that a one-way flow of current is obtained. The three-phase alternating-current generator, which is sometimes used in automobiles instead of the direct-current generator, has no commutator. Rectification of the alternating current is, instead, effected by means of semiconductor diodes (see Vol. I, page 80) which prevent the flow of current in one direction and permit it in the other direction. These diodes are accommodated in the generator. The latter differs in construction from the direct-current generator in that the exciting current is here fed—through bushes and slip rings—to the exciting coil which is in the rotor of the machine and establishes a rotating magnetic field (Fig. 12). The magnetic lines of force intersect the fixed coils in the stator, the current thus being generated by induction in these coils. The residual magnetism of the rotor core is not sufficient for self-excitation, and for this reason the magnetic field initially has to be built up by an energizing current supplied by the battery. The three-phase generator has the advantage that it is of lighter and more compact construction than a direct-current generator of comparable output and that it supplies electricity even when the engine is idling.

It is regulated by appropriate control of the exciting current, just as in the case of the direct-current generator.

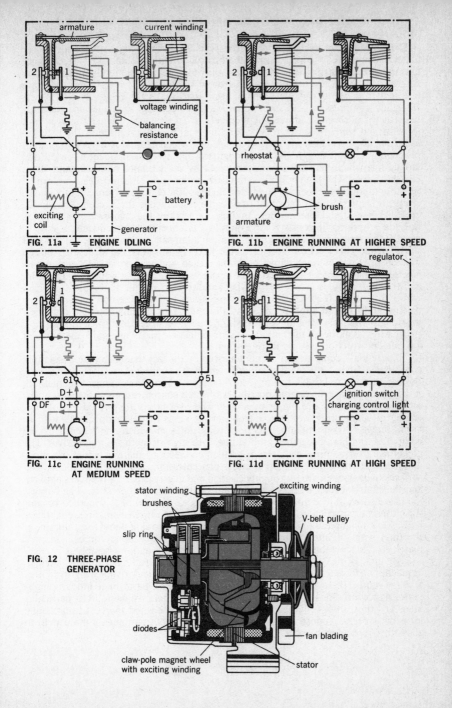

FIG. 11a ENGINE IDLING

armature
current winding
voltage winding
balancing resistance
2 1
exciting coil
generator
battery
– +

FIG. 11b ENGINE RUNNING AT HIGHER SPEED

2 1
rheostat
brush
armature
– +

FIG. 11c ENGINE RUNNING AT MEDIUM SPEED

2 1
F 61 51
D+
DF D+ D–
– +

FIG. 11d ENGINE RUNNING AT HIGH SPEED

regulator
2 1
ignition switch
charging control light
– +

FIG. 12 THREE-PHASE GENERATOR

stator winding
brushes
exciting winding
slip ring
V-belt pulley
diodes
fan blading
claw-pole magnet wheel with exciting winding
stator

PRESENT-DAY METHODS OF AIRCRAFT CONSTRUCTION

The construction methods in current use for aircraft vary according to the materials employed: wood, composite construction, aluminum alloys, steel, glass-fiber-reinforced plastics. The most commonly used materials are aluminum alloys. Aircraft structures are usually constructed in the form of "shells," and the main problem from the designer's point of view is to make the relatively thin sheets withstand compression and shear loads without buckling. Fuselages are generally constructed of sheets reinforced by members called stringers. These structures have to withstand bending, shear, torsion, and internal pressure. Care must be taken to avoid stress concentrations at rivet holes, etc. Metal fatigue is a very important problem in aircraft design and it is recognized that every component has a definite fatigue life, which may be seriously reduced by the presence of cracks and stress concentrations. The fatigue life is estimated on the basis of fatigue tests performed on the components concerned and in test flights of the completed aircraft. Obviously, stringent inspection and detection of possible cracks while the aircraft is in service are further essential precautions. In this connection, the "fail-safe" principle is nowadays applied. It consists, for example, in deliberately introducing a joint to prevent the spread of a fatigue crack. Thus, in a fail-safe wing structure a crack that starts in one panel cannot spread to the next. In this way the structure, though weakened, is not at once dangerously weakened and will be able to perform its function until the crack is detected at the next inspection. Steel, as well as a magnesium and titanium alloy, is used in parts where additional strength or rigidity combined with lightness are required. For supersonic aircraft, which become heated in their passage through the atmosphere, light alloys undergo an unacceptable reduction in strength. Stainless steel must be used for the whole structure of such aircraft.

Broadly speaking, "differential," "semi-integral" and "integral" methods of aircraft construction can be distinguished.

"Differential" construction is characterized by the fact that each major assembly comprises a fairly large number of units which are connected together by riveting, bolting or spot welding, riveting being the most important connecting method. A drawback is the ever-present possibility of stress concentrations at the rivet holes, which are formed by drilling or punching. The rivet heads are formed by cold up-setting or hammering with the aid of various kinds of power-operated tools. In "semi-integral" construction, the numerous units of which the assemblies are composed are joined together by bonding with high-strength resin adhesives. The advantage over riveting is that with bonded connections there is better and more uniform stress distribution, so that stress concentrations are largely obviated. The surfaces to be bonded are thoroughly cleaned and degreased, the adhesive is applied, and the components are held pressed together under uniform pressure. The bonded joints are cured at a temperature of around 150° C. Typical of this method of construction, more particularly for wings, are so-called sandwich structures, characterized by low weight and high strength. They consist of two skin plates (e.g., aluminum sheeting) with an intermediate core of sheet-aluminum "honeycombs," foamed plastics, or balsa wood. In "integral" construction each major assembly is in effect a single unit, fabricated in one piece—e.g., by casting, extruding, stamping, etching, or machining from a solid block. It is employed more particularly for wing structures and tail units, these being the most severely stressed parts of a modern high-performance aircraft. Etching techniques are used to remove metal from particular areas and thus produce "integral" components of the desired shape. A nonmetallic form of "integral" construction consists in the use of components molded from glass-fiber-reinforced plastics.

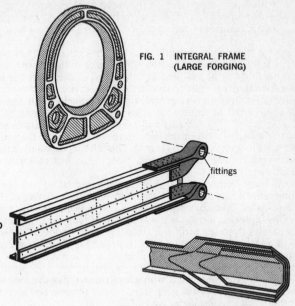

FIG. 1   INTEGRAL FRAME
(LARGE FORGING)

fittings

FIG. 2   SPAR COMPOSED
OF INDIVIDUAL
SECTIONS AND
PLATES

FIG. 3   GLUED NOSE SPAR ROOT

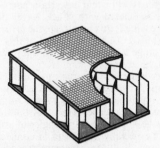

FIG. 4   SANDWICH CONSTRUCTION
COMPRIZING HONEYCOMB CORE
WITH GLUED-ON SKIN PLATES

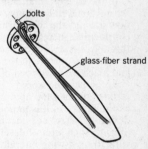

bolts

glass-fiber strand

FIG. 5   COMPONENT MADE FROM
GLASS-FIBER-REINFORCED
PLASTIC

An airfoil (or aerofoil), such as the wing of an aircraft, develops a resultant force acting upwards, i.e., transversely to the direction of flight. How this upward force, the "lift," is developed by the airfoil section is explained in Vol. I, page 554. The forces involved are dependent on the speed of the aircraft (the dynamic pressure increases proportionally to the square of the speed) and more particularly also on the geometric features of the wing—i.e., its cross-sectional shape and its shape in plan.

The term "drag" in a general sense denotes a resistant force acting in a direction opposite to the direction of motion and parallel to the relative airstream. The wing of an aircraft flying at less than half the speed of sound encounters, in addition to surface-friction drag and pressure drag, a resistance called induced drag, which is the part of the drag associated with the development of lift and is proportional to the square of the lift coefficient. This coefficient represents the relative lift of a particular airfoil. The induced resistance can be kept down to a low value by providing a large ratio of wingspan to mean chord—i.e., the ratio of the length of the wing to its average width (Fig. 1). This ratio is especially important in aircraft whose wings are required to have a high lift coefficient and therefore develop a large lift—e.g., gliders.

The "sweep" denotes the slant of the wing in relation to a line perpendicular to the longitudinal axis of the aircraft. It influences the behavior of the flow conditions in the so-called boundary layer immediately adjacent to the wing surface. In the case of a swept-wing aircraft with sweepback of the wing (Fig. 2), the streamline flow first separates from the surface of the wing in the region of the wing tips when the angle of attack is very large (see Vol. I, page 554). Since the ailerons are located in that part of the wing they are liable to become ineffective under such conditions, so that any minor disturbance may cause stalling and uncontrolled rolling motion ("roll-off"). On the other hand, with a forward sweep of the wing (Fig. 3), separation of the flow starts nearer the wing root, so that the ailerons continue to be effective and the aircraft remains under control.

In the case of an aircraft flying at more than half the speed of sound (but still in the subsonic range), supersonic speeds will occur in certain parts of the streamline flow around the wing, which are associated with shock waves that give rise to a considerable increase in drag. If the wing is given a sweep, the critical speed at which this objectionable effect occurs is shifted to a higher value. For this reason all aircraft designed to operate in the speed range between mach 0.5 (half the speed of sound) and mach 1.0 (the speed of sound) have swept wings.

For supersonic speeds, i.e., exceeding mach 1.0, the optimum wing shape in plan is different from the optimum shape for subsonic speeds. The ratio of span to mean chord is now of less importance; on the other hand, a more pronounced sweep is desirable. These considerations led to the development of the delta-wing aircraft (Fig. 4). Such aircraft develop a large lift only at high angles of attack as compared with straight-wing aircraft. This is especially important in connection with takeoff and landing. Landing speeds are very high. At supersonic speeds above mach 1.5 the nonswept short-span wing (Fig. 5) is, in terms of drag, more favorable than any other shape. However, though important, this is not the only consideration that governs the geometry of aircraft wings. In modern military aircraft the "variable-geometry" wing has been introduced in order to obtain relatively favorable conditions for takeoff and landing (long wingspan, position "a" in Fig. 6) and for supersonic flight (position "b"). The changeover is effected by swinging the wings back when the aircraft is in flight ("swing-wing" aircraft).

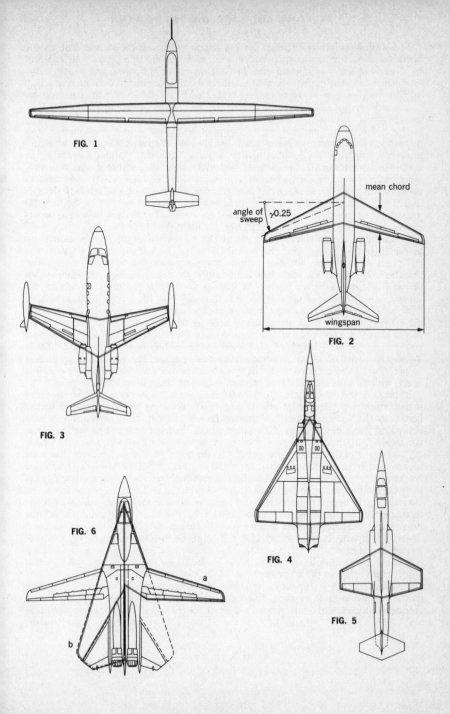

FIG. 1

angle of
sweep γ0.25

mean chord

wingspan

FIG. 2

FIG. 3

FIG. 4

FIG. 5

FIG. 6

a

b

331

The airflow behavior depends on the shape of the airfoil section and on the phenomena occurring at the *boundary layer*, which is the thin layer of air adjacent to a solid surface—such as an airfoil—over which air is flowing and which is distinguished from the main airflow by distinctive flow characteristics of its own set up by friction. The flow in the boundary layer (see Fig. 1, where the thickness of this layer is shown greatly exaggerated) may be *laminar* or *turbulent*. In laminar flow the velocity distribution in the layer shows a steady increase from zero at the surface of the airfoil—more particularly, the wing of an aircraft—to a maximum corresponding to the velocity of the main airflow. The flow is relatively smooth and moves in layers parallel to the surface; hence the term "laminar." In turbulent flow, the fairly regular motion of the laminar boundary layer is destroyed: the boundary layer undergoes transition; it becomes thicker and is characterized by large random motions (turbulence). These effects may give rise to *separation*, a term which denotes that the flow in the boundary layer detaches itself from the surface of the wing (at the separation point) and that immediately adjacent to the surface, flow even occurs in a direction opposite to the direction of the main flow (Fig. 2).

The airflow around the wing starts at the stagnation point and is laminar up to the transition point, where turbulence sets in. The latter point is located near the point of minimum pressure, approximately where the wing has its greatest thickness. Normally the turbulent boundary layer detaches itself (separates) from the trailing edge of the wing, where eddies develop. If this separation occurs too far forward toward the leading edge, there is serious loss of lift and an increase in drag. This is liable to happen when the angle of attack exceeds the critical value called the stalling angle or when the airspeed becomes too low. Some aircraft, especially sports planes, are equipped with a stall-warning device which may consist of a short triangular plate or a length of wire fitted to the leading edge of the wing (Fig. 3). When the angle of attack becomes dangerously large, separation of the airflow commences at this plate or wire. There is an immediate (but not yet dangerous) loss of lift, which warns the pilot that he is approaching the stalling angle.

Regions of the wing where laminar separation is liable to occur may be provided with devices for producing turbulence (Fig. 4). The resulting turbulent flow "adheres" better to the surface than the laminar flow and premature separation is thus prevented. Such devices may, for example, take the form of small projecting plates which break up the laminar flow. "Swept" wings are provided with so-called fences, which are plates or vanes placed parallel to the main airflow and prevent flow (and separation) in the direction from wing root to tip, this subsidiary flow being promoted by the sweep of the wing. A similar effect is obtained by forming the leading edge of the wings with "sawtooth" notches (Fig. 6). At the tail of the aircraft, interference of the boundary layers of the horizontal and the vertical stabilizers produces interference drag. To diminish this, the so-called T tail has been developed, in which the horizontal surfaces are placed at the top of the vertical fin (Fig. 7), while the junction of these components is provided with a fairing—i.e., a streamlined casing designed to reduce drag.

Steering an aircraft in three directions is effected by means of the rudder (which guides the aircraft in the horizontal plane), the elevator (which controls the pitch—i.e., makes the tail go up or down), and the ailerons (which control the rolling motion of the aircraft by their differential rotation). The rudder is attached to a vertical stabilizer, while the ailerons are set at the trailing edges of the wings (Fig. 5).

(more)

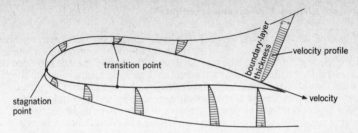

FIG. 1

boundary-layer thickness

velocity profile

transition point

velocity

stagnation point

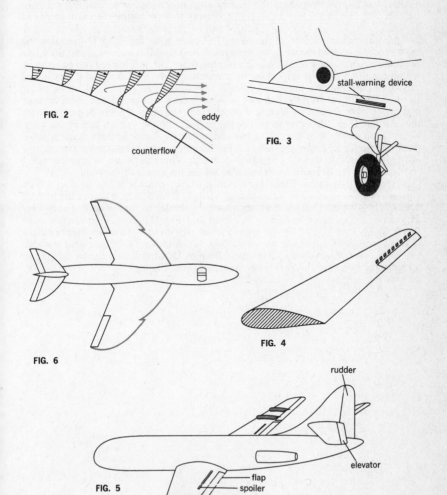

FIG. 2

eddy

counterflow

FIG. 3

stall-warning device

FIG. 6

FIG. 4

FIG. 5

rudder

elevator

flap

spoiler

fence

aileron

Sometimes the horizontal and vertical stabilizers are not provided with separately movable attachments (elevator and rudder), but can each be moved as a whole so as to alter the angle of attack. The trailing edge of the rudder may be provided with a small subsidiary rudder called a trimming tab (Fig. 7) by means of which the pilot adjusts the trim of the aircraft—i.e., the condition of static balance in pitch during rectilinear flight, with the main control surfaces seeking their neutral positions.

Further adjustments are achieved by means of flaps, these being control surfaces which serve to control the speed by increasing the drag and thus acting as a brake or to increase the lift or aid in recovery from a dive. The slat (Fig. 8) is a movable auxiliary airfoil running along the leading edge of a wing; in normal flight it is in contact with the latter, but it can be lifted away to form a slot at certain angles of attack, so that air flows through the slot and reenergizes the boundary layer on the low-pressure upper surface of the wing. The plain wing flap (Fig. 9) increases the camber (curvature) of the wing, with the result that the lift is improved and the angle of attack at which separation occurs is increased, so that the airspeed can be reduced without stalling. This is important in connection with the takeoff and landing of high-speed aircraft. An improved form is the slotted flap (Fig. 8); the flow of air through the slot between the flap and the wing gives a further increase in lift without separation of the boundary layer. In contrast with the other types of flap mentioned, the split flap (Fig. 10) serves to reduce the pressure on the suction face of the wing, whereby an increase in the lift is likewise achieved. Landing flaps (Fig. 9) serve primarily to slow down the aircraft for landing; they break down the airflow around the aircraft and thus function as brakes. Such flaps are sometimes called spoilers, more particularly when installed on the underside of the wing.

A special "dynamic" device for boundary-layer control is the jet flap which consists of a flat jet of air expelled at high velocity from a narrow slot at the trailing edge of the wing and which exercises an action similar to that of an ordinary flap. The same principle is embodied in the blown flap (Fig. 11), an ordinary trailing-edge flap in which separation from the upper surface is delayed by blowing. This principle is also successfully applied to elevators (Fig. 12). Another modern control method, still in the experimental stage, consists in keeping the airflow laminar by sucking in air from the boundary layer through numerous small holes.

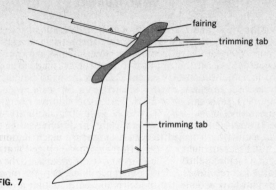

fairing

trimming tab

trimming tab

FIG. 7

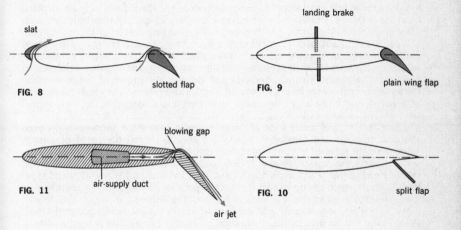

slat

slotted flap

FIG. 8

landing brake

plain wing flap

FIG. 9

blowing gap

air-supply duct

air jet

FIG. 11

split flap

FIG. 10

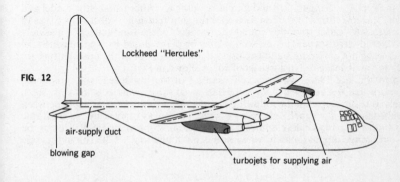

Lockheed "Hercules"

FIG. 12

air-supply duct

blowing gap

turbojets for supplying air

Many of the forces and moments to which an aircraft is subjected by the airflow cannot be accurately determined by purely theoretical calculations. The aircraft designer must therefore have recourse to experimental aerodynamics, which from the earliest days has contributed much to the progress made in aeronautical science. One of the most important experimental aids is the wind tunnel, a device whereby the reactions of a carefully controlled airstream on scale models of airplanes or their component parts can be studied. The first condition that a model for testing in the wind tunnel must satisfy is that of geometric similarity with the full-scale prototype. In addition, certain other important conditions relating to flow conditions and velocity must be satisfied (Reynolds number; mach number; see page 330) to enable valid measurements to be performed on the model. The Reynolds number is a correction factor applied to the analysis of the flow around the model; it corrects for the scale effect resulting from the difference in size between model and prototype. When the fluid flow around the model is the same as that around the prototype, there is said to be dynamic similarity. For complete similarity between the full-scale airplane and a model that is, say, one-tenth its linear size, the air velocity in the wind tunnel would have to be ten times as high as the speed for which the airplane is to be designed. For high-speed aircraft this would require impracticably high wind velocities in the tunnel and impracticably strong models to withstand the high pressures associated with such velocities. For these reasons, the tests are usually made on models at Reynolds numbers well below those for the full-scale conditions; in the interpretation of the results, due allowance is made for this difference in dynamic conditions. Various methods and devices are employed for performing the measurements of the forces, moments, torques and pressures to which the models, attached to special balances or rigidly supported, are subjected in the wind tunnel (Fig. 1). The airflow pattern can be made visible by a number of methods.

There are several categories of wind tunnels: low-speed tunnels, high-speed subsonic tunnels, and transonic tunnels (for speeds around the speed of sound), supersonic tunnels (up to about mach 5—i.e., five times the speed of sound), hypersonic tunnels (for higher speeds). A distinction is further to be made between the open-circuit tunnel (Eiffel type, Fig. 2) and the closed-circuit tunnel (Prandtl type, Fig. 3). In the open-circuit tunnel the duct is large at its entrance, but reduces to a much smaller area at the test section (or working section), in which the desired wind velocities are produced and the measurements on the model are performed. Beyond the test section, the duct expands gradually. The airflow is produced by a powerful fan. The closed-circuit tunnel enables much better control to be maintained over the pressure, temperature and humidity of the air. After passing the model and the fan, the air returns to the test section through a return-flow duct with banks of guide vanes to lead it smoothly round the bends. Modern wind tunnels are usually of the closed-circuit type. The test section may be of the closed-jet type (in which the tunnel walls are continuous and enclose the test section) or of the open-jet type (in which the model is tested in a free jet of air, as shown in Fig. 3).

Up to the late 1920s, wind tunnels were all of the low-speed type, producing maximum air speeds of about 120 mph. High-speed subsonic tunnels and supersonic tunnels were developed in the following decade. For a time there was a gap between the subsonic and the supersonic speed ranges, which was bridged by the transonic wind tunnel, a postwar development, enabling tests to be made right through the transonic range (approximately between mach 0.8 and mach 1.2). The hypersonic

(more)

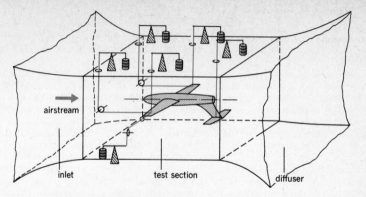

FIG. 1   MODEL SUSPENDED IN WIND TUNNEL

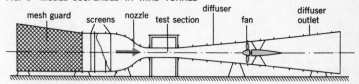

FIG. 2   OPEN-CIRCUIT SUBSONIC WIND TUNNEL
         (EIFFEL TYPE)

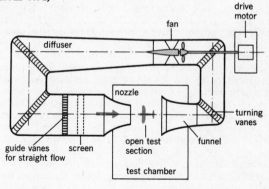

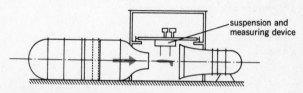

FIG. 3   CLOSED-CIRCUIT SUBSONIC WIND TUNNEL
         (PRANDTL TYPE)

337

wind tunnel, the most recent development, is used for studying the conditions associated with the launching and flight of rocket-propelled missiles and earth satellites.

In the subsonic wind tunnel, as described in the foregoing, the test section is located at the narrowest part of the duct, where the highest speeds—below the speed of sound—are produced. In the supersonic wind tunnel (Fig. 4), the test section is preceded by a constriction, a so-called convergent-divergent nozzle, in which the very high speeds are attained. Each different supersonic speed requires the use of a differently shaped nozzle; in some tunnels the nozzle has a flexible wall so that it can be varied in shape by hydraulic adjusting equipment instead of having to be exchanged for another. Beyond the test section is a second constriction, in which the ultrasonic speed diminishes to subsonic values. The wind is produced by a multi-stage axial-flow compressor or by the high-speed jet from a set of gas turbines. The friction of the wind against the tunnel walls generates heat, which is removed by a cooler incorporated into the circuit, so as to maintain a constant temperature in the test section. The power requirement to maintain a continuous flow of air at supersonic speeds is very high. For very high speeds this becomes a very un-economical method of operation, and to overcome this problem intermittently operated wind tunnels have been developed. Power is stored in the form of com-pressed air or vacuum, the wind being produced in short blasts, whereby a con-siderable saving in power input for operating the tunnel is effected. Broadly speaking, there are two types of intermittent wind tunnel. In one type the measurements are performed during the time when a valve between the test section and the pressure storage vessel (e.g., vacuum vessel, Fig. 5) is open. This vessel is connected to the wind tunnel through a quick-closing valve; the actual tunnel comprises a convergent-divergent nozzle, the test section, and a second constriction (the diffuser). Before the test commences, the vacuum vessel is evacuated; when the valve is opened, air rushes into the vessel so that a supersonic speed is attained in the test section, depending on the shape of the nozzle and the degree of vacuum in the vessel. So long as this vacuum is sufficient to maintain sonic speed in the throat of the nozzle, the supersonic speed in the test section remains constant. An air drier is installed at the intake to intercept any moisture that might condense into droplets in the test section, where they could disturb the flow conditions. The second type of intermittent wind tunnel (Fig. 6) is a tube along which gas is driven by various means for a very short time (a fraction of a second) during which the force acting on the model is measured by special techniques. The tube, which is of constant cross section and closed at both ends, is divided into a high-pressure and a low-pressure part by a gastight diaphragm. The test section, located behind a convergent-divergent nozzle, is in the high-pressure part. Before the test is started, the appro-priate pressures are produced in the two parts of the tube by pumping in and pumping out air respectively. When the diaphragm is ruptured, a constant airflow speed will very briefly exist in the test section.

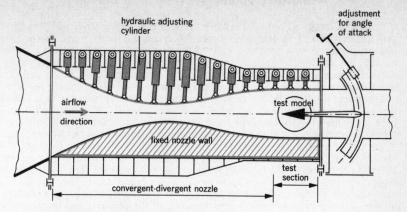

FIG. 4   CONVERGENT-DIVERGENT NOZZLE AND TEST-MODEL
SUSPENSION IN SUPERSONIC WIND TUNNEL

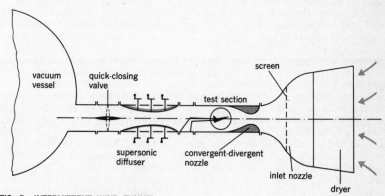

FIG. 5   INTERMITTENT WIND TUNNEL

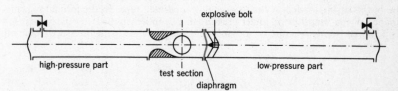

FIG. 6   INTERMITTENT WIND TUNNEL, SHOCK-WAVE TYPE

In most aircraft, actuation of control surfaces (rudder, elevator, ailerons, flaps) and retractable landing gear is performed by hydraulic power transmission. One important reason why hydraulic systems have come to be widely used in aeronautical engineering is that they can transmit and exert large forces without taking up much space. The hydraulic working fluid is high-pressure oil, which also acts as a lubricant for the moving parts of the system.

The *pumps* used in the hydraulic system are of the displacement type. These pumps deliver a pulsating output which is, however, quite satisfactory for the purpose. They are of the rotary type (vane pumps; gear pumps) or the reciprocating type (plunger pumps); see Fig. 1. The plunger pump is the type most widely employed in aircraft because it can produce the highest pressures, mainly, up to about 350 kg/cm^2, which is about twice as high as the maximum attainable by gear pumps. Plunger pumps are of the radial-cylinder or the axial-cylinder type and comprise more than one cylinder, in order to equalize the pulsations in the output. The axial-cylinder pump is driven by a swash plate which drives the plungers by performing a wobbling rotary motion. The stroke of the plungers and therefore the output flow rate can be varied by tilting the swash plate (variable-stroke pump).

The hydraulic power supplied by the pumps is converted back into rotary motion by *hydraulic motors*. They are similar to pumps in construction in that they may be of the gear, vane or plunger type. In the last-mentioned type the plungers drive the swash plates, to which the power output is connected. *Hydraulic cylinders* convert hydraulic power into linear reciprocating motion by the admission of oil to one or the other side of the piston (Fig. 2). *Valves* serve to control the direction and rate of flow of the hydraulic oil, and to regulate the pressure. A relief valve bypasses the oil flow back to the tank as soon as the oil pressure exceeds a certain preset value. When it falls below this value, the valve automatically closes again. A pressure-reducing valve reduces the oil pressure of the system to a lower value which may be required for operating a second hydraulic system. Control valves are of various kinds, more particularly rotary valves and piston valves. Fig. 3 shows a section through a four-way/three-way piston valve: depending on the position to which the piston is moved, certain ports in the cylinder are opened and closed, causing the oil to flow as selected. Such selector valves may be operated manually, mechanically, electrically or hydraulically. Nonreturn valves (Fig. 4) cut off the oil flow in one direction, but allow it to pass in the other direction. Throttle valves are used to establish a particular oil-flow rate, which is dependent on the pressure, however. A flow-control valve, on the other hand, maintains the flow constant, regardless of the pressure. It comprises an adjustable throttle and a differential-pressure piston installed before or after it. Hydraulic accumulators in which oil is stored under constant pressure are employed in cases where oil-flow rates exceeding the delivery rate of the pump are required for short periods—e.g., when a number of controls in the aircraft have to be simultaneously actuated. Accumulators may be of the spring-loaded or the gas-pressure-operated type. In the former a spring-loaded piston keeps the oil under pressure in a cylinder; in the latter the same result is achieved by gas pressure on a diaphragm which transmits the pressure to the oil.

(more)

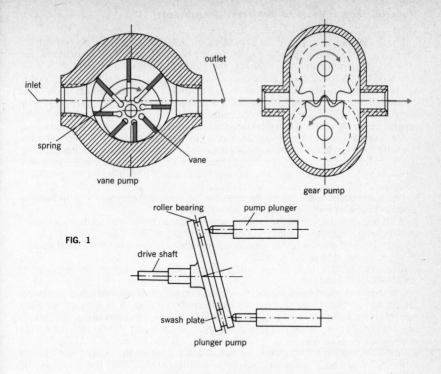

FIG. 1

roller bearing   pump plunger

drive shaft

swash plate

plunger pump

inlet

spring

vane pump   vane

outlet

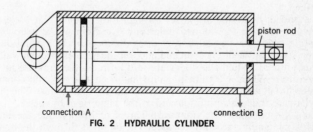

connection A   connection B

FIG. 2 HYDRAULIC CYLINDER

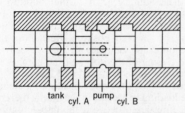

tank cyl. A pump cyl. B

FIG. 3 FOUR-WAY/THREE-WAY PISTON VALVE

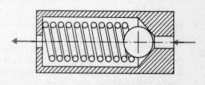

FIG. 4 NONRETURN VALVE

The hydraulic installation in a modern aircraft comprises the equipment for supplying the oil under pressure (pumps), the pipeline system for distributing the oil, and the equipment for utilizing the hydraulic power (cylinders, hydraulic motors). In addition, there are devices for controlling the airflow and pressure (valves of various kinds); a monitoring system may also be incorporated. The control surfaces of airliners and military aircraft are actuated by means of two or three systems which are entirely independent of one another and are also quite separate from the systems for operating the landing gear, brakes, etc. This separation of the systems is a safeguard to obtain maximum functional reliability. The hydraulic working fluid is usually a mineral oil. However, for systems where fire hazard is an important consideration, a solution of glycol in water may be used, while synthetic fluids or silicic acid esters are used for systems in which high working temperatures occur.

The accompanying diagram represents the hydraulic system operating the landing gear and flaps of a six-seater passenger aircraft.

1. Landing gear. Lowering the landing gear is initiated by means of a hand-operated control lever, whereby the following operations are set in motion:

(a) the "landing gear" selector valve is mechanically shifted from the locked neutral position to the "lowering" position;

(b) the loading valve cuts in the pump (closes the bypass), so that the pressure in the system builds up;

(c) by means of a limit switch, the energizing current of an electromagnet actuating the "doors" selector valve is interrupted; the piston moves to its spring-controlled central position, in which the oil passages for opening the doors are connected to one another. While the aircraft is in flight, the electromagnet is constantly energized in order to keep the piston in its end position, against the action of the centering springs. The reason for this arrangement is that it enables the doors to be opened in the event of a fault in the electric-power supply, because when that happens, the piston of the selector valve automatically moves to the central position, as already described.

The flow of hydraulic oil passes through the "landing gear" selector valve and nonreturn valve to the "doors" selector valve, from where it flows to the door-actuating cylinders. The oil first releases the mechanical locking devices of these cylinders, so that the doors open and uncover the hatches through which the landing gear is lowered. When the doors are fully opened, the pressure in the hydraulic system rises. When a certain pressure has been reached, the connector valve 3 allows oil to flow to the cylinders for releasing the landing gear. The pistons of these cylinders are thus set in motion and at the same time, by means of a mechanical tripping device, cause the nonreturn valves to open, while the locking devices securing the landing gear are released. The oil flows through the nonreturn valves into the actuating cylinders which lower the landing gear. A further buildup in pressure now occurs; the "landing gear" selector valve moves to its central position and thereby relieves the pressure in the system. The landing gear has meanwhile automatically been locked mechanically in the lowered position. Retracting the landing gear after takeoff proceeds in the opposite sequence to that described, except that now a time-lag valve delays the closing of the doors of the hatches to ensure that the landing gear has had ample time to be retracted and locked.

2. Flaps. The "flaps" selector valve is moved to the "lowering" position by actuation of a hand-operated lever. Now again the loading valve cuts in the pump, thereby causing pressure to build up in the system. Interposed between the flap-actuating cylinders and "flaps" selector valves is a throttle which prevents too-rapid lowering of the flaps after the locking devices have been released.

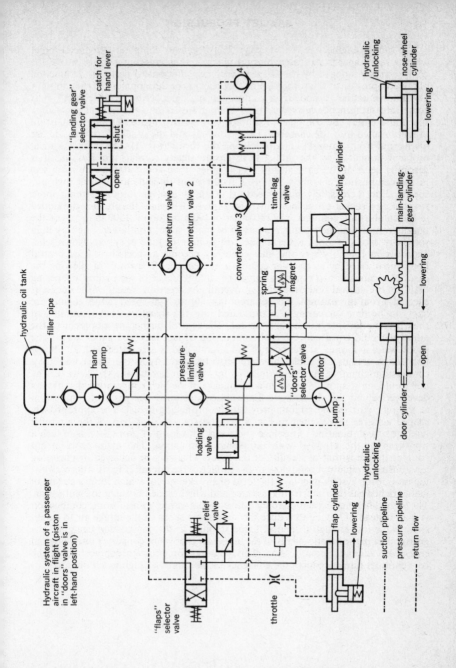

Hydraulic system of a passenger aircraft in flight (piston in "doors" valve is in left-hand position)

hydraulic oil tank

filler pipe

hand pump

pressure-limiting valve

nonreturn valve 1

nonreturn valve 2

converter valve 3

"landing gear" selector valve

catch for hand lever

shut

open

4

time-lag valve

locking cylinder

hydraulic unlocking

nose-wheel cylinder

lowering

main-landing-gear cylinder

lowering

spring

magnet

"doors" selector valve

motor

pump

loading valve

door cylinder

open

lowering

"flaps" selector valve

relief valve

throttle

flap cylinder

hydraulic unlocking

lowering

suction pipeline

pressure pipeline

return flow

343

The ramjet engine (see Vol. I, page 562) began to compete with the turbojet when aircraft speeds increased beyond mach 2, thus entering a range where the compression produced by the airspeed becomes sufficient to perform the function of the compressor of the turbojet. For this reason the development of the ramjet—also known as the "athodyd," a contraction of "aerothermodynamic duct"—has come into prominence in recent years. The air rushes into the inlet at supersonic speed and enters the combustion chamber, where it is heated by the combustion of fuel injected into the chamber. The heated air and the gases of combustion are discharged from a nozzle, thus producing the thrust (Fig. 1). The main technical problem presented by the ramjet is to ensure steady combustion. For this it is generally necessary to have airflow speeds of less than about 100 m/sec (330 ft/sec) in the combustion chamber. This is a requirement difficult to fulfill at high airspeeds. For increasingly high speeds the ramjet evolves into something more resembling a rocket-propulsion unit (Fig. 2). In such engines the pressure developed in the combustion chamber is of the order of 100 atm. (about 1500 lb/in.^{2}), and the nozzle from which the jet emerges has to be made larger and larger. For very high speeds, in excess of mach 6, the engine evolves into the kind of system shown schematically in Fig. 3, where the inlet cone has become a specially shaped central body surrounded by an annular combustion chamber. As a result of allowing the gases of combustion to expand around the circumference of the conically tapering "tail" of the central body, a saving in overall construction weight of the engine is effected. From this example it is apparent how future high-speed ramjet engines are likely to become increasingly incorporated into the structure of the aircraft and thus become an integral feature thereof. The logical further development of the athodyd would consist in external combustion of fuel behind a shock wave. The shock wave is formed at the nose of the aircraft and is associated with an abrupt increase in pressure. It could therefore serve theoretically as the "front wall" of a combustion chamber, fuel being injected into the air behind the shock wave. The fuel would ignite spontaneously in consequence of the high temperature that always develops behind the shock wave. External expansion of the gases of combustion at the rear part of the aircraft provides the propelling thrust. The appropriately shaped surfaces represented in Fig. 4 may be conceived as part of the aircraft's fuselage or combination of fuselage and wing. This form of propulsion is in turn a transition to the athodyd with ultrasonic combustion. The main problem encountered here is that of stability of the flame. This may be achieved by enclosing it within a recirculation zone close to the surface of the aircraft (Fig. 5). Alternatively, the propulsion system may take the form of a rocket motor which emits a stream of fuel-enriched gas into which air is injected and which is then brought to combustion. The main difference in relation to the conventional ramjet with subsonic combustion is that, instead of having to reduce the supersonic speed of the intake air to a subsonic value low enough to permit flame stability in the combustion chamber, the greater part of the kinetic energy of the intake air is now not converted into potential energy by adiabatic compression. This compression prior to combustion in the conventional ramjet reduces the efficiency of the ramjet at high mach numbers.

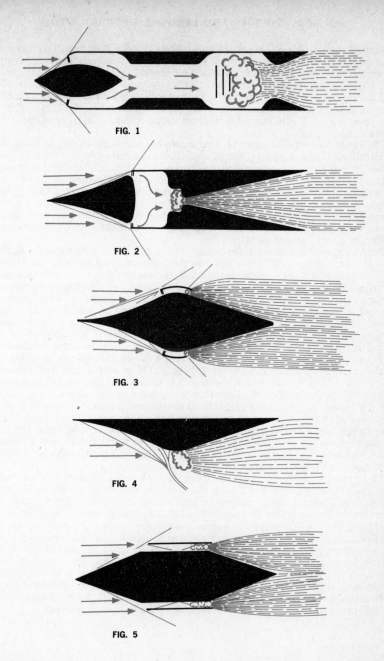

FIG. 1

FIG. 2

FIG. 3

FIG. 4

FIG. 5

VERTICAL-TAKEOFF-AND-LANDING AIRCRAFT (VTOL)

Aircraft capable of vertical takeoff and landing (usually abbreviated as VTOL) have important military and civil potentialities and are under active development by aircraft firms in a number of countries. One general advantage claimed for these aircraft is their greater safety in that the hazards associated with conventional takeoff and landing are eliminated. A further advantage is that they do not require airfields with long and expensive runways (see Fig. 1), so that, despite the higher direct operating costs of VTOL aircraft, an overall saving is effected in comparison with conventional aircraft.

Of course, the helicopter (see Vol. I, page 560) has essentially solved the problem of vertical takeoff and landing, but its applications are limited by its relatively low speed. In recent years a large number of new VTOL aircraft types have emerged. Vertical takeoff without the use of large rotors became practicable with the advent of the gas-turbine engine because it could generate much more thrust for a given weight than the piston engine. Many new problems have had to be solved: e.g., in connection with stability and control during hovering and the transition from vertical to forward flight, and vice versa, since conventional control surfaces are ineffective at low forward speeds.

Depending on the position of the aircraft during the takeoff, a distinction is made between "tail sitters" and "flat risers." In the first-mentioned category the whole aircraft rests with its tail on the ground and its nose pointing vertically upwards. After takeoff, it is gradually brought into the normal flying position by operation of the controls. The "flat riser" takes off in the normal position, i.e., with the fuselage parallel to the ground. In this last-mentioned category of VTOL aircraft, the propulsion engines may be swiveled from the vertical position for takeoff and landing to horizontal for forward propulsion. This principle is illustrated for a propeller engine in Fig. 2. With turbojet propulsion, the propulsion engines can be used for takeoff and landing by suitably directing the jets downwards. In addition to the propeller VTOL aircraft and the turbojet VTOL aircraft, a third type is based on the ducted fan, this being a propeller or fan within a duct or shroud, which in some types can be tilted in the same manner as the propeller engine in Fig. 2. In Fig. 3 a ducted-fan propulsion unit of the dual-propulsion type is illustrated. It is a combination of a ducted fan and a jet engine. Each of the two wings of the aircraft may be provided with such a fan, "buried" in the thickness of the wing. The jet engine provides the propulsion in the normal way when the aircraft is in forward flight. For takeoff and landing, the jet exhaust is deflected to drive the fan, which thus develops a powerful vertical thrust.

The present trend of development is toward the direct utilization of the thrust developed by turbojet (see Vol. I, page 564). In a case where separate lift engines are provided in addition to the propulsion engine there is of course the problem of extra weight due to having two sets of engines, only one of which is in use at any particular time. In this respect the arrangement where only one set of engines is provided, which can be swiveled from vertical to horizontal, and vice versa, or where the jets themselves can be deflected to produce a thrust in the desired direction (Fig. 4) is advantageous. This is especially true in high-speed fighter aircraft, whose engines produce a large thrust which can be utilized for vertical takeoff. On the other hand, separate lift engines (as in Fig. 5) or a combination of swiveling jet engines and a set of auxiliary lift engines (as in Fig. 6) may be more advantageous for other types of aircraft, such as civil aircraft, with lower cruising speeds.

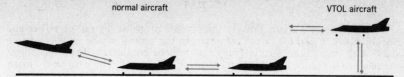

normal aircraft　　　　　　　　　　　VTOL aircraft

FIG. 1

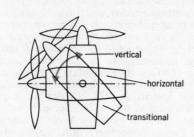

vertical

horizontal

transitional

FIG. 2　SWIVELING ENGINE (PRINCIPLE)

FIG. 3

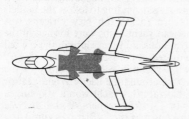

FIG. 4a

FIG. 4b

FIG. 5

lift engines

engine for
forward propulsion

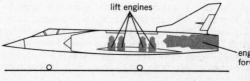

FIG. 6

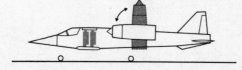

"Yacht" denotes a comparatively small vessel propelled by sail or power and used for pleasure and for racing purposes. In the present article only sail-driven boats—with or without auxiliary engines—will be considered. Thus "yachting" here refers to the sport of cruising and racing in sailing craft, which may be of many and widely varying types, ranging in size from small boats to large oceangoing yachts.

The term "yacht" is of Dutch origin—"jacht," meaning "hunt" or "chase." From the 14th century onward the seafaring people of Holland had built small, highly maneuverable craft for hunting down pirates and smugglers. At a later period these boats came to be used for pleasure cruising or racing, and it is in this context that the term is now used, to the exclusion of all craft primarily serving commercial or other purposes (fishing boats, cargo ships, etc.). In 1660 the King of England, Charles II, was presented with a Dutch yacht. This marked the beginning of yacht building in England. Tradition has it that the first regatta—a race between small sailing craft—took place in 1662. The year 1720 saw the foundation of the world's first yacht club, the Cork Harbour Water Club, in Ireland. The second yacht club was an English one: the Cumberland Fleet, founded in 1775, of which the Royal Thames Yacht Club is the descendant. Yachting in the modern sense, however, started with the establishment of the Yacht Club (now the Royal Yacht Squadron) at Cowes, Isle of Wight, in 1812. The Yacht Racing Association (Y.R.A.) was founded in 1875 with the object of establishing rules for regatta racing.

American yachting can be said to have started in the early part of the 18th century, but it was not until early in the next century that sailing for pleasure came to be more widely practiced. The New York Yacht Club (N.Y.Y.C.) was founded in 1844. The yachts of that period were usually schooner-rigged, tending to follow the models developed for commercial purposes. The Southern Yacht Club, New Orleans, was founded in 1859, and a number of other American yacht clubs were formed in the next two decades. At present there are close to a thousand yacht clubs in the United States, a high proportion of which are concentrated in the New York area. A very important event in the history of international yachting was the victory of the 100-ft. 170-ton schooner *America* in a 53-mile race around the Isle of Wight against a large number of British yachts in 1851. In that race this yacht won a 100-guinea cup offered by the Royal Yacht Squadron. In 1857 the owners of the *America* presented the cup to the N.Y.Y.C. (henceforth known as the "America's cup") as a perpetual challenge trophy in a race open to competitors from all countries. These contests have had considerable influence on the evolution of modern yacht design.

Figs. 1 and 3 show two basic types of small sailing craft. The small boat illustrated in Fig. 1 has a drop keel which can be raised and lowered. When the boat is sailing before the wind, the keel is raised to reduce resistance and thus increase speed. With side wind, however, the keel is lowered to counteract lateral drift and assist in obtaining a forward-propelling component of the wind force on the sails (Fig. 2; see also Vol. I, page 544). Most larger yachts have fixed keels, lead-ballasted to reduce the risk of capsizing (Fig. 3). However, the drop keel is not confined entirely to very small craft; it is used in certain boats up to about 30 ft. in length because of the more versatile performance it provides.

(more)

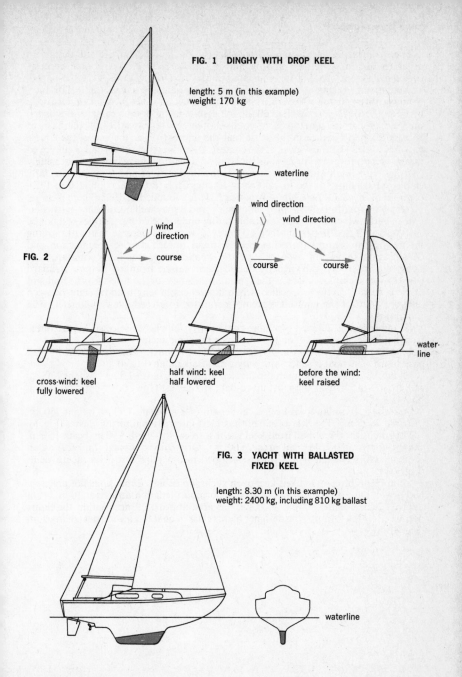

FIG. 1 DINGHY WITH DROP KEEL

length: 5 m (in this example)
weight: 170 kg

waterline

wind
direction

course

FIG. 2

wind direction

wind direction

course

course

water-
line

cross-wind: keel
fully lowered

half wind: keel
half lowered

before the wind:
keel raised

**FIG. 3 YACHT WITH BALLASTED
FIXED KEEL**

length: 8.30 m (in this example)
weight: 2400 kg, including 810 kg ballast

waterline

The early racing yachts differed widely from one another in size and design. In order to obtain something approaching a fair basis of comparison in contests between different types, measurement formulas and rating rules were developed. The competing yachts were measured and appropriate allowances (time allowance, handicap) were made to compensate for the inherent differences in performance between the various types. Essentially the results of races thus came to be judged on the basis of the time taken to complete the course instead of on the basis of "absolute" performance in the sense that the first arrival at the end of the course must necessarily be the winner of the contest. Over the years the formulas employed in various countries and by various yachting organizations have become increasingly complex. For international ocean racing, the British rules and formulas generally employed are those of the Royal Ocean Racing Club (R.O.R.C.), founded in 1925, or those of the Cruising Club of America (C.C.A.), founded in 1922. There is as yet no single internationally adopted set of rules, and it may well occur that a particular type of yacht which receives a favorable rating under one country's rules will not be so favorably rated under another's. However, with the increase in regatta sailing in the present century there has been a trend toward precise specification and classification of yachts, especially in the smaller sizes, so that contests on an "absolute" basis between virtually identical craft can be arranged. In these standard classes even such particulars as the type and thickness of the wood for the hull and other parts are carefully specified, while differences in length and weight between different boats of the same class are not permitted to exceed a few millimeters or a few grams.

Of international interest are more particularly the classes selected by the International Yacht Racing Union (I.Y.R.U.) for Olympic racing:

Finn Dinghy: one-man dinghy, length 4.50 m, beam 1.51 m, weight 105 kg, sail area 10 m^2. The boat is sailed with only a mainsail, no jib (Fig. 4).

Flying Dutchman: two-man dinghy, length 6.05 m, beam 1.80 m, weight 170 kg, sail area 15 m^2. This is the most up-to-date and fastest type of conventional sailing boat (Fig. 5).

Star class: two-man fixed-keel yacht, length 6.92 m, beam 1.73 m, weight 750 kg, sail area 26.13 m^2. The Star is one of the oldest international racing classes (Fig. 6).

Dragon class: three-man fixed-keel yacht with cabin, length 8.90 m, beam 1.90 m, weight approx. 2000 kg, sail area 26.60 m^2. Originally designed (in 1929) as an ordinary cruiser-racer, it has since developed into a pure first-class racing yacht (Fig. 7).

5.5 m class: three-man fixed-keel open yacht (no cabin); dimensions not precisely laid down, but must conform to a specific formula; length a little over 10 m, beam approx. 2 m, weight approx. 2000 kg, sail area approx. 29 m^2. Within the limits permitted by the formula, the designer of this class of boat has a good deal of freedom (Fig. 8).

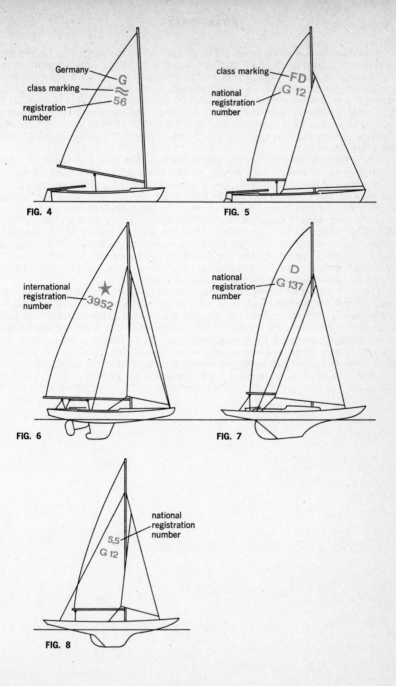

Germany

class marking

registration number

G

≈

56

FIG. 4

class marking

national registration number

FD

G 12

FIG. 5

international registration number

★

3952

FIG. 6

national registration number

D

G 137

FIG. 7

national registration number

5,5

G 12

FIG. 8

A submarine is a naval vessel that can operate on the surface or underwater. A distinction must accordingly be made between the vessel's displacement on the surface and its greater displacement when submerged. For example, these displacements are 1300 tons and 1575 tons, respectively, for the British "T" class submarine, which is a typical conventional submarine developed in World War II. It is 273 ft. in length. Modern submarines range in size from relatively small vessels intended for use in coastal waters (e.g., about 140 ft. long, with 359/430 tons displacement) to large oceangoing long-distance vessels (e.g., about 420 ft. long, with 7900/9000 tons displacement).

When floating on the surface, the submarine displaces a tonnage of water equivalent to the vessel's own weight. In the submerged condition the submarine is in a state of neutral buoyancy, i.e., it can then, with only minor changes in the quantity of water ballast, be made to float at any desired level. Continual adjustment is necessary to maintain this condition, as the submarine uses up fuel when cruising underwater. Consumption of food and water by the crew, and the firing of torpedoes, must also be compensated. Since the center of gravity of the submarine never exactly coincides with the center of buoyancy, motionless "hovering" is hardly possible in actual practice : the submarine must always have a certain amount of forward motion to enable it to maintain its depth with the aid of the hydroplanes, which moreover keep the vessel at the desired angle.

The main part of the submarine is the pressure hull, which usually has a circular shape in cross section (see Figs. 1a to 1e). This cylindrical shape is most suitable for withstanding the water pressure to which it is subjected at great depths. The hulls of modern submarines are always of welded steel construction and are divided into compartments separated by strong bulkheads. Approximately amidships is the superstructure, comprising the conning tower and bridge. The tower is provided with upper and lower hatches giving access to the interior of the submarine.

Under the tower is the control room. Enclosing the pressure hull is a casing; water has free access (through suitable apertures) to the space between this casing

(more)

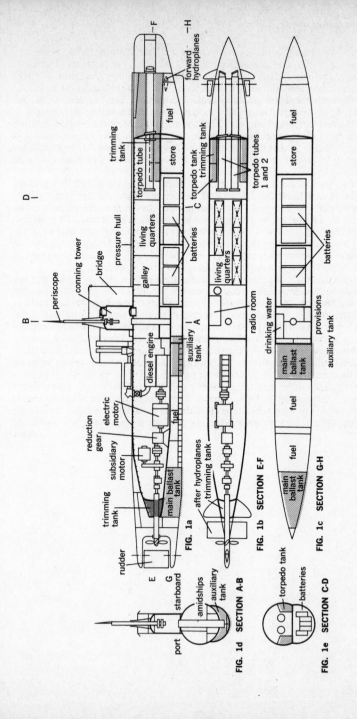

periscope

conning tower

bridge

pressure hull

B

galley

living quarters

batteries

reduction gear

electric motor

diesel engine

subsidiary motor

fuel

trimming tank

rudder

main ballast tank

E

G

starboard

FIG. 1a

auxiliary tank A

after hydroplanes

trimming tank

FIG. 1b SECTION E-F

F

H

forward hydroplanes

fuel

trimming tank

torpedo tube

store

torpedo tank

trimming tank

torpedo tubes 1 and 2

C

D

living quarters

radio room

drinking water

auxiliary tank

main ballast tank

fuel

provisions

fuel

FIG. 1c SECTION G-H

fuel

store

batteries

port

amidships

auxiliary tank

FIG. 1d SECTION A-B

torpedo tank

batteries

FIG. 1e SECTION C-D

353

and the hull. Contained within or outside the hull are various tanks, more particularly the fuel tanks and the water-ballast tanks (see Figs. 2a–2e). The latter comprise the main ballast tanks (which are empty when the submarine is on the surface and are completely flooded when it dives), the trimming tanks (forward and after trimming tanks) and the auxiliary tanks (amidships). The two last-mentioned sets of tanks serve to keep the vessel "in trim" by compensating for changes in weight due to fuel consumption, etc. The trimming tanks correct any alteration in the longitudinal distribution weight, while the auxiliary tanks are for control of overall weight. Rapid flooding of the main ballast tanks is essential to enable the submarine to dive quickly; they are provided with large power-operated vents and flood valves.

If the submarine has some forward motion when it dives (as is usually the case), the forward tanks are flooded a few seconds before the after tanks, so that the bow dips and the vessel's motion helps it to glide quickly under the surface, assisted by the hydroplanes. The submerged submarine is controlled by means of the hydroplanes (depth fins) at front and rear. These help to maintain depth and keep the vessel at the desired angle. The control effect of the hydroplanes is greater according as the submarine's forward speed is greater. Surfacing is effected by admitting compressed air to the ballast tanks and thus expelling the water.

The main propulsion of a conventional submarine is provided by diesel engines. As these engines require air, they could formerly be used only for propelling the submarine on the surface. Underwater it was always necessary to use battery-powered electric motors. A large submarine may have four main propulsion diesel engines, each with an output of 1600 hp. In a modern submarine the diesel engines are not, as a rule, directly connected by mechanical means to the propellers. Instead, they drive generators which supply electricity to the motors that drive the propellers. A World War II development is the "schnorkel," a breathing tube which is raised while the submarine is operating at periscope depth. Air for the diesel engines is drawn in through this tube, and the exhaust gases are expelled from a second tube. Diesel-powered propulsion is thus possible even when the submarine is almost totally submerged. For operating at greater depths the diesels and generators are not used, the power for the motors being supplied by storage batteries, which are

(more)

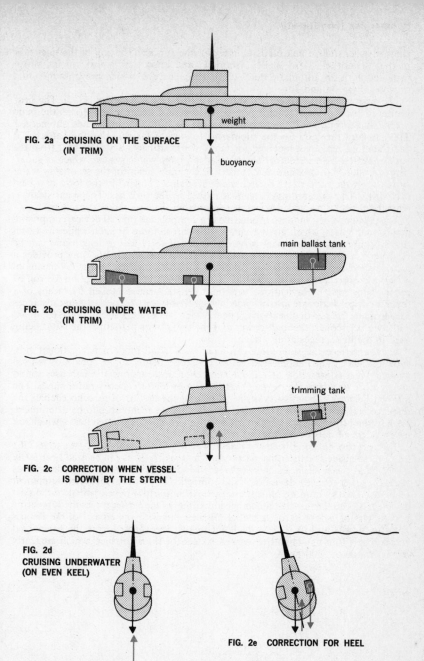

FIG. 2a CRUISING ON THE SURFACE
(IN TRIM)

weight

buoyancy

FIG. 2b CRUISING UNDER WATER
(IN TRIM)

main ballast tank

FIG. 2c CORRECTION WHEN VESSEL
IS DOWN BY THE STERN

trimming tank

FIG. 2d
CRUISING UNDERWATER
(ON EVEN KEEL)

FIG. 2e CORRECTION FOR HEEL

charged when the submarine is cruising on the surface. The top of the inlet tube of the "schnorkel" (also spelled "snorkel" and known as "snort" in Britain) is provided with an automatic quick-closing intake valve which prevents the entry of water (Figs. 3a and 3b).

The first nuclear-powered submarine was the U.S.S. *Nautilus* (1955). The heat produced by a nuclear reactor is used to generate steam which drives the main propulsion turbines and the turbogenerators, both on the surface and submerged. This obviates the need for the submarine to surface for charging the batteries. It can therefore remain submerged for many weeks, if necessary. The heat from the reactor is first transferred to the so-called primary water system, which is super-heated under high pressure and is used to generate steam in the secondary water system, in which the water is circulated through a "boiler" in the form of a heat exchanger. The dry saturated steam is passed to the turbines. The spent steam is condensed back into water, which is recirculated through the heat exchanger.

The periscope of a modern submarine is a complicated optical device, comprising prisms and lenses, which gives a view of the surrounding horizon while the vessel remains submerged. A typical periscope is about forty feet in length and can be raised or lowered by telescoping action (Figs. 4a and 4b). It not only provides a view around the horizon, but is equipped with a tilting head prism by means of which the observer can also scan the sky and which permits correction for roll or pitch of the vessel. Navigational sightings on stars can be taken by means of a sextant device. Scales are also provided which enable the observer to gauge distances and estimate the size of objects in the field of view. As a rule, a submarine is equipped with two periscopes, the high-power and the low-power periscope, the latter being used in the final stages of an attack.

On the surface the submarine can avail itself of all modern navigational aids, such as radar. When submerged below periscope depth, sonar apparatus is used ("sonar" is a contraction of "sound navigation and ranging"), which uses sound waves more or less in the manner that radar uses high-frequency radio pulses. The interval between the emission of the pulse and the arrival of its echo enables the distance between the submarine and, for example, an enemy ship to be determined. Large submarines are further equipped with inertial navigation systems, which are independent of outside signals (see page 374).

The torpedo has long been the classic weapon of the submarine (see page 236), for both offensive and defensive purposes. The torpedo tubes can be fired electrically or by hand when surfaced or submerged. Modern developments include various types of rocket-torpedo devices, remote-controlled and/or provided with equipment for automatically homing onto the target. An important development of the past decade has been the missile-launching submarine. In the earlier stages the submarine had to surface before it could fire its missiles, but with the advent of the Polaris ballistic missile, and its successors, launching is now effected from the submerged vessel. Each Polaris submarine carries sixteen of these missiles, which currently have a range of 2500 miles.

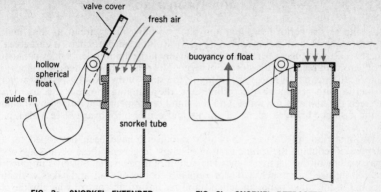

FIG. 3a SNORKEL EXTENDED

FIG. 3b SNORKEL RETRACTED

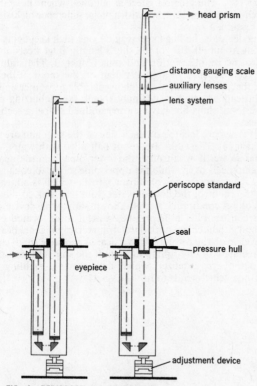

FIG. 4 PERISCOPE: (a) RETRACTED;
(b) EXTENDED (RAISED)

A ship at sea performs rolling and pitching movements about its longitudinal and its transverse axis respectively (Fig. 1). Rolling, in particular, is disagreeable to crew and passengers because of its relatively large amplitude, besides presenting problems with regard to the storage of cargo. The rolling motion depends on various factors: the wave movement according to the state of the sea, the vessel's moment of inertia with respect to the rolling axis, the damping moment due to friction between the hull and the water, and the stability moment, determined by the horizontal distance between center of gravity and center of buoyancy (see Vol. I, page 542).

Various kinds of devices, known as stabilizers, have been developed for the purpose of reducing the rolling motion of ships. In general these devices are of the "passive" or of the "active" type. The action of a passive stabilizer is initiated by the rolling itself, i.e., such a device *responds* to the motion and takes corrective action. On the other hand, an active stabilizer has preset control whereby the corrective action in the form of a counteracting movement is programed to take place simultaneously with the occurrence of the disturbing movement that causes the rolling of the ship. The wave movements, in particular, are never quite regular, but it is nevertheless possible, by means of appropriately designed active stabilizers, to reduce rolling by at least 75%. The greatest effect is obtained when the stabilizer operates at the natural frequency of the ship, but with a phase difference of 90 degrees in relation to the ship's motion.

The simplest stabilizing device is the bilge keel (Fig. 2), one such keel being fitted on each side and extending about 30–50% of the ship's length. Bilge keels develop considerable resistance to the rolling motion and thus reduce it. The stabilizing effect achieved by these keels depends to a great extent on the speed of the ship. They have the drawback that they present a not inconsiderable resistance and thus slow down the vessel. Instead of being a continuous keel, the stabilizing device may take the form of a series of short fins having a streamlined shape in section so as to reduce the resistance.

Stabilizing (or antiroll) tanks are located on each side of the ship and are interconnected by two pipes (Fig. 3). The tanks are about half filled with water, oil or some other suitable liquid. Water flows through the lower pipe from the upper to the lower side when the ship heels over, while the upper pipe serves to equalize the air pressure in the tanks. This upper pipe contains a throttle which is adjusted to regulate the airflow and thus control the flow of water from one tank to the other in accordance with the rolling conditions. A well-known stabilizing device is the gyrostabilizer (Fig. 4). It consists of a large and heavy steel rotor located on the center line of the ship and mounted in horizontal transverse gudgeon bearings. When the ship is on an even keel, the rotor axis is vertical. A sensitive small control gyroscope responds immediately to any rolling motion of the ship and transmits a counteracting motion to the gyrostabilizer, which thus exerts a righting force against the action of the wave which tends to roll the vessel over.

(more)

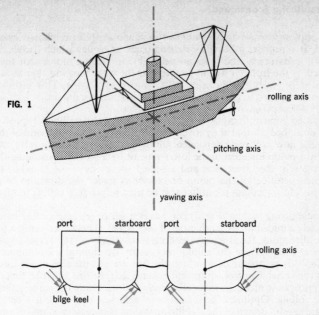

FIG. 1

rolling axis

pitching axis

yawing axis

port   starboard   port   starboard

rolling axis

bilge keel

FIG. 2 BILGE KEEL (PASSIVE)

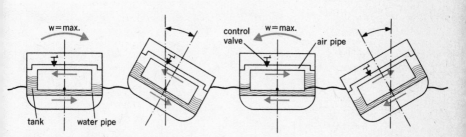

w = max.   control valve   w = max.   air pipe

tank   water pipe

FIG. 3 STABILIZING TANKS (PASSIVE)

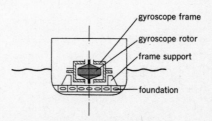

gyroscope frame

gyroscope rotor

frame support

foundation

FIG. 4 GYROSTABILIZER (PASSIVE)

The simplest form of active stabilizer is the antiroll device shown schematically in Fig. 5. It comprises a heavily ballasted truck or trolley which travels on a track extending transversely and is so propelled by an electric motor that the moment developed by the trolley's weight counteracts the wave moment that causes rolling. The motor is under the direction of a control gyroscope. This installation can alternatively be used to give the ship a rolling motion in calm water, as is sometimes necessary for experimental and testing purposes.

"Activated" antiroll tanks (Fig. 6) differ from the ordinary "passive" type, already described, in that the movement of the water from one side of the ship to the other is now not made dependent on the action of a throttle valve, but is controlled by a pump installed in the lower pipe or by a blower controlling the airflow and pressure in the upper pipe and air-filled space above the liquid in the tanks. Again the operation of the pump or blower is under the direction of a control gyroscope. This installation can likewise be used to produce rolling motion in calm water.

A fin stabilizing system (Fig. 7) comprises a set of retractable fins mounted approximately amidships on each side of the vessel. These fins can be pivoted in opposite directions about axes extending transversely to the vessel's longitudinal center line. They reduce the rolling motion by developing a counteracting effect which depends on the angle at which the fins are set, the size of the fins, and the speed of the vessel. Control equipment ensures that the fins are at all times swiveled to the appropriate angle for most effectively counteracting the wave action tending to cause rolling. Optimum performance is achieved only within a certain speed range, and careful design of the stabilizing system is necessary to make it as effective as possible. When not in use, the fins can be retracted into the hull or swung back into recesses provided for them.

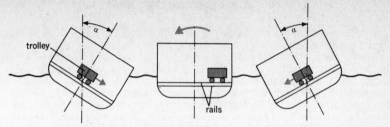

FIG. 5   STABILIZING TROLLEY (ACTIVE)

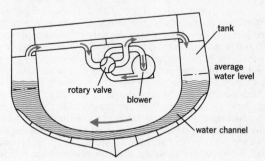

FIG. 6   ACTIVATED ANTIROLL TANKS

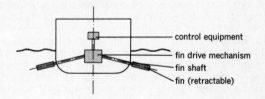

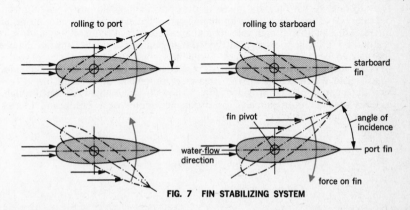

FIG. 7   FIN STABILIZING SYSTEM

LIQUID-PROPELLANT ROCKET SYSTEMS

A liquid-propellant rocket engine or motor comprises a combustion or thrust chamber, one or more storage tanks for the liquid combustibles, a supply system for feeding the combustibles to the combustion chamber, and a suitable power unit for driving the propellant pumps (Fig. 1). Most liquid-propellant rockets use two combustibles (bipropellant system), such as kerosene (the fuel) and liquid oxygen (the oxidant). With modifications, this arrangement, first successfully applied in the German V2 rockets of World War II, has been used in postwar rocket developments everywhere. However, efforts to simplify modern rockets and make them more reliable, particularly for space travel, have led to the utilization of gas pressure as the operating agent for the propellant-supply system, as in Fig. 2. Here the propellants (fuel and oxidant) are fed to the combustion chamber by the pressure of an inert gas such as helium and nitrogen. The drawback of gas pressurization is that the high pressures necessitate heavy tanks. In turbopump supply systems (Fig. 1) the turbine wheel is driven by a high-velocity gas stream and in turn drives single-stage or multistage centrifugal pumps. More recently, axial-flow pumps have also been employed, more particularly for liquid hydrogen. The gas to drive the turbine may be produced in a gas generator by the decomposition of hydrogen peroxide by means of a catalyst. The decomposition products, oxygen and steam, impinge on the turbine blades. Alternatively, the gas for driving the turbine for the propellant pumps may be obtained direct from the combustion chamber, or gaseous hydrogen from the regenerative cooling system of the combustion chamber may be utilized for the purpose. The propellants are injected at high pressure into the combustion chamber, in which the propellants are atomized, mixed and burned. The hot gaseous combustion products are expelled through the exhaust nozzle. The combustion chamber of a liquid-propellant rocket has to be cooled, for which purpose the regenerative principle is generally applied : one of the propellants flows through a jacket around the combustion chamber and cools it before passing to the injector. (The other propellant flows directly to the injector head of the combustion chamber.) An alternative method is so-called ablation cooling, operating on the same principle as the ablation shield of a rocket capsule on reentry into the atmosphere (see page 370). The combustion chambers of small rockets may be made of a high-melting-point metal such as molybdenum or tungsten, the heat being dissipated by radiation from the external surface of the chamber. Efficient injection of the propellants into the combustion chamber is very important, as it must ensure thorough mixing and complete combustion. The injector head atomizes the propellants and controls the feed rate so that they are mixed in the correct ratio. Various types of injector have been developed, such as the impinging spray type in which streams of propellant intersect one another at high velocity and are thereby broken up into small droplets ("atomized"). In the shower-head type, the propellant is sprayed into the chamber through concentric rows of holes. Sometimes concentric slots are employed to produce intersecting conical sheets of propellant; or the injector may embody a splash plate which breaks up the streams of propellant directed against it; or a swirl-type spray may be provided. Fig. 3 shows a longitudinal section through a modern rocket motor, namely, the third-stage motor of a European "Eldo-A" satellite launch vehicle.

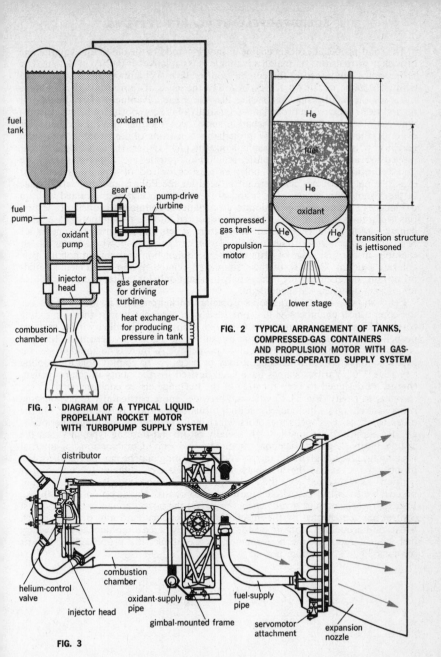

fuel tank

oxidant tank

gear unit

pump-drive turbine

fuel pump

oxidant pump

injector head

gas generator for driving turbine

combustion chamber

heat exchanger for producing pressure in tank

FIG. 1 DIAGRAM OF A TYPICAL LIQUID-PROPELLANT ROCKET MOTOR WITH TURBOPUMP SUPPLY SYSTEM

He

fuel

He

oxidant

compressed-gas tank

propulsion motor

He

He

transition structure is jettisoned

lower stage

FIG. 2 TYPICAL ARRANGEMENT OF TANKS, COMPRESSED-GAS CONTAINERS AND PROPULSION MOTOR WITH GAS-PRESSURE-OPERATED SUPPLY SYSTEM

distributor

helium-control valve

injector head

combustion chamber

oxidant-supply pipe

gimbal-mounted frame

fuel-supply pipe

servomotor attachment

expansion nozzle

FIG. 3

The solid-propellant rocket engine or motor embodies the oldest known principle of rocket propulsion. Its major advantage is its relative simplicity in design. The propellant is contained in the combustion (or thrust) chamber, which is provided with an exhaust nozzle for the expulsion of the gases of combustion. For military purposes the solid-propellant rocket has the great advantages of readiness and reliability. Modern solid propellants are usually of the composite (or heterogeneous) type: i.e., they consist of two separate substances, the fuel and the oxidant—for example, the American GALCIT propellants consisting of a mixture of potassium perchlorate (the oxidant) and asphalt oil (the fuel). Other oxidants similarly employed are ammonium perchlorate, ammonium nitrate, etc. Superior results are obtained with certain modern polymer plastics instead of asphalt as fuel. For example, polyurethane fuels were introduced for the Polaris and the second stage of the Minuteman rockets. The liquid polymers are catalyzed and mixed with an oxidant, e.g., ammonium perchlorate, to produce a composite propellant. At first this is of a thickly liquid consistency and is poured into the rocket motor case. After being cured at a slightly elevated temperature, it sets and acquires the appearance of a stiff rubber. Other modern fuels of this category are polybutadienes as, for example, in the first stage of Minuteman, in conjunction with an ammonium perchlorate oxidant. Because of their simplicity, reliability and convenience, solid-propellant rockets are being increasingly used not only for military rockets but also for rockets designed for space flight.

Fig. 1 shows a typical large solid-propellant rocket motor. The motor case—i.e., the combustion chamber—is first provided with a rubber lining; then a molding core is placed in the chamber and the remaining space is filled with a castable propellant. When the latter has set as a result of curing or polymerization, the core is withdrawn, so that a cavity is left in the propellant. As there is no liquid propellant available for regenerative cooling, it may be necessary to employ ablation cooling (see page 370). Various shapes have been devised for the cavity in the propellant charge, all designed to keep the area of the burning surface constant, such as the star-shaped cavity (Fig. 2). Cavities are provided more particularly in fast-burning propellant charges which develop a high thrust over a short period of time. On the other hand, solid-propellant motors which have to operate for longer periods—e.g., for surface-to-air missiles—are usually of the end-burning type, in which the charge burns "cigarette fashion" from one end only. Combinations of internal burning (to give a high initial thrust) and end burning may be employed. The solid-propellant rocket has the drawback that it is relatively difficult to control after ignition; as contrasted with the rocket propelled by liquid combustibles, it is not practicable to swivel the whole motor so as to control the direction of the thrust. Instead, rather complex arrangements to swivel the nozzle are necessary, or thrust control may be effected by the injection of a secondary fluid into the exhaust jet. Most solid propellants contain an admixture of powdered aluminum as an auxiliary fuel, which improves the performance. For example, the propellant of Polaris contains 20% of aluminum powder.

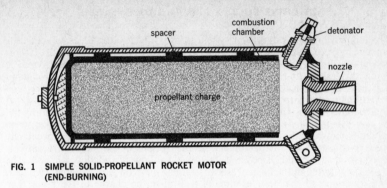

FIG. 1   SIMPLE SOLID-PROPELLANT ROCKET MOTOR
(END-BURNING)

FIG. 2   LARGE SOLID-PROPELLANT ROCKET MOTOR WITH
STAR-SHAPED CAVITY

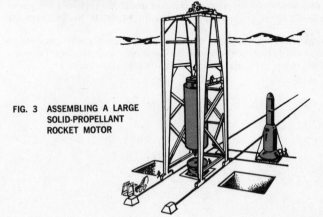

FIG. 3   ASSEMBLING A LARGE
SOLID-PROPELLANT
ROCKET MOTOR

ION-DRIVE ROCKET PROPULSION SYSTEMS

The ion-drive rocket propulsion system utilizes electrostatic fields to accelerate positively charged particles (ions), which are ejected rearwards. Ions are atoms which have acquired a positive charge by the removal of one or more electrons. The ions may be formed by passing a working fluid, such as cesium vapor, through an ionizing device (electrically heated tungsten grids) whereby the atoms lose electrons and are thus turned into positively charged ions. These ions are first concentrated into a beam by repulsion from positive electrodes and are then accelerated by the attraction exercized by negative electrodes (Fig. 1). To maintain the rocket in an electrically neutral state, it is necessary also to discharge the electrons (negatively charged particles) into space—otherwise the rocket would become negatively charged, so that a cloud of positive ions would follow it and slow it down. The electrons are ejected in the form of a beam from an electron gun and are mixed with the positive ions so that they eventually neutralize the charge of the latter. The velocity attainable by the ions is governed by the difference in voltage along the path they have to traverse in the propulsion motor and by the charge and mass of the ions themselves. Since the acceleration process does not constitute an electric arc or some form of combustion, this is a cold drive system. Very high ejection velocities can be attained without giving rise to the difficulties associated with high temperatures at the exhaust. Typical velocities range from 30 to 300 km/sec.

Various sources of ions can be used. The simplest method of producing ions, as already described, is by direct contact, e.g., between cesium and heated ionizing grids made of tungsten, a metal which can operate at elevated temperatures and has a high affinity for electrons. Another very effective method is provided by the Kaufman system, which embodies a device resembling a magnetron and produces ions by electron bombardment of a metal vapor (Fig. 2). With the aid of a thermionic cathode located at the axis of the ionization chamber and of a magnetic field between the cathode and the chamber wall, the paths of the electrons are so curved by the magnetic field that no anode current will flow until the electrons are scattered by collisions with gas atoms or molecules inside the chamber.

The "ions" employed in an ion drive system may alternatively consist of electrically charged particles other than atomic or molecular ions—namely, dust particles, liquid droplets, or colloidal particles. In the last-mentioned case, half the number of colloidal particles employed are given a positive and the other half a negative charge. These particles are respectively accelerated in two separate chambers and ejected.

Fig. 3 is a schematic representation of an ion-drive system of rocket propulsion. The ion drive is a low-thrust system and can function only in a vacuum. For these reasons it is suitable more particularly for interplanetary space flight.

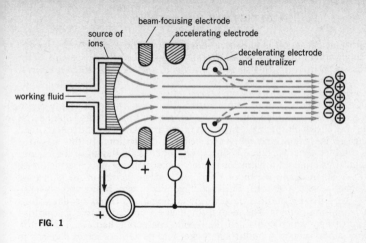

FIG. 1

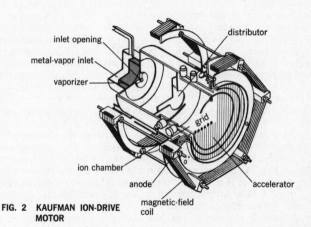

FIG. 2 KAUFMAN ION-DRIVE
MOTOR

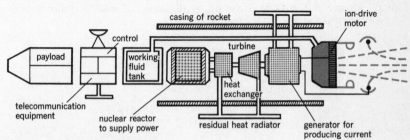

FIG. 3 ION DRIVE SYSTEM FOR
ROCKET PROPULSION (SCHEMATIC)

With multistage rockets it is in general possible to attain higher final velocities than with single-stage rockets of equal overall weight. When the propellant in the first stage has been used up, this stage is jettisoned, and the next stage is ignited. The final stage, usually the smallest, carries the payload. Its final velocity is the sum of the final velocities attained by all the rocket stages.

As Fig. 1 shows, there are in principle four different ways of constructing multistage rockets. The system hitherto most widely employed is that of tandem or series staging (Fig. 1a), in which the successive stages are arranged one above the other, the first stage to be ignited and jettisoned being at the bottom. In this way it is possible to combine different types of propulsion and rocket design in the various stages. A typical example of a giant multistage rocket is the Saturn 5, which took America's first astronauts to the moon (Fig. 2e). It stood 263 ft. on the launching pad and weighed about 3000 tons. The first stage was of very heavy and powerful construction, with five F 1 motors powered by kerosene and liquid air, developing a thrust of 7.5 million pounds. The second stage was propelled by five T 2 motors, while the third stage, which contained the Apollo spacecraft, had one T 2 motor. The optimum subdivision of a multistage rocket into its various stages depends to a great extent upon the chosen combination of propellants. A disadvantage of the series-type multistage rocket is that the propulsion systems of the various stages are ignited and operate consecutively, so that they cannot act simultaneously in accelerating the rocket. For this reason the booster-rocket principle has been applied in the Atlas intercontinental ballistic missile (Fig. 1b). The main rocket is essentially a single-stage liquid-propellant vehicle powered by a sustainer motor. In addition there are two booster units, burning the same fuels and developing a very high thrust. The boosters are jettisoned at burnout, and the sustainer accelerates the missile to maximum velocity and is then shut off. In this method only one liquid-propellant supply system is required, whereas separate stages arranged in series each require their own supply system. A third possibility is the parallel-stage rocket (Figs. 1c and 2c), comprising a main rocket and a number of jettisonable solid-propellant booster units for high lift-off thrust and initial acceleration. This arrangement is regarded as most suitable for future space-flight projects. Another possible combination is illustrated in Fig. 1d: a small manned spacecraft is carried into orbit by a launching rocket to which it is attached "piggyback" fashion. Proper separation of the stages at burnout is an important operation, generally carried out in a program-controlled sequence. Actual separation of the stages is effected by means of explosive bolts or similar devices. It is essential that the burned-out stage should become detached simultaneously at all points of connection and that the ignition of the next stage should take place, not immediately upon separation, but with a few seconds' delay. Booster units may be jettisoned by the action of small side-thrust rockets which release the units from the main rocket.

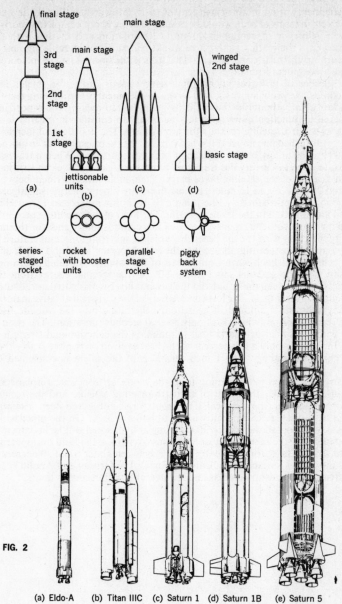

FIG. 1

final stage

3rd stage

2nd stage

1st stage

(a)

series-staged rocket

main stage

jettisonable units

(b)

rocket with booster units

main stage

(c)

parallel-stage rocket

winged 2nd stage

basic stage

(d)

piggy back system

FIG. 2

(a) Eldo-A　(b) Titan IIIC　(c) Saturn 1　(d) Saturn 1B　(e) Saturn 5

369

The kinetic energy of a satellite or spacecraft is many times greater than the amount of energy which, in terms of heat, would be needed to bring about complete vaporization of the satellite or spacecraft in question. Retardation and friction with the air on reentry into the earth's atmosphere would release a considerable amount of heat and burn up the reentering body. Indeed, this is the normal fate of meteorites entering the atmosphere from outer space.

After a number of unsuccessful attempts with various protective shields and cooling methods, the problem was solved by the development of the ablating reentry shield. Made of ablative material, the shield dissipates heat by melting and vaporizing. To understand its function more fully, it is necessary to consider in greater detail what happens when a satellite reenters the atmosphere. The first signs of increased air resistance (drag) become noticeable when it enters the thermosphere at an altitude of about 100 km (60 miles), but no significant heating occurs here. When it traverses the mesopause, as yet in a very flat trajectory almost parallel to the earth's surface, some slight heating of the satellite takes place. Then, in an altitude range between about 40 and 25 km (25 and 15 miles), almost the entire kinetic energy is dissipated within a period of approximately one minute. It is in this range that the satellite and its crew are subjected to the severest strains. The front of the satellite is protected by a shield made of synthetic resin or plastic of low combustibility and low thermal conductivity. Friction with the air heats this ablative material to a temperature of several thousand degrees centigrade, so that the resin becomes liquid and "boils off." The layer behind the shock-wave front ahead of the satellite is heated to around $6000°$ C and is in the gaseous state. About 80% of the thermal energy from the intermediate layer between the liquid and the gaseous layer is dissipated as radiation to the surrounding air (Fig. 1). The low conductivity of the still-solid ablative material prevents any substantial amount of heat from penetrating into the satellite itself during the reentry period (which is of only 50–100 seconds' duration). The resin is reinforced with glass fiber or quartz fiber to maintain the cohesion and strength of the shield. In this way only a few percent of the heat evolved on reentry is absorbed by the satellite; only about $\frac{1}{2}$ to 1 inch thickness of the shield is consumed by ablation.

In the case of a spacecraft returning from the moon or from an interplanetary flight, the speed is higher than that of an earth-orbiting satellite and the reentry problem correspondingly more critical. If the reentry trajectory is too steep, frictional heating will be too severe to be dealt with by ablative cooling. On the other hand, if the trajectory is too flat, the spacecraft traveling at high speed will be in danger of being "bounced off" the earth's atmosphere, thus overshooting it and disappearing forever into space. The optimum reentry angle is between about 5 and 10 degrees in relation to the earth's surface, as indicated in Fig. 2. The Gemini and Apollo space capsules have ablating reentry shields and are steerable within certain limits.

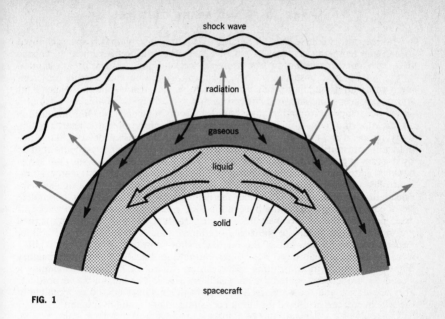

shock wave

radiation

gaseous

liquid

solid

spacecraft

FIG. 1

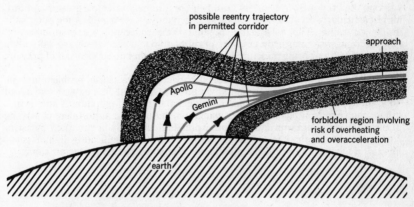

possible reentry trajectory
in permitted corridor

approach

Apollo

Gemini

forbidden region involving
risk of overheating
and overacceleration

earth

**FIG. 2 REENTRY CORRIDOR FOR APOLLO SPACECRAFT
(HYPERBOLIC SPEED) AND GEMINI SPACECRAFT
(CIRCULAR ORBITAL SPEED)**

SPACE SUITS AND SPACE CAPSULES

Man cannot, of course, exist in the near vacuum of interplanetary space, exposed to extreme temperatures, the hazard of meteorites, and bombardment of cosmic rays (very energetic radiation from outer space) and dangerous radiation emitted by the sun. For man to survive in space, he must carry his normal "earthly" environment with him. In the first place, the astronaut wears a space suit and helmet. This equipment keeps him supplied with oxygen for breathing, maintains his body at a controlled temperature, and removes moisture from the air enclosing him. In addition all the important data of his vital functions, such as blood pressure, heart rhythm, body temperature, etc., are continually monitored by sensing devices and automatically transmitted by radio back to earth, where they are supervised and studied by doctors and scientists. Besides wearing this self-contained space suit, the astronaut is additionally enclosed within a fully air-conditioned and temperature-controlled capsule. He can, however, leave the capsule to perform particular tasks outside it ("walk in space"). The capsule is furthermore provided with a complete set of equipment for sustaining human life over a period of days or even weeks, as exemplified by the Apollo spacecraft which performed the first American manned moon flight. With regard to the waste products of the human body this is still an "open" system, on account of the relatively short duration of the space flight—a matter of about eight days. For longer journeys in space—e.g., an interplanetary flight or a sojourn in an orbiting space station—it will be necessary to employ a closed circuit, as represented schematically in Fig. 2. The human waste products are dehydrated, and the water recovered from them is purified and recirculated in the system and can be reused for human consumption over and over again.

The walls of a space suit and of a space capsule are of multilayer construction. An important component of a space suit is a layer of plastic provided with a coat of metal (aluminum or gold) applied to it in vapor form. This metallic coating reflects the intensive solar radiation to which the astronaut is exposed when he ventures outside the capsule. Fig. 1 illustrates the Mercury capsule, which was used in earlier American space flights. A large proportion of the equipment carried in the capsule serves to maintain the controlled conditions in the crew's space suits and in the cabin of the capsule itself. In the Apollo capsule and in future earth-orbiting space stations the crew will live in less cramped conditions and enjoy greater comfort, more particularly because they will be able to remove their cumbersome outer suits inside the cabin. This is possible because these spacecraft are of double-wall construction, with intermediate layers of rubberlike material which are self-sealing in the event of their being punctured by meteorites, so that no sudden loss of pressure within the capsule will occur.

The choice of atmosphere in the cabins of spacecraft is a special problem in itself. The natural air we breathe is essentially a mixture of about 20% oxygen and about 80% nitrogen. Since considerations of weight saving are of paramount importance in all space flights, it has been investigated whether a cabin atmosphere consisting of pure oxygen at appropriately reduced pressure can safely be breathed by astronauts. Another method is to use helium—which is a significantly lighter gas—instead of nitrogen to produce a breathable atmosphere in the spacecraft cabin.

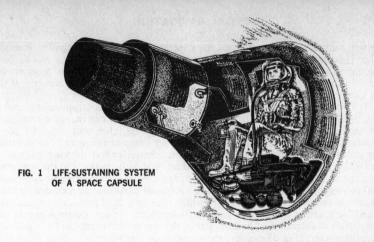

FIG. 1 LIFE-SUSTAINING SYSTEM OF A SPACE CAPSULE

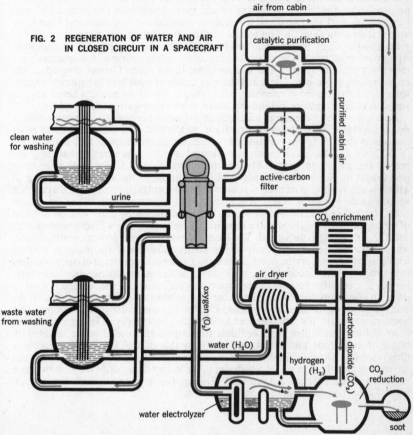

FIG. 2 REGENERATION OF WATER AND AIR IN CLOSED CIRCUIT IN A SPACECRAFT

air from cabin

catalytic purification

purified cabin air

clean water for washing

urine

active-carbon filter

CO_2 enrichment

oxygen (O_2)

air dryer

waste water from washing

water (H_2O)

hydrogen (H_2)

carbon dioxide (CO_2)

CO_2 reduction

water electrolyzer

soot

Inertial guidance systems are used for controlling the flight of intercontinental missiles. This method of guidance makes use only of the laws of classical mechanics to calculate the flight path of the missile and compel it to follow a programed trajectory from launching site to target. A simple inertial system was incorporated in the German V2 rockets which bombarded London in the final stages of World War II. The basis of a modern inertial guidance system is an arrangement, comprising three mutually perpendicular accelerometers, which can measure forces in any direction in space, coupled with three gyroscopes, also with mutually perpendicular axes, which constitute an independent frame of reference. The inertial system may be supplemented by an independent celestial, or star-tracking, guidance system which maintains the missile in a fixed attitude with reference to the sun or a particular star and thus serves as a check on the inertial system (stellar-monitored inertial guidance). The inertial principle has acquired particular importance for the navigation of spacecraft. Reference axes which take up a fixed position in space are provided by a set of gyroscopes and are independent of any external points of reference such as a lighthouse, beacon, star or the sun, as in conventional navigation. The accelerometers (Fig. 1) measure the acceleration components in three mutually perpendicular directions. By means of successive integration (performed by computer equipment), the speed and the distance traveled can be determined from these acceleration data (Fig. 2). The results of this computation are compared with the precalculated flight path which has been fed into the data-processing computer equipment in advance. The requisite corrections to the course of the spacecraft are then automatically applied. The satellites used in the Lunar Orbiter program relied on astronavigation, based on using the star Canopus (visible in the southern hemisphere) as a fixed point of reference, but in the vicinity of the moon, where Canopus was concealed from view behind the moon, inertial navigation was employed.

An accelerometer measures acceleration or, more particularly, the force that is exerted when a body possessing inertia is accelerated. The inertia tends to resist the acceleration. It is this resistance to a sudden change in speed that is the origin of the force that thrusts a motorist backward into his seat when he suddenly depresses the accelerator pedal. This reaction force is equal to the product of mass (the weight of the motorist's body) and acceleration. In the accelerometer (Fig. 3) a mass is maintained in its neutral position by two springs; it remains in this position so long as the system is at rest or is in motion at constant speed. When the system is accelerated in the arrowed direction, the spring-mounted mass will, on account of its inertia, at first lag behind the movement; the front spring will thus be stretched, and the rear spring compressed. The inertia mass controls a potentiometer contact whose zero position corresponds to the neutral position of the mass. A positive acceleration thus causes the potentiometer to give a positive voltage of corresponding magnitude to the acceleration; a negative acceleration (deceleration) similarly produces a negative voltage of corresponding magnitude. If the voltage is used to drive an electric motor, the total number of revolutions will be proportional to the average speed of the missile or spacecraft over any particular period of time. This combination of the potentiometer and motor thus constitutes an integrating device. A second integrator uses the speed data to compute the total distance traveled (in terms of magnitude and direction). In Fig. 4 this double integration process is illustrated schematically for one coordinate direction (x). Fig. 5 represents an inertial navigation system for a vessel which can move in two dimensions only; in this case only two accelerometers are required. Finally, Fig. 6 shows a three-dimensional inertial guidance system in the nose of a rocket.

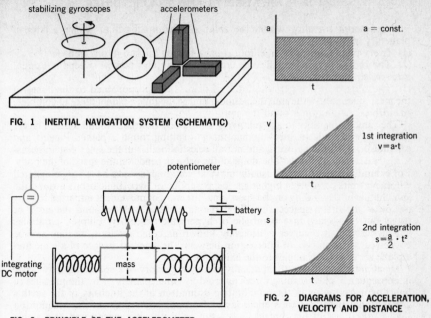

FIG. 1 INERTIAL NAVIGATION SYSTEM (SCHEMATIC)

FIG. 3 PRINCIPLE OF THE ACCELEROMETER

$a = $ const.

1st integration
$v = a \cdot t$

2nd integration
$s = \dfrac{a}{2} \cdot t^2$

FIG. 2 DIAGRAMS FOR ACCELERATION, VELOCITY AND DISTANCE

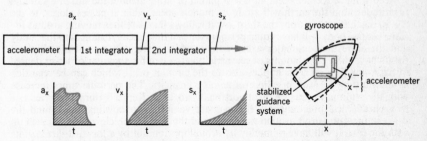

FIG. 4 INERTIAL NAVIGATOR (ONE-DIMENSIONAL, SCHEMATIC)

FIG. 5 TWO-DIMENSIONAL INERTIAL NAVIGATION

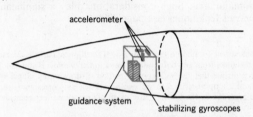

FIG. 6 NOSE OF ROCKET WITH THREE-DIMENSIONAL INERTIAL GUIDANCE SYSTEM

A spacecraft traveling within the solar system is subject to Kepler's laws of planetary motions (Fig. 1):

(1) Every planet moves in an elliptical orbit having the sun in one of its foci.*

(2) The radius vector (i.e., the line connecting sun and planet) sweeps out equal areas in equal times.

(3) The squares of the periodic times of planets are proportional to the cubes of the mean distances to the sun (the periodic time is the time a planet takes to complete one orbital revolution around the sun).

The same laws apply to any relatively small body orbiting around a much larger one—e.g., a satellite (natural or man-made) orbiting round a planet. Planets and satellites have elliptical orbits, and not all celestial bodies in the solar system necessarily travel in such orbits. The shape of the orbit depends on the speed of the body. For example, comets, which usually move at very high speeds, have very elongated elliptical orbits or, at even higher speeds, parabolic or hyperbolic orbits around the sun. Similarly, depending on the speed and direction of movement imparted to it by its rocket motors, a spacecraft will pursue an elliptical orbit around the earth or travel out into space in a parabolic or hyperbolic trajectory. Ellipses, parabolas and hyperbolas are curves collectively known as conic sections, as they can be conceived as the curves of intersection between the external surface of a cone and a plane set at various angles to the axis of the cone.

Deviations from the mathematically precise trajectory of a spacecraft may occur in consequence of disturbing forces (exerted by other planets or by the pressure of light) or inaccuracies in the quantitative estimation of the intensity of the earth's gravitational field (the theoretical laws of motion are based on the assumption that all the mass of the central body—sun or earth—is concentrated at a single point, whereas in reality the central body is of somewhat flattened spherical shape).

According to Kepler's second law, a planet orbiting around the sun or a satellite orbiting around the earth will attain its highest speed when it passes closest to the sun or earth (Fig. 1). From the third law it follows that satellites moving around the same central body will have equal periodic times if their mean distances to that body and therefore the semimajor axes of their orbits are equal, even if the elliptical orbits have different amounts of eccentricity—i.e., exhibit a greater or lesser degree of flattening (Fig. 2). This applies also to the circular orbit, which can be regarded as the limiting case of an ellipse of zero eccentricity. The periodic times increase with the semimajor axis "a" (proportionally to $a^{3/2}$). Kepler's third law gives rise to a paradoxical phenomenon. If two space vehicles are traveling one behind the other in the same orbit around the earth, and the second vehicle wishes to catch up with the first, it will have to modify its orbital motion not by a forward thrust of its rocket motors, but by means of a retroactive thrust. A thrust to accelerate the spacecraft in its forward motion would cause it to move into a larger orbit, i.e., an ellipse with a larger semimajor axis. Such considerations play a significant part in connection with rendezvous techniques (see page 380).

(more)

---

*Every ellipse has two points called foci, located on the major axis. The elliptical curve is characterized by the fact that the sum of the distances from any point on the curve to the two foci is constant for that particular ellipse. It is on this principle that the well-known method of drawing an ellipse is based, namely, by means of a pencil and a loop of string around two pins. Each pin corresponds to a focus of the ellipse. A circle may be regarded as the limiting case of an ellipse whose foci coincide at one point, the center of the circle.

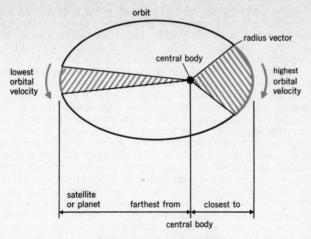

**FIG. 1 THE RADIUS VECTOR SWEEPS OUT
EQUAL AREAS IN EQUAL TIMES**

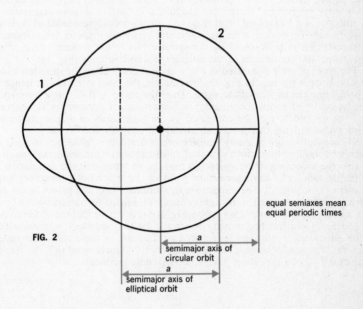

FIG. 2

equal semiaxes mean
equal periodic times

a
semimajor axis of
circular orbit

a
semimajor axis of
elliptical orbit

As explained in the foregoing, the orbit of a nonpropelled body moving around a central body in space will be a curve belonging to the family of conic sections. Planets and satellites travel in elliptical orbits, i.e., closed curves, whereas bodies moving at high speeds may pursue "open" (parabolic or hyperbolic) orbits. By appropriately firing the rocket motors of a spacecraft it is possible to vary the speed and thus alter the shape of its trajectory in space. Fig. 1 shows some theoretically possible trajectories drawn in red for a spacecraft on a journey from, for example, the earth to Mars. The orbits of these two planets are drawn in black. The part of the trajectory actually traveled by the spacecraft is shown as a full red line. Fig. 1a represents a so-called Hohmann transfer orbit, which is tangential to the inner and outer planetary orbits respectively. The spacecraft, which is first assumed to be moving along with the earth (at A) around the sun, is given an initial thrust which accelerates it in a direction tangential to the earth's orbit. It will then continue to travel, without additional propulsion, along the red trajectory to Mars (at B). In the absence of any further propulsive intervention the spacecraft would then return along the dotted trajectory constituting the other half of the elliptical transfer orbit. If this spacecraft could, on arrival at B, be given a slowing-down thrust in a direction parallel to the surface of Mars, it would go into orbit around that planet. Hohmann orbits are referred to as "minimum-energy" transfer orbits because they have the advantage of requiring relatively low rocket-fuel consumption. A disadvantage, however, is that the spacecraft approaches the destination planet (Mars) at very low speed. Fig. 1c shows an elliptical orbit which permits a more rapid transfer between the two planetary orbits. In addition to these transfer orbits which require two separate propulsive thrusts to put the spacecraft into orbit around the destination planet, there are other possible orbits which would require three or more of such "kick maneuvers" (corrective propulsive thrusts) to achieve that result. Thus, in the case of a bielliptical orbit (Fig. 1d), thrusts would be needed at A, B and C. Figs. 1e and 1f are examples of parabolic and hyperbolic transfer orbits respectively. A special case is represented by the spiral orbits of ion rockets which develop a low-power but continuous thrust during the whole journey (Fig. 1g).

The types of orbit illustrated in Fig. 1 are not confined to the situation where the central body is the sun. In Fig. 1b, for example, the inner circle can be conceived as representing the surface of the earth. Then the curve A–B–C might correspond to the trajectory of an intercontinental ballistic missile. Furthermore, the curves A–B in Figs. 1b and 1c may be conceived as the trajectories of a spacecraft rising to rendezvous with another spacecraft already in orbit around the earth.

Theoretically a great many transfer orbits are conceivable, but in reality their number is limited by various practical considerations. One important requirement is that the launching of the spacecraft must be so timed that the destination planet is in the vicinity of the intersection point of the spacecraft's trajectory and that planet's orbit (Fig. 2). The conditions of illumination of the planet at the time of arrival of the spacecraft is also important with regard to transmission of pictures back to earth. Thus, for a lunar probe (Fig. 3) it is desirable that the observed portion of the moon's surface be situated close to the shadow boundary, where small details show up clearly by the shadows they cast. Because of these and other restricting factors, the actually available time interval ("launch window") for successful launching may be very short (hours, perhaps only minutes).

(more)

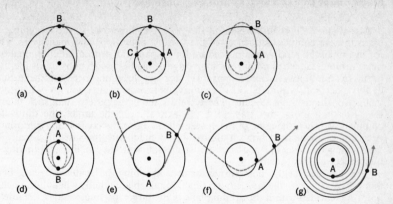

FIG. 1

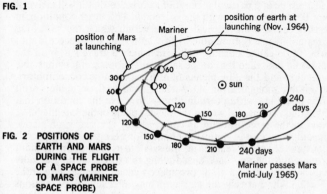

FIG. 2 POSITIONS OF EARTH AND MARS DURING THE FLIGHT OF A SPACE PROBE TO MARS (MARINER SPACE PROBE)

position of earth at launching (Nov. 1964)

Mariner

position of Mars at launching

⊙ sun

30
60
90
120
150
180
210
240 days

Mariner passes Mars (mid-July 1965)

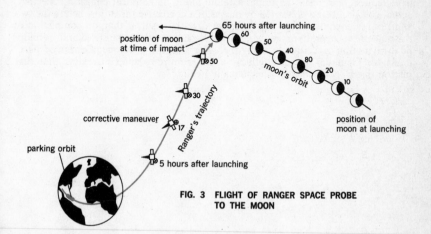

65 hours after launching

position of moon at time of impact

corrective maneuver

parking orbit

Ranger's trajectory

5 hours after launching

moon's orbit

position of moon at launching

FIG. 3 FLIGHT OF RANGER SPACE PROBE TO THE MOON

Actual space flights are based on the principles described in the foregoing, but often involve complex sequences of the various operations and maneuvers. Thus, for example, the rendezvous procedure for a spacecraft with a satellite already in orbit around the earth comprises four stages. The first stage consists in the flight of the spacecraft and its launching rocket through the atmosphere and the injection of the spacecraft into its elliptical rendezvous trajectory that will carry it into the vicinity of the orbiting satellite. The second stage is the purely "ballistic" flight of the spacecraft along this trajectory. On arrival near the target, the third stage commences, when the crew of the spacecraft take over its control and perform the precision maneuvers to bring it almost into contact with the satellite. Finally, the fourth stage consists in "docking" the spacecraft with the satellite and linking the two space vehicles together.

Fig. 1 shows various possibilities for accomplishing the "ballistic" phase of the flight. In one case (a) the launching is timed to take place exactly when the launching point A is located within the orbital plane of the target satellite. Besides, the satellite itself must, at that instant, be at a particular point of its orbit so that it will arrive at the rendezvous point B simultaneously with the spacecraft. This latter condition, in particular, considerably restricts the interval of time within which the launching must take place. Greater latitude is attainable by first launching the spacecraft into an intermediate orbit (a so-called parking orbit) in which it travels around the earth in a shorter periodic time than that of the satellite. The spacecraft is thus able to "overtake" the satellite and then, by means of a thrust from its rocket motors, be transferred at a suitable moment from the parking orbit to the satellite's orbit (b). Alternatively, it is possible—within certain limits—to waive the condition that the point of launching must be located within the orbital plane of the satellite (c).

The final approach to the target satellite is controlled by the spacecraft's crew themselves, with the aid of radar and optical methods. For example, the spacecraft can be held on a constant course in relation to the satellite as seen against distant stars in the background and the approach speed along this direction be gradually slowed down so that the relative speed of the two space vehicles becomes zero at the instant of contact (Fig. 2a). Alternatively, the spacecraft may, after suitable corrections in its course, approach the target satellite along an elliptical collision trajectory, actual collision being prevented by means of a corrective thrust just before the two vehicles make contact (Fig. 2b). In either case the result is that, on completion of the rendezvous operation, the two space vehicles have the same speed and therefore have no motion in relation to each other. The rendezvous between Gemini 6 (spacecraft) and Gemini 7 (target satellite)—a rather more complex procedure than that just outlined—is illustrated schematically in Fig. 3.

FIG. 1 EXAMPLES OF HOW THE FIRST
PHASE OF A RENDEZVOUS
OPERATION (BALLISTIC PHASE)
CAN BE PERFORMED:
(a) COPLANAR LAUNCHING
INTO A BALLISTIC ORBIT
(b) LAUNCHING INTO A PARKING ORBIT
(c) NONCOPLANAR LAUNCHING

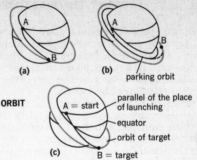

(a)

(b) parking orbit

A = start

parallel of the place
of launching

equator

orbit of target

(c) B = target

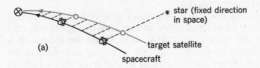

(a)

* star (fixed direction
in space)

target satellite

spacecraft

FIG. 2 POSSIBLE METHODS OF ACHIEVING
THE FINAL PHASE OF A RENDEZVOUS

(b)

FIG. 3 RENDEZVOUS OF GEMINI 6 (BLACK)
WITH GEMINI 7 CAPSULE (RED)
SERVING AS TARGET

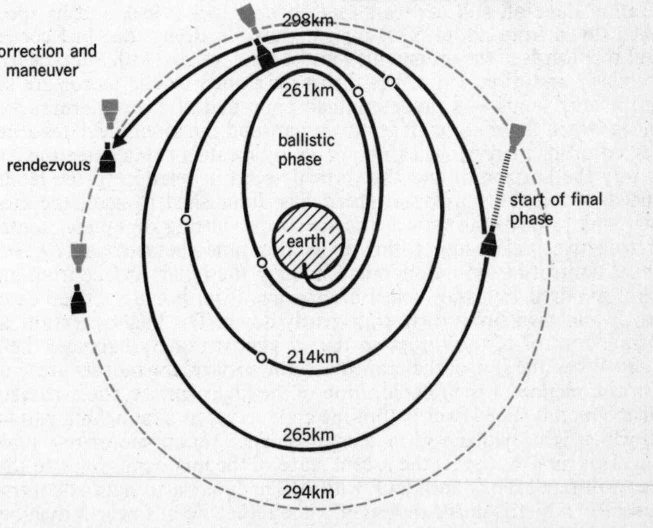

o = correction and
thrust maneuver

298km

261km

ballistic
phase

rendezvous

start of final
phase

earth

214km

265km

294km

LANDING THE FIRST MEN ON THE MOON

Landing a space vehicle on a celestial body devoid of atmosphere, such as the moon, presents special technical problems of braking the vehicle's descent and accurately controlling the braking action. The conditions of entry are different from those presented by a major planet enveloped in a relatively dense atmosphere. For a moon landing the entire kinetic energy of the vehicle must be braked by a counteracting thrust developed by rockets.

On arrival in the vicinity of the moon, the spacecraft is first slowed down by firing its retro-rocket motors, so that it goes into a circular parking orbit round the moon. This applies more particularly to a manned spacecraft such as the Apollo which the Americans have used for landing men on the moon. In the case of the Apollo XI project the actual descent onto the moon's surface was made by two astronauts in a special mooncraft, the "lunar (excursion) module" (abbreviated as "LEM" or "LM"), which was detached from the orbiting spacecraft, in which one astronaut remained awaiting the return of the mooncraft.

The entire procedure of releasing the mooncraft, landing it at a predetermined site on the moon and then linking it once again to the spacecraft is one which requires precise control of the direction and velocity of both vehicles. When the spacecraft is accurately in orbit and in the correct position on its orbit to ensure a landing in the desired area, the mooncraft briefly fires its rocket motor so that it moves away from the spacecraft and goes into an elliptical orbit whose point nearest the moon is located some 10 to 20 miles before and six or seven miles above the planned landing area. The periodic time of the mooncraft in this elliptical orbit must be the same as that of the spacecraft in its circular parking orbit. This particular requirement is for the astronauts' safety; should the landing rockets fail to fire, the mooncraft will simply continue in orbit and automatically encounter the spacecraft; the latter can then be maneuvered into a docking position with the mooncraft, so that the two astronauts in it can return to the spacecraft that will take them back to earth.

When the mooncraft is in the correct position in its orbit, the actual landing maneuver can commence. The landing rocket motor of the mooncraft must be able to develop a thrust that can be suitably varied, because at the start of the landing operation the craft still carries its full load of rocket fuel, and its speed has to be slowed down from about 5000 mph to zero. In doing this, fuel corresponding to about two-thirds of the mooncraft's initial total weight (with full tanks) is consumed. The power and direction of the thrust developed by the motor are so controlled that the craft lands at a predetermined point and at a predetermined speed. If the orbit in which the mooncraft is moving around the moon deviates a little from the specified orbit, corrections can be made by means of small steering rocket jets. In this way the horizontal and the vertical speed in relation to the landing area are reduced. When the horizontal speed has diminished to zero, the mooncraft will slowly sink towards the surface, the actual speed being kept under control by means of retroactive rocket motor thrust. By this time the astronauts have taken over manual control of the mooncraft. Scanning the lunar surface from an altitude of several hundred feet, they select a zone free from boulders, deep cracks or other hazards and then bring their craft gently down. The final operation calls for very accurate control of the thrust so that it almost exactly balances the mooncraft's weight. When the feet of the craft touch the surface, the motors are shut off.

On completion of their exploration of the lunar surface, the astronauts return to their mooncraft. The lower half of the craft serves as a launching pad for the upper half, which is provided with a second, smaller rocket motor just under the crew cabin. This motor propels the ascent stage of the mooncraft back to the spacecraft, which will rendezvous and dock with it. The lunar astronauts then transfer to the spacecraft and jettison the mooncraft; the return flight to earth then begins.

The sequence of operations for the Apollo XI moon landing project was as follows (see illustrations on pp. 384–385):

1. Saturn rocket is launched, carrying the Apollo spacecraft with the mooncraft enclosed within it.

2. First stage of the rocket is jettisoned, second stage is fired.

3. Second rocket stage is jettisoned, third stage goes into orbit around the earth (1½ revolutions).

4. Third stage is fired, thereby increasing the speed from 17,500 mph to the so-called "escape velocity" of almost 25,000 mph.

5a–5d. Third stage burns out. Apollo spacecraft is released. Mooncraft (LEM) and command module are now docked together nose to nose by a complex maneuver. They then reconnect with the third stage and continue the flight. Third stage is then finally jettisoned, and Apollo spacecraft starts up its own rocket motors. Apollo comprises the command module (i.e., the crew capsule), the service module, and the mooncraft.

6. Braking rockets are fired, spacecraft goes into orbit around the moon at an altitude of about 70 miles.

7. Two astronauts enter mooncraft, which is now detached from the spacecraft.

8. Mooncraft goes into its own orbit bringing it over the landing area.

9. Main braking rocket motor of mooncraft is fired. Telescopic legs of mooncraft are extended and it lands on the lunar surface.

10. Ascent stage of mooncraft launched for return flight to orbiting spacecraft.

11. Rendezvous with spacecraft. Lunar astronauts transfer themselves from mooncraft to spacecraft.

12. Mooncraft is jettisoned.

13. Spacecraft starts return flight to earth.

14. Command module is detached from service module.

15. Command module is maneuvered so that the heat shield is facing forward on entering the earth's atmosphere (see page 370).

16. Final parachute descent to earth.

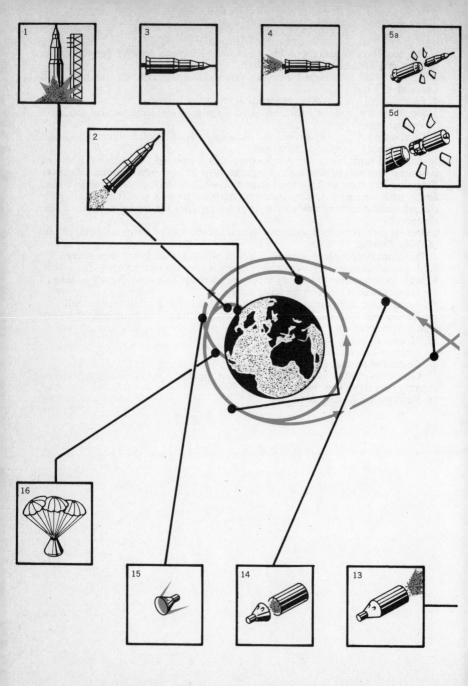

STAGES OF THE MANNED APOLLO SPACECRAFT FLIGHT TO THE MOON

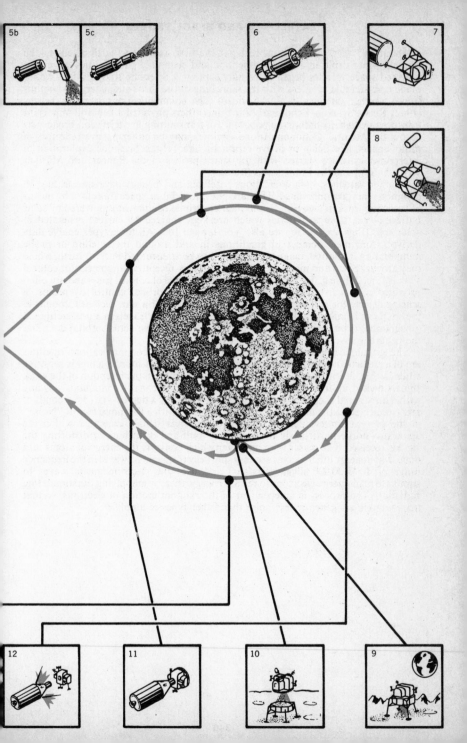

About five years after Russia had put the first man-made earth satellite—the Sputnik—into orbit, in 1957, the scientific and economic potentialities of rocket-powered space vehicles began to be fully exploited. Scientific study of the weather made considerable progress with the launching of the Tiros and, later, the Nimbus (1964) weather satellites (both American). The communications satellites Telstar (1962), Relay, Syncom, Echo, Early Bird, and others provided a new and important transoceanic link, including the possibility of transmitting live television broadcasts across the Atlantic. Communications satellites may be either passive reflectors of radio signals, like Echo, or active repeaters, like Telstar. Scientific exploration of interplanetary space started with the space probes of the Ranger and Mariner series (U.S.A.) and the Lunik series (Russia).

Weather satellites, communications satellites and probes have one feature in common: they are unmanned. They are put into orbit or space trajectory by means of launching rockets (see Fig. 1), and their electronic equipment is powered by solar batteries—i.e., storage batteries which are kept charged by current generated in solar cells (Figs. 2a, 5a, 5b; see also Vol. I, page 102). All three types enable data derived from measurement of conditions in and around the satellite or probe (temperature, radiation, magnetic field, etc.) to be transmitted back to earth, where they are picked up and monitored. Also, they carry reception equipment for control signals (Fig. 2a)—e.g., for orbital correction by means of control jets—and are often provided with star-tracking systems for automatic orbital control by means of sensing devices for guidance with reference to the sun or a star, such as Canopus in the southern hemisphere. In most cases the space vehicle travels in a predetermined (programed) orbit, so that only the difference between the actual orbital data and the program have to be monitored and corrected.

The ground tracking stations for space probes and communications satellites are of a particular design (Fig. 2b). Their most important feature is a highly sensitive antenna (Fig. 3) with a gain of 60 db (one million times) in the reception of the weak signals picked up from the probe or satellite. The antenna is continuously swung with a directional accuracy of about one-thousandth of a degree—i.e., 3.6 seconds of arc. Connected to the antenna is a maser amplifier with a low noise factor. "Noise" in the present context denotes unwanted electrical signals generated within the apparatus (more particularly the amplifier) itself and disturbing or distorting the signals received. The maser is an amplifier which converts the energy of atoms into microwave energy (microwaves are very short electromagnetic waves in the frequency range of 1000–30,000 megacycles/sec.) and uses the electromagnetic waves to stimulate high-energy electrons to release energy, thereby amplifying the stimulating radiation. The process is independent of the random motion of electrons, so that maser amplifiers generate less noise than other types of amplifier.

(more)

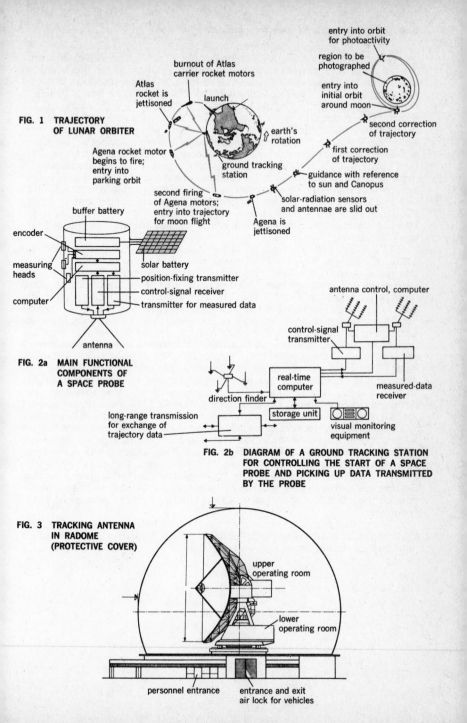

FIG. 1 TRAJECTORY OF LUNAR ORBITER

burnout of Atlas carrier rocket motors

Atlas rocket is jettisoned

launch

earth's rotation

ground tracking station

Agena rocket motor begins to fire; entry into parking orbit

second firing of Agena motors; entry into trajectory for moon flight

Agena is jettisoned

solar-radiation sensors and antennae are slid out

guidance with reference to sun and Canopus

first correction of trajectory

second correction of trajectory

entry into initial orbit around moon

entry into orbit for photoactivity

region to be photographed

buffer battery

encoder

measuring heads

computer

solar battery

position-fixing transmitter

control-signal receiver

transmitter for measured data

antenna

FIG. 2a MAIN FUNCTIONAL COMPONENTS OF A SPACE PROBE

antenna control, computer

control-signal transmitter

direction finder

real-time computer

measured-data receiver

long-range transmission for exchange of trajectory data

storage unit

visual monitoring equipment

FIG. 2b DIAGRAM OF A GROUND TRACKING STATION FOR CONTROLLING THE START OF A SPACE PROBE AND PICKING UP DATA TRANSMITTED BY THE PROBE

FIG. 3 TRACKING ANTENNA IN RADOME (PROTECTIVE COVER)

upper operating room

lower operating room

personnel entrance

entrance and exit air lock for vehicles

One of the main functions of a *weather satellite* is to report back on cloud formations to the ground tracking stations; in addition, the physical conditions such as temperature are measured and transmitted. The satellite is equipped with a television camera equipped with a vidicon tube (see Vol. I, page 124). In a typical camera of this kind, light is admitted to the single plate in the tube for 1/25 second, and the image is then scanned by an electron beam in 800 lines in a period of 200 seconds. In all, $800 \times 800 = 640,000$ picture points are thus scanned; the requisite band width is 1600 cycles/sec.; the transmitter frequency is 136 megacycles/sec.; the output is 5 watts. The antennae for picking up the television pictures from weather satellites need not be so elaborate as those used for communications or for the tracking of space probes far out in the solar system, and ordinary amplifiers instead of masers are generally employed. These satellites orbit at heights varying from about 500 to 1500 miles above the earth. One complete picture is scanned and transmitted every 208 seconds with the type of camera envisaged here; as already stated, the actual scanning time is 200 seconds, which is of course very much longer than in an ordinary television camera.

Space probes: The moon probes of the Ranger series were designed with the object of obtaining high-quality close-up pictures of the lunar surface (from an altitude of about 1000 ft.) and transmitting them back to earth. The three Ranger probes (1964) together sent back 17,000 pictures, some of which provide a thousand times better resolution of detail than could be obtained with moon photographs obtained through telescopes on earth. Ranger was equipped with two television cameras producing 1125-line images with a scanning time of 2.5 seconds. Light was admitted to the signal plate of the special vidicon tube for 1/250 of a second. Each of the two cameras, operating alternatively, took pictures at a rate of one every 5 seconds. They were transmitted at a frequency of 960 megacycles/sec., output 60 watts. The signals were picked up on earth by a parabolic antenna of about 90 ft. diameter and recorded on 35 mm photographic film and on Polaroid film with the aid of a television picture tube. Four other cameras in the probe obtained lower-definition close-up pictures (exposure time 1/500 sec., 282 lines, scanning time 1/5 second). The six cameras operated only during the last 20 minutes before impact.

Although the first pictures of the hidden side of the moon were transmitted back to earth by the Russian moon probe Luna 3 in 1959, serious mapping began with the Lunar Orbiter program. Each of these spacecraft was equipped with two cameras, one for wide-angle, the other for telephoto photography. The spacecraft traveled at a speed of 4500 mph in relation to the surface of the moon, and to prevent blurring, the film was moved slightly between exposures. Lunar Orbiters 1, 2 and 3 photographed the visible face of the moon; Lunar Orbiters 4 and 5 concentrated on the far side.

In 1965 the space probe Mariner IV (Fig. 5) took 21 pictures of the planet Mars from 6000 miles and transmitted them back to earth, a distance of about 140 million miles, using a transmitter with an output of only 10 watts. The spacecraft took 228

(more)

FIG. 4 SPAIN, PORTUGAL AND THE STRAITS OF GIBRALTAR AS PHOTOGRAPHED BY NIMBUS-A WEATHER SATELLITE FROM A HEIGHT OF ABOUT 550 MILES; PICTURE RECEIVED AT A TRACKING STATION AT MUNICH, GERMANY

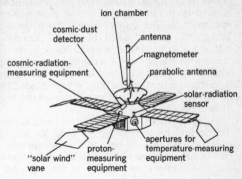

cosmic-dust detector
ion chamber
antenna
magnetometer
cosmic-radiation-measuring equipment
parabolic antenna
solar-radiation sensor
apertures for temperature-measuring equipment
"solar wind" vane
proton-measuring equipment

FIG. 5a

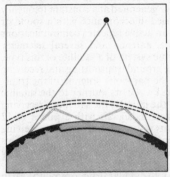

FIG. 6 UNDERSEA CABLE, SHORT-WAVE RADIO AND COMMUNICATIONS SATELLITE AS MEANS OF INTERCONTINENTAL COMMUNICATION

solar-radiation sensor
radiation telescope
television camera
Mars sensor
Canopus sensor

FIG. 5b DETAILS OF MARINER IV SPACE PROBE TO MARS

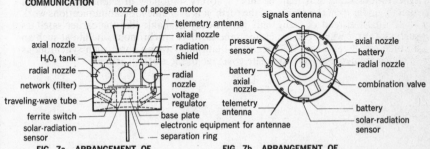

nozzle of apogee motor
telemetry antenna
axial nozzle
axial nozzle
radiation shield
H_2O_2 tank
radial nozzle
network (filter)
radial nozzle
traveling-wave tube
voltage regulator
ferrite switch
base plate
solar-radiation sensor
electronic equipment for antennae
separation ring

FIG. 7a ARRANGEMENT OF COMPONENT UNITS IN A SATELLITE (SIDE VIEW)

signals antenna
pressure sensor
axial nozzle
battery
battery
radial nozzle
axial nozzle
combination valve
telemetry antenna
battery
solar-radiation sensor

FIG. 7b ARRANGEMENT OF COMPONENT UNITS IN A SATELLITE (REAR VIEW)

days to reach the vicinity of Mars. Its vidicon camera took 200-line pictures, exposure time 1/5 second, scanning time 24 seconds. Each of the $200 \times 200 = 40,000$ picture points was assigned a numerical brightness value in a scale of 64 possible values by an analogue-digital converter and these values were stored in binary form on a magnetic tape. For transmission to earth, this tape was subsequently "played back" at a greatly reduced speed (because of the relatively low transmitter power) so that it took 8 hours and 48 minutes to transmit the data of a single picture that had been scanned and taped in 24 seconds (the recording speed of the tape was 13 in./sec., whereas the playback speed was only 0.01 in./sec.). At the tracking station the data were stored on a magnetic tape. When decoded, the numerical values were reproduced in their appropriate brightness intensities in a television picture tube and recorded on 35 mm photographic film.

Communications satellites, more particularly the active repeater type such as Telstar and its successors, are used for long-distance transmission of signals in circumstances that are beyond the performance range of other systems. This applies particularly to band width, which for television transmission is about 5 megacycles/sec. and is thus over a thousand times greater than the band width required for ordinary radiotelephonic communication. ("Band width" denotes the frequency limits of a given wave band; more specifically, it indicates the frequency range occupied by a modulated carrier wave—i.e., a wave generated at a constant frequency whose amplitude, frequency or phase is then varied in accordance with a sound or video signal to be transmitted.) Fig. 7 illustrates an active repeater communications satellite. Externally it carries a communications antenna and several telemetry antennae. Fig. 8 is a block diagram of the electronic system of a satellite of this type, while Fig. 9 represents the electronic system of the ground transmitting and receiving station. The signal beamed to the satellite from the parabolic antenna of the transmitter is equivalent to a power of about 2 million kw. On its journey to the satellite this signal is weakened about 10^{20} times, so that the input at the satellite's receiving antenna is 20×10^{-12} watt. The satellite transmits its signals with a power of 12 watts, which is weakened by a factor of 5×10^{19} on its way to the transatlantic receiving station, so that the input at the receiving antenna is only 2.4×10^{-19} watt. This extremely weak signal is amplified with the aid of maser amplifiers as already described.

Communications satellites of this kind are put into circular orbit around the earth at a distance of about 23,000 miles radius. The periodic time of the satellite in its orbit is 24 hours—i.e., synchronous with the earth's rotation. The advantage of the synchronous satellite is that, because it hovers stationary in relation to the point directly under it on the surface of the earth, it is available for communications work at all times, instead of being only periodically available like the earlier satellites such as Telstar which orbited close to the earth. A synchronous satellite at such an altitude can "see" about 120 degrees of latitude. At least three such synchronous satellites are necessary for complete coverage of the earth's surface.

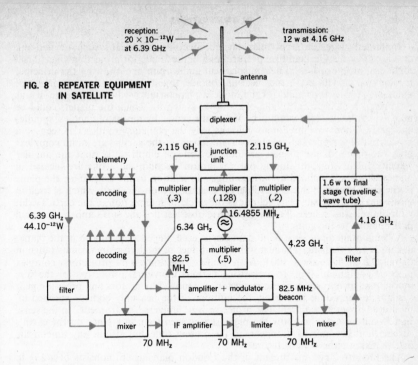

reception:
20 × 10⁻¹²W
at 6.39 GHz

transmission:
12 w at 4.16 GHz

antenna

FIG. 8 REPEATER EQUIPMENT
IN SATELLITE

diplexer

2.115 GHz | junction unit | 2.115 GHz

telemetry

encoding | multiplier (.3) | multiplier (.128) | multiplier (.2)

6.39 GHz
44.10⁻¹²W

1.6 w to final stage (traveling-wave tube)

16.4855 MHz

6.34 GHz

4.16 GHz

decoding

82.5 MHz

multiplier (.5)

4.23 GHz

filter

filter

amplifier + modulator

82.5 MHz beacon

mixer | IF amplifier | limiter | mixer

70 MHz | 70 MHz | 70 MHz

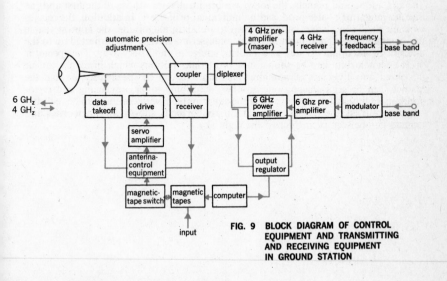

automatic precision adjustment

4 GHz pre-amplifier (maser) | 4 GHz receiver | frequency feedback | base band

coupler | diplexer

6 GHz
4 GHz

data takeoff | drive | receiver

6 GHz power amplifier | 6 Ghz pre-amplifier | modulator | base band

servo amplifier

antenna-control equipment

output regulator

magnetic-tape switch | magnetic tapes | computer

input

FIG. 9 BLOCK DIAGRAM OF CONTROL
EQUIPMENT AND TRANSMITTING
AND RECEIVING EQUIPMENT
IN GROUND STATION

PLANETARIUM

In modern usage the term "planetarium" denotes an optical system, devised and developed by the German firm of Carl Zeiss in the 1920s, for projecting an artificial night sky on the inside of a hemispherical auditorium and showing the principal motions of planets and other celestial bodies, together with other astronomical phenomena. The forerunner of the modern planetarium was the device known as the orrery, a mechanical model of the solar system in which the planets could be made to move in orbit around the sun in the center by means of handles, spindles and gears. There are three main reasons why the planetary motions as seen by a geocentric observer—i.e., an observer stationed on the earth—are quite complex: first, the orbits of the planets around the sun are elliptical, so that the angular velocity is not constant during one revolution; second, the orbits are located in planes which are inclined at various angles in relation to the ecliptic (or the plane of the earth's orbit); third, the observer is stationed on the earth which is itself in motion around the sun. The ecliptic is the circle of intersection of the earth's orbit with the celestial sphere. To the geocentric observer it is the sun's apparent annual path relative to the stars.

When a dome of at least 50 ft. diameter is used, the effect produced in the planetarium is very realistic. The hemispherical domes of modern planetariums range in diameter from 50 to about 100 ft. At the center of the dome is the projector, a complex piece of equipment (Fig. 1) comprising various individual projectors for the sun, moon, planets and stars. These projectors are moved by motors and precision gear systems to reproduce accurately the motions of the heavenly bodies—speeded up or slowed down, if required. The projector is a "universal instrument" in the sense that it can be made to show the night sky as seen from any place on the earth's surface, and not only as it appears at the present time, but also at any time in the distant past or the distant future.

The modern Zeiss instrument in the London planetarium consists of a $13\frac{1}{2}$ ft. long dumbbell-shaped assembly, weighing over two tons and comprising nearly two hundred optical projectors. It contains about 29,000 individual parts. Besides the sun, moon and planets, the forty-two brightest stars—those of the first and the second magnitude—are produced by individual projectors. In addition, the images of some 8850 other stars, ranging down to the sixth magnitude (the faintest visible to the unaided eye), and including some of lesser magnitude, are projected on to the interior of the dome, as well as a realistic image of the Milky Way. The spectator in the planetarium can be shown in a short time motions which in reality can be observed only by years of watching. The motions can be speeded up, so that the planets can be seen chasing one another, describing loops, advancing and retrograding. Eclipses of the moon and of the sun (complete with corona) can be realistically reproduced, the phases of the moon are correctly shown, and meteorites and comets can be seen racing across the night sky.

(more)

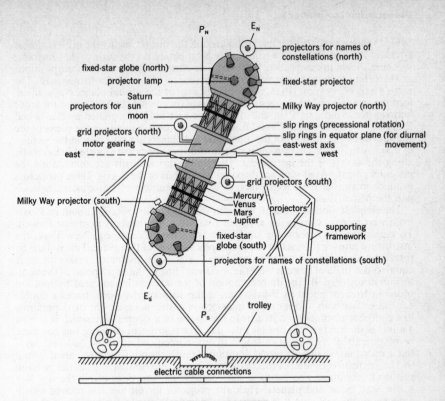

projectors for names of constellations (north)

fixed-star globe (north)

projector lamp

fixed-star projector

Saturn
projectors for sun
moon

Milky Way projector (north)

grid projectors (north)

slip rings (precessional rotation)

motor gearing

slip rings in equator plane (for diurnal movement)

east

east-west axis

west

grid projectors (south)

Milky Way projector (south)

Mercury
Venus
Mars
Jupiter

projectors

supporting framework

fixed-star globe (south)

projectors for names of constellations (south)

E_S

P_S

trolley

electric cable connections

FIG. 1 THE ZEISS PLANETARIUM PROJECTOR

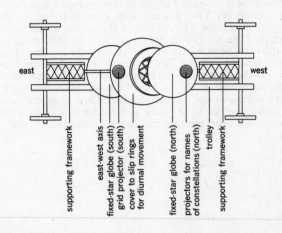

east

west

supporting framework

east-west axis

fixed-star globe (south)

grid projector (south)

cover to slip rings for diurnal movement

fixed-star globe (north)

projectors for names of constellations (north)

trolley

supporting framework

Fig. 3 shows the earth's annual orbit around the sun. (In reality the orbit is almost circular; its elliptical shape would hardly be apparent on the scale of this diagram; it is here shown in perspective, however.) The earth is represented in four positions in the orbit, corresponding to the four seasons of the northern hemisphere. The earth's axis of rotation (P) is not perpendicular to the orbital plane, but is tilted, forming an angle of 23.5 degrees with the axis of the ecliptic (E). This is also the angle between the polar axis $P_n P_s$ and the ecliptic axis $E_n E_s$ (designated as P-axis and E-axis) of the instrument. Viewed from the earth, the sun moves in the plane of the ecliptic, while the motions of the moon and the planets always occur in the vicinity of that plane. For this reason the projectors for these bodies are located individually along the E-axis of the instrument. The two spherical heads accommodating the projectors for the fixed stars are disposed at the ends of the E-axis. These projectors produce images of some 8900 stars, as well as a number of star clusters, nebulae and the Milky Way.

The simplest motion is due to the earth's diurnal (daily) rotation about its P-axis from west to east, causing an apparent rotation of the heavens from east to west. To reproduce this motion for any particular place on earth, e.g., New York, the instrument must first be set to the geographic latitude of that place. This is done by rotating it about its east-west axis until the P-axis of the instrument makes an angle equal to the angle of latitude with the horizontal line to the north point of the audi-torium dome (Fig. 4). The diurnal motion of the heavens is simulated by rotation of the instrument about its P-axis. During this rotation the E-axis traces a double conical surface with its apex at the intersection of the two axes. The earth performs one complete revolution on its axis in the course of a sidereal day which is almost 4 minutes shorter than the (mean) solar day. The instrument performs one complete revolution about its P-axis in a length of time corresponding to the "sidereal day," but speeded up, so that its actual duration is much shorter (it can be varied from 3 to 12 minutes). At the end of this period the instrument returns to its initial position relative to the dome, and the fixed stars occupy the same positions in the sky. Not so the sun, moon and planets. Thus, the projector for the sun has rotated nearly 1 degree to the east, and it is not until 4 "minutes" later (actually a much shorter length of time corresponding to the reduced time scale), when the diurnal rotation motors have rotated the instrument 1 degree farther westward, that the sun regains its initial position in the sky. The "diurnal" motion of the moon, as viewed in the planetarium, is very noticeable, whereas that of the planet Saturn, for example, is almost undetectable. The position of the instrument, in performing the diurnal rotation, must therefore be characterized not by the ordinary time of day, but by sidereal (or stellar) time, as in Fig. 4. This is because of the above-mentioned differ-ence in time between the sidereal day and the solar day. Fig. 3 indicates the positions taken up by the instrument in the course of demonstrating the earth's diurnal rotation in winter (far right) and summer (far left). These diagrams show how the position of the instrument has to be varied with the rotation of the earth in order to keep the E-axis of the instrument parallel to the E-axis of the ecliptic.

(more)

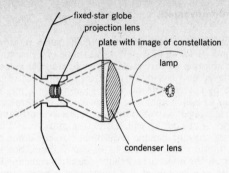

FIG. 2 FIXED-STAR PROJECTOR

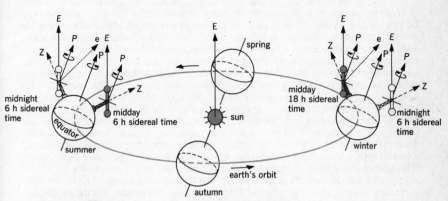

P = polar axis of earth
P = polar axis of instrument
Z = zenith direction of place
where instrument is located

E = axis of ecliptic (perpendicular to earth's orbital plane)
E = axis of ecliptic of instrument
e = direction of E-axis prior to half turn of instrument about P-axis

**FIG. 3 ANNUAL MOTION OF THE EARTH AROUND THE SUN
AND DIURNAL ROTATION OF THE EARTH AROUND ITS POLAR AXIS**

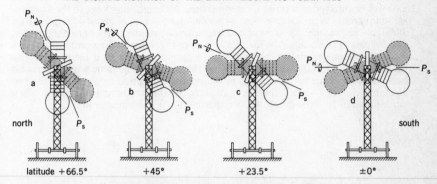

**FIG. 4 POSITIONS OF INSTRUMENT AT FOUR DIFFERENT
GEOGRAPHIC LATITUDES AND FOR 6 H (RED) AND
18 H (BLACK) SIDEREAL TIME**

The principle of the demonstration of the sun's motion in the planetarium is shown in Fig. 5. Because of the elliptical shape of the earth's orbit, the sun as seen by a geocentric observer moves with varying angular velocity in the course of the year. Since the eccentricity of the orbit is only slight, this variation can adequately be simulated by causing the earth, represented by a pin, to rotate with constant angular velocity on a circular disc whose axis of rotation corresponds to the E-axis, while the sun is given a fixed amount of eccentricity. Earth and sun are interconnected by a guide rod, at one end of which is the sun projector. When the disc is rotated at constant speed, the sun's image projected on to the dome moves at varying speed. The moon's motion is simulated in similar fashion, but the mechanism is more complex because the lunar orbit around the earth is inclined at an angle of 5 degrees to the ecliptic and moreover itself performs a precessional rotation once every $18\frac{1}{2}$ years.

Because of the eccentricity of planetary orbits, the accurate geocentric representation of the motion of a planet is by no means simple, and a full explanation would be outside the present scope. Fig. 6 merely represents the simplified case where the earth's orbit and the planet's orbit are concentric circles situated in the same plane, with the sun at the center. The pins representing the earth and the planet rotate in their respective orbits around the sun. The two pins are interconnected by means of a guide rod, so that the image of the planet (like that of the sun in the mechanism described above) is always projected in the direction away from the earth, and the image is thus always viewed geocentrically. With the planetarium instrument it is also possible to demonstrate the precessional motion of the earth's axis, or "precession of the equinoxes," a phenomenon already known in ancient times and subsequently explained by Newton as being caused by the equatorial bulge (or the flattened shape at the poles) of the earth. It is characterized by the fact that the earth's axis revolves in a small circle about the pole of the ecliptic in a period of 26,000 years. This causes the equinoctial points to move retrograde in the ecliptic by an annual amount of a little over 50 seconds of arc, so that the coordinates of the stars are gradually altered in various ways. The equinox is the instant, occurring twice a year, at which the sun apparently crosses the celestial equator. The two points, diametrically opposite each other, in which the ecliptic cuts the celestial equator are called the equinoctial points. In practical terms the precessional motion means that the earth's axis is not steady in space, but performs a slow "wobble," like the axis of a spinning top. At present the axis points approximately in the direction of the Pole Star (Polaris), in the constellation of Ursa Minor, but in 13,000 years' time the axis will be directed at a point located between the bright star Vega in the Lyre and the star gamma in the Dragon—i.e., at an angular distance of about 47 degrees from Polaris. After another 13,000 years the axis will have returned to its present position, and Polaris will once again be the North Star. In Fig. 7 (upper diagram) the direction P_A is the present position of the earth's axis at midsummer; P_B is the position that the axis will have 13,000 years hence, when the earth is at the same point of its orbit, but—because of the tilt of the axis in the opposite direction as a result of the precessional motion—it will then be midwinter in the northern hemisphere. These variations can be reproduced, greatly speeded up, in the planetarium.

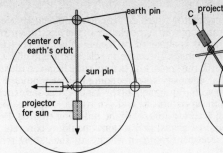

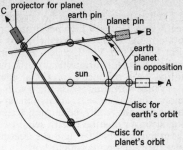

FIG. 5 THE SUN'S ANNUAL ORBIT SEEN BY A GEOCENTRIC OBSERVER

FIG. 6 GEOCENTRIC MOTION OF A PLANET (MARS)

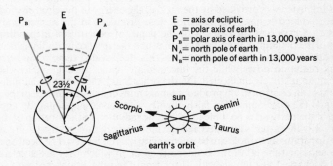

E = axis of ecliptic
P_A = polar axis of earth
P_B = polar axis of earth in 13,000 years
N_A = north pole of earth
N_B = north pole of earth in 13,000 years

FIG. 7 PRECESSION OF THE EARTH'S AXIS

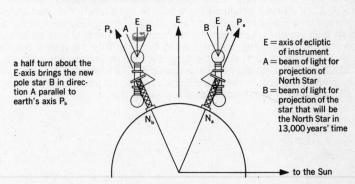

a half turn about the E-axis brings the new pole star B in direction A parallel to earth's axis P_b

E = axis of ecliptic of instrument
A = beam of light for projection of North Star
B = beam of light for projection of the star that will be the North Star in 13,000 years' time

MEDICAL APPLICATIONS OF THE BETATRON

The betatron is an electron accelerator which can produce high-energy electron beams and X-rays of high penetrating power. Acceleration of the electrons traveling in a circular orbit of constant radius is achieved by means of a rapidly changing electric field (see Vol. I, page 108). In medical science the electron beam and X-rays are employed for various therapeutic purposes. For example, the betatron illustrated in Fig. 1 can produce X-rays of an intensity corresponding to that produced in an X-ray tube operating at 42 million volts. Equipment of this kind is large and expensive. Important requirements are that the apparatus be easily movable, simple to operate, safe, and—above all—permitting accurate control of the radiation emitted. Protection against radiation hazard requires special precautions, including concrete walls about five feet thick around the treatment room, in which only the patient remains once the radiation is switched on.

The biological action of electron beams and X-rays is based on the splitting of macromolecules, which is effected through the medium of ionization processes. Cells in pathological growths (tumors, etc.) are more severely affected by the radiation. The actual effect of the radiation on tissues is due to electrons set in motion within them, these electrons being either directly supplied in the radiation (electron beams) or released as a secondary phenomenon from atoms bombarded by electromagnetic radiation of very short wavelength (X-rays). A measure for the radiation dose administered is provided by the ratio of the energy of the electrons to the mass of the irradiated region. The radiation dose produced in the human body by the action of X-rays first increases with the depth of penetration (depending on the energy of the radiation), attains a maximum, and then gradually diminishes according to the attenuation of the rays. Because of this, deep-seated tumors can be subjected to high radiation doses while the intermediate tissues through which the X-rays pass remain relatively unaffected. On the other hand, the radiation administered in the form of an electron beam produces a maximum dose at the surface of the patient's body, and the dose diminishes fairly rapidly with increasing depth. By appropriate choice of the energy it is possible to adjust the radiation to the location of the tumor, so that only the latter is attacked and the more deeply situated tissues are spared.

The apparatus as a whole can be swung in various directions. The electrons injected to the annular tube (a) of the betatron (Fig. 2), in which they are accelerated, are either discharged at high velocity direct from the tube or directed on to an anticathode ("target") (b) to produce X-rays (see Vol. I, page 106). The betatron is fed with alternating current which generates a magnetic field (c). The electrons are injected at (d) into the annular tube, which is enclosed in a vacuum jacket (e). An alternating electric field is produced in the tube by the magnetic field and maintains the electrons on a predetermined circular path (f), whereby they are speeded up to extremely high velocities. The magnetic field and the electric field must be accurately interadjusted. The electrons can be emitted from the tube through a thin metal "window" at the periphery, when the magnetic field is weakened (Fig. 3). On the other hand, when the intensity of the magnetic field is increased, the radius of the electron path is reduced, so that the electrons now strike an anticathode (b) at the inner wall of the tube and therefore produce X-rays (Fig. 4).

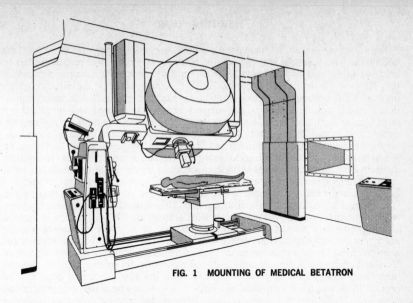

FIG. 1 MOUNTING OF MEDICAL BETATRON

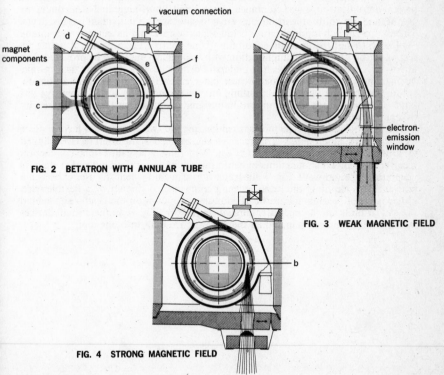

vacuum connection

magnet
components

d

e

f

a

b

c

FIG. 2 BETATRON WITH ANNULAR TUBE

electron-
emission
window

FIG. 3 WEAK MAGNETIC FIELD

b

FIG. 4 STRONG MAGNETIC FIELD

Fig. 1 shows a section through the human ear, comprising the outer ear, the middle ear, and the inner ear. Beyond the eardrum is the air-filled middle ear cavity bridged by a chain of three small bones (auditory ossicles) forming a mechanical link between the drum and the so-called oval window, which is the entrance to the inner ear. The semicircular canals in the inner ear are concerned with bodily equilibrium, while the fluid-filled spiral, shaped like a snail's shell and divided by a partition, serves for hearing. Located on the partition is the organ of Corti, a complex structure in which the auditory receptor cells are embedded. These cells are connected to the ends of the auditory nerve fibers. Defective hearing may be caused by a functional disorder, due to an accident or disease, at any point in this system, including the auditory nerve and indeed the auditory center in the brain. In some cases an impairment can be cured or alleviated by medical treatment or surgery. In other cases it may be possible to obtain improvement by means of a hearing aid.

In general, a hearing aid is a sound amplifier. The earliest form was the ear trumpet, which amplifies sound by collecting it with a large mouth and leading it down a tapering tube to a narrow orifice which is inserted into the ear. A modern hearing aid is a transistorized electronic device serving to amplify sound by means of electrical amplification. The sound is picked up by a microphone, which converts it into weak electrical currents. These are amplified and passed to the receiver, which converts them back into sound of greater loudness than the original sound (Fig. 2). The power for the amplifier is supplied by a battery, which may be of the ordinary dry-cell type or a rechargeable storage battery. Present-day hearing aids may comprise several amplification stages combined into a single unit (integrated semiconductor circuit) and are adjustable in various ways to suit them to individual requirements. The user can switch the apparatus on and off, as desired, and he can vary the volume (loudness) to suit the acoustic conditions. Tone control, i.e., a choice of frequency response (the variation of amplification with frequency), may also be provided. The intensity range in which speech is understood may be wide for some deaf people, but narrow for others. In some of these latter cases a hearing aid with automatic volume control may help by smoothing out the variations in sound intensity; this kind of control varies the amplification automatically, so that the output intensity is kept constant.

The user of the early type of electrical hearing aid wore earphones held in place by a headband. Later, this external receiver was replaced by the insert receiver clipped to a molded plastic insert in the outer ear. Sometimes a so-called bone-conduction receiver is used, which functions by vibrating the bones of the skull rather than by generating a sound wave. Fig. 3 illustrates a hearing aid which the user carries in a pocket of his clothing, the receiver being connected to the aid by a flexible wire. Other types of hearing aid are small enough to be worn on the head—e.g., behind the ear, or built into an eyeglass frame, or in the ear (Fig. 4). In these small devices all the component parts, including the receiver, are built into one unit.

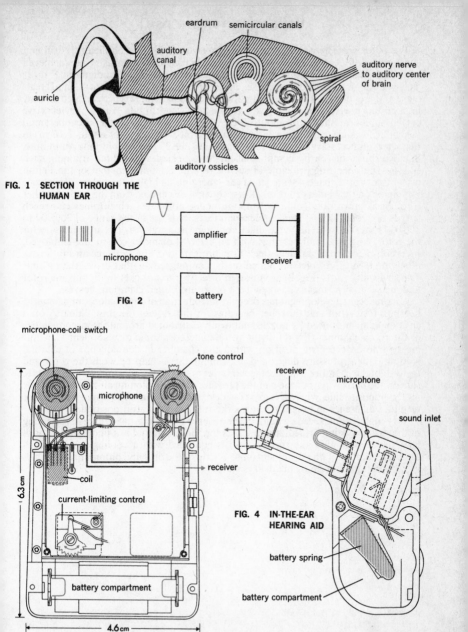

FIG. 1 SECTION THROUGH THE HUMAN EAR

auricle

auditory canal

eardrum

semicircular canals

auditory nerve to auditory center of brain

spiral

auditory ossicles

FIG. 2

microphone

amplifier

receiver

battery

FIG. 3 POCKET HEARING AID (REAR COVER OPENED)

microphone-coil switch

tone control

microphone

coil

current-limiting control

receiver

battery compartment

6.3 cm

4.6 cm

FIG. 4 IN-THE-EAR HEARING AID

receiver

microphone

sound inlet

battery spring

battery compartment

The outer protective layer of a tooth, the enamel (Fig. 1), is difficult to drill and cut because of its great hardness. Considerable progress in dentistry was achieved by the introduction of small drilling tools tipped with hard metal or diamond (Fig. 2). To obtain high cutting efficiency, i.e., faster preparation of cavities for dental treatment and therefore less pain and discomfort to the patient, it is necessary to employ high cutting speeds. Diamond and hard-metal-alloy drill bits exert a very effective cutting action at speeds of about 60 or 70 ft./sec. To achieve the circumferential velocities for such cutting speeds with a drill bit which is only about 1/16 in. in diameter it is necessary to rotate it at something like 250,000 revolutions per minute. Small turbines driven by compressed air have proved suitable for attaining such speeds. The operating principle of such a turbine is similar to that of the Pelton wheel (see Vol. I, page 48): a jet of air emerging at high velocity from a nozzle impinging on the blades of a rotor; the spent air is discharged from the rotor chamber through a return duct (Fig. 3). Running under no load, such turbines can reach speeds of about 400,000 rpm; under load the speeds are in the range of 200,000 to 300,000 rpm. Very carefully manufactured high-precision bearings are necessary for the rotor. As a rule, ball bearings with an external diameter of $\frac{1}{4}$ in. are employed, precision-made from stainless steel. For lubrication of the bearings an oil "mist" (finely divided oil droplets) is added to the compressed air that drives the turbine. The small drill tools (Fig. 2) are inserted directly into a socket in the turbine rotor and are secured by means of screw-type or spring-loaded clamping sleeves.

A more recent development has been the introduction of air-cushion, or pneumatic, bearings (Fig. 4) instead of roller bearings. In this system the rotor "floats" on a cushion of air produced by nozzles uniformly distributed around the circumference of the rotor chamber (Fig. 5). If the rotor undergoes some displacement and thus moves into an eccentric position within the chamber in consequence of external loading (lateral pressure developed in drilling), the air gap between the rotor and the wall of the chamber will become narrower on one side and wider on the opposite side. In the narrow part of the gap the pressure of the cushioning air builds up, while on the opposite side, where the gap is widest, a decrease in air pressure occurs. The resulting difference in pressure returns the rotor to its central position.

Very high cutting speeds would, in the absence of cooling, cause excessive heating of the surfaces being drilled, inasmuch as the tooth material is a poor conductor of heat. To prevent this, the turbine drill is equipped with a cooling system which directs a number of fine water jets on to the tip of the drill and automatically comes into operation as soon as the drill is started (Fig. 1).

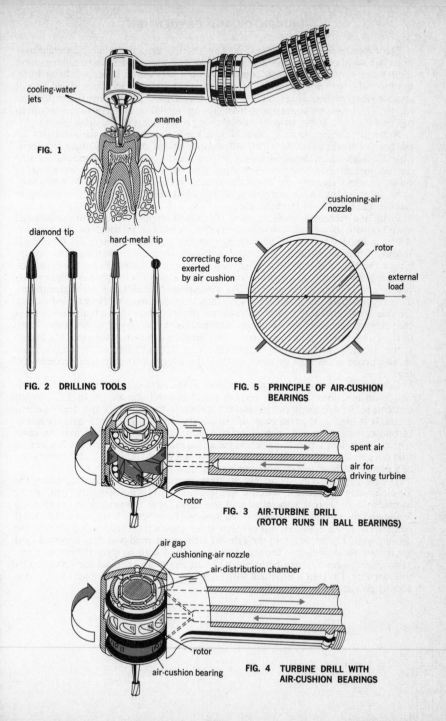

cooling-water jets

enamel

FIG. 1

diamond tip

hard-metal tip

FIG. 2   DRILLING TOOLS

cushioning-air nozzle

rotor

correcting force exerted by air cushion

external load

FIG. 5   PRINCIPLE OF AIR-CUSHION BEARINGS

spent air

air for driving turbine

rotor

FIG. 3   AIR-TURBINE DRILL (ROTOR RUNS IN BALL BEARINGS)

air gap

cushioning-air nozzle

air-distribution chamber

rotor

air-cushion bearing

FIG. 4   TURBINE DRILL WITH AIR-CUSHION BEARINGS

INCUBATOR AND OXYGEN TENT

The function of an incubator (Fig. 1) is to help sustain the life of prematurely born babies who, because of their low weight and weak general constitution, have diminished viability. For such babies, weighing perhaps as little as 2 lb. at birth, the incubator provides an ideal life environment, capable of being varied to suit specific requirements, characterized by: sterility (freedom from germs), temperature (up to 37°C), relative atmospheric humidity (up to 100%), an oxygen concentration (up to 40%—i.e., about twice the normal oxygen content of the atmosphere).

Normally, the prematurely born baby is kept continuously in the incubator for the first few weeks of its life. All nursing manipulations are done by means of special holes through which the nurse inserts her arms and which are provided with sealing devices that fit closely around the arms. The baby is fed, weighed (by means of built-in scales), cleaned and medically treated inside the incubator. Minor operations, transfusions and X-ray photography can likewise be done without having to remove the baby from the incubator.

From the technical point of view the incubator is a special air-conditioned chamber with its own climate-control system. A built-in motor draws in fresh air, which is freed from germs by passing through a bacterial filter and then flows through a thermostatically controlled heating system and a water evaporator (to give it the required humidity). If necessary, oxygen is added. Because of the continuous inflow of air—at a rate of about 6 liters (0.2 cu. ft.) per minute—a slight excess pressure in relation to the surrounding atmosphere is maintained in the incubator. This ensures that the airflow is always directed outwards and the exhaled carbon dioxide is carried away and no external air can penetrate through any small leaks that may be present in the enclosure. The incubator is equipped with various safety devices which sound a warning and/or take preventive action in the event of a failure in the electric power, water or oxygen supply, or if the temperature in the incubator should become too high or too low, or if the oxygen concentration becomes too high.

Comparable in principle to the incubator is the oxygen tent (Fig. 2). It performs two main functions: to supply oxygen-enriched breathing air (30 to 50% oxygen content) to the occupant and to cool the air within the tent by 5 to 8 degrees centigrade. It is used in the treatment of respiratory diseases and heart diseases and in certain cases for the care of persons recovering from serious operations. In cases where the patient's breathing functions are impaired, the oxygen tent can ensure that the normal oxygen content in his blood is maintained, and the body temperature of a patient with very high fever can be kept down.

The oxygen tent consists mainly of a transportable cabinet (accommodating the air-circulating equipment, electric cooling unit, oxygen-dispensing unit, water atomizer, and control apparatus) and the actual tent comprising the supporting frame and the envelope, which is fitted around the patient's bed so as to enclose it completely. A fan draws air out of the tent and passes it through a dust filter to the cooling unit. The moisture in the exhaled air is condensed and thus removed. The air is then recirculated to the tent and is enriched with oxygen. If necessary, the atmospheric humidity can be increased by means of an artificial fog produced by the atomizer. The tent is provided with large openings, closed by zippers (zip fasteners), to give access to the patient.

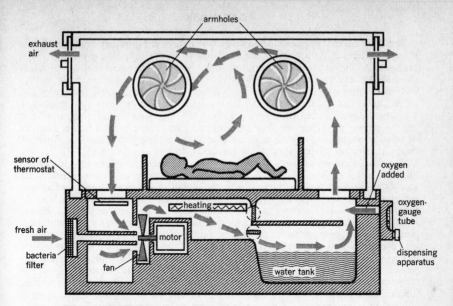

armholes

exhaust air

sensor of thermostat

fresh air

bacteria filter

fan

heating

motor

water tank

oxygen added

oxygen-gauge tube

dispensing apparatus

FIG. 1 INCUBATOR

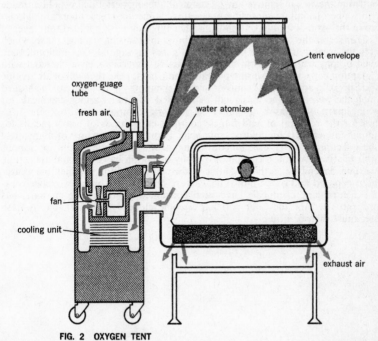

tent envelope

oxygen-guage tube

fresh air

water atomizer

fan

cooling unit

exhaust air

FIG. 2 OXYGEN TENT

405

The function of a gas mask is to protect the wearer's respiratory organs and eyes from the effects of poison gases, fumes and dust. Various types of protective mask are used for industrial and for military purposes. For the successful use of a gas mask that filters the air through chemicals in a canister, the basic condition is that the toxic fumes or gas are present in relatively low concentrations (generally not exceeding about 2% by volume) in the air and that the air must also contain a sufficiently high content of oxygen (at least 15% by volume, and at least 17% if carbon monoxide is present). For exposure to atmospheres with a higher content of toxic constituents, a self-contained type of breathing apparatus—i.e., with its own independent oxygen supply—has to be worn.

Gas masks are widely used in industry—e.g., in chemical plants and in certain mining operations where fumes of an injurious character occur. Firemen and rescue squads are also normally equipped with gas masks. A gas mask consists of a face-piece, straps for attaching the mask to the wearer's head, and a canister for filtering the inhaled air and absorbing gases and fumes from it. In one type of mask the exhaled air is discharged through the canister, i.e., air inlet and outlet are combined. Another type is equipped with a separate outlet valve for discharging the exhaled air. The facepiece is molded to fit closely around the wearer's face so as to form a gastight seal around mouth, nose and eyes, thus ensuring that only air which has passed through the canister is inhaled. In the type of mask illustrated in Fig. 1, the canister is screwed to the inlet opening located approximately at chin level. The inhaled air is purified in the canister and thus made safe to breathe. Purification is effected by a combination of physical and chemical processes. Particles or droplets suspended in the air are removed by mechanical filtering performed by a filter made of various fibers (cellulose, glass fibers, asbestos). Sometimes these fibers are of loose texture in the form of a thick felt pad; in other types of mask a folded thin layer of filter paper serves the same purpose (Fig. 2 shows the canister of a dust mask, which protects the wearer from dust but not from gases or vapors). Gas molecules are removed by physical adsorption on surface-active materials (active charcoal with high retention capacity); this principle can be utilized for the removal of all organic vapors (Fig. 3). In addition, the canister may—depending on the nature of the hazard to which the wearer of the mask will be exposed—contain various chemicals for absorbing particular gases or fumes by forming compounds with them—e.g., alkalies for the removal of acid fumes, complex compounds of heavy metals for ammonia, copper salts for hydrocyanic acid. Hopcalite (a mixture of manganese dioxide and cupric oxide) is used for converting carbon monoxide, a highly poisonous gas, into relatively harmless carbon dioxide by oxidation based on catalytic action. After a time, depending on the gas, fume or dust concentration to which the wearer has been exposed and on certain other factors, the canister becomes ineffective— i.e., the neutralizing chemicals have been consumed, or the active charcoal has become saturated, or the filter pad has become clogged with dust, etc. A fresh canister must then be fitted.

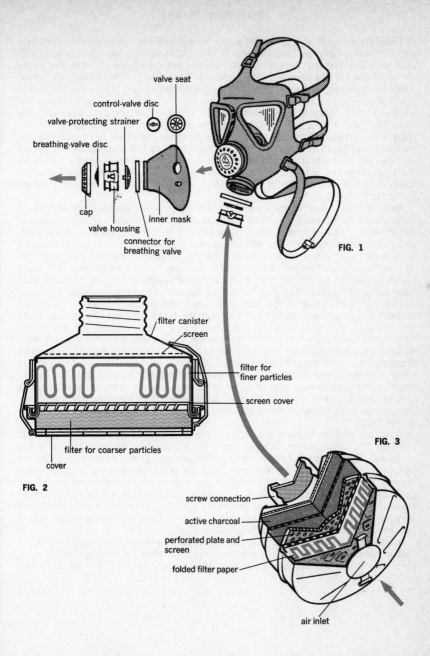

valve seat

control-valve disc

valve-protecting strainer

breathing-valve disc

cap

valve housing

connector for
breathing valve

inner mask

FIG. 1

filter canister

screen

filter for
finer particles

screen cover

filter for coarser particles

cover

FIG. 2

FIG. 3

screw connection

active charcoal

perforated plate and
screen

folded filter paper

air inlet

The human heart is a muscular pump comprising four separate cavities and a series of valves allowing blood to pass in one direction only. Man, like other mammals, has a double circulatory system. Blood that has parted with oxygen to the tissues (9 and 13) and absorbed carbon dioxide from them (venous blood) is returned to the heart through the superior and the inferior venae cavae (11 and 10). This blood enters the right auricle (3), whose contractions cause the blood to pass through the tricuspid valve (16) into the right ventricle (1). The contractions of the right ventricle pass the blood through the pulmonary semilunar valves (17) and along the two pulmonary arteries (5) to the lungs (6). In the lungs the blood is oxygenated and returns to the heart through the pulmonary veins (7) and thus enters the left auricle (4). This chamber contracts and passes the blood through the bicuspid, or mitral, valve (15) into the left ventricle (2), whose contractions force the blood through the aortic semilunar valves (18) into the aorta (12 and 13), which is the biggest artery of the body (Fig. 1).

Thus the right side of the heart serves mainly to pump deoxygenated blood through the lungs, while the left side pumps oxygenated blood throughout the rest of the body. This is represented schematically in Fig. 2. The output, or rate of flow, of the blood pumped into the arteries by the ventricles varies to suit the body's requirements. Physical effort—manual labor, running, climbing, etc.—increases the output of blood; when the body is at rest, the output diminishes. The heart varies the output by varying the volume of blood admitted into the ventricles each time the latter are filled and also by varying the rate of contraction (faster or slower heartbeat). The left side of the heart (left auricle and ventricle) has to circulate the blood through all parts of the body, except the lungs, and has thicker and more strongly muscular walls than the right side, which has to perform the pulmonary blood circulation only. For proper functioning, the left side and the right side must be accurately interadjusted, both with regard to the contraction rate of the respective chambers and with regard to the output of blood. In certain diseased conditions there is an imbalance between the two sides of the heart because one side fails to function adequately (e.g., due to valve malfunctioning or other causes) and thus cannot cope properly with the flow of blood pumped by the other (healthy) side. This causes congestion of blood in the blood vessels, especially in the lower part of the body.

What can be done to correct such functional disorders of the heart? For example, consider the relatively simple case where the bicuspid or the tricuspid valve has sustained damage and has ceased to function properly. If it is a case of constriction (stenosis), so that the valve does not open properly, modern surgery can often remedy the defect: dilation of the valve can restore its function and enable the patient to lead a normal life again. In a case where the valve fails to close completely, the valve opening can be reduced in size by stitching. However, if the valves have been destroyed—as a result of disease, for example—the only remedial action consists in the insertion of an artificial valve (Fig. 3, p. 411).

A ball valve is used for the purpose. It comprises a small metal cage in which a plastic ball can move up and down so as to open and close the passage. Such a valve is secured in position in the heart or in a blood vessel (to replace a semilunar valve, for example).

(more)

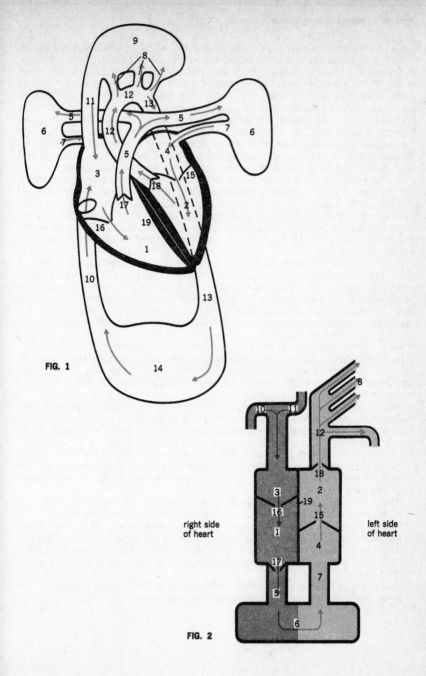

FIG. 1

FIG. 2

right side
of heart

left side
of heart

Fig. 4 shows the mitral valve of the heart replaced by an artificial valve of this type. In Fig. 4a the left auricle has been filled with oxygenated blood from the lungs. At the instant when the heart muscle slackens and expands, suction is developed in the left ventricle, so that the ball is lifted off its seat and thus opens the valve passage, allowing the ventricle to fill with blood. Simultaneously, the aortic semi-lunar valves (at the entrance to the aorta, or main artery) close. This rhythmic expansion of the heart whereby blood is drawn into the ventricle is called the diastole. In the next stage, called the systole, the heart contracts and pumps the blood through the arteries, the aortic semilunar valves now being open (Fig. 4b). When this happens, the pressure of the blood in the contracting ventricle forces the ball of the artificial valve back on to its seat, so that no blood can flow back into the auricle.

During operations on the heart it is essential to keep the patient's blood circulating. The best way is to let a machine temporarily perform all the functions of the heart and also of the lungs. The human organism, especially the brain, is very sensitive to oxygen deficiency. If the supply of oxygen (via the blood) to the tissues is stopped for more than ten minutes, permanent damage to the organism or death will ensue. For this reason a device called a heart-lung machine is used in major heart operations (see Vol. I, page 446).

A sensational development in modern cardiac surgery has been the heart transplant —i.e., the replacement of the diseased or damaged heart by the heart of another person, the donor, who usually is the victim of an accident or has died from some cause other than a heart disease. The problems of tissue rejection have not yet been fully overcome, but it is likely that the rejection mechanism, which causes the transplanted heart to be treated as an "intruder" by the body, will be better understood in due course, so that effective measures can be taken to improve the patient's chances of survival and resumption of a more or less normal life.

In a more distant future it may become possible to replace a defective heart by an artificial heart—a wholly man-made pumping machine that can be implanted permanently into the body. An artificial heart of this kind would have to fulfill the following requirements:

1. The "right" and "left" sides must be accurately interadjusted, so that their respective outputs (flow rates) are properly in balance.
2. The pump would have to be double-acting in a case where both sides of the heart are replaced.
3. In order to cope with the body's increased demand for blood under conditions of physical exertion, both the frequency (number of "beats" per minute) and the output (rate of flow in cubic centimeters per minute) must be capable of variation within wide limits.
4. The pressure in the auricles must be kept accurately within specific limits, as it is this pressure that governs the output of the respective ventricles.
5. The output of the right ventricle must not vary too greatly in relation to that of the left ventricle; otherwise congestion of blood in the blood vessels would occur.
6. The artificial heart must not heat up or, alternatively, the heat must be quickly carried away by the bloodstream and dissipated in the lungs and other parts of the body.
7. The artificial heart must be small, and capable of being easily installed in the body, and ensure reliable maintenance-free operation.

(more)

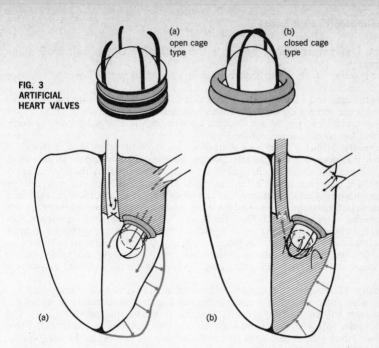

FIG. 3
ARTIFICIAL
HEART VALVES

(a)
open cage
type

(b)
closed cage
type

(a)

(b)

FIG. 4

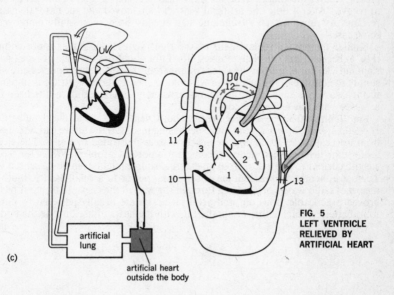

(c)

artificial
lung

artificial heart
outside the body

11
3
10
1
2
4
12
13

FIG. 5
LEFT VENTRICLE
RELIEVED BY
ARTIFICIAL HEART

Implanting an artificial heart into a living human body presents some major problems:

1. The material of which the heart consists must not call forth a defensive rejection from the body.
2. The material must be smooth so as not to damage the blood corpuscles, as that could have serious effects (thrombosis: clotting of the blood).
3. The artificial heart must not become electrostatically charged inside the body, as this too might cause clotting.

Hence the artificial heart would have to be made of a synthetic material—a plastic—that does not undergo rejection by the body. Such materials already exist. Also, it is possible to cancel changes in the electrical potential by the application of a corrective current from an external source. The artificial heart can moreover be coated with a synthetic substance that can absorb heparin, an anticoagulant which prevents clotting of the blood. And these are by no means all the requirements that the artificial heart must fulfill. Thus, the heart—or rather, the pump—must not suck in too much blood from the veins, as this would cause the latter to collapse and bring the circulation to a standstill. The many factors and variables that must be supervised and controlled require an automatic monitoring device which, for example, regulates the "heartbeats" (i.e., the pulse rate), keeps a check on temperature and output rate, etc.

Because of the functional division of the human heart into two systems—left and right—which operate separate circulations (see Fig. 2), it is, in cases where one side of the heart fails to function properly, possible to bypass that side for a time. This can be done with the aid of relatively simple and inexpensive equipment. Fig. 5 (on p. 411) shows the bypassing of part of the left side of the heart by the interposition of an artificial ventricle between the left auricle and the aorta. This arrangement greatly relieves the strain on the natural ventricle, so that it can rest and get a chance to recover. After a time, the artificial ventricle is removed and the natural ventricle resumes its function. This technique has already been successfully employed in some cases.

Similar temporary replacement of the right ventricle is rather more complex (Fig. 6) because this ventricle receives blood not only from the venae cavae (via the right auricle) but also from the veins coming from the coronary blood vessels, which supply the muscle of the heart wall with blood. The artificial ventricle is implanted in the two venae cavae and the pulmonary artery. Encouraging results have been achieved with this technique, too.

Fig. 7 illustrates the replacement of both ventricles by artificial substitutes. It would appear, however, that more promising results in the future can be expected from artificial parts of compact construction, as illustrated in Fig. 8. The electric motors for driving such hearts present an important problem. Wires for feeding electric current to the motor from an external source might even be dispensed with. There are theoretical possibilities of supplying power to a small electric motor by means of radio waves penetrating through the wall of the chest. Hitherto it has not proved practicable, however, and artificial hearts are usually powered by current from batteries which either are grafted in with the heart or are outside the body.

(more)

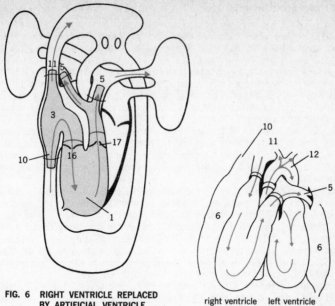

FIG. 6 RIGHT VENTRICLE REPLACED BY ARTIFICIAL VENTRICLE

right ventricle   left ventricle

FIG. 7 BOTH VENTRICLES REPLACED BY TWO SEPARATE ARTIFICIAL VENTRICLES

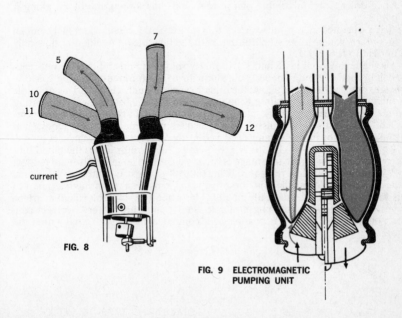

current

FIG. 8

FIG. 9 ELECTROMAGNETIC PUMPING UNIT

The main problem, however, is presented by the design and construction of the actual pumping unit with its drive system. The following systems have been developed:

1. Devices powered by atomic energy. A device of this kind comprises a capsule containing radioactive isotopes and a miniature steam engine. Such capsules have already been tried experimentally on animals. Weighing about 400 grams, including the radiation-protection shield, they function continuously, and without requiring attention, for about two years.

2. Piezoelectric devices (see Vol. I, page 86). A pumping chamber is located between two piezoelectric crystals, which have the property of expansion and contraction when subjected to an alternating electric field and can thus be made to develop a rhythmical pumping action. So far, however, this promising principle has not been put into actual effect.

3. Electromagnetic drive (Fig. 9). The electromagnet consists of a coil and a core which is connected to a diaphragm. The space between the coil and the diaphragm is filled with a liquid. When energized, the coil pulls the core into it, and the diaphragm, actuated by the motion of the core, forces the operating liquid into the spaces around the two flexible "ventricles," thereby compressing them and causing them to pump the blood into the arteries. When the coil is deenergized, the diaphragm springs back to its initial position, so that the liquid returns to the space it originally occupied and blood can flow into the now distended ventricles.

4. Pumping units driven by electric motors. There are several types of artificial heart based on electric motor drive. The first three of these are used mainly for the "extracorporeal hearts" (i.e., outside the body) employed in heart-lung machines:

(a) Screw pump comprising a screw rotating within a flexible-walled chamber (Fig. 10).

(b) Finger pump, in which a series of "fingers" perform a rhythmic wavelike motion, pressing down on to a flexible plastic tube and thus forcing the blood along it (Fig. 11).

(c) Valve-operated pump, in which a ram squeezes the blood alternately out of plastic tubes, while valves synchronized with the ram allow the blood to flow only from right to left (Fig. 12).

(d) Oscillating pump (Figs. 8 and 13). In this type of pumping unit the two ventricles are compressed not simultaneously but alternately. Each time the oscillating "piston" squeezes one "ventricle," it allows the other to expand, and vice versa. This principle of alternate action has certain advantages. Thus, there is better energy utilization in the course of the operating cycle, so that a smaller and lighter motor can be employed. There is also better dissipation of heat from the motor to the bloodstream because the motor in this arrangement is directly adjacent to the ventricles.

(e) Rotary pump (Fig. 14). In this type of pump a lobed rotor forces the blood into the "ventricles." The great advantage of this system is that no inlet valves are needed, such valves with low flow resistance being difficult to construct for artificial hearts. The rotor also performs the function of valves. A cross section through a pump of this kind is shown schematically in Fig. 14a, while Fig. 14b is a diagram which serves to illustrate the pumping principle. Such pumps are of very compact construction, the more so as it is possible to accommodate the motor itself within the rotor.

(more)

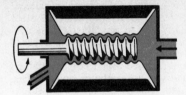

FIG. 10   SCREW PUMP

FIG. 11   FINGER PUMP

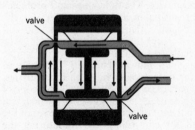

valve

valve

FIG. 12   VALVE PUMP

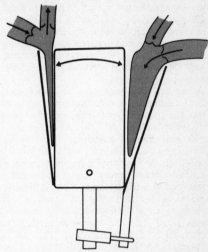

FIG. 13   OSCILLATING PUMP

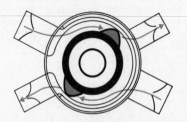

FIG. 14a   ROTARY PUMP

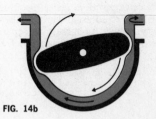

FIG. 14b

5. Pumping units powered by compressed gas or air. The source of power for these artificial hearts is located outside the chest and can be regulated by means of external controls. Such pumps are especially suitable as replacements for one or the other side of the natural heart, as the two sides are fairly easy to separate in this system. This also makes for easier implantation in the patient's chest. There are two types of artificial heart operating on this principle:

(a) Bag-type heart (Fig. 15). A flexible plastic bag is enclosed within a rigid container. Compressed air is admitted into the space between the container and the bag, causing the blood to be squeezed out of the latter. Then, when the air pressure is decreased, the bag expands and automatically fills up with blood again. This type, illustrated in Fig. 15a, is intended for use in situ within the human body, whereas Fig. 15b schematically shows the same principle applied to the external "heart" in a heart-lung machine.

(b) An artificial heart may be constructed on the principle represented in Fig. 16: compressed air is admitted, through a pressure regulator and a three-way valve, into the space in front of the piston of the artificial heart. With this arrangement the frequency of the heart's rhythm can be regulated at will. The development of a dependable and efficient pumping unit is an important step in solving the problem of devising a successful and viable artificial heart, but it is by no means the complete solution. The fact that it has hitherto not proved possible to keep an experimental animal alive for more than a couple of days by means of an artificial heart is in itself sufficient indication of the difficulty of the problem. However, some measure of success has been achieved—even in human subjects—with the temporary replacement of one side of the heart by an artificial device. The difficulties and problems associated with the complete surgical substitution of an artificial heart for the diseased or defective natural heart are so complex that success in this field still appears to be a long way off.

The difficulties can be illustrated with reference to an in itself relatively simple problem, which could be solved by the arrangement indicated in Fig. 17. The natural heart can vary the power of its beat and the output of blood from its ventricles, according to the body's requirements at any particular time. Technically this control function can be reproduced by installing a piezoelectric transducer (P), which is a sensing device that measures the blood output rate and controls each stroke of the pump accordingly. The equipment needed for controlling and monitoring the functioning of an artificial heart in an experimental animal in the laboratory is very elaborate. Fig. 18 shows—in a schematic and greatly simplified form—an electronic control system for a heart pump powered by compressed air. The diagram shows three interlinked control loops, each of which reports the individual data (pressures, valve position, piston position) back to a higher monitoring center and causes the latter to adjust and modify the next stroke of the pump. The various stages of the system have to operate as separate computers. (The illustrations for this article were supplied by Professor W. J. Kolff, Cleveland.)

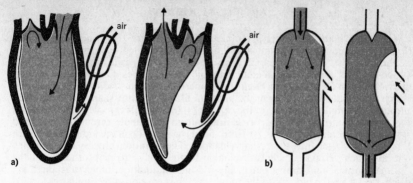

FIG. 15 BAG-TYPE ARTIFICIAL HEART

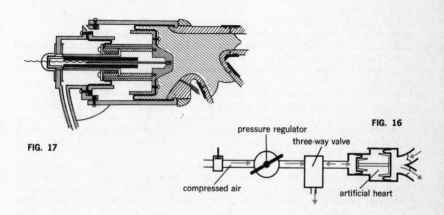

FIG. 17

FIG. 16

pressure regulator

three-way valve

compressed air

artificial heart

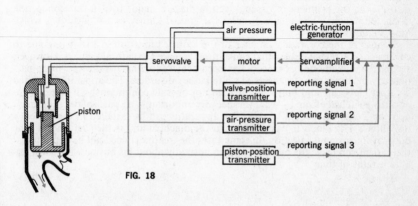

piston

air pressure

electric-function generator

servovalve

motor

servoamplifier

valve-position transmitter

reporting signal 1

air-pressure transmitter

reporting signal 2

piston-position transmitter

reporting signal 3

FIG. 18

ARTIFICIAL KIDNEY

The kidneys are two excretory organs which produce the urine. Each kidney (Fig. 1a) is composed of about a million tiny units called nephrons (Fig. 1b). A nephron consists of a tuft of capillaries (fine blood vessels) called the glomerulus (1). Blood enters it through the afferent arteriole (2) and leaves it through the efferent arteriole (3). In its passage through the glomerulus it is purified of waste products, which drain out with fluid from the blood (a filtration process) and are discharged through the tubule leading from the capsule (4) (Bowman's capsule) which encloses the glomerulus (Fig. 1c). The urine drains from the tubule through a branched system of collecting ducts (5, 6, 7, 8) into the pelvic space (9) of the kidney and thence through the ureter into the urinary bladder. The filtration process is a complex one in which certain substances valuable to the body are reabsorbed from the "primary urine" that drains out of the capsule. In addition, other substances are discharged from the blood into the tubule (which is enveloped by capillaries formed by the efferent arteriole). The process in the glomerulus consists of "passive" filtration, whereas the excretion and reabsorption of substances in the tubule is an "active" selective process. An "artificial kidney" can only reproduce the passive filtration process.

The kidneys are important not only as organs of excretion, but also as regulating devices for the body's internal environment—the precise conservation of the salt and water content. Nearly one-fifth of the blood pumped out of the heart (5000–6000 cc per minute) goes to the kidneys; urine is produced at a rate of 1 to 2 cc per minute. If the kidneys fail to function properly, other parts of the body—stomach, bowels, lungs, skin—can, for a time, compensate for the deficiency by taking over part of the excretory function. In the long run, however, the balance will be disturbed, and a diseased condition known as uremia will develop—i.e., the waste products of protein metabolism are retained in the blood. As a result, the content of urea and other waste products in the blood increases. These cause the grave symptoms of toxemia which accompany uremia and are due to the action of poisonous substances in the blood (headache, dizziness, diminished or suppressed urine, rapid pulse; in far advanced cases: nausea, vomiting, coma and death). Remedial treatment may take various forms; the most effective method is by means of the "artificial kidney," which permits more rapid correction of the blood chemistry. In general, the aims of treatment are to give rest to the kidneys, so that they may recover. The artificial kidney can render valuable assistance in this respect. Of course, modern surgery has developed the kidney transplant—replacement of a diseased kidney by a healthy kidney supplied by a suitable donor—as a radical form of treatment. As with heart transplants, the problem of tissue rejection plays a major role in this connection—i.e., the patient's body tends to treat the grafted kidney as an "intruder" which must be got rid of.

Because of the complex character of the kidney's function, it is not possible to construct an artificial kidney that can entirely replace the natural kidneys. However, a considerable amount of success has been achieved with such devices, more particularly in combating the effects of acute kidney disorders. The apparatus required is rather elaborate and bulky, but it need not operate continuously; it can be temporarily switched off, and—what is especially important—no parts of the equipment need be inside the patient's body. The apparatus can be installed in the patient's own home. His blood is passed through the machine and is then returned to the body by intravenous injection. In some cases the treatment need not be continuous: the patient can leave the machine and go normally about his business for considerable lengths of time.

(more)

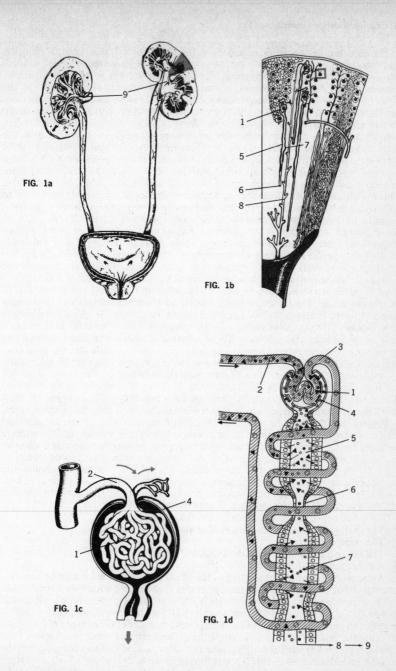

FIG. 1a

FIG. 1b

FIG. 1c

FIG. 1d

The function of the artificial kidney is based on the physical laws of diffusion and osmosis. The term osmosis denotes the diffusion of a solvent through a semipermeable membrane—a membrane that allows the solvent to pass but is impermeable to the dissolved substances—into a more concentrated solution. In the artificial kidney the semipermeable membrane is similar to the walls of the glomerulus capillaries in some essential respects (pore width, total pore area, filtration effect) and is referred to as the dialyzer. The blood is subjected to a treatment called hemodialysis (Fig. 2): a very thin film of flowing blood (1) is separated from the surrounding rinsing solution (2) by an approximately 0.08 mm (0.003 inch) thick semipermeable membrane (10). Formerly, natural membranes (fishes' air bladders, animals' intestinal membranes) were used for the purpose; at the present time, suitable plastics such as cellophane are employed for the semipermeable membrane. This membrane allows substances in normal molecular solution and the solvent to pass through its pores, but it prevents the passage of very large molecules (proteins, in particular) and of the cellular constituents of the blood (corpuscles, etc., 7, 8, 9). An important advantage is that bacteria and viruses (5, 6) cannot penetrate this membrane either, so that sterilization of those parts of the apparatus which are located outside the membrane is unnecessary. Since the dialyzer obeys the laws of diffusion and osmosis, the rinsing liquid must contain in physiological concentration all those membrane-passing dissolved normal constituents of the blood (electrolytes, in particular) which are required to maintain in the blood and not be lost to the rinsing liquid. On the other hand, this liquid must not contain any of the substances present in above-normal (harmful) concentration in consequence of defective functioning of the natural kidneys—e.g., potassium, urea, harmful acid radicals such as phosphate and sulphate. These substances must be diffused through the membrane into the rinsing liquid. The latter may also contain suitable high concentrations of substances which it is desirable to introduce *into* the blood stream, by diffusion in the opposite direction through the membrane—e.g., drugs, dextrose, etc. The rinsing liquid must furthermore be of the same temperature as the blood it encloses, this being ensured by thermostatic control (4). Also, oxygen (3) must be allowed to bubble through it, so that the stream of blood can absorb this oxygen.

Present-day dialyzers have membranes that function more or less as superfine strainers which allow molecules below a certain size to pass and prevent the passage of larger ones. Efforts are being made to develop membranes that exercize a selective action, not with regard to the size of the molecules, but with regard to the chemical constitution thereof. To be able to pass through the pores in the membrane, the molecule must have a distinctive chemical property that will act like a key to unlock a door. In this way it might be possible to produce membranes that perform selective filtration in somewhat the same manner as the natural kidney does.

The principal components of an artificial kidney are:
1. The dialyzer containing the membrane and the rinsing liquid;
2. The pumping system;
3. The heating equipment with thermostat;
4. The oxygen supply system;
5. Arrangements to prevent clotting of the blood (drugs such as hirudin or heparin; smooth surfaces on all parts of the apparatus that are in contact with the blood).

The prototype of the dialyzer commonly employed at the present time is illustrated in Fig. 3: the blood diverted from an artery (1) has hirudin (2) added to it and is supplied to the apparatus. The latter comprises a number of tubes made of collodion (a cellulose tetranitrate), which constitutes the semipermeable membrane. The rinsing liquid is supplied through a tube (5). After passing through the dialyzer, the blood is introduced into a vein (3) and is thus returned to the patient's body. Constant blood temperature is ensured by a thermostat (6), and oxygen is supplied (7) to the rinsing liquid.

(more)

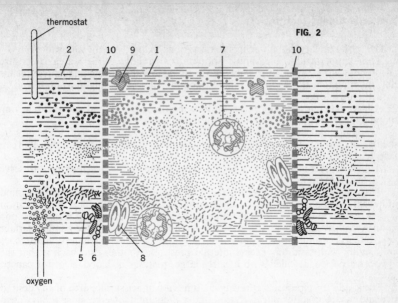

FIG. 2

thermostat

2 10 9 1 7 10

5 6 8

oxygen

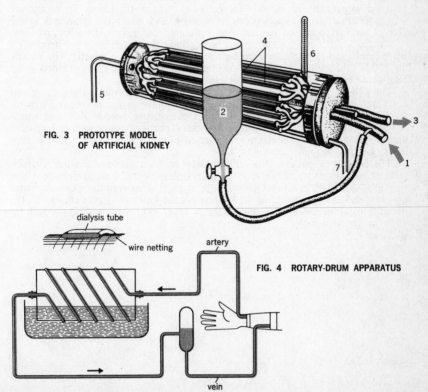

FIG. 3 PROTOTYPE MODEL
OF ARTIFICIAL KIDNEY

dialysis tube

wire netting

artery

FIG. 4 ROTARY-DRUM APPARATUS

vein

The requirements that an artificial kidney has to fulfill are the following:

1. The semipermeable membrane should have a large surface area to ensure adequate osmotic interchange between the blood and the rinsing liquid. This means that the blood must flow in the thinnest possible film, providing maximum contact with the membrane bathed by the liquid.

2. A large difference in concentration between the rinsing liquid and the inflowing blood should be maintained by frequent renewal of the liquid.

3. Any rough surfaces or turbulence in the flow of blood must be avoided to obviate the risk of clotting. As already mentioned, the use of anticoagulins (hirudin, etc.) can help to achieve this object.

Medical science has developed several types of blood-dialysis equipment to serve as "artificial kidneys":

1. The rotary-drum apparatus (Fig. 4). A blood-filled flat cellophane tube is wound in helical fashion around a rotating drum made of wire mesh. The rotary motion causes the blood to flow along the tube in the same way that water is moved along in an Archimedean screw. This device does indeed provide a large contact-surface area, but it is unsuitable as a means of ultrafiltration, as the cellophane tube stretches and cannot withstand any excess internal pressure. Another drawback is that the apparatus has to be filled with a fairly large quantity of blood before it can be started.

2. The sandwich-type apparatus (Fig. 5). It has a contact-surface area of any desired size and can be used for ultrafiltration under pressure. However, since it has "dead corners," it presents an inherent danger of clotting of the blood. To counteract this, large amounts of anticoagulins have to be used, which is a drawback in the case of recently operated patients and sufferers from gastric disorders, as these substances are liable to cause serious hemorrhage. In this form of construction a flat cellophane bag is sandwiched between grooved plates of plastic. The rinsing liquid flows through the grooves in the opposite direction to the flow of blood in the cellophane bag (countercurrent principle).

3. The Allwall apparatus (Fig. 6). In this system a tube through which the blood flows and which forms the semipermeable membrane is wound around a wire-mesh drum and is itself enclosed within a close-fitting wire-mesh jacket which restrains the expansion of the tube, so that excess pressure can be used and effective ultrafiltration applied. The drum is immersed in the rinsing liquid, which is kept in motion by a propeller device.

4. The Moeller apparatus (Fig. 7) comprises a drum made of plastic in which grooves are cut and which is enclosed within a plastic outer shell, likewise provided with grooves. The blood-filled dialysis tube is gripped between the inner and outer grooves, and the rinsing liquid flows through the latter in countercurrent to the blood. This arrangement, too, enables the pressure to be made sufficiently high for effective ultrafiltration.

(more)

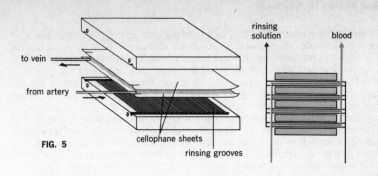

to vein

from artery

cellophane sheets

FIG. 5

rinsing grooves

rinsing solution

blood

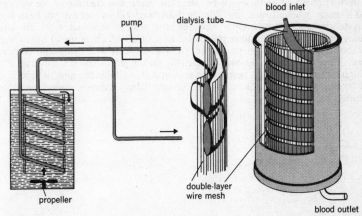

pump

dialysis tube

blood inlet

propeller

double-layer wire mesh

blood outlet

FIG. 6

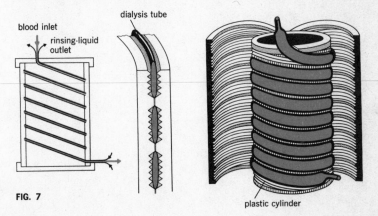

dialysis tube

blood inlet

rinsing-liquid outlet

FIG. 7

plastic cylinder

5. The twin-coil apparatus (Fig. 8a), in which the semipermeable membrane is in the form of a flat, wide cellophane tube which is coiled in two tiers around a hollow core. The coils are mounted inside a container through which the rinsing liquid is passed. A very thin film of blood flows through the coils, which present a large area of contact with the surrounding liquid. Fig. 8b shows the actual dialyzer. It has the advantage that it is supplied sterilized and ready to use inside its container; it can moreover be quickly and easily exchanged.

The artificial-kidney treatment procedure is shown schematically in Fig. 9. The blood is diverted from an artery (1), passes through the dialyzer tubes, and is returned to a vein (2). Circulation of the blood through the apparatus is maintained by a pump. Another pump circulates the rinsing liquid. Before the purified blood enters the vein, it passes through a device in which any air or clots are trapped. The whole procedure takes about six hours and has to be repeated at intervals, depending on the results of blood analyses. In between the successive treatments the patient can lead a more or less normal life. The ideal artificial kidney has not yet been devised. Despite certain shortcomings, however, the results achieved with an artificial kidney are sometimes quite remarkable: as a result of the removal of the noxious metabolic substances from the blood, the patient feels well again and can resume work. Thus his active life can be greatly prolonged. Also, the artificial kidney, though it cannot provide a permanent substitute for nonfunctioning natural kidneys, can be of great value in the relief of reversible kidney damage, giving the kidneys a chance to recover.

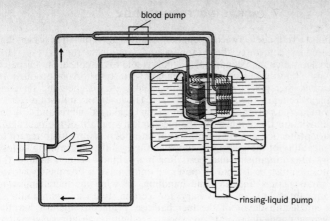

FIG. 8a

FIG. 8b

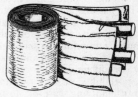

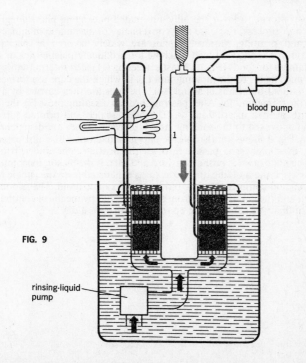

FIG. 9

A distinction is to be made between consumer packaging and dispatch packaging: The former category comprises the wrappers and containers (boxes, tins, jars, bottles, bags, etc.) in which merchandise of all kinds is sold to the consumer. Dispatch packaging serves primarily to protect the merchandise in transit from producer to wholesaler or consumer (packing cases, drums, sacks, etc.). Depending on their purpose and their contents, the requirements vary. However, the following factors are usually of major importance: protection of the packaged merchandise from pressure and knocks (in this connection the mechanical strength of the packaging is all-important), protection against drying out or entry of moisture (with regard to this the impermeability of the packaging to water vapor is the determining factor), protection from microorganisms and pests (for many kinds of merchandise the packaging material must be devoid of pores; it must have a completely closed surface), protection against fingering and handling, allowing the merchandise to "breathe" (permeability to air—necessary for certain types of commodities), preserving the smell and flavor of the merchandise, protection against harmful effects of light (some goods deteriorate on exposure to light of certain wavelengths), etc. The packaging materials must be suitable for use in packaging machines. Hence they should, for example, be suitable for printing upon, they should be suitable for gluing, welding or sealing, and should in many cases moreover withstand destructive chemical action (due to acids, alkalies, oil, grease, etc.). The customer or consumer also makes certain demands as to the quality of the packaging; of course, these vary for different types of merchandise and may include, for example: good transportability, ease of opening, easy inspection of the contents (transparency), ease of emptying, ease of closing, good stackability (e.g., in a refrigerator), easy disposability, etc.

Modern industry tends to favor the fully automated packaging plant integrated into the production process. For ease and convenience of automatic manufacture, filling, closing and printing, various plastics are widely favored as packaging materials. An example is afforded by a machine for producing plastic tubs or cups to serve as containers for foods (Fig. 1). Plastic sheeting (1) is fed intermittently to a heating unit (2) and thence to the molding unit (3), in which the cups are formed by pressing and separated by means of a knife (4). The cups are then carried by a belt conveyor (5) to the filling unit (6). After passing through a sealing press (7), the cups are provided with printed inscriptions (8). The filled, sealed and printed cups are ejected at (9) and conveyed to the packing table (10), whence they are dispatched to their destination. Fig. 2 shows another plastic-cup-manufacturing machine based on thermoforming. Plastic sheet is passed through the heating zone and fed to the forming zone. The male mold is pushed forward and carries the plastic sheet into the female mold. To assist the molding operation, air is exhausted from the female mold (vacuum molding) and compressed air is blown into the male mold. The two molds then move apart, and the cup is ejected from the male mold by means of compressed air. A machine of this type can produce up to a million cups a day.

(more)

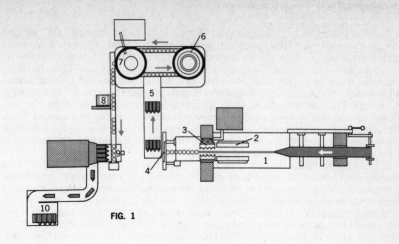

FIG. 1

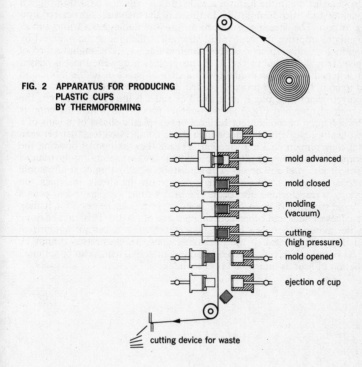

**FIG. 2 APPARATUS FOR PRODUCING
PLASTIC CUPS
BY THERMOFORMING**

mold advanced

mold closed

molding
(vacuum)

cutting
(high pressure)

mold opened

ejection of cup

cutting device for waste

In other processes, packaging consists merely in putting a "second skin" around the goods concerned. More particularly, this takes the form of a shrunk-on heat-sealed wrapping of plastic placed around various foods (Fig. 3). The merchandise to be packaged—i.e., poultry—is first put into a plastic bag (a), and then the air is extracted from the bag (b). The closed bag (c) is then passed through a heated tunnel (d) in which the plastic is heated to above its softening point (above 90° C) by means of hot air. The softened plastic sheet shrinks around the contents of the bag and closely envelops it. A short cooling period completes the packaging process. With this method of preservation it is possible to keep poultry intact and in good condition for human consumption for long periods. Extraction of the air from the package is a major factor in the prevention of putrefaction.

Packaging problems impose basic requirements as to the materials used. Most kinds of merchandise have special properties and call for special packaging methods. The simplest method consists in enclosing the merchandise with a single protective layer of material. However, it is becoming increasingly common practice to employ composite materials (laminates)—e.g., paper combined with plastic sheet, or paper with metal foil, or metal foil sandwiched between sheets of plastic. Plastic sheet or film is manufactured as follows. Granules of plastic are fed into an extruder, in which they are subjected to kneading and mixing, in conjunction with heating. The material is extruded from a slot die in the form of a wide ribbon, which is passed through a series of stretching and calendering rolls. In this way the material is processed into thin sheeting or film. The sheeting produced in this way undergoes various subsequent treatments to improve its mechanical strength: stiffening of the material by grooving, bonding to other plastic sheets (lamination), etc. The manufacture of laminated sheeting is illustrated in Fig. 4. In this process propylene sheet is heated and is, when softened, pressed in contact with a strip of paper or textile fabric and then allowed to cool. In another process—called the "molten bead" process—a hot filament of polyethylene emerging from a small extruder is introduced between the sheets to be bonded together. Fig. 5 shows another process, in which the bond is established by softening of the contact surfaces of two plastic sheets by means of a wedge-shaped heating element, the sheets then being compressed together between rollers so that they remain sticking together. Yet another method of bonding the sheets for forming laminates is by ultrasonic welding: the two sheets are introduced between a "sonotrode" and an "anvil." The sonotrode, which vibrates at ultrasonic frequency, exerts a rhythmically oscillating pressure on the sheets, in which the ultrasonic energy is transformed into heat. The heat softens the plastic and causes the sheet to bond together (Fig. 6). Similar bonding processes are used for sealing (welding) the edges of bags and other containers made of plastic. This may be done by inserting the material between heated jaws which exert a certain amount of pressure and thus effect a seal. If the packaging materials themselves cannot be sealed by direct bonding, a layer of suitable bonding material, which can be softened by the application of heat, is interposed (hot sealing).

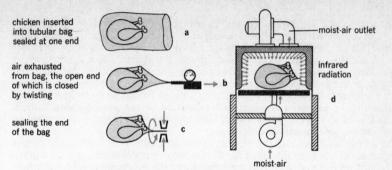

chicken inserted into tubular bag sealed at one end — a

air exhausted from bag, the open end of which is closed by twisting — b

sealing the end of the bag — c

moist-air outlet

infrared radiation

d

moist-air

FIG. 3 WRAPPING CHICKENS IN SHRUNK-ON HEAT-SEALED PLASTIC

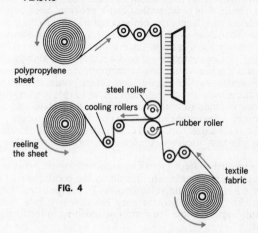

polypropylene sheet

steel roller

cooling rollers

rubber roller

reeling the sheet

textile fabric

FIG. 4

laminated-sheet manufacture

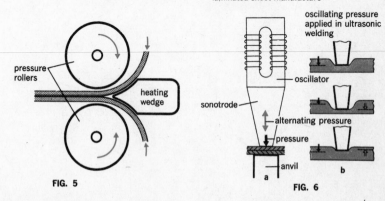

pressure rollers

heating wedge

FIG. 5

oscillating pressure applied in ultrasonic welding

oscillator

sonotrode

alternating pressure

pressure

anvil

a

b

FIG. 6

429

The use of silos for the storage and preservation of green fodder (forage) by a process of controlled fermentation is described by the term "ensilage." The fermented foodstuff for farm animals is called "silage." Various crops can be processed in this way: grass crops, leguminous crops (peas, beans), cereals, etc. There are many types of silo. Here only one type, the metal-tower silo, will be dealt with. It is a gastight structure, so designed that fresh forage can be introduced from the top and fermented silage extracted from the bottom at any time, without detriment to the quality of the product. In good silage the composition of the ensiled food crop is largely unchanged, and its feeding value is high if it is made from a protein-rich crop, besides providing a useful source of minerals and vitamins.

The green forage put into the silo has a moisture content of 30–40%. The crop is chopped and elevated to the top of the silo; in modern installations, as shown in the accompanying illustration, this is done pneumatically by means of a blower which provides a stream of air to carry the forage to the top of the silo.

The silage is extracted from the bottom by means of a built-in cutting and conveying device which consists of an endless chain provided with cutter blades. This device removes silage from the bottom of the column and conveys it to the discharge gate at the center. The silage sinks down in the silo under its own weight. As heat is evolved in the fermentation process, there is no risk of the silo contents' freezing and thus remaining blocked in the silo in cold weather. To maintain a fairly constant pressure in the silo despite variations in temperature, the silo is equipped with pressure compensators—baglike flexible containers, communicating with the external air, which take in air at low temperatures and expel it when the temperature rises. The silo wall is made of aluminum alloy or steel plate and is of bolted or welded construction. Advantages of ensilage are that the process is labor-saving because it can be mechanized and that it is independent of weather conditions. Against this, the silo and ancillary equipment involve considerable initial expenditure.

If air is freely admitted to fresh green forage, the biological process of respiration continues for a considerable time. This results in the breakdown of nutritive constituents, so that a loss in feeding value occurs. To prevent this, it is necessary to exclude air from the forage; hence the need for storing it in a closed container. Plant enzymes and bacteria cause fermentation, resulting in the formation of acids, mainly lactic acid.

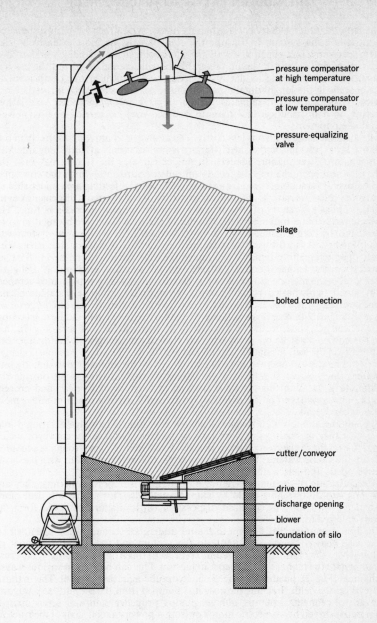

pressure compensator
at high temperature

pressure compensator
at low temperature

pressure-equalizing
valve

silage

bolted connection

cutter/conveyor

drive motor

discharge opening

blower

foundation of silo

THE MODERN CATTLE BARN (COW SHED)

Accommodation for dairy cows is mainly of two types: the stall-barn system and the loose-housing system. In Europe the stall barn is the most extensively used system: each cow is confined in a stall, the stalls being arranged side by side in rows facing on a feedway, which should preferably be wide enough to permit the passage of a wagon for delivering fodder to the troughs. A typical layout comprises two rows of stalls in the longitudinal direction of the barn or shed, with three alleys, or service passages. In cold climates the barns have insulated walls. Ventilation, watering, feed distribution and manure removal may be carried out by power-operated equipment.

The loose-housing system is less costly. The animals can move about freely in one or more large pens inside the shed. Manure is removed only at fairly long intervals by means of power manure loaders. In milder climates the system may take the form of a semiopen shelter adjacent to an outside yard into which the cows may freely move. Hayracks and feed troughs are installed in a feeding area under shelter. Sometimes the cows have access to a self-feeder with an automatic dispensing device which supplies a quantity of hay or grain at predetermined intervals of time. The cows are milked in a milking room adjacent to the shed. They are brought in to be milked and are then returned to the shed or the yard. The loose-housing system with semiopen shelters is for livestock fattening (beef cattle). In modern stall barns with power-operated manure-removal equipment the stalls are usually made 4 to 8 inches shorter than the animals' bodies. Adjacent to the standing area and at a slightly lower level is the manure gutter, which is cleared by means of rope-hauled scrapers or by small two-wheeled power scoops. Alternatively, in some installations the manure mixed with straw is removed by a conveyor system which discharges it on to a manure heap. In recent years, however, manure removal in liquid form has come into increasingly widespread use, because of its convenience and economy. In that case the cows do not lie on straw which is cleared out along with the manure but lie directly on the well-insulated floor of the barn (timber, plastic or ceramic flooring). The liquid mixture of manure and urine produced by the animals flows directly into a gutter which slopes slightly and discharges its contents into a pit outside the building (Fig. 1). With this system it is easier to keep the stalls clean and prevent disease. The cows stand on a clean floor, so that milking is carried out under more hygienic conditions.

Feeding the animals with silage is a simple operation. The silage is brought into the barn either in small wagons on rails or by means of a screw conveyor which brings it straight to the troughs from the silo. Each animal's feed trough is provided with an automatic waterer. When the cow prods the water container with her nose, a valve opens and lets water flow out.

Mechanization of the manure-removal operations is important because they are laborious and disagreeable to do by manual methods. This article will deal more particularly with fully mechanized disposal of cattle manure in liquid form, as already referred to.

The liquid manure that flows into the pit undergoes sedimentation when left to stand. The heavier constituents settle to the bottom and the lighter ones float to the top where they form a surface scum. To prevent this from happening, it is necessary to agitate the manure—i.e., keep it in motion. This can be done in various ways: mechanical (Fig. 2), pneumatic (Fig. 3) or hydraulic agitation (Fig. 4). The manure is spread on the fields. For this purpose it is pumped from the pit into tank wagons by means of centrifugal pumps, plunger pumps, propeller pumps or screw pumps. The pump is driven by an electric motor or from a power-takeoff shaft of the tractor.

(more)

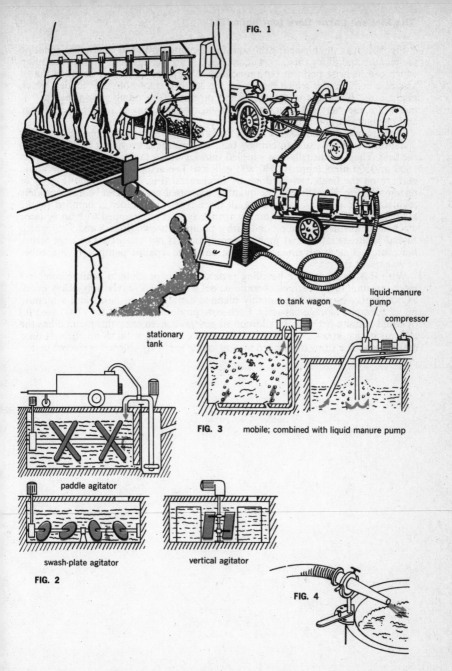

FIG. 1

to tank wagon

liquid-manure pump

compressor

stationary tank

FIG. 3   mobile; combined with liquid manure pump

paddle agitator

swash-plate agitator

FIG. 2

vertical agitator

FIG. 4

Fig. 5 shows a double-acting (lift-and-force) pump which can be used for agitating the manure and filling the tank wagon. Fig. 6 is an illustration of a propeller pump which can likewise perform both functions. This pump can be so designed that it acts as a comminuter which breaks up the solids (remains of cattle fodder) which might cause blockages in the pump and pipelines. The pump shown in Fig. 7 is merely for the "straight" handling of liquid manure, without having an agitating and/or comminuting action.

Manure may be pumped on to the fields directly through pipelines. For greater distances it is more usual to employ tank wagons for conveying it from the pit to the field. These are usually two-wheeled vehicles with a tank capacity ranging from 1500 to 3500 liters (approx. 330–800 gallons). The actual spreading of the liquid manure on the fields can be done by mechanical (Fig. 10), hydraulic (Fig. 9) or pneumatic (Fig. 8) methods. The quantity dispensed per acre can be appropriately regulated. The power for operating all three types of spreader is supplied by the takeoff shaft of a tractor. The tank wagon may also be equipped with an agitator mechanism for the manure pit and with a pump of its own (Figs. 8 and 9), in which case separate agitating and pumping equipment is not required. On large farms, however, it is more advantageous to use one large separate pump to fill a number of wagons at the pit.

With this liquid-manure-handling procedure it is possible to achieve complete mechanization of the various operations, but it does involve relatively heavy initial capital outlay in providing suitably planned cattle sheds or barns and a manure pit of adequate storage capacity. Each cow produces about 20 m^3 (700 cubic ft.) of liquid manure per year. In addition to general convenience, this method has the advantage that straw-storage facilities can be dispensed with, as no straw is used in the sheds. The straw is chopped up and left on the fields, where it is plowed under.

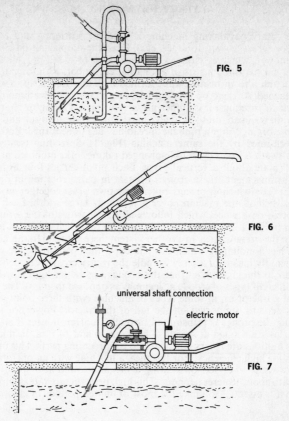

FIG. 5

FIG. 6

universal shaft connection

electric motor

FIG. 7

FIG. 9   TANK WAGON WITH MANURE SPRAYING PUMP

float valve

pressure gauge

shutoff valve

dome

four-way valve

stributor

air compressor

agitator suction hose

FIG. 8

tank
three-way valve

agitator hose

eccentric screw pump

suction hose

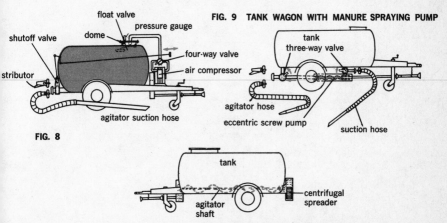

tank

agitator
shaft

centrifugal
spreader

FIG. 10   TANK WAGON WITH CENTRIFUGAL
MANURE SPREADER

ROTARY HAYMAKING MACHINE

The rotary haymaking machine is used for scattering and turning the mown hay crop (grass, clover, legumes, etc.). First, the crop is cut off close to the ground by a mower, which is either a tractor-drawn or a self-propelled machine whose cutting mechanism consists of a long flat cutter bar, with forward-pointing slotted fingers, and a reciprocating knife formed by a thin steel strip to which knife blades are attached. As the crop is cut, it falls in a continuous swath on the ground, where it is left to cure until it is sufficiently dry to rake. During or shortly after mowing, the swath is spread out so as to give maximum exposure to sun and wind and thereby promote drying. After a time the spread hay has to be turned. Both these operations are performed by the same machine (Fig. 1). Spreading (scattering) the freshly mown swath is done by double-pronged raking forks mounted at the ends of arms which rotate about a vertical pivot. Each pivot carries four arms in a horizontal cross arrangement; these crosses operate in pairs, rotating in opposite directions (Fig. 2). The whirling prongs rake the mown grass together in front and fling it fanwise behind the machine over a relatively large width. Each rotating cross is supported on a wheel which follows all irregularities of the ground. The machine shown in Fig. 1 has an articulated frame, so that each cross can individually adapt itself to the irregularities. In another system there is a rigid frame, and every alternate wheel has a mounting which can telescope, thus permitting the wheel to move up or down. By means of a screw spindle the rotary crosses can be varied in height in relation to the ground, so that the prongs can be set higher or lower for dealing with different types of crop (e.g., legumes as opposed to grass). The machine is towed behind a tractor or, in some types of machine (with three-point support), the front end is actually supported on the tail of the tractor. Power for rotating the arms carrying the prongs is transmitted through a universal-joint shaft from the tractor. The arms are rotated at speeds of between 10 and 14 rpm. In this type of machine all the motions are rotary; there are no reciprocating parts. This operating principle makes for high efficiency. For instance, a machine with an operating width of 16 ft. (six "crosses"—i.e., sets of rotating arms) and towed at a speed of 9–10 mph can deal with about 18 acres per hour. This high performance rate is important in climates where it is essential to make the most of limited periods of favorable weather.

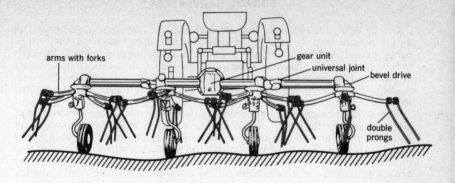

FIG. 1 ROTARY HAYMAKING MACHINE

arms with forks

gear unit

universal joint

bevel drive

double prongs

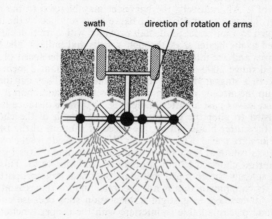

swath

direction of rotation of arms

**FIG. 2 ROTARY MACHINE SPREADING THE HAY
(VIEWED FROM ABOVE)**

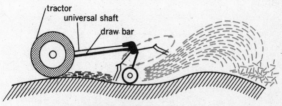

tractor

universal shaft

draw bar

**FIG. 3 ROTARY MACHINE TURNING THE HAY
(SIDE VIEW)**

FORAGE HARVESTERS

Certain crops, such as grass, legumes, and corn, are often ensiled (see page 430) for use as cattle food. To make silage, they must be chopped up so that they can be tightly packed in the silo, whereby anaerobic fermentation is promoted. The crop is cut in the field with a machine called a forage harvester which chops it immediately or picks up and chops a windrow which has been cut and raked earlier. One such machine, the so-called economy-type forage harvester, is illustrated in Fig. 1, and its mode of functioning will be described here.

This versatile machine can perform various operations: cutting and loading of green forage crops for the day-to-day feeding of cattle (Fig. 2); cutting grass, clover and other crops and placing them in windrows (Fig. 3); cutting and chopping of forage for ensilage (Fig. 4); picking up and loading of crops which have been placed in windrows to wilt or dry a little before being chopped for ensilage; picking up and chopping straw from windrows (left by a combined harvester) and scattering the chopped straw on the fields to make it easier to plow it under; cutting and chopping of remains of crops left in the fields, such as potato plants, corn, tobacco or cabbage stalks.

When towed by a tractor, the forage harvester moves along behind, but somewhat to the right of it. Alternatively, the harvester may be fixed to the right-hand side of the tractor. The power for driving the harvester is supplied by the tractor's takeoff shaft connected to a universal-joint shaft provided with a friction clutch. The power is transmitted to the beater shaft through V-belts and pulleys. The belt pulleys are of various sizes and are interchangeable so as to vary the speed of rotation of the beaters (speed range 1000–1700 rpm). The latter are about 2 inches in width and are mounted in a staggered arrangement on the shaft. They cut the growing crop off, or pick up the already cut crop from the windrow, and chop it up by whirling the stalks and leaves past a cutter blade (Fig. 1b) whose distance from the beaters can be adjusted to alter the cutting action. The length of the chopped material depends on the cutter blade setting, the speed of rotation of the beater shaft, and the speed of forward travel of the harvester. When the blade is set close to the beaters and the beaters rotate at high speed, while the machine moves forward at a slow pace, the shortest chopped length is obtained, and vice versa. This length is only an average; actually the chopped material contains a large proportion of oversize pieces—i.e., its chopping action is not accurate, and for this reason the material is not suitable for ensilage in gastight fermentation silos because an excessive proportion of long pieces is liable to interfere with the proper functioning of the discharging mechanism. The chopped material is conveyed up the curved duct by a combination of throwing and blowing. It can thus be discharged into a wagon. Alternatively, it can be deposited in a windrow; in that case the outlet of the duct is fitted with a discharge flap and guide plates which permit varying of the width of the windrow.

(more)

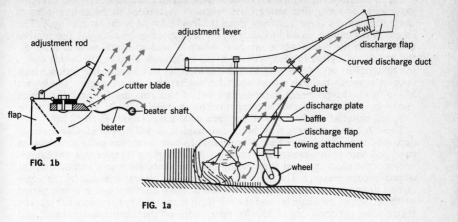

adjustment rod

adjustment lever

cutter blade

discharge flap

curved discharge duct

duct

discharge plate

baffle

discharge flap

towing attachment

wheel

flap

beater shaft

beater

FIG. 1b

FIG. 1a

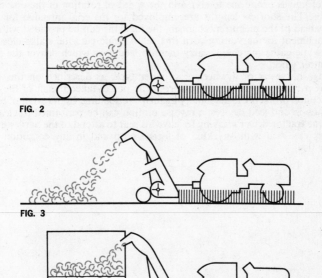

FIG. 2

FIG. 3

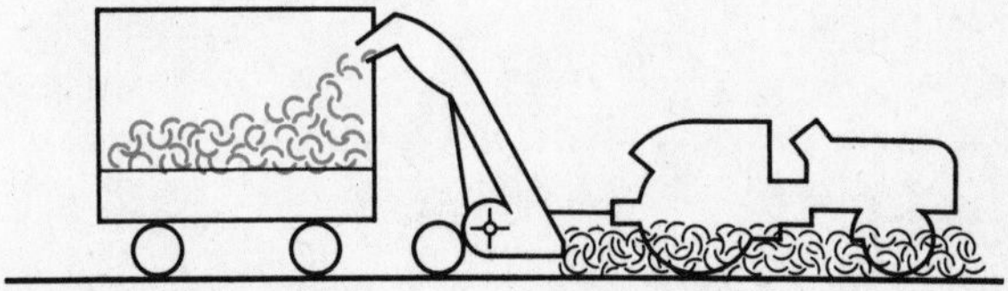

FIG. 4

Fig. 5 illustrates a forage harvester which enables the length of the chopped material to be controlled with greater precision. According to the arrangement of the cutting knives, these machines are of the flywheel type or the cylinder type. The length of cut can be varied by changing the feed speed or by changing the number of knives. In the cylinder-type machine, which is predominantly used in the U.S., the knives are bolted to a rapidly revolving cylindrical knife holder which chops the material in very short lengths and additionally has an impeller action which flings it upwards through a duct and thus discharges it into a wagon. The knives of the flywheel-type forage harvester are bolted to a rotating flywheel disc which carries up to six knives, each accompanied by an impeller blade. The length of the chopped material can be adjusted between wide limits (about $\frac{5}{16}$ to 12 in.).

These forage harvesters may be tractor-hauled or self-propelled. Cut material deposited in windrows is picked up by means of a cam-controlled pickup mechanism equipped with spring-mounted prongs (a). A cover plate (b) over this mechanism prevents the short material from falling back and subjects it to a certain amount of preliminary pressure. The material compacted in this way is seized by the auger (c) and draw-in chain (d); together they feed it into a duct in which it is compacted by the action of the chain (e) and the roll (f); it is then fed to the cutter bar (g). The cutter wheel (h), revolving at 300–700 rpm, chops off the compressed material, and impeller blades fling it through the curved discharge duct into the wagon. The length of the chopped material can be varied by changing the intake speed, the number of knives (from one to six), and the speed of rotation of the disc carrying the knives. The shortest lengths are employed for material intended for making silage. Instead of the pickup mechanism, the machine can be provided with a row-crop attachment for harvesting corn (maize). A knife cuts the stalks close to the ground, while gathering chains carry them to feed rolls which deliver the material to the cutter wheel.

A forage harvester of flywheel or cylinder type, as described on this page, is driven by means of a universal shaft from the power-takeoff shaft of the tractor. A safety clutch protects the machinery against overloading. In the event of blockage of the draw-in and feed devices, a reverse motion can be performed to clear them, without the tractor driver's having to leave his seat to attend to the harvester. These harvesters can deal with any kind of forage crop and in any condition—green, wilted or dry.

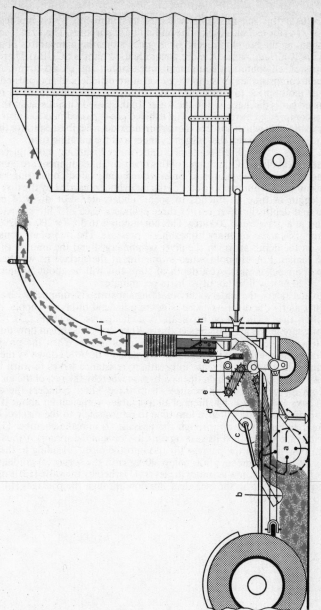

FIG. 5

SELF-CONTAINED UNDERWATER BREATHING APPARATUS (SCUBA)

Apparatus of this general category is used in skin diving, which is practiced as a sport as well as for technical, scientific and military purposes. The term "Aqualung" is sometimes applied to skin divers' or frogmen's breathing apparatus of any kind, but is actually a trade name. A more generally used term is SCUBA. There are two main types of self-contained equipment: open-circuit and closed-circuit. An open-circuit unit comprises an air supply (one or more cylinders of compressed air) and a mask or mouthpiece to which air is fed through a tube via a demand regulator. The exhaled gas is discharged into the water. In the closed-circuit system the breathing gas is compressed pure oxygen. The exhaled gas is passed into a canister of soda lime or caustic soda which absorbs the carbon dioxide; a flexible respirator (breathing bag) is interposed between the oxygen cylinder and the inhaling tube.

Fig. 1 shows a typical self-contained underwater breathing mouthpiece as used for amateur skin diving. Other items of this open-circuit equipment are compressed-air cylinders, demand regulator or pressure-reducing valve, connecting tubes, and pressure gauge. The number and size of the air cylinders carried by the diver depend on the length of time he intends to spend underwater. For dives of maximum duration and depth the diver carries three cylinders, each of 7 liters capacity, containing air at a pressure of 200 atm. This corresponds to $3 \times 7 \times 200 = 4200$ liters of "free" air—i.e., air at ordinary atmospheric pressure. The rate of air consumption depends on the depth at which the diver is submerged and the amount of physical effort he makes. For example, when swimming at the surface he will need about 25 liters of air per minute. At a depth of 10 m this will be about 50 liters, and at a depth of 20 m it will be about 75 liters per minute.

For greater safety, the underwater breathing apparatus is equipped with a warning device that alerts the diver when he is nearing the end of his air supply. A reserve air supply is then switched on by means of a changeover valve (Fig. 3). So long as the air pressure in the air cylinders is above 40 kg/cm^2, the air can flow unhindered to the demand regulator and thence to the mouthpiece. When the pressure falls below that value, the flow to the regulator is gradually throttled down by the pressure of a spring. The resulting increase in breathing resistance serves to warn the diver, who now actuates a lever which opens a bypass whereby the rest of the air can flow freely from the cylinders. For example, with three 7-liter cylinders the diver will then still have $3 \times 7 \times 40 = 840$ liters of air available—sufficient for about 10 minutes at a depth of 20 m. This will give him time to return safely to the surface. In Fig. 3, the normal air supply flows through the passage (2) into the chamber (3); the air pressure lifts the valve (4) off its seat, against the force of the spring (5), thus allowing the air to flow through the passage (6) and into the pipe (7) leading to the demand regulator. When the pressure falls below 40 kg/cm^2, the spring (5) gradually closes the valve (4). The diver now actuates the lever (11) whereby the valve (13) is opened, so that the remaining air in the cylinders can flow through the passages (14, 15, 16, 6 and 7) to the regulator.

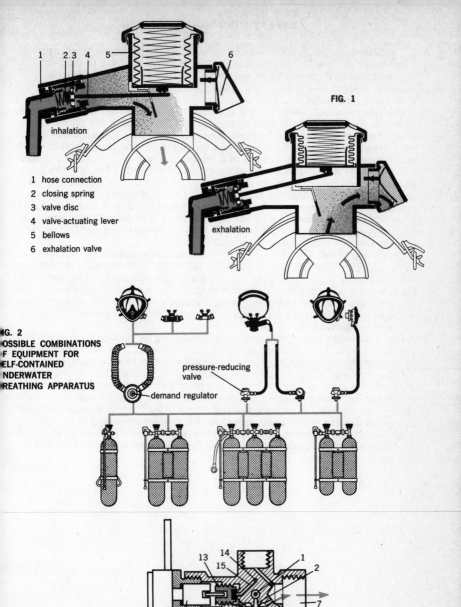

FIG. 1

1 hose connection
2 closing spring
3 valve disc
4 valve-actuating lever
5 bellows
6 exhalation valve

inhalation

exhalation

FIG. 2
POSSIBLE COMBINATIONS
OF EQUIPMENT FOR
SELF-CONTAINED
UNDERWATER
BREATHING APPARATUS

pressure-reducing valve

demand regulator

FIG. 3 SECTIONAL DRAWING OF CHANGEOVER VALVE

SAFETY BINDING FOR SKIS

The skier wears heavy boots with stiff soles, which enable him to exercise precise control over the skis. The boots are secured by means of bindings. In the event of a severe fall, the rigid connection between boot and ski is liable to cause injuries. This hazard is reduced by so-called safety bindings (or release bindings), which free the skier's foot when he falls.

The safety binding illustrated in Fig. 1 comprises a swiveling plate (1), which is rotatably secured to the ski with screws (2), and the gripping and release device which is attached to the swiveling plate by means of a cable (3). In a fall, the swiveling plate, which functions in the manner of a "turntable," cooperates by rotating and thus providing an additional degree of freedom. Each end of the cable is provided with a nipple (4 and 5) which bears against a coil spring (6 and 7). The two springs are accommodated in the trapezium-shaped frame (8), to which the middle flap (9) is pivotably connected. A spring catch (14) holds the middle flap (9) in the locked position. The coil spring on the left-hand side is guided by a hole in the nipple and by a pin (10) which is riveted to a rocker (11). The latter is cushioned by a rubber pad (12) and is provided, on the right-hand side, with a hole in which a stud (13), screwed into the nipple (5), can slide longitudinally.

On the right-hand side of the binding is the mechanism which releases it in the event of overloading and thus protects the skier's legs and especially the Achilles tendon. The seating is so designed that the loaded spring tends to swing outwards. However, it is prevented from doing this by the stud (13), which engages with a hole in the rocker. The coil spring can swing outwards only after the stud has been withdrawn downwards out of the hole in the rocker. When this happens, the heel of the boot is automatically released.

Fig. 3 shows the movements that occur in fastening the boot to the binding. If it swung about the cable, the binding would move along a curve with radius r_1; but the middle flap bears against a groove at the back of the heel and compels the binding to swing upwards along the curve with radius r_2. As a result, the two coil springs are tensioned and thus keep the boot firmly pressed against the ski. Fig. 2 shows the binding in its "locked" position, gripping the boot. The catch (14) engages with the crosspiece (16), thus preventing accidental release of the binding. Release is effected as a result of a jerk which disengages the catch and dislodges the binding from its "top dead center" position, so that it returns to its bottom position as shown in Fig. 3. To reset the released binding, the stud (13) is reinserted into its hole and the coil spring pressed back into its seating.

(more)

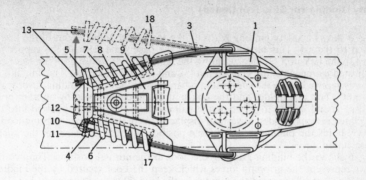

FIG. 1

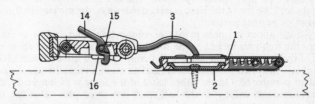

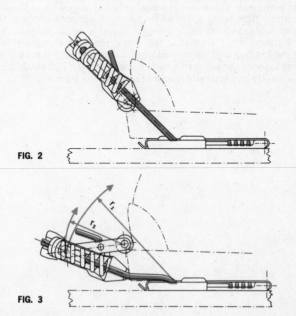

FIG. 2

FIG. 3

Fig. 5 shows a safety binding whereby the toe of the boot instead of the heel is secured to the ski. This binding comprises the base plate (1) with the supporting pillar (2), the link unit (3) incorporating the ball catch (4) whose gripping force can be adjusted by means of the knurled nut (5), and the sole holder (6). The link unit (3) is loosely screwed on to the supporting pillar, so that the entire binding is vertically adjustable. The sole holder and the link unit are interconnected by another pivot (the stud 7) and locked together by the ball catch. A stop pin (8) screwed into the base plate prevents accidental vertical displacement of the binding. A spring-loaded bush (9) holds the binding in the middle position, but has no effect on the safety-release action.

Fig. 6 shows the binding gripping the boot in the normal position. The two red arrows represent the gripping forces which keep the boot secured by the binding. The sole holder balances on the two teeth; the stud (7) is the fulcrum. In Fig. 7 the binding is shown in the "release" position. The device functions as follows: when the skier falls and the toe of the boot is jerked sideways, the movement dislodges the binding from its normal position.

It first swivels about its main pivot (the supporting pillar), while as yet the ball catch keeps the link unit locked to the sole holder. If the twisting force is relatively small and of short duration, the catch will merely "give" a little but remain engaged. The binding will then swing back to its normal position, so that the stable equilibrium of the system is restored. On the other hand, if the force is of longer duration and attains the maximum value that the ball catch can resist (this value can be varied by adjusting the gripping force by means of the nut 5), the catch will disengage and release the sole holder, so that now the link unit and the sole holder can also swivel in relation to each other. From Fig. 7 it is apparent that the binding swivels farther about the main pivot and that release is longer deferred according as the ball catch engages more firmly (this being adjusted by means of the nut). Between the main pivot, the loaded tooth, and the ankle a toggle action is developed, which ensures that release will always occur, no matter how firmly the catch engages. Hence there is no risk that the binding will fail to perform its releasing action even if the skier sets the gripping force of the catch to a high value. Once the ball catch disengages, the binding completely and instantly releases the toe of the boot (Fig. 8).

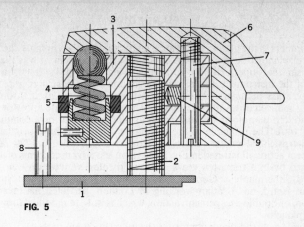

FIG. 5

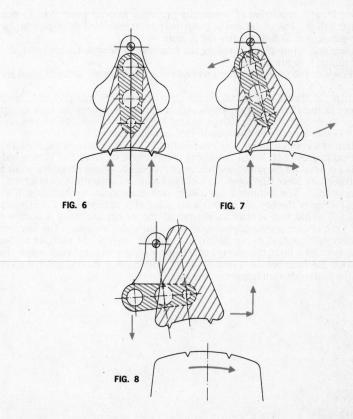

FIG. 6

FIG. 7

FIG. 8

Forerunners of the piano include such instruments as the virginals, the harpsichord and the clavichord. A more advanced form of the harpsichord was the spinet. The mechanism that operated the harpsichord involved quills that "twanged" the strings of the instrument. In the clavichord, however, the strings were struck from below by brass hammers. While both instruments were played through the 18th century, the mechanism of the clavichord was the immediate ancestor in principle of the modern pianoforte (the full name for the instrument now commonly known as the piano). The principle of the pianoforte appears to have originated with an Italian harpsichord maker, Bartolommeo Cristofori, who in 1720 produced the escapement action illustrated in Fig. 2, based on less advanced forms of a mechanism introduced by him some ten years earlier. Cristofori's invention was perfected by Gottfried Silbermann, a German, and his pupil Johann Andreas Stein. In 1821 Sebastian Erard, an Alsatian working in London, perfected his invention of the repetition, or double-escapement, action, which is now, in modified forms, employed in most pianos.

The modern grand piano comprises the following main parts:

1. The cast-iron frame, which has to resist the pull of about 18 tons exerted by the more than 220 strings.

2. The soundboard, consisting of a carefully prepared wooden panel 9 to 11 mm (about $\frac{3}{8}$ inch) thick. The vibrations of the strings are transmitted through a bridge to the soundboard, which intensifies the sound.

3. The casing and wrest plank (fixed to the iron frame) provided with holes into which the tuning pins are inserted.

4. The action, i.e., the system of levers whereby the hammers are actuated, and the keyboard.

5. The strings: in present-day grand pianos the over- or cross stringing introduced by Steinway in the 19th century is employed. In this arrangement the bass strings are made to cross over the tenor strings, so that longer bass strings can be used and the area of bridge pressure on the soundboard can be increased (Fig. 1).

The action of a grand piano is illustrated in Fig. 3. This is essentially the Steinway action introduced in 1884. When the key t_1 is depressed, its other end t_2 is raised. The damper is lifted off the string, and the lever system abc pivots about the point p until the bell-crank lever (the lower arm of the hopper c) comes into contact with the setoff button s_1. The hammer h has now moved much closer to the string. With further depression of the key, the hopper must (since c_1 is against the button) rotate to the right. The hammer is thus further raised and strikes the string. Continued pressure on the key causes the hopper to escape from under the roller. The hammer automatically falls back, allowing the string to vibrate freely. If the pressure on the key is now reduced a little, the spring f causes the hopper c to come back under the roller, so that the hammer can again be struck without first having to let the key return to the initial position (repetition action).

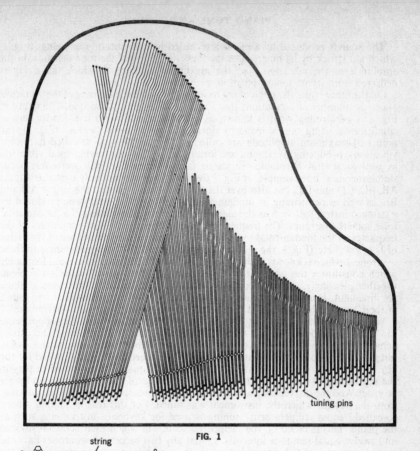

tuning pins

FIG. 1

string

hammer

check

key

FIG. 2

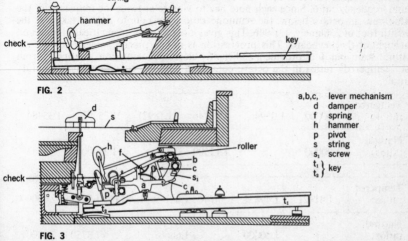

d

s

h   f

roller

b

c

s₁

check

p   a

c

t₁

t₂

a,b,c,	lever mechanism
d	damper
f	spring
h	hammer
p	pivot
s	string
s₁	screw
t₁	} key
t₂	

FIG. 3

The sounds produced by a piano are caused by the vibration of strings (Fig. 1) which are struck by hammers. The loudness (intensity) of the tone depends on the amplitude of the vibration—i.e., the maximum distance to which the string is deflected from its stationary position.

On the other hand, the pitch of the tone depends on the frequency of the vibration —i.e., the number of oscillations that the string performs per second. The string in Fig. 1 is performing what is known as its fundamental vibration. Under certain conditions a string can be made to vibrate in two or more "waves" (Fig. 2); the points of maximum amplitude are called nodes. In this way so-called harmonic vibrations, producing overtones, are formed. In reality the fundamental vibration as well as several harmonic vibrations (and therefore several overtones) occur simultaneously. For example, in Fig. 2 the string vibrates not only over the lengths AB, BC, CD and DE, but also over the length AC and EC and the length AD and BE, as well as performing its fundamental vibration. Thus, each tone produced by a stringed instrument such as the piano is a compound, consisting of a fundamental tone and its overtones. The frequencies of the overtone are whole multiples of the frequency of the fundamental tone. Thus, the overtone frequencies are in the ratios of $1:2:3:4:5$ etc. (Fig. 3: the notes marked correspond only approximately to these overtones), which is known as the "harmonic series." The notes of various frequency which constitutes the harmonics of, for example, the tone of "middle C" blend together pleasingly, with the exception of the seventh and ninth harmonics, which are dissonant. The best tone quality is produced by striking the strings at one-eighth of their length; this is believed to discourage the seventh and ninth harmonics.

With "exact" tuning, the frequencies of the note within an octave are proportional to 24, 27, 30, 32, 36, 40, 45 and 48. This is known as the "natural scale" and can be achieved only in instruments of continuously variable pitch, such as the violin, but not in keyboard instruments. A piano could indeed be "exactly" tuned for the key of C major, for example, as in Fig. 4; but if the pianist then wished to play in the key of, say, D major, the ratios of the frequencies of the notes would no longer be exact. And the discrepancy would be greater according as he moved farther away from the key for which the instrument was tuned. For this reason, among others, so-called "equal temperament" tuning is used for keyboard instruments such as the piano. This is based on the "tempered scale," in which each octave is divided into twelve equal semitone intervals, so that any two successive semitones have the same frequency ratio. Since each note has to vibrate at twice the frequency of the same note an octave below, the semitone ratio from note to note is taken as the twelfth root of 2, namely, 1.05946. This gives a continuous geometrical progression throughout the keyboard. This progression is not in precise agreement with the natural scale, but does provide a sufficiently close approximation, as appears from the "tempered" ratios in the octave compared with the corresponding "natural" ratios:

Tempered ratios:	1.00000	1.05946	1.12246	1.18921	1.25992	1.33484
Natural ratios:	1.00000		1.12500		1.25000	1.33333

Tempered ratios:	1.41421	1.49831	1.58740	1.68179	1.78180	1.88775	2.00000
Natural ratios:		1.50000		1.66667		1.87500	2.00000

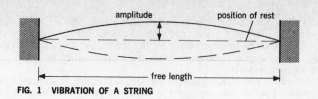

FIG. 1 VIBRATION OF A STRING

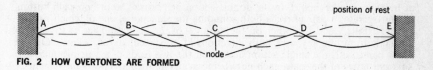

FIG. 2 HOW OVERTONES ARE FORMED

FIG. 3 OVERTONES ASSOCIATED WITH THE FUNDAMENTAL TONE C

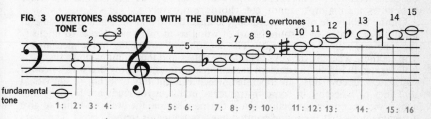

frequency ratios between overtones and fundamental tone

FIG. 4 "EXACT" TUNING IN C MAJOR

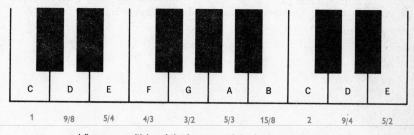

red figures = multiples of the frequency of the fundamental tone C

C	D	E	F	G	A	B	C	D	E
1	9/8	5/4	4/3	3/2	5/3	15/8	2	9/4	5/2
1	2	3	4	5	6	7	8	9	10

tone of the scale

ACCORDION

The accordion is a free-reed instrument which was developed in the first half of the 19th century. Each reed consists of a metal tongue over a slot in a metal plate, the pitch of the reed being determined by the length and thickness of the tongue. The tongue is "sprung up" above its frame and vibrates when air flows around the reed from its upper side.

Fig. 1 represents a section through a modern accordion. It comprises a bellows secured between two wooden end units in which the reeds are accommodated. The wind is admitted selectively to the reeds through valves controlled by finger buttons or by keys on a keyboard. In the double-action or "piano" accordion each button or key operates a pair of reeds both sounding the same note, one of which sounds on the "press," and the other on the "draw," of the bellows. Thus every note is available from one key with both directions of movement of the bellows. On the other hand, with the single-action accordion the two reeds of a pair produce two adjacent notes of the scale, each note being available with a bellows movement in one direction only. For the left hand there are usually two keys or buttons to produce bass notes and major or minor, etc., chords respectively.

On the treble side of the accordion, the fingerboard and the frame of the instrument are separated by a soundboard of wood or metal, which serves to intensify the tone and improve its quality. The soundboard covers the treble mechanism. The "action" whereby the keys operate the valves and thus admit wind to the reeds is illustrated in Fig. 2. The modern accordion is furthermore provided with so-called registers, whereby slides are operated which can cut out or bring in individual tonalities or whereby, in some instruments, extra sets of reeds are brought into action, one set pitched an octave below the main set and the other off-tuned to give a tremulant, so that the sound emitted has a pulsating or tremolo effect. The reeds are screwed or riveted over slots in metal plates. These rectangular plates each have two slots and cover the air ducts through which the wind for blowing the reeds is admitted. The reeds themselves are made of watch-spring steel (brass was used in earlier accordions) and can vibrate freely in their slot ("free reeds"). To minimize the air consumption, the slots are closed by flexible flaps of leather.

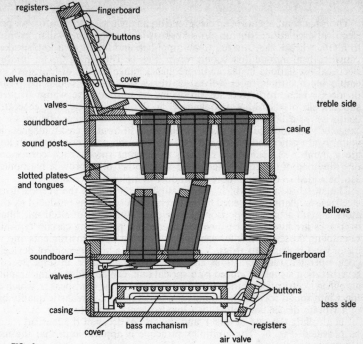

registers — fingerboard
buttons
valve machanism — cover
valves
soundboard — treble side
sound posts — casing
slotted plates and tongues
bellows
soundboard
valves — fingerboard
casing — buttons
bass side
cover — bass machanism — registers
air valve

FIG. 1

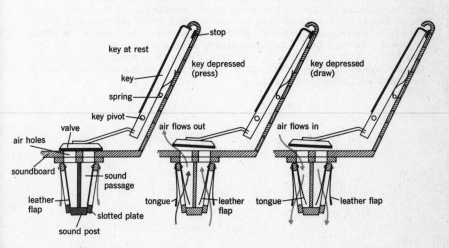

key at rest — stop
key depressed (press)
key depressed (draw)
key
spring
key pivot
valve — air flows out — air flows in
air holes
soundboard
sound passage
leather flap — tongue — leather flap — tongue — leather flap
slotted plate
sound post

FIG. 2

453

The main components of an electronic organ are a set of generators for producing electrical oscillations, one or more keyboards whereby the oscillations are passed to filters, whence they are fed to an amplifier, and finally to a loudspeaker which converts them into audible sound waves (Fig. 1). The generators for producing the electrical oscillations in the audio-frequency range consist of feedback oscillators employing electronic tubes (valves) or transistors, and are connected to frequency dividers, which are devices for producing an output wave whose frequency is a submultiple of the input frequency (Fig. 2). Alternatively, the generators may comprise so-called reed oscillators (vibrating metal reeds acting as variable capacitors) or toothed iron discs which rotate in front of electromagnets in which voltages are generated which are used to form the pitch note for each key; there are a number of such discs rotating at different speeds, each corresponding to one semitone of the equally tempered scale; other discs provide harmonics which can be added to the pitch note to produce complex tone colors.

The switch contacts which are actuated by the key of the keyboard cause the electrical oscillations to be fed to the filters. The oscillations produced by the generators contain a high proportion of overtones. By means of electronic filters those overtones are filtered out which are needed for producing certain typical sounds resembling those emitted, for example, by various string instruments and wood or metal wind instruments (Figs. 3 and 4). Through a volume-control device, usually pedal-operated, the oscillations are passed to the amplifier. In some instruments a reverberation signal, produced by a special unit, can be added to the amplifier. The amplified electrical oscillations are finally passed to the loudspeaker, which converts them into sound waves. The sounds can be given a more realistic quality by means of a vibrato device controlled by a low-frequency generator (Fig. 5). Also, devices for controlling the rate of attack and the rate of decay of the musical sounds may be provided (Fig. 6). The multiplex principle is applicable in that the individual oscillations of the generators can be utilized in multiple fashion by suitably combining and adding together the tones.

In the electronic organ provided with frequency dividers, as envisaged here, there are twelve oscillators for the top octave (twelve semitones). All the lower octaves are obtained from these oscillators by circuits which divide the frequency by two in each successive octave. In the so-called free phase system, however, one oscillator is provided for each note. The earlier electronic organs were comparatively large and bulky. With the advent of semiconductor components (transistors) it has become possible to build very compact instruments, some of which are reduced to suitcase size for convenient portability.

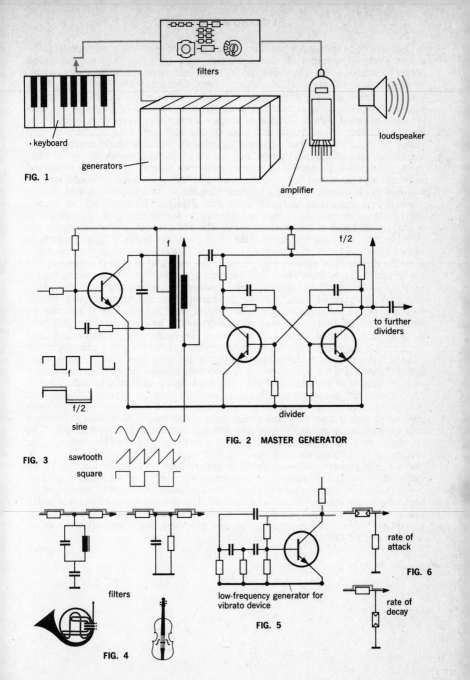

keyboard

generators

amplifier

loudspeaker

filters

FIG. 1

FIG. 2 MASTER GENERATOR

to further
dividers

divider

sine

sawtooth

square

FIG. 3

filters

FIG. 4

low-frequency generator for
vibrato device

FIG. 5

rate of
attack

rate of
decay

FIG. 6

Oscillations in the audio-frequency range are produced and processed with the aid of electronic equipment, stored on tape, and converted into audible sound waves by loudspeakers. The "material" of electronic music can, in acoustical terms, be subdivided into tones, sounds, tone mixtures, sound mixtures, and noises. *Tones* are the simplest acoustic effect and are based on sinusoidal oscillations. They do not occur in conventional music. *Sounds*, which in musical terminology are conventionally called "tones," consist of harmonic overtones (sinusoidal tones whose frequencies bear whole-numbered ratios to one another). *Tone mixtures* contain sinusoidal tones of different, arbitrary frequencies. *Sound mixtures* correspond to the chords of conventional music and consist of sounds. *Noises* are either tone mixtures of very high density or have a continuous acoustic spectrum (sounds and tone mixtures have line spectra). An acoustic effect is essentially defined by four parameters: frequency (pitch), amplitude (loudness), time (duration), and quality (timbre). Frequency is measured in cycles/second (Hertz, Hz), amplitudes in decibels (db), duration in seconds or in a sound track in centimeters (cm). Timbre (tonal quality) is the result of the overtone composition of a sound. In a sound studio it is possible to produce any frequency in the range from 1 to 20,000 cycles/second (range of audibility: 16 to 20,000 cycles/sec.), any loudness (sound intensity) in the range from the threshold of audibility (0 phon) to the threshold of pain (130 phons), and any sound duration from about 1/500 second upwards. Suitable measuring instruments are available for measuring these quantities (frequency, sound, level, etc.). The various types of apparatus for producing (generators), processing (filters, modulators, etc.), storage (magnetic recorders), and reproduction (loudspeakers) of the sounds, as well as the measuring instruments, have for the most part been adopted and adapted from communication engineering, electroacoustics, and electrical measurement technology. Electromechanical and electronic (electrophonic) musical instruments serve merely to supplement the apparatus. Fig. 1 schematically shows the layout of the equipment for an electronic music studio. The precise nature and the number of the units can be varied. Each unit is manually operated, but in some studios semiautomatic or fully automatic control equipment is additionally employed. An electronic composition generally consists of a number of "layers," which are first produced separately and stored and are then mixed. Composing electronic music comprises numerous individual operations and is quite time-consuming. The composer monitors and controls each step directly by means of loudspeakers; he can immediately intervene and make such corrections as he considers necessary. When the composer has completed the composition, it is ready for playing by electronic reproduction equipment—no human interpreter of the music is needed.

Production of tone, sound and noise: The composer of electronic music always begins by producing the "audio" raw material. This consists essentially of sinusoidal, square-wave (rectangular) and sawtooth tones of various frequencies and intensities, as well as "white" (all-frequency) noise. The material either is immediately used or is first stored on tape.

(more)

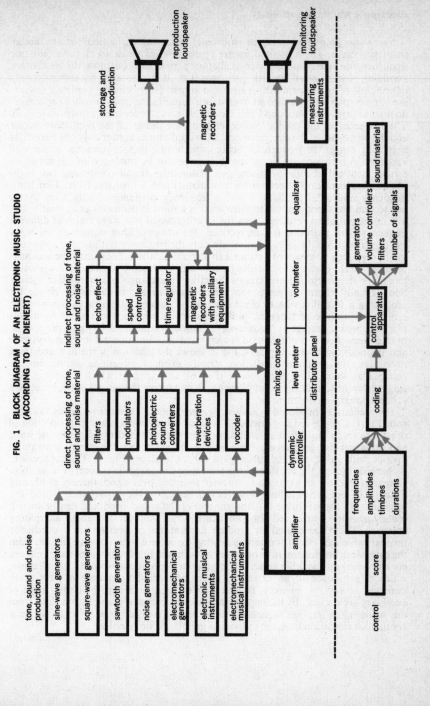

FIG. 1 BLOCK DIAGRAM OF AN ELECTRONIC MUSIC STUDIO
(ACCORDING TO K. DIENERT)

457

Sine-wave generators produce sinusoidal electrical oscillations in the audio-frequency range. These oscillations are reproduced as tones which have only one frequency and one amplitude per oscillation period and constitute the elementary "building blocks" of all acoustic processes (Fig. 2). A sine-wave generator consists in principle of an electron tube (valve) and an oscillatory circuit. Feedback circuits are often employed. So-called RC (resistance-capacitance) generators (or oscillators) do not use coils, their function being based solely on resistors and capacitors (Fig. 3). The frequency of the oscillations depends on the design of the oscillatory circuit and the properties of the electron tubes. Each oscillator can cover only one particular frequency range, the frequencies being capable of either continuous or stepwise variation. All frequencies in the audible range can be produced by means of an arrangement called a *beat-frequency oscillator* (Fig. 4) and comprising two high-frequency generators, a rectifier (or modulator) and a low-pass filter. Two high-frequency sinusoidal oscillations with a frequency difference of only a few cycles per second are superimposed to form what is known in acoustics as a "beat"—i.e., a periodic pulsation resulting from the interference of two wave trains of different frequency. This pulsation is then rectified; the low-pass filter suppresses the high-frequency portions. In all generators not only the frequency but also the amplitude can be regulated by means of continuously variable controls. Sometimes the building-up time and the decay time can also be adjusted. With some generators it is possible to change the sinusoidal oscillations directly into rectangular square-wave oscillations. Finally, there are generators which have an additional sweep device for producing a so-called vibrato effect (periodic variation of frequency). *Square-wave generators* produce rectangular oscillations which, when made audible by means of a loudspeaker, emerge as "square-wave" sounds containing a very high proportion of overtones. A sound of this kind is composed of a fundamental (or primary) tone and the odd-numbered overtones. Fig. 5 shows the addition of the first and third overtones; the resultant oscillation already exhibits something approaching a rectangular shape. The *multivibrator* (Fig. 6), an oscillator which comprises two stages coupled so that the input of each is derived from the output of the other, is a suitable square-wave generator.

The oscillations produced by *sawtooth generators* are, when converted into audible sound waves, perceived as "sawtooth" sounds, which contain a high proportion of overtones (even-numbered as well as odd-numbered). In contrast with sine-wave and square-wave generators, these sawtooth generators (or ramp generators) operate with gas-discharged tubes (glow tube, thyratron; Fig. 7). The capacitor C is charged through the resistor R; the glow tube discharges it as soon as its ignition voltage is reached; this process is repeated over and over again, thereby producing a sawtooth oscillation whose frequency depends on the capacitance C and the resistance R. The overtone content (timbre) of the sound produced can be varied. *Noise generators* are used for producing noise voltages (irregular alternating voltages). These are caused by random fluctuations of electrons in a conductor through which an electric current is flowing (resistance noise) or on emission from the cathode in an electron tube or valve (valve noise). In a loudspeaker such voltages are converted into audible noise. Noise generators produce so-called "white" noise, which is composed of an extremely dense nonperiodic sequence of very short impulses, of varying intensity, in which all the frequencies of the entire range of audibility are present. By means of suitable equipment it is possible to isolate, from this white noise, impulse sequences whose density can be varied so as to obtain a continuous transition from the individual impulse to white noise.

(more)

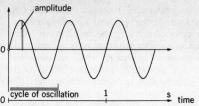

2 cycles per second (2 Hz);
duration 1.5 sec.

amplitude

0

cycle of oscillation   1   s time

FIG. 2 SINE-WAVE OSCILLATION

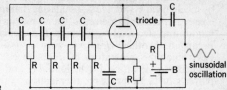

triode

sinusoidal
oscillation

FIG. 3 SINE-WAVE GENERATOR

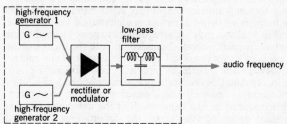

high-frequency
generator 1

G ~

high-frequency
generator 2

G ~

rectifier or
modulator

low-pass
filter

audio frequency

FIG. 4 BEAT-FREQUENCY OSCILLATOR

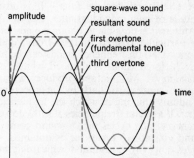

amplitude

square-wave sound
resultant sound
first overtone
(fundamental tone)
third overtone

0   time

FIG. 5 ADDITION OF FIRST AND THIRD OVERTONES

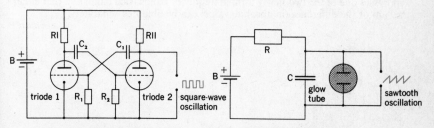

B +

RI   C₂   C₁   RII

triode 1   R₁   R₂   triode 2   square-wave
oscillation

**FIG. 6 SQUARE-WAVE GENERATOR
(MULTIVIBRATOR)**

B +

R

C   glow
tube   sawtooth
oscillation

**FIG. 7 SAWTOOTH GENERATOR
(DIAGRAM ILLUSTRATING PRINCIPLE)**

The audio-frequency oscillations from *electromechanical generators* are produced by mechanical systems (e.g., oscillating diaphragms, rotating discs). Electromagnetic, electrostatic or photoelectric sound-pickup devices convert the mechanical oscillations into electrical oscillations. The oscillator systems are employed in electronic organs and other *electronic musical instruments* (see page 454), as also are electronic-tube generators in feedback circuits, glow-tube generators, and beat-frequency generators. Pitch, loudness and timbre can be controlled to suit the player's or composer's requirements.

The "audio" material can be directly utilized by connecting the generators, through amplifiers, to the appropriate equipment (filters, modulators, etc.). The "processing" of the material is concerned primarily with producing the tonal qualities (timbres) to give the desired musical effects. It is then stored on tape.

Electric *filters* are composed of capacitors and choke coils and are used for intensifying or suppressing certain frequency ranges (frequency bands). A low-pass filter (Fig. 8) allows only frequencies below a certain predetermined value to pass, whereas a high-pass filter (Fig. 9) allows only high frequencies to pass. The band-pass filter (Fig. 10) allows a certain frequency band—bounded on both the high- and the low-frequency side—to pass and suppresses all other frequencies. Tone-control (actually timbre-control) devices and equalizers (antidistortion devices) are also forms of filter. In the studio, combinations comprising a number of band-pass filters connected in parallel are mainly employed, which together cover approximately the whole range of audible frequencies. The band width of the individual filters is usually equivalent to a third or an octave—hence the designations "one-third octave filter" and "octave filter." It is advantageous to be able to control the amplitude of each individual filter, e.g., the Albis variable filter (Fig. 11). With the aid of filters it is possible to obtain sounds having a particular overtone structure—i.e., a particular desired timbre—from "audio" material with a high content of overtones. Certain frequency bands can be filtered out of "white" (colorless) noise so that "colored" noise is obtained, which possesses timbre ("sound color").

Modulators: Modulation produces a major change in the given frequency spectrum—more particularly by the process termed multiplicative mixing, as distinct from additive mixing in which the frequencies are merely superimposed upon one another. If a sound (frequencies 50, 100, 150 cycles/sec.) is modulated with a carrier frequency of 160 cycles/sec., a sound combination with the frequencies 210, 260, 310 and 110, 60, 10 cycles/sec. is obtained. All the frequencies are displaced to a different range (upwards and downwards); intervals between the several frequencies remain unchanged, but the ratios of the frequencies are altered. Multiplicative mixing is often done by means of a so-called ring modulator (or double-balanced modulator). The *frequency converter* (Fig. 12) functions on the same principle, except that it suppresses one of the two newly formed frequency bands (side bands), so that either the sum or the difference (in absolute value) can be obtained separately.

(more)

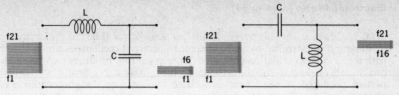

FIG. 8 LOW-PASS FILTER (PRINCIPLE)   **FIG. 9 HIGH-PASS FILTER (PRINCIPLE)**

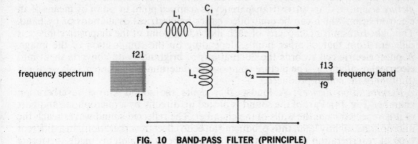

FIG. 10 BAND-PASS FILTER (PRINCIPLE)

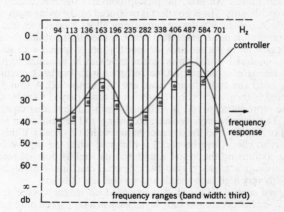

FIG. 11 ALBIS VARIABLE FILTER (PORTION OF SCALE)

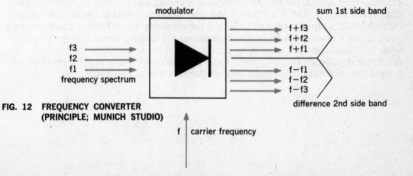

FIG. 12 FREQUENCY CONVERTER
(PRINCIPLE; MUNICH STUDIO)

Photoelectric sound converters: In the *photomodulator* (Fig. 13) the cathode-ray oscillograph is controlled by audio-frequency electrical oscillations which, together with the sawtooth oscillation generated in the apparatus, produces a corresponding oscillogram on the screen of the cathode-ray tube. This oscillogram (i.e., the luminous pattern produced on the screen) is focused by lenses on the photocathode of a photomultiplier, a device which converts the image into electrical oscillations (see Vol. I, page 118). The image can be varied by means of control devices on the oscillograph; in this way the timbre of a given sound can be continuously varied. In a *picture scanner* a diapositive (transparency) is scanned point by point by means of an electron beam which can be controlled either by electrical oscillations or by hand. The light-transmitting capacity of each individual point of the diapositive image is different from that of other points, depending on the composition of the image. A photoelectric cell records the fluctuations of brightness and converts them into electrical oscillations. The picture composition determines the contents of overtones (timbre) of the sound produced.

Reverberation devices: A loudspeaker emits the sound into a reverberation chamber (Fig. 14). Part of the sound is picked up directly by a microphone and part of it is reflected from the walls of the chamber. The reflected sound waves reach the microphone slightly later, thus producing an echo effect (reverberation). In a different type of reverberation device a metal plate (*reverberation plate*) is made to vibrate by electrical means. An acoustic pickup converts the mechanical vibrations into electrical oscillations. The echo effect is produced by the dying away of the vibrations of the plate. A *vocoder* is a device comprising filters, modulators and two generators (impulse generator, noise generator) which is used in the studio for special modulation processes, more particularly for modulation of speech with sound and noises.

"Audio" material stored on magnetic tape can be modified to produce an *echo effect* by means of a magnetic tape recorder with feedback circuit (Fig. 15); the recording head and reproducing head are interconnected. A sound is recorded on the tape and is then, after a short interval of time, played back by the reproducing head. The reproduced sound is recorded again on the same tape; this process is repeated. The time interval between recording and reproduction is determined by the speed of the tape and the distance between the two heads. If this interval is very short, an echo effect (reverberation) is obtained when the tape with the composite recording comprising the initial sound and its slightly delayed second and subsequent recordings is finally played back. Special machines embodying this principle operate with very high tape speeds and several sets of heads.

Tape-speed control: If a sound is recorded on magnetic tape at a speed of, for example, $7\frac{1}{2}$ inches per second and played back at a tape speed of 15 inches per second, the played-back sound will have half the duration and twice the frequency of the original sound. The intervals between the various frequencies composing the sound are likewise doubled, the frequency band is widened, and the buildup and dying-out effects are altered. On the other hand, the frequency ratios remain unchanged. Playback of the tape at reduced speed has the opposite result. Special tape-recording and playback equipment embodies continuous ("infinitely variable") speed control from 0 to 60 inches per second by means of a suitable amplifier with a sine-wave generator (frequency range approx. 15 to 100 cycles/sec.). The synchronous motors driving the tape vary their speed in accordance with the frequency.

(more)

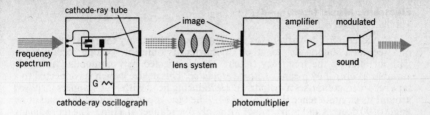

FIG. 13 PHOTOMODULATOR (DARMSTADT STUDIO)

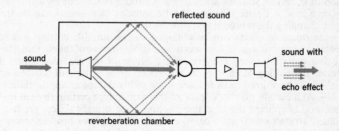

FIG. 14 REVERBERATION CHAMBER

FIG. 15 ECHO EFFECT

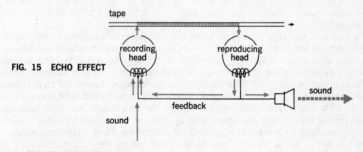

**FIG. 16 REPRODUCING HEAD
WITH FOUR SCANNERS
FOR TIME REGULATOR
(ACCORDING TO SPRINGER)**

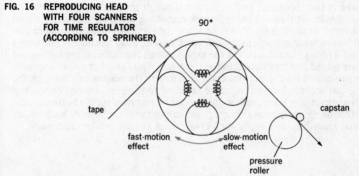

As an alternative to varying the speed, it is possible to vary the reproduction time of a recorded sound—i.e., lengthen or shorten its duration—within certain limits without altering the frequency (pitch). This is effected by continuous ("infinitely variable") control by means of a *time regulator*. This device, which is connected to a tape recorder, comprises a cylindrical reproducing head with four scanners disposed around its circumference (Fig. 16, p. 463). The tape is passed around the head at an angle of 90 degrees and is in contact with only one scanner at a time. The reproducing head can be rotated in both directions; the scanning rate can thus be varied independently of the tape speed. To increase the reproduction time, any particular section of the tape is scanned a number of times; conversely, to shorten the reproduction time, certain sections are skipped. A variant of the time regulator is the tone-pitch regulator, a device whereby the frequencies of a sound can be transposed without changing the reproduction time.

A wide range of effects can be obtained by the suitable cutting and editing of magnetic tape recordings. In this way it is possible to control the rhythm, the buildup and dying-out effects, and the enveloping curves of the recorded sound. Fig. 17a illustrates the editing of an impulse sequence: strips of tape of selected length, on which sounds have been recorded, are stuck to blank tape at appropriately selected intervals. At a certain playback speed of this assembly, a certain rhythm is obtained. The method of splicing the tape also has an effect: straight splices produce abrupt, or "hard," transitions (Fig. 17b), whereas oblique splices produce more gradual, or "soft," transitions (Fig. 17c), in respect of both the buildup and the dying out of particular sound sequences. The enveloping curves (or envelopes) can likewise be modified to some extent by cutting the tape to vary its width (Fig. 17d). (The envelope is the curve enclosing the peak values of a high-frequency oscillation.) Looped tapes can be used to give any desired number of repetitions of recorded sound sequences.

Using "audio" material stored on tape, the composer of electronic music determines the timing and overall dynamic structure of his composition and applies such corrections as may be necessary (blending in of sounds, adjusting the volume, intensifying or attenuating certain frequency ranges by means of equalizers, fading in and fading out of particular sounds). *Mixing* the various "layers" of an electronic musical composition is done by means of several synchronized tape recorders or by means of a four-track instrument whereby the recordings on three tapes are recorded on the fourth tape, the operations being controlled from a mixing console. Fig. 18 schematically shows that re-recording procedure. First, the sound sequence is recorded, through the stereo reproducing head, on to the bottom track of the tape. The tape is then rewound and again run through the recorder; this time the recording is transferred through the full-track recording head, amplifier and stereo reproducing head to the top track. At the same time another sound sequence can be recorded on the bottom track, as the full-track erasing head has meanwhile wiped the tape clean. During the third run of the tape through the machine the two sound recordings are mixed and transferred to the top track, while another sound sequence can again be recorded on the bottom track. Each time the tape is run through the machine, the last recorded sound sequence is added to the previous one on the top track (additive mixing), while a fresh sequence can be recorded on the bottom track. During the recording operations the composer can monitor the sounds both before and after final recording and apply such corrections as he considers necessary.

(more)

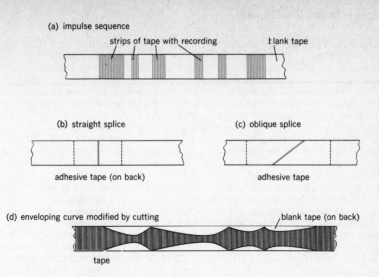

(a) impulse sequence

strips of tape with recording   blank tape

(b) straight splice   (c) oblique splice

adhesive tape (on back)   adhesive tape

(d) enveloping curve modified by cutting   blank tape (on back)

tape

FIG. 17 TAPE CUTTING AND EDITING

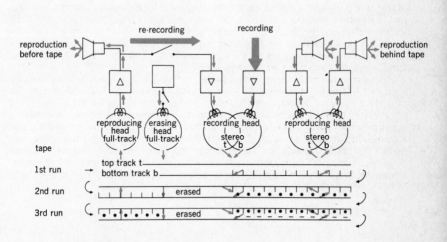

reproduction before tape

re-recording   recording

reproduction behind tape

reproducing head full-track   erasing head full-track   recording head stereo t b   reproducing head stereo t b

tape

1st run   top track t   bottom track b

2nd run   erased

3rd run   erased

FIG. 18 SPECIAL TAPE-RECORDING APPARATUS (PRINCIPLE) (ACCORDING TO HEISS/VOLLMER)

Storage and reproduction: Recording the electronic composition on magnetic or perforated magnetic film is the final stage of the process. Reproduction of the composition may be monaural (single-channel), in which only one loudspeaker (or set of speakers) is used; with stereophonic (or binaural) reproduction two channels are used; some systems use four reproduction channels. With these multichannel systems two or more loudspeakers are suitably positioned in the room or hall, whereby a "three-dimensional" effect can be obtained.

Control: In recent years, electronic data-processing equipment has been utilized for automatically controlling the production process. Computers and punched-tape equipment are used for the purpose. Such apparatus comprises generators, volume regulators and filters which are controlled by punched tape; for this it is necessary first to convert the frequencies, amplitudes, timbres and durations into a suitable code system. By means of automatic control systems it is possible to produce highly differentiated musical compositions; for instance, four hundred variations in pitch, volume and timbre per second can be achieved by means of punched-tape control.

Notation: The score of an electronic musical composition contains precise instructions for the technical production, represented in the form of diagrams provided with numerical data, symbols and explanatory notes. The method of presentation varies and depends primarily on the structure of the composition. The example of a score presented in Fig. 19 is read from top to bottom (time values in centimeters, referred to a tape speed of 19 cm or $7\frac{1}{2}$ inches per second) and from right to left (loudness in decibels; 0 db = minimum damping, i.e., maximum loudness). The frequencies are indicated at the beginning of the respective sounds. The line diagram indicates the loudness as a function of time.

In the score illustrated in Fig. 20, only the enveloping curves (volume-time curves) of the sound sequence are indicated, distributed over four channels (tracks). Each of the four channels is played back over a separate loudspeaker. The figures relate to the "audio" material for which the frequencies, timbres, amplitudes and time values are separately recorded in another score.

Fig. 21 indicates data for preparing a punched tape which will automatically control the frequency ranges of the filters (timbre). Similarly, the frequencies and amplitudes are controlled by other punched tapes.

Electronic music may also, within limits, be produced without a score—i.e., by a process of free improvisation by the composer, who records his work directly on tape. There are no hard-and-fast rules or clear-cut standards for the composition of electronic music. The character of an electronic composition depends not only on the composer's ideas, but also on the facilities provided by the studio; artistic and technical factors have an equal share in the process of composing. True "electronic music" is based entirely on electronically produced sound, whereas so-called "musique concrète" uses sounds actually occurring in nature and in human environment (train whistles, factory sirens, automobile horns, industrial sounds, etc.) as its "raw material."

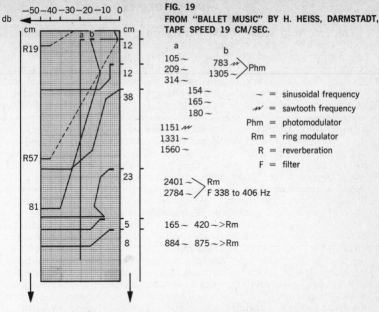

FIG. 19
FROM "BALLET MUSIC" BY H. HEISS, DARMSTADT,
TAPE SPEED 19 CM/SEC.

a
105 ~
209 ~
314 ~

b
783 ⋀ ⟩Phm
1305 ~

154 ~
165 ~
180 ~

1151 ⋀
1331 ~
1560 ~

2401 ~ ⟩Rm
2784 ~ ⟩F 338 to 406 Hz

165 ~   420 ~ >Rm

884 ~   875 ~ >Rm

~ = sinusoidal frequency
⋀ = sawtooth frequency
Phm = photomodulator
Rm = ring modulator
R = reverberation
F = filter

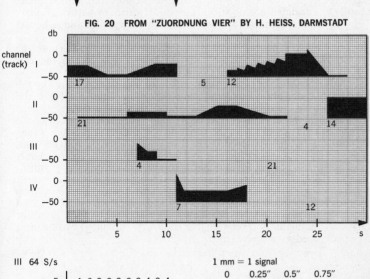

FIG. 20   FROM "ZUORDNUNG VIER" BY H. HEISS, DARMSTADT

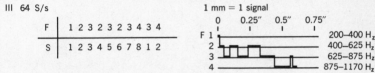

III 64 S/s

| F | 1 2 3 2 3 2 3 4 3 4 |
| S | 1 2 3 4 5 6 7 8 1 2 |

1 mm = 1 signal

F 1   200–400 Hz
2   400–625 Hz
3   625–875 Hz
4   875–1170 Hz

FIG. 21   FROM "COMPOSITION NO. 2" BY J. A. RIEDL, MUNICH

III = punched-tape loop III   64 S/s = 64 signals per second
F = filter   S = signal
F1 = filter 1   S2 = 2 signals

The term "high fidelity" ("hi-fi") is applied to sound recording and reproducing equipment whereby a reasonably faithful reproduction of the quality of the original sound can be obtained. Oscillations, however complex they may be, can always be conceived as the sum of various pure sine and cosine oscillations (the latter are of precisely the same shape as sine oscillations, but are displaced one-quarter of a wavelength in relation to them). This possibility of splitting up any oscillation into an assembly of sine (and cosine) oscillations (Fig. 1) is important because the electronic components which are used for the transfer of oscillations can retransmit pure sine oscillations as oscillations which are again at least approximately sinusoidal in form, but amplified by a factor depending on the frequency and delayed by a length of time likewise depending on the frequency (Fig. 2). Such components are referred to as "linear" transfer elements; their performance is fully characterized by the amplification and the delay (phase displacement) as a function of the frequency. Together these two data determine the frequency response of the equipment (in Fig. 2 only the amplification response, not the phase response, is indicated). The "overall frequency response" of a high-fidelity system is determined by the product of the amplification responses and by the sum of the phase responses of all the transfer elements. The system should provide constant amplification and phase displacement (delay) over the entire range of audibility (approx. 16 to 20,000 cycles/sec.). Since the frequency dependence of the amplification is generally linked to the frequency dependence of the phase displacement, the amplification must not suddenly decrease at the limits of the audibility range, for otherwise phase displacements will occur within that range. This is especially important for stereophonic reproduction, inasmuch as the "three-dimensional" effect is produced partly by particular phase relations between the two channels corresponding to the hearer's left and right ears respectively. Between the amplification response and the phase response of the system there must exist a mathematical relationship which ensures that in the "overall frequency response" a constant amplification over a certain range is associated with a constant phase displacement within that range.

Rigorously linear behavior of transfer elements is an idealization. In reality the pure sine oscillations will always undergo some distortion. Thus, if a sinusoidal voltage is applied to a loudspeaker, the diaphragm responds accurately to the voltage only at low amplitudes. With increasing amplitudes the deflection of the diaphragm reaches a maximum value; the variation of the acoustic pressure is then no longer sinusoidal, but has flattened peaks (Fig. 3). Such nonlinear distortions, once they have developed, can—in contrast to linear distortions—be corrected only in exceptional cases. This is done, for example, in certain disc-recording procedures: the oscillations to be recorded are so distorted prior to recording that the nonlinear distortions which subsequently occur are largely compensated. The oscillation which results from nonlinear distortion of a sinusoidal oscillation can be resolved into a number of purely sinusoidal oscillations in the manner shown in Fig. 1. For each transfer element, the proportion of higher frequencies produced by nonlinear distortion can be stated as a percentage.

(more)

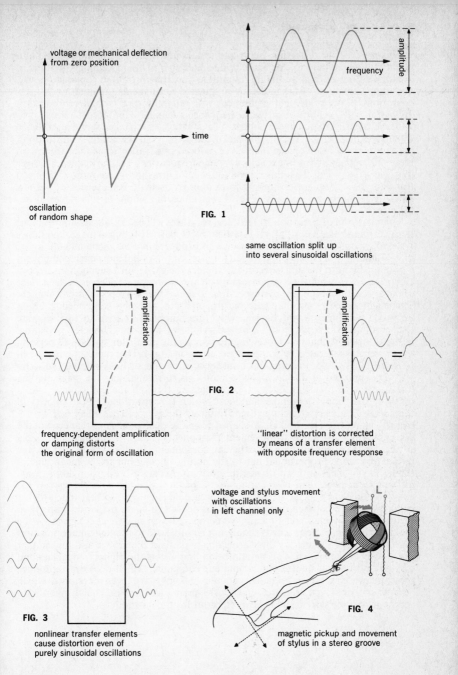

voltage or mechanical deflection
from zero position

time

oscillation
of random shape

FIG. 1

amplitude

frequency

same oscillation split up
into several sinusoidal oscillations

amplification

FIG. 2

frequency-dependent amplification
or damping distorts
the original form of oscillation

amplification

"linear" distortion is corrected
by means of a transfer element
with opposite frequency response

FIG. 3

nonlinear transfer elements
cause distortion even of
purely sinusoidal oscillations

voltage and stylus movement
with oscillations
in left channel only

L

L

FIG. 4

magnetic pickup and movement
of stylus in a stereo groove

This is called the distortion factor and should not exceed 1% in the case of high-fidelity systems. The whole object of hi-fi technique is to attain faithful reproduction over the entire range of audibility (minimum linear distortion) in conjunction with minimum distortion of the individual frequency components (minimum nonlinear distortion). Simpler amateur equipment will generally give faithful reproduction only of a narrower range of audio frequencies (standard broadcast radio up to 9000 cycles/sec.; record players approx. 50 to 5000 cycles/sec.). The sound reproduced by such equipment is "darker" and lacking in clarity, but the nonlinear distortions, which are much more objectionable, are reduced in advance, since they are caused more particularly at low frequencies and high frequencies—i.e., the frequency ranges that such equipment does not reproduce at all. Ultrashort-wave radio (VHS) can transmit audio frequencies ranging from approx. 30 to 15,000 cycles/sec., but few amateur receivers are actually able to reproduce the whole of this range of frequencies.

The reduction of distortion by technical measures will be explained with reference to the example of a record player (see also Vol. I, page 314). Fig. 4 (page 469) shows a stereo groove and the needle movement it produces in a magnetic pickup. In the case of a monaural phonograph record (or disc), only a radial movement component within the plane of the disc is recorded, whereas the groove of a stereophonic record comprises two movement components which are inclined at 45 degrees to the surface of the disc, each component corresponding to one of the two reproduction channels. The magnetic or electromagnetic pickup separates the two components: i.e., it transmits two independent signals, which are amplified and fed to two separate loudspeakers. This separation is achieved as follows: if the left channel is oscillating, whereas the right channel is receiving no signals, the needle (or stylus) will perform the movement indicated by the thick red arrow in Fig. 4. This causes the black coil to rotate in its plane; in doing this it intersects no magnetic lines of force and so no voltage is induced in this coil. The red coil L is so rotated that the magnetic flux (the number of magnetic lines of force) passing through the coil changes. As a result of this change a voltage is induced which is proportional to the velocity of the needle. The crystal pickup (Fig. 5) utilizes the piezoelectric effect (see Vol. I, pages 86 and 316). The resilient W-shaped coupling element ensures that each of the two mutually perpendicular movement components causes flexing movement only of the crystal assigned to that particular component. In this type of pickup, the output voltage is proportional not to the needle velocity but to the magnitude of the needle's deflection from its neutral position. Accurate proportionality is more difficult to achieve than with the magnetic pickup. Besides, with the crystal the counteracting force to the deflection of the needle is greater, so that much higher contact pressure between disc and needle is needed to guide the latter properly in the groove: between 5 and 10 grams, as compared with as little as 0.5 gram for the magnetic pickup. Stereo records should not be subjected to needle-contact pressures exceeding about 5 grams, as these are liable to damage the groove and thus impair the quality of the record. This deterioration first affects the higher frequencies and may at first remain undetected in amateur equipment which does not reproduce these frequencies. More recently, ceramic pickups have been developed which operate with contact pressures of about 3 grams and are not much inferior to magnetic pickup systems as regards freedom from distortion.

(more)

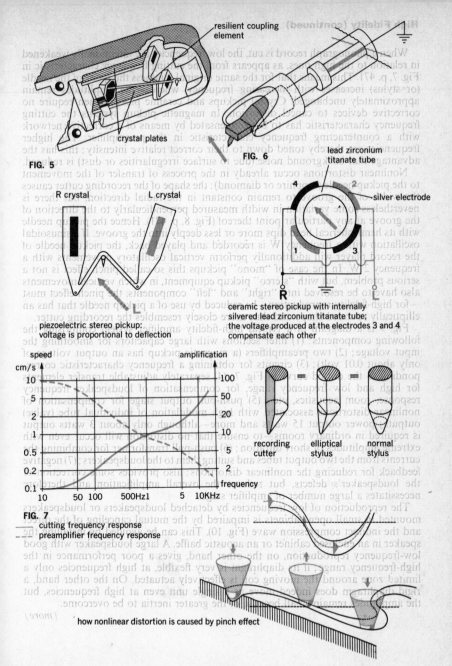

resilient coupling element

crystal plates

FIG. 5

FIG. 6

lead zirconium titanate tube

R crystal   L crystal

silver electrode

R   L

ceramic stereo pickup with internally silvered lead zirconium titanate tube; the voltage produced at the electrodes 3 and 4 compensate each other

piezoelectric stereo pickup: voltage is proportional to deflection speed

amplification

cm/s

recording cutter   elliptical stylus   normal stylus

FIG. 7

cutting frequency response
preamplifier frequency response

how nonlinear distortion is caused by pinch effect

FIG. 8

471

When a phonograph record is cut, the low frequencies are intentionally weakened in relation to the high ones, as appears from the cutting frequency characteristic in Fig. 7, p. 471. This means that for the same original loudness the speed of the needle (or stylus) increases with increasing frequency, whereas the amplitudes remain approximately unchanged. Crystal pickups and ceramic pickups then require no corrective devices to cancel distortion. In magnetic pickup systems the cutting frequency characteristic has to be compensated by means of a corrective network with a counteracting frequency characteristic in the preamplifier. The higher frequencies are thereby toned down to their correct relative intensity; this has the advantage that background noise (due to surface irregularities or dust) is reduced.

Nonlinear distortions occur already in the process of transfer of the movement to the pickup needle (sapphire or diamond): the shape of the recording cutter causes the width of the groove to remain constant in the radial direction, but there is nevertheless some variation in width measured perpendicularly to the direction of the groove at any particular point thereof (Fig. 8, p. 471). Hence the pickup needle with its hemispherical head dips more or less deeply into the groove. If a sinusoidal oscillation with a frequency W is recorded and played back, the pickup needle of the record player will additionally perform vertical oscillatory movements with a frequency 2 W. In the case of "mono" pickups this so-called pinch effect is not a serious problem, but with "stereo" pickup equipment, in which vertical movements also have to be resolved into "right" and "left" components, the pinch effect must —for high-fidelity reproduction—be reduced by use of a pickup needle that has an elliptically rounded shape and thus more closely resembles the recording cutter.

Fig. 9 is a diagram of a typical high-fidelity amplifier system comprising the following components: (1) filter sections with large capacitors for smoothing the input voltage; (2) two preamplifiers (a magnetic pickup has an output voltage of only about 0.01 volt); (3) circuits for obtaining a frequency characteristic corresponding to the dotted curve in Fig. 7; (4) separately adjustable transfer elements for high and low frequency range, for compensation of loudspeaker frequency response, room acoustics, etc.; (5) push-pull output stage for compensation of nonlinear distortions associated with high modulation of individual tube (valve) outputs; power output 15 watts and more—although only about 3 watts output is required in ordinary rooms—to ensure that no distortion will occur even with extreme amplitudes of short duration; (6) push-pull transformer for combining the currents from the two output tubes and suiting them to the loudspeakers; (7) negative feedback for reducing the nonlinear distortions; also provides some correction of the loudspeaker's defects, but reduces the overall amplification and therefore necessitates a large number of amplifier stages.

The reproduction of low frequencies by detached loudspeakers or loudspeakers mounted in small open cabinets is impaired by the mutual canceling of the direct and the indirect compression wave (Fig. 10). This can be avoided by installing the speaker in an enclosed cabinet or an acoustic baffle. A large loudspeaker with good low-frequency reproduction, on the other hand, gives a poor performance in the high-frequency range; if its diaphragm is very flexible, at high frequencies only a limited zone around the moving coil is effectively actuated. On the other hand, a rigid diaphragm does indeed move as a single unit even at high frequencies, but the amplitudes remain small because of the greater inertia to be overcome.

(more)

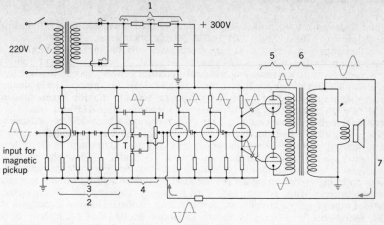

FIG. 9

input for
magnetic
pickup

+ 300V

220V

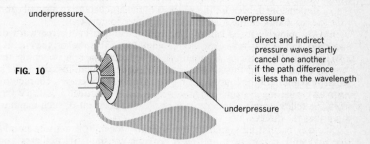

underpressure

overpressure

FIG. 10

direct and indirect
pressure waves partly
cancel one another
if the path difference
is less than the wavelength

underpressure

FIG. 12

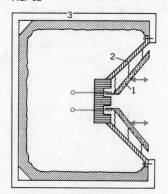

modern high-fidelity loudspeaker
for low and medium frequencies
(approx. 30–1500 Hz)

FIG. 11

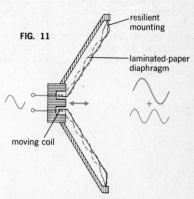

resilient
mounting

laminated-paper
diaphragm

moving coil

formation of harmonic oscillations
due to nodes in the diaphragm
of an ordinary loudspeaker

In a wide-range loudspeaker the frequency range is increased by subdivision of the diaphragm into a rigid inner and a flexible outer zone, with a very flexible transition region in between. At high frequencies only the inner part vibrates; at low frequencies the whole diaphragm is actuated. A more consistent solution consists in dividing the incoming frequency mixture between special loudspeakers for low and for high frequencies with the aid of electrical filter circuits. The laminated-paper diaphragm (cone) of a simple loudspeaker tends to develop harmonic components when energized by a sinusoidal oscillation (Fig. 11, p. 473) and thus to emit additional higher frequencies. Fig. 12 (p. 473) illustrates a high-fidelity loudspeaker for low and middle frequencies; in this speaker the above-mentioned defects have been reduced by the following arrangements: (1) rigid diaphragm made of thick but light material (1 cm thick foam plastic coated with tinfoil, total weight only about 10 grams) to obviate harmonic components; (2) very flexible triple suspension for exact centering and ensuring accurately axial movement and for obtaining a resonance frequency at the bottom end of the frequency range (approx. 25 cycles/sec.); (3) thick cabinet stiffened by ribs and corner blocks, to obviate harmonic components, with sound-absorbing lining to absorb the sound emitted into the interior of the cabinet and to prevent stationary waves from forming; the cabinet is moreover amply dimensioned, because the stiffness of the air cushion in a small cabinet would shift the resonance of the loudspeaker to higher frequencies.

In a completely closed cabinet the energy emitted in the rearward direction is absorbed. In the bass-reflex speaker cabinet (Fig. 13), however, the energy is utilized also in the low frequencies. By enlarging or reducing the reflex opening, the enclosed volume of air is tuned to such a resonance frequency that, instead of the steep resonance maximum of the loudspeaker, two flat resonance curves, characteristic of two coupled oscillating systems, are obtained. Also, the indirect pressure wave emerging from the reflex opening undergoes a phase displacement of such magnitude that it intensifies the direct wave.

Like any oscillating system, a loudspeaker does not exactly respond to a force that suddenly commences and then remains constant. It oscillates about a new middle position (Fig. 14b, black curve). In the schematic model (Fig. 14a) such an oscillation induces a current in the damping coil. At low damping resistance this current becomes very strong, and its magnetic field damps the oscillation. In a loudspeaker the moving coil functions both as driving coil and as damping coil. The output circuit of the amplifier serves as a damping resistance. With correct design of the equipment this damping resistance is of just the right magnitude to provide optimum damping (continuous red curve in Fig. 14b) and thus ensure faithful reproduction of steplike modes of oscillation.

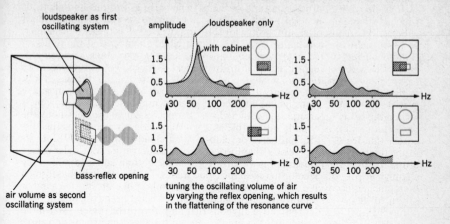

loudspeaker as first oscillating system

amplitude — loudspeaker only

with cabinet

air volume as second oscillating system

bass-reflex opening

tuning the oscillating volume of air by varying the reflex opening, which results in the flattening of the resonance curve

FIG. 13   BASS-REFLEX SPEAKER

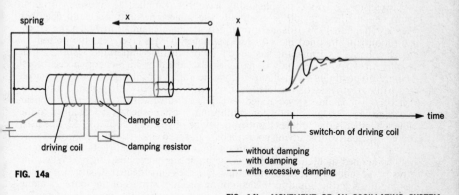

spring

driving coil

damping coil

damping resistor

FIG. 14a

x

switch-on of driving coil

time

—— without damping
—— with damping
- - - with excessive damping

FIG. 14b   MOVEMENT OF AN OSCILLATING SYSTEM WITH AND WITHOUT DAMPING

CLOUD CHAMBER AND BUBBLE CHAMBER

Important contributions to the discovery and investigation of elementary particles such as electrons and positrons and of the particles such as protons and neutrons which are the components of atomic nuclei have been made by experimental methods of detection. An important apparatus for this kind of research is the cloud chamber, of which there are two main types: the Wilson cloud chamber and the diffusion cloud chamber. By means of such equipment it is possible to observe and photograph the paths of high-speed electrically charged particles. When such particles pass through a gas, they collide with, and dislodge electrons from, the gas atoms, so that the latter become positively charged ions. The dislodged electrons (negative particles) and the gas ions function as centers of condensation upon which droplets form if the gas is supersaturated with a vapor. The succession of droplets along the path of a particle is visually observable as a so-called cloud track. It was this discovery that led C. T. R. Wilson to develop the first cloud chamber in 1912. In its original form, the Wilson cloud chamber consists of a circular glass cylinder closed at the top by a glass observation window (Fig. 1). The bottom and side wall of the chamber are of a black color to reveal the cloud tracks. The bottom, which is formed by a plunger, is covered with wet absorbent paper to keep the air in the chamber saturated with water vapor. An electric field is applied between the top and bottom of the chamber in order to remove any ions that may already be present. When the plunger is suddenly withdrawn (lowered) a certain distance, the expansion causes the gas and vapor in the chamber to cool, whereby the gas becomes supersaturated with vapor. When this happens, the electrically charged particles introduced into the chamber leave characteristic cloud tracks formed by tiny droplets of condensed vapor, as exemplified in Fig. 2, where the longer and finer track is due to the smashing of an atom, causing a proton to be liberated. A limitation of the Wilson cloud chamber is that it allows only instantaneous observations, since the supersaturated conditions suitable for the formation of cloud tracks exist only very briefly during the expansion. Various modifications have been applied to Wilson's original chamber, but the principle is essentially the same. The diffusion cloud chamber was developed in 1950. It is a continuously sensitive chamber in which the supersaturation is maintained more or less permanently in a layer near the bottom of the chamber. This apparatus has acquired considerable importance as a research tool, particularly in the study of particles discharged by high-energy accelerators. The prototype of this cloud chamber consisted of an air-filled glass cylinder placed in a pan of methyl alcohol cooled by dry ice. Warm methyl alcohol in a tray near the top of the chamber evaporated and diffused downward. Cloud tracks appear in a zone where a state of diffusion equilibrium is established. A device allied to the cloud chamber is the bubble chamber, developed by D. A. Glaser in 1952. It is used more particularly for studying collision of high-speed charged particles with atomic nuclei. A pressure-tight vessel contains a liquid heated far above its boiling point, but maintained at high pressure so that boiling is prevented (Fig. 3). When the pressure is suddenly reduced, the liquid becomes superheated. Charged particles passing through the liquid in this condition produce strings of tiny bubbles along their paths (Fig. 4). These bubble tracks are recorded by high-speed photography; precision measurements on the photographs provide information on the nuclear processes concerned. One important advantage over the cloud chamber is that in the bubble chamber rare nuclear events are of relatively frequent occurrence because of the high density in the liquid.

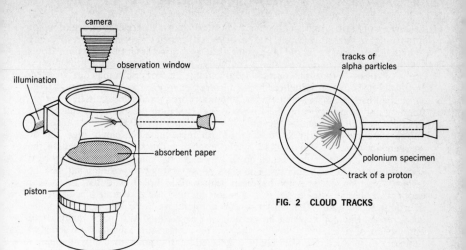

camera

observation window

illumination

absorbent paper

piston

FIG. 1   WILSON CLOUD CHAMBER (SCHEMATIC)

tracks of alpha particles

polonium specimen

track of a proton

FIG. 2   CLOUD TRACKS

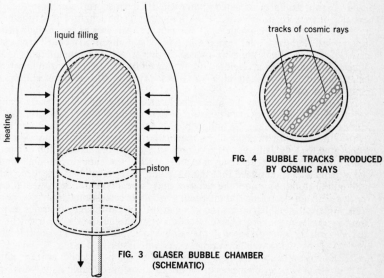

liquid filling

heating

piston

FIG. 3   GLASER BUBBLE CHAMBER (SCHEMATIC)

tracks of cosmic rays

FIG. 4   BUBBLE TRACKS PRODUCED BY COSMIC RAYS

LINEAR ACCELERATOR

Devices of various types collectively known as particle accelerators are of major importance in nuclear research. The particles accelerated to velocities corresponding to many millions of volts, and used as high-energy projectiles for bombarding atoms and for other purposes, are usually the nuclei of light atoms such as the proton from hydrogen or the alpha particle from helium; heavier nuclei may also be used. In the accelerator the particle acquires a kinetic energy equal to its electric charge multiplied by the difference in electric potential through which it falls. The corresponding unit of energy is the electron-volt (ev), based on the electronic charge and the volt as a unit of potential; the unit Mev represents 1,000,000 electron-volts.

The principle of multiple acceleration whereby these extremely high kinetic energies are attained is illustrated by the mechanical model in Fig. 1, where the "particle" (the red ball) is speeded on its way at an increasingly high velocity by a succession of rotating hammers. The class of apparatus comprising what may be termed "circular accelerators" has been dealt with in Vol. I, page 108. In a linear accelerator the particles travel in a straight line and are accelerated by a rapidly alternating potential. The earlier linear accelerators were so-called resonance accelerators (Fig. 2a), in which the particles are accelerated in steps by the repeated application of a relatively small voltage. The accelerator consists of a series of tubular units which are alternately connected to the poles of a high-frequency generator. Acceleration of the particles (e.g., low-energy ions) occurs at the gaps between the units, the frequency of the accelerating voltage being so adjusted that the correct polarity to speed the particle on its way is applied at the correct instant when the particle arrives at the gap (Figs. 2b and 2c). In simple terms: when a positive particle enters a gap just when the next tubular unit is negative and the preceding unit is positive, it will be attracted by the former and repelled by the latter and thus accelerated; by the time the particle reaches the next gap, the polarity has reversed, and it is again similarly attracted and repelled. To satisfy this condition there must be the following relation between the frequency f, the length l of a tubular unit, and the velocity v of the particle: $f l = v$. Since v increases as the particle proceeds along the accelerator, while f remains unchanged, this means that l must progressively increase—i.e., the gaps must be spaced farther apart (the tubular units must be longer). The ions to be accelerated are produced by an ion source and injected into the accelerator at a suitable high initial velocity with the aid of an appropriately applied voltage.

In its present-day form the linear accelerator makes use of an electromagnetic (radio) wave, with a frequency of around 3000 megacycles/second, traveling along an evacuated waveguide: i.e., a hollow metal conductor (see page 526) through which high-frequency microwaves—more particularly, very short radio waves—are propagated (Fig. 4). This type of apparatus is used for the acceleration of electrons. These are injected into the waveguide and travel in the same direction as the wave. The apparatus is so designed that the wave and the electron have the same velocity at all points along the guide; the electron thus travels synchronously with the wave. The latter has an electric field component which is directed along the axis of the waveguide. An electron which enters the guide at the correct time (or phase) is propelled along by a force arising from the interaction of its charge and the electric field. Since the wave and the electron travel at the same velocity, the electron is subjected to this force all along the waveguide and thus travels faster and faster. To achieve the desired synchronization, the phase velocity of the high-frequency wave is adjusted to the electron velocity by means of suitably dimensioned diaphragms (with circular openings in them) spaced at intervals equal to one-quarter of the wavelength (Fig. 3).

The world's largest linear accelerator is at Stanford University, in Palo Alto, California. It is nearly 2 miles long and can, in its present stage of development, accelerate electrons to energies of 20 million electron-volts (20,000 Mev). The electron beam can easily be brought out from the accelerator, so that certain precision experiments can be performed which are not possible with circular accelerators.

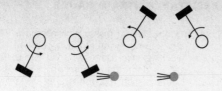

FIG. 1 MECHANICAL ANALOGY OF ACCELERATION PRINCIPLE

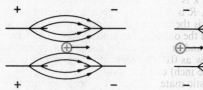

ions V_0

acceleration gap

high-frequency generator

FIG. 2a SCHEMATIC DIAGRAM OF LINEAR ACCELERATOR

FIG. 2b IONS IN THE ACCELERATION GAP

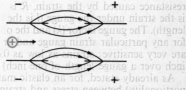

FIG. 2c IONS IN THE FIELD-FREE INTERIOR OF THE TUBE

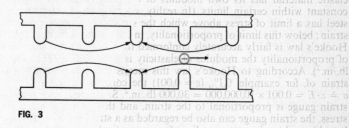

FIG. 3

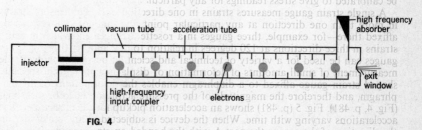

collimator   vacuum tube   acceleration tube

injector

high-frequency input coupler

electrons

high frequency absorber

exit window

FIG. 4

ELECTRIC-RESISTANCE STRAIN GAUGE

Strain denotes the specific deformation, often expressed as a percentage, caused in a material by the action of a stress—i.e., a force per unit area (e.g., lb./in.^{2}, kg/cm^2, kg/mm^2, etc.). In an elastic material such as steel there is direct proportionality between stress and strain. Because of this definite relationship it is possible to determine the magnitude of the stresses in a structure if the strains are known. The latter can be measured by devices termed strain gauges, which are extensively used in science and engineering for purposes of research and testing. There are various types of strain gauge. One very widely applied type is the bonded electric-resistance strain gauge, first developed some thirty years ago. In its simplest form it consists of a length of fine wire in a zigzag or grid pattern cemented between two thin sheets of paper (Figs. 1 and 2). The gauge is bonded to the part being tested, and the wire grid participates in the deformations thereof, either in compression (shortening) or tension (elongation) so that the length and cross-sectional area of the wire undergo changes, which cause changes in the electrical resistance of the wire which are proportional to the changes in length (elongation causes an increase, shortening causes a decrease in resistance). These changes in resistance, though usually very small, can be measured by means of suitable electrical circuits and instruments. The relationship between the unit change in resistance and strain is expressed by the formula: $\Delta R/R = k \cdot \Delta l/l$, where k is the so-called gauge factor, ΔR is the change in resistance caused by the strain, R is the original resistance of the gauge, and $\Delta l/l$ is the strain under the gauge (l is the original gauge length and Δl is the change in length). The gauge factor (k) and the original resistance (R) are predetermined values for any particular strain gauge and are supplied by the manufacturer. Such gauges are very sensitive; strains as low as 0.001% (corresponding to one millionth of an inch over a gauge length of one inch) can be detected.

As already stated, for an elastic material there is a linear relationship (i.e., proportionality) between stress and strain. This is known as Hooke's law and can be expressed by the following formula: $\sigma = \varepsilon \cdot E$, where σ is the stress, ε is the strain, and E is a coefficient termed the modulus of elasticity (or Young's modulus). Every elastic material has its own modulus of elasticity, which—for that material—is constant within certain limits. (In reality, no material is completely elastic; even steel has a limit of stress above which the stress is no longer proportional to the strain; below this limit of proportionality, in the so-called elastic range of behavior, Hooke's law is fairly accurately conformed to, however.) For steel below the limit of proportionality the modulus of elasticity is about 2,100,000 kg/cm^2 (30,000,000 lb./in.^{2}). According to Hooke's law this means that when steel is subjected to a strain of, for example, 0.1% ($\varepsilon = 0.001$) the corresponding stress in the steel is: $\sigma = \varepsilon \cdot E = 0.001 \times 30,000,000 = 30,000$ lb./in.^{2}. Since the electrical resistance of the strain gauge is proportional to the strain, and this in turn is proportional to the stress, the strain gauge can also be regarded as a stress-measuring device and it can be calibrated to give stress readings for any particular material.

A single strain gauge measures strains in one direction only. To measure strains in more than one direction at any particular point, two or more gauges can be affixed there—for example, three gauges in a rosette arrangement which measure strains in three directions at 120 degrees in relation to one another (Fig. 3). Strain gauges can be used for a variety of technical and scientific purposes involving the measurement of small amounts of deformation or displacement. Thus, a specially shaped strain gauge affixed to a diaphragm enables the deformations of the diaphragm, and therefore the magnitude of the pressure acting on it, to be measured (Fig. 4, p. 481). Fig. 5 (p. 481) shows an acceleration pickup for the measurement of accelerations varying with time. When the device is subjected to an acceleration in the direction of the arrow, the part A with the bonded-on strain gauge undergoes a deflection (in the opposite direction) which is directly proportional to the acceleration.

(more)

FIG. 1 ELECTRIC-RESISTANCE STRAIN
GAUGE WITH ZIGZAG WIRE

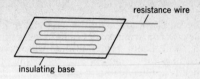

resistance wire

insulating base

FIG. 2 ELECTRIC-RESISTANCE STRAIN
GAUGE WITH WIRE GRID

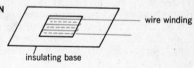

wire winding

insulating base

FIG. 3 THREE STRAIN GAUGES
IN ROSETTE ARRANGEMENT

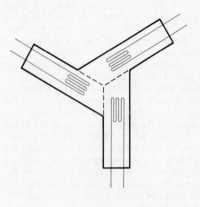

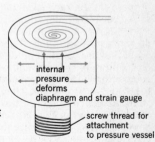

internal
pressure
deforms
diaphragm and strain gauge

screw thread for
attachment
to pressure vessel

FIG. 4 LOAD CELL WITH STRAIN-GAUGE
WIRE IN SPIRAL PATTERN

FIG. 5 ACCELERATION PICKUP WITH
ELECTRIC-RESISTANCE STRAIN GAUGE

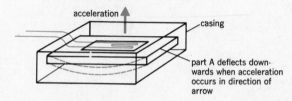

acceleration

casing

part A deflects down-
wards when acceleration
occurs in direction of
arrow

The small variations in the electrical resistance of a strain gauge are measured by means of an arrangement called a measuring bridge which embodies the principle of the Wheatstone bridge (Fig. 6) which comprises two resistance branches connected in parallel, each branch consisting of two resistances in series. The supply voltage is applied at the two diagonal points I and II. The measuring instrument (galvanometer) connected between the other two diagonal points III and IV will give a zero voltage reading only when the same voltage drop occurs between I and III as between I and IV—for example, when $R_1 = R_4$ and $R_2 = R_3$, and also when $R_1 = R_2$ and $R_3 = R_4$. In general, the bridge will be "in equilibrium"—i.e., the measuring instrument will give a zero reading, when the mathematical relationship $R_1 : R_2 = R_4 : R_3$ is satisfied. If three of the resistances are of known value, e.g., R_2, R_3 and R_4, then the (unknown) fourth resistance R_1 can be determined by "balancing" the bridge (by varying the resistance R_2) and calculating it from $R_1 = R_2 R_4/R_3$.

For measurements performed with strain gauges, all four resistances of the bridge may be replaced by strain gauges, all of equal resistance (Fig. 7); alternatively, only two equal strain gauges may be connected into the bridge circuit, while the other two resistances are incorporated in the measuring-bridge apparatus itself (Fig. 8). When the bridge has been "balanced," i.e., the measuring instrument has been brought to a zero reading, a variation in the magnitude of one or more of the resistances will cause the needle of the instrument to deflect from the zero position. The arrangement shown in Fig. 7 may, for example, be applied in measuring the strains occurring at the top and at the underside of a beam loaded in bending (Fig. 9). When there is no load on the beam, the bridge is in equilibrium; when load is applied, the two gauges R_1 and R_3 undergo an equal tensile strain (elongation), while the two gauges R_2 and R_4 undergo an equal compressive strain (shortening). The purpose of the arrangement in Fig. 10 is to measure the strain at the top of the beam subjected to a combination of flexural and tensile loading. Only the strain gauges R_1 and R_3 are affixed to the beam; the gauges R_3 and R_4 are dummies in that they do not participate in the strain of the beam, but provide temperature compensation —i.e., they are installed in the vicinity of the other two, so that any changes in the temperature of the surroundings (whereby the resistances are altered) affect all the gauges equally. Similar compensation can be provided in the arrangement shown in Fig. 8 and applied to the strain measurement on a beam as shown in Fig. 11. With strain gauges of equal resistance the measuring arrangement in Fig. 10 is twice as sensitive as that in Fig. 11, because in the former case the change in resistance of both R_1 and R_3 causes twice the deflection of the measuring instrument as does the change in resistance of R_1 alone in the latter case. Fig. 12 shows strain gauges for measuring the strains in a shaft subjected to a torque (twisting moment). The largest strains occur in directions at 45 degrees to the center line of the shaft.

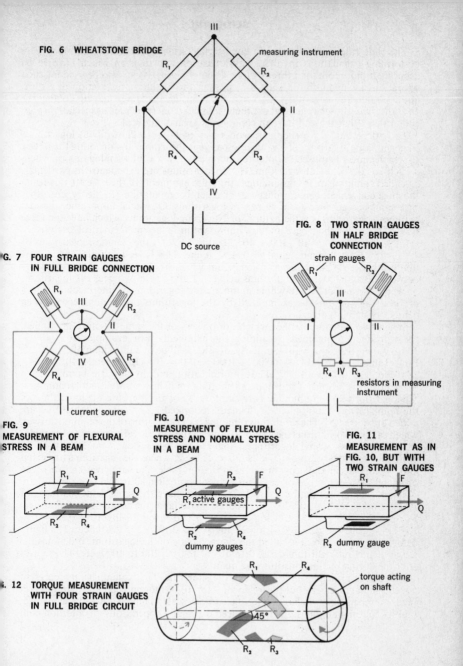

FIG. 6 WHEATSTONE BRIDGE

measuring instrument

R_1

R_2

III

I

II

R_4

R_3

IV

DC source

FIG. 7 FOUR STRAIN GAUGES IN FULL BRIDGE CONNECTION

R_1

III

R_2

I

II

IV

R_3

R_4

current source

FIG. 8 TWO STRAIN GAUGES IN HALF BRIDGE CONNECTION

strain gauges

R_1

R_2

III

I

II

R_4

IV

R_3

resistors in measuring instrument

FIG. 9 MEASUREMENT OF FLEXURAL STRESS IN A BEAM

R_1 R_3 F

Q

R_2 R_4

FIG. 10 MEASUREMENT OF FLEXURAL STRESS AND NORMAL STRESS IN A BEAM

R_3 F

R_1 active gauges

Q

R_2 R_4

dummy gauges

FIG. 11 MEASUREMENT AS IN FIG. 10, BUT WITH TWO STRAIN GAUGES

R_1 F

Q

R_2 dummy gauge

FIG. 12 TORQUE MEASUREMENT WITH FOUR STRAIN GAUGES IN FULL BRIDGE CIRCUIT

R_1 R_4

torque acting on shaft

45°

R_2 R_3

The slide rule is widely used in engineering, science and commerce for rapidly performing calculations involving multiplication and division which have to be accurate to not more than three or four decimal places. It can also be used for such operations as involution (raising to a power) and evolution (extraction of a root) and for calculations with trigonometric functions (sine, cosine, tangent, cotangent). In addition to those for general use there are many different types of special-purpose slide rules. What they all have in common is logarithmic scales.

To understand how a slide rule works it is essential to know about logarithms. In general, the logarithm of a value a to base b is the exponent n denoting the power to which b must be raised in order to obtain a; i.e., if $a = b^n$, then $\log_b a = n$, where a itself is called the antilogarithm. Hence logarithms are exponents. In particular, so-called common, or Briggsian, logarithms are exponents of the base 10, this being the most convenient base for purposes of general computation. (In theory, any other base could be adopted for a system of logarithms. Of special importance, besides common logarithms, are the natural, or Naperian, logarithms, which have the value $e = 2.718\ldots$ for their base and are important in higher mathematical analysis.)

Since $1 = 10^0$, $10 = 10^1$, $100 = 10^2$, $1000 = 10^3$, etc., the common logarithms of 1, 10, 100, 1000, etc., are equal to 0, 1, 2, 3, etc. This is set forth in Table 1. Any number between 1 and 10 has a logarithm which lies between 0 and 1; numbers between 10 and 100 have logarithms between 1 and 2; etc. Whereas the logarithms of whole powers of 10 are whole numbers, the logarithms of intermediate values are decimal fractions, as exemplified by the logarithms of the numbers between 0 and 10 listed in Table 2.

When two (or more) values which are powers of the same base are multiplied, their product is obtained by adding the exponents. For example: $10^2 \times 10^3 = 100 \times 1000 = 100,000 = 10^5$. Similarly, division is done by subtraction of exponents. For example: $10^5 \div 10^3 = 100,000 \div 1000 = 100 = 10^2$. In general: $10^m \times 10^n = 10^{m+n}$ and $10^m \div 10^n = 10^{m-n}$. Thus multiplication and division can be reduced to the simpler operations of addition and subtraction if, instead of the actual numbers, the logarithms of those numbers are employed. For this purpose the logarithms of the numbers to be multiplied or divided are looked up in suitable tables and are added or subtracted. The result thus obtained is the logarithm of the required answer, which can be looked up (as the antilogarithm) in the same logarithm tables. Example: to calculate 2×3 using logarithms: $\log 2 \times 3 = \log 2 + \log 3 = 0.301 + 0.477 = 0.778$; the antilogarithm of 0.778 is 6 (see Table 2, where the logarithms are given in only three decimal places; in actual logarithm tables they are given in five, seven or sometimes even more places, thus providing greater accuracy).

The logarithms of the numbers 1, 2, 3, 4, etc., have the same values as those of 10, 20, 30, 40, etc., and 100, 200, 300, 400, etc., in so far as the decimals are concerned. For example: $\log 2 = 0.301$; $\log 20 = \log 2 \times 10 = \log 2 + \log 10 = 0.301 + 1 = 1.301$; $\log 200 = \log 2 \times 100 = \log 2 + \log 100 = 0.301 + 2 = 2.301$; etc. The figure before the decimal point is called the characteristic of the logarithm; it denotes the range of the antilogarithm—e.g., 0 to 10, 10 to 100, 100 to 1000, etc. The figures after the decimal point constitute the mantissa.

The above principles are utilized in the slide rule. The scale divisions represent not the actual numbers, but the logarithms (or, to be more precise, the mantissas) of those numbers. Fig. 1 shows the logarithmic scale. Thus, as already explained, multiplication and division are reduced to simple operations of addition and

(more)

Table 1

power	exponent	value of the power
10^0	0	1
10^1	1	10
10^2	2	100
10^3	3	1000
10^4	4	10,000
10^5	5	100,000
10^6	6	1,000,000
...	...	......

Table 2

number anti logarithm	power	exponent (logarithm)
1	10^0	0
2	$10^{0.301...}$	0.301...
3	$10^{0.477...}$	0.477...
4	$10^{0.602...}$	0.602...
5	$10^{0.699...}$	0.699...
6	$10^{0.778...}$	0.778...
7	$10^{0.845...}$	0.845...
...	...	......

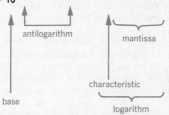

$$\log_{10} 5 = \log 5 = 0.699...$$

$$\log_{10} 50 = \log 50 = 1.699...$$

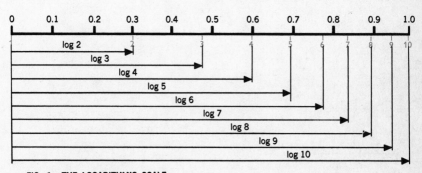

FIG. 1   THE LOGARITHMIC SCALE

subtraction, performed by sliding one part of the rule in relation to the other. It should be noted that in logarithmic calculations the position of the decimal point is determined by the characteristic of the logarithm. The slide rule, however, gives only the mantissa. It is up to the user to determine the position of the decimal point independently. In actual practice this is not a serious difficulty, as it is usually a simple matter to estimate the approximate magnitude of the answer and thus know where to place the decimal point.

An ordinary slide rule consists of the actual rule, the slide, and the transparent cursor with a hairline. Various logarithmic scales are engraved on the rule and the slide (Fig. 2). When the rule is "closed," the pairs of scales A and B, and C and D, respectively, coincide. The divisions on A and B extend from 1 to 100; those on C and D extend from 1 to 10. To determine the square of a value, the hairline is moved to that value on scale C or D, and the square is indicated by the hairline on scale B or A (Fig. 3). Square roots are obtained by the reverse procedure (Fig. 4). The scale K (above A in Fig. 2) gives the cubes of the values on scale D (Fig. 3). Conversely, scale D gives the cube roots of the values on K (Fig. 4).

For multiplication, the scales C and D are preferably employed (Fig. 5). It may occur that the slide has to be moved too far to the right to give a reading of the answer. In that case the slide should be moved back to the left until the figure 10 instead of the figure 1 on scale C is opposite the first factor of the multiplication. The result is then obtained in the usual way on scale D, opposite the second factor on scale C. Alternatively, the scales A and B may be used, though with reduced accuracy of the readings.

Division is performed as indicated in Figs. 7 and 8. Operations involving both multiplication and division may be carried out in accordance with Fig. 9. On a standard slide rule there is, between the scales C and D, another scale, usually marked CI. It gives the reciprocals of the values on scale C (Fig. 10). For the values on scale B the scale CI gives the reciprocals of the square roots; for example, a value a on scale B corresponds to $1/\sqrt{a}$ on scale CI. Reciprocals can be used advantageously for multiplying (dividing by the reciprocal value) and also for complex multiplications and divisions involving several factors (Fig. 11).

FIG. 2

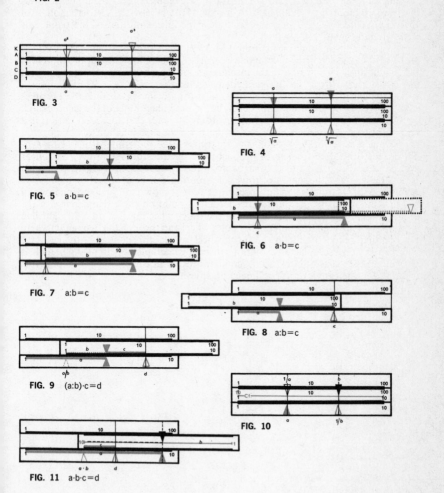

FIG. 3

FIG. 4

FIG. 5   a·b=c

FIG. 6   a·b=c

FIG. 7   a:b=c

FIG. 8   a:b=c

FIG. 9   (a:b)·c=d

FIG. 10

FIG. 11   a·b·c=d

A variety of link mechanisms, or linkages, enable movements to be produced which are exactly or at least approximately rectilinear—i.e., directed along a straight line. The applied motion may be circular or rectilinear. A four-bar linkage (see pp. 200 and 202) can be so contrived that certain points of the mechanism describe substantially straight paths. One such contrivance, proposed by Hoecken, is shown in Fig. 1. Particularly favorable dimensions are obtained when the stationary link d is made twice as long as the crank a and all the other dimensions (b, c, e) are made 2.5 times as long as a. In the case of the oscillating crank (Fig. 2), the point C travels in an approximately straight path. A mechanism whereby the circular motion of the point A is converted into the accurately rectilinear motion of the point B is illustrated in Fig. 3.

In the straight-line link mechanism devised by Watt (Fig. 4a), the point S travels along a so-called lemniscoidal curve, two parts of which are close approximations of straight lines. Fig. 4b shows the mechanism for a particular type of gas-pressure indicator used with a piston-operated machine; the pressure is exerted upon a spring-loaded measuring piston K in a measuring cylinder, so that the piston is raised a greater or lesser amount, depending on the magnitude of the gas pressure. This motion is so transmitted to the recording stylus S that the latter performs a rectilinear vertical motion and records the pressure as a function of the position of the machine's working piston, the rotation of the recording drum being synchronized with the movement of this last-mentioned piston. Accurately rectilinear motion can also be obtained by a pantograph-type mechanism (Fig. 5; see also page 202) whereby the length of travel of a motion can be enlarged or reduced as required. In the cardan gears (invented by Cardano) shown in Fig. 6 the inner wheel has half the radius of the outer. Any particular point on the circumference of the inner wheel, as the latter rotates within the outer wheel, moves on a straight line which passes through the center of the outer wheel and through the two points where that circumferential point on the inner wheel comes into contact with the outer wheel in the course of each revolution. Instead of completely circular wheels it is possible to use parts of circles—i.e., circular arcs, as in Fig. 7, where this principle is utilized in the cam lever (or rolling contact lever).

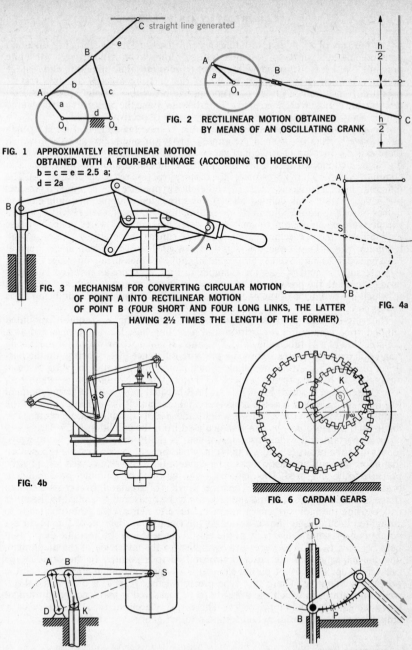

C straight line generated

FIG. 2 RECTILINEAR MOTION OBTAINED
BY MEANS OF AN OSCILLATING CRANK

FIG. 1 APPROXIMATELY RECTILINEAR MOTION
OBTAINED WITH A FOUR-BAR LINKAGE (ACCORDING TO HOECKEN)
$b = c = e = 2.5\ a$;
$d = 2a$

FIG. 3 MECHANISM FOR CONVERTING CIRCULAR MOTION
OF POINT A INTO RECTILINEAR MOTION
OF POINT B (FOUR SHORT AND FOUR LONG LINKS, THE LATTER
HAVING 2½ TIMES THE LENGTH OF THE FORMER)

FIG. 4a

FIG. 4b

FIG. 6 CARDAN GEARS

FIG. 5 INCREASING THE LENGTH OF TRAVEL
OF A RECTILINEAR MOTION

FIG. 7 CAM LEVER BASED ON
CARDAN GEAR PRINCIPLE

489

The pressure of a fluid (gas or liquid) is defined as the force it exerts in a direction perpendicular to a surface of unit area. A distinction is to be made between "absolute pressure," which is measured with respect to zero (absolute vacuum), and "gauge pressure," which is the amount by which the pressure exceeds the atmospheric pressure. Hence: gauge pressure + atmospheric pressure = absolute pressure. This relationship is further clarified in Fig. 1 for the case where the pressure to be measured is higher and lower than the atmospheric pressure respectively.

The simplest form of pressure gauge is the U-tube manometer (Fig. 2). It may either have both arms open to the atmosphere (a) or one sealed arm (b) in which there exists a vacuum (Torricellian vacuum) over the sealing liquid. With type (a) one arm is connected to the pressure p_1 to be measured and the other arm is in communication with the pressure of the atmosphere (the reference pressure p_b). The difference in level H is a measure of the difference in pressure $p_1 - p_b$; i.e., H represents the gauge pressure (as defined above) measured, for example, in millimeters or inches of sealing-liquid column. If the sealing liquid is water, then H will represent the pressure in units (mm or inches) of water column (or "water gauge"). If the sealing liquid has a specific gravity γ, then H can be converted to "water gauge" by multiplying it by γ; hence: $p_1 - p_6 = H \gamma$ (w.g.). The U-tube manometer open at both ends can be employed as a differential gauge for measuring difference between two pressures p_1 and p_2—as, for example, in Fig. 2c, where an inverted U tube is used to measure the pressure difference $p_1 - p_2$ in millimeters or inches of the liquid to which the two arms of the gauge are connected, the sealing medium being a gas in this case.

The ring-balance pressure gauge (Fig. 3), comprising a pivotably mounted annular tube containing a partition and a sealing liquid, may be regarded as a combination of a U-tube gauge and a balance. It is particularly used for measuring small differential pressures. Thus the pressure difference $p_1 - p_2$ acting on the partition produces a rotating movement which causes the partition to swing through an angle α in relation to the vertical, so that a state of equilibrium is established, when the turning moment $M_p = (p_1 - p_2)$ AR is equal to the counteracting moment $M_G = G \sin\alpha$. a, where A is the cross-sectional area and R is the mean radius of the annular tube, while G is a known weight and a is its distance to the center. The differential pressure $p_1 - p_2$ can be calculated from the condition $M_p = M_G$.

An important and widely used instrument is the Bourdon-tube pressure gauge (or spring-tube pressure gauge, Fig. 4) in which pressure measurement is based on the deformation of an elastic measuring element (in this case a curved tube) by the pressure to be measured. The deformation is indicated by a pointer on a dial calibrated to give pressure readings. The tube, which is of circular or oval cross-sectional shape, is closed at one end, and the pressure to be measured is applied to the other end, causing the radius of curvature of the tube to increase (i.e., the tube tends to straighten itself out, as shown dotted in the right-hand diagram of Fig. 4). In the diaphragm-pressure gauge (Fig. 5) the elastic element is a stiff metallic diaphragm held between two flanges; pressure is applied to the underside of the diaphragm, and the movement of the latter is transmitted to a pointer. In the capsule-type pressure gauge (Fig. 6) the elastic element is a capsule to the interior of which the pressure is admitted. The piston-type pressure gauge (Fig. 7) is a so-called dead-weight apparatus in which the pressure to be measured is balanced by adjustment of the weight G placed on the piston. This is a very accurate type of gauge, usually employed for the calibration and testing of other gauges.

(more)

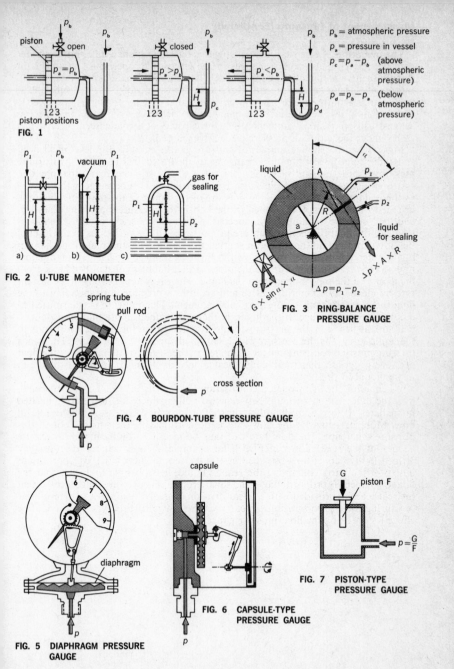

FIG. 1

p_b = atmospheric pressure
p_a = pressure in vessel
$p_c = p_a - p_b$ (above atmospheric pressure)
$p_d = p_b - p_a$ (below atmospheric pressure)

piston positions 1 2 3

FIG. 2 U-TUBE MANOMETER

FIG. 3 RING-BALANCE PRESSURE GAUGE

$\Delta p = p_1 - p_2$

$\Delta p \times A \times R$

$G \times \sin \alpha \times \alpha$

FIG. 4 BOURDON-TUBE PRESSURE GAUGE

spring tube
pull rod
cross section

FIG. 5 DIAPHRAGM PRESSURE GAUGE

diaphragm

FIG. 6 CAPSULE-TYPE PRESSURE GAUGE

capsule

FIG. 7 PISTON-TYPE PRESSURE GAUGE

piston F

$p = \dfrac{G}{F}$

491

The operation of electric pressure gauges is based on quite a different principle: namely, that pressure causes changes in the electrical properties of various substances such as manganin (a copper-base alloy containing manganese and nickel), carbon, etc. These devices are called resistance pressure gauges. Other substances, e.g., quartz, acquire an electric charge when subjected to pressure (piezoelectric effect). This principle is utilized in pressure-measuring instruments (piezoelectric gauges), as is also the phenomenon that the capacity of a condenser (or capacitor) varies under the action of pressure (capacitive gauges).

Figs. 1, 2 and 3 show examples of pressure gauges employed in level measuring and indicating devices. In the arrangement illustrated in Fig. 1 the hydrostatic pressure of the liquid, or its level, is indicated by the gauge—a mercury-float pressure gauge—illustrated in Fig. 4. This device is essentially a U-tube gauge with mercury as the sealing liquid. The variable pressure of the liquid in the tank (in which the level is to be measured) is applied to the "positive" pressure-measuring chamber, and the "negative" chamber is connected to the atmospheric pressure. The movements of the float on the mercury are proportional to the variations in the level of the liquid in the tank and are transmitted through a rack-and-pinion mechanism to a pointer. Another system is shown in Fig. 2: gas (air, nitrogen, carbon dioxide) under pressure is introduced into the pipes so that bubbles constantly emerge from the mouth of the pipe immersed in the liquid. The gas is kept flowing at a constant rate by means of a metering device and acquires a pressure corresponding to the liquid level in the tank at any particular moment. This pressure is transmitted to the float pressure gauge. If the liquid in the tank is under more than atmospheric pressure, as in Fig. 3, the pressure acquired by the gas in the pipes corresponds to the liquid level plus the pressure of the saturated vapor over the liquid. The vapor pressure must be compensated; it is applied to the "negative" chamber of the float pressure gauge, so that the latter indicates only the liquid pressure or the depth of the liquid in the tank.

When a fluid flows through a constriction (orifice, diaphragm, nozzle) in a pipe-line, the difference in pressure between two points which are respectively located immediately before and after the constriction provides a measure of the rate of flow. More particularly, the flow rate is proportional to the square root of this pressure difference. The flow rate can thus be measured by means of a pressure gauge—of the type shown in Fig. 4, for example—whose scale is appropriately divided to give direct flow-rate readings (in ft.^{3}/sec., m^3/min., etc.). By appropriate design of the parts containing the sealing liquid, it is possible to ensure that the float movement is proportional to the square root of the pressure difference, so that the scale can be provided with a linear division (Fig. 5). Fig. 6 shows a rate-of-flow measuring system comprising a float pressure gauge of this type and a U-tube gauge for checking the float pressure gauge.

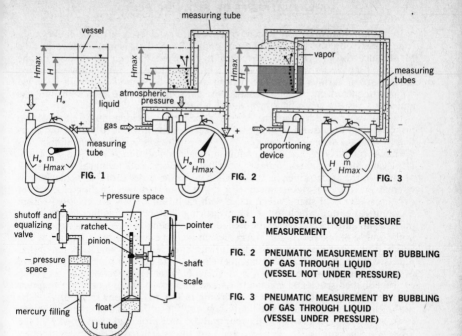

FIG. 1 HYDROSTATIC LIQUID PRESSURE MEASUREMENT

FIG. 2 PNEUMATIC MEASUREMENT BY BUBBLING OF GAS THROUGH LIQUID (VESSEL NOT UNDER PRESSURE)

FIG. 3 PNEUMATIC MEASUREMENT BY BUBBLING OF GAS THROUGH LIQUID (VESSEL UNDER PRESSURE)

FIG. 4 MERCURY-FLOAT PRESSURE GAUGE

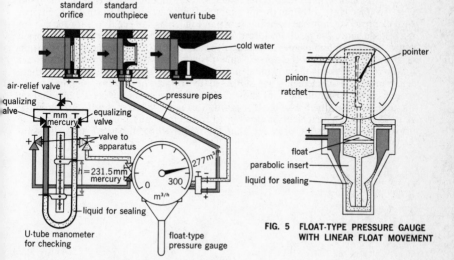

FIG. 6 RATE-OF-FLOW-MEASURING EQUIPMENT

FIG. 5 FLOAT-TYPE PRESSURE GAUGE WITH LINEAR FLOAT MOVEMENT

In modern industrial technology, pipelines are extensively used for conveying a wide variety of gases, liquids and even solids. Obviously it is important to be able to measure the rate of flow of materials conveyed through pipelines—i.e., the quantity that passes in the unit of time (ft.^{3}/sec., m^3/min., etc.). Flow-measuring devices are of various kinds. The present article is concerned only with so-called volumetric meters. Such meters—used more particularly for the measurement of gas and water, respectively—have been dealt with in Vol. I, pp. 224–227.

Flow-measuring devices of the "direct" type for liquids function on either of two principles: (1) a measuring chamber of known capacity is repeatedly filled and emptied (Fig. 1; see also Vol. I, page 226); (2) the rotating measuring element displaces a known quantity of liquid in performing each revolution (Figs. 2 to 5). These displacement-type meters, though differing in design and technical features, all operate on the same principle, which will here be described more particularly with reference to the oval-runner meter (Fig. 5). It comprises two rotating elements of oval cross-sectional shape which mesh with each other. They are enclosed within a cylindrical casing which forms the measuring chamber and is provided with an inlet and an outlet. Fig. 6 shows the oval rotating elements in four successive positions in the course of one revolution, during which each crescent-shaped space at the top and bottom of the measuring chamber is twice filled. The total volume of liquid that is passed through the measuring chamber from inlet to outlet during each revolution of the oval elements is equal to 4 $F_s h$, where F_s is the cross-sectional area of each crescent-shaped space and h is the transverse dimension of the measuring chamber (perpendicularly to the plane of the drawing in Fig. 6). The power for driving the oval elements is supplied by the liquid flow itself (Fig. 7a). The pressure difference Δp across the meter acts upon the major and minor projected areas f and F of the lower oval element in Fig. 7a, which areas are thus subjected to the resultant forces P_F and P_f respectively (Fig. 7b). Since P_F is larger than P_f and moreover has a larger lever arm with respect to the center of rotation of the oval element, the latter is thus subjected to a torque (turning moment) which causes it to rotate. The upper oval element, when in the position shown in Fig. 7b, is subjected to a torque of zero magnitude, since the resultant forces acting on each side of the center balance each other. When the two elements have each rotated through 90 degrees, the situation is reversed, in that now the upper element is subjected to its maximum torque and on the lower element there is momentarily no torque acting at all. In intermediate positions, the two elements are subjected to torques of varying magnitude: the torque on the lower element (as shown in Fig. 7a) diminishes to zero, while that on the upper element increases to its maximum value. The meter is self-starting in the sense that the oval elements will begin to rotate from any position as soon as the liquid flow commences, provided that the liquid to be measured has a pressure

(more)

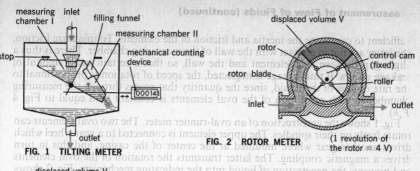

FIG. 1 TILTING METER

FIG. 2 ROTOR METER
(1 revolution of the rotor = 4 V)

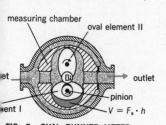

G. 3
ANET-WHEEL
ETER
(1 revolution of the rotor = 4 V)

FIG. 4 CYLINDRICAL-PISTON METER
(1 REVOLUTION OF PISTON = $V_1 + V_2$)

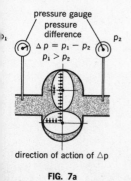

FIG. 5 OVAL RUNNER METER
$V = F_s \cdot h$

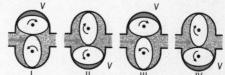

FIG. 6 PRINCIPLE OF OVAL-RUNNER METER

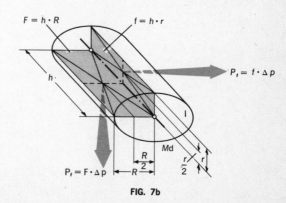

pressure gauge
pressure difference
$\Delta p = p_1 - p_2$
$p_1 > p_2$

direction of action of Δp

FIG. 7a

$F = h \cdot R$ $f = h \cdot r$

$P_t = f \cdot \Delta p$

$P_t = F \cdot \Delta p$

Md

FIG. 7b

495

sufficient to overcome the inertia and friction of the elements. To minimize friction, the latter are not in contact with the wall of the measuring chamber. There is thus a slight gap between each element and the wall, so that a certain small amount of leakage occurs. If this leakage is neglected, the speed of rotation is proportional to the rate of flow of the liquid, since the quantity that passes through the measuring chamber in each revolution of the oval elements is constant (and equal to Fig. 4 $F_s h$).

Fig. 1 shows the construction of an oval-runner meter. The two oval elements can rotate freely on their spindles. The upper element is connected to a gear wheel which drives another gear wheel, mounted at the center of the casing, and this in turn drives a magnetic coupling. The latter transmits the rotation of the oval elements and prevents the penetration of liquid into the indicating mechanism. Fig. 2 shows a single-pointer dial mechanism with a pointer which rotates continuously while measuring is in progress. The pointers of the two-pointer dial mechanism illustrated in Fig. 3 can be reset to zero on completion of the measuring operation.

As already stated, a certain amount of leakage occurs in consequence of the clearances between the rotating elements and the wall of the chamber. The amount of leakage, besides obviously being dependent on the precision of manufacture of the meter and the rate of flow, is dependent on the viscosity of the fluid passing through the meter. Fig. 4 shows an installation for testing the performance of an oval-runner meter, more particularly with liquids of different viscosity. The object of the gas separator is to remove any gas dissolved in the liquid, as its presence produces errors in the measurements. With this equipment the accuracy of the meter can be checked for different flow rates and different liquids.

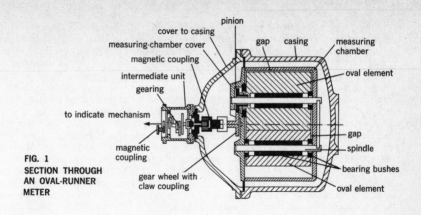

FIG. 1 SECTION THROUGH AN OVAL-RUNNER METER

pinion
cover to casing
measuring-chamber cover
magnetic coupling
intermediate unit
gearing
to indicate mechanism
magnetic coupling
gear wheel with claw coupling
gap
casing
measuring chamber
oval element
gap
spindle
bearing bushes
oval element

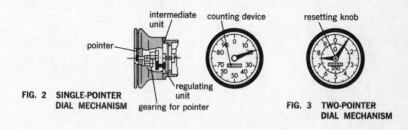

FIG. 2 SINGLE-POINTER DIAL MECHANISM

intermediate unit
pointer
counting device
regulating unit
gearing for pointer

FIG. 3 TWO-POINTER DIAL MECHANISM

resetting knob

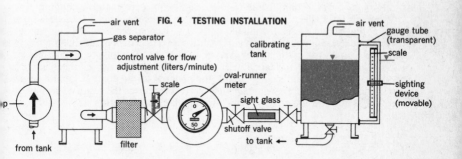

FIG. 4 TESTING INSTALLATION

air vent
gas separator
control valve for flow adjustment (liters/minute)
scale
oval-runner meter
sight glass
shutoff valve
to tank
from tank
filter
calibrating tank
air vent
gauge tube (transparent)
scale
sighting device (movable)

Operations involving the mixing of solid, liquid and gaseous substances occur in innumerable industrial manufacturing processes. Each industry has developed mixers unique to its own use and has in most cases done this chiefly on an empirical basis, which has given rise to considerable diversification of the equipment employed. The main requirement applicable to any mixing operation, however, is to achieve as homogeneous a mixture as possible. In many cases some kind of chemical reaction and/or physical change of the materials concerned is required to take place during mixing—e.g., heating, cooling, dissolving, aeration, de-aeration, change of state (e.g., liquid to solid or vice versa), agglomeration, granulation, dispersion (suspension, emulsion), wetting, coloring, change of viscosity, etc. The intimacy or degree of mixing achieved is directly related to the homogeneity of the mixture. Absolute homogeneity would correspond to theoretically perfect mixing; in actual practice only a certain degree of homogeneity, sufficient to fulfill the requirements of the process concerned, is aimed at. The individual components of a mixture sometimes offer considerable resistance to the attainment of uniform distribution and dispersion in the specified proportions (by weight or by volume). This may be due to difference in specific gravity or bulk density of the component materials, the action of adhesive or cohesive forces, surface features of the particles, etc. In particular, when solids are mixed with liquids, undesirable agglomerations are liable to occur which prevent uniform wetting of the solid particles unless such agglomerations are broken up in the course of the mixing operation. This can be achieved by appropriate design of the mixing elements and the mixing container, appropriate speed of rotation of the mixing elements, and adequate driving-power input. In difficult cases it may be necessary to perform the mixing operation in two or more stages—e.g., a coarse preliminary mixing operation (macromixing) followed by a homogenizing process (micromixing) in which a high power input is applied to the premixed material so that large shearing forces are produced whereby the agglomerations are broken up.

Friction is an attendant phenomenon of every mixing operation. Intensification of the frictional effect in conjunction with an increase in power input causes the actual mixing effect to diminish in importance as compared with the comminuting (or disintegrating) effect associated with friction. Machines which thus develop a comminuting action are called mixing mills. An essential requirement applying to every mixing operation is that both horizontal and vertical flow of sufficient intensity occur and that all the material is moved frequently into the zone of intense mixing action in the vicinity of the mixing element (i.e., the mixing paddle, impeller, etc.). Stratification, settling and segregation of the material in the container must on no account take place. These phenomena are liable to occur as the result of gravity or centrifugal force and must be prevented by suitable mixing action.

(more)

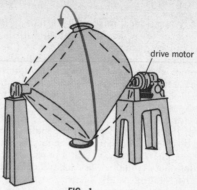

FIG. 1

drive motor

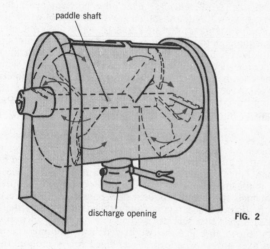

paddle shaft

discharge opening

FIG. 2

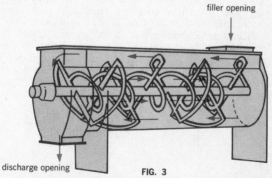

filler opening

discharge opening

FIG. 3

499

As already mentioned, mixers for various purposes present a wide diversity of types, only a few of which can be dealt with here. Four main classes of mixing appliances can be distinguished, however: (1) flow mixers: these are used in circulating systems for the mixing of miscible fluids, the mixing effect being produced by interference with the flow (jet mixers, injectors, turbulence mixers, etc.); (2) paddle mixers: one or more blades rotate on a shaft within the container so that the material to be mixed is moved around in a circular path; (3) propeller mixers: mixing is effected by revolving helical blades which constantly push the material along; (4) turbine (or centrifugal impeller) mixers: broadly speaking, these operate on the principle of the centrifugal pump, the material being accelerated by the impeller vanes and discharged tangentially. The mixers of classes (2), (3) and (4) are positive-action mixers, characterized in that a power-driven mixing element moves (rotates) within a stationary container, as exemplified by the paddle mixer in Fig. 2 (p. 499), in which the rotating paddle blades move the material toward the center of the container, while the arms on which the blades are mounted move it back toward the periphery (counterflow principle). A different class of mixing equipment, not included in the above classification, is formed by the so-called gravity mixers, of which there are various types, one of which is illustrated in Fig. 1 (p. 499). The container is constantly rotated, so that the material inside it is tumbled about. The interior of the container may be fitted with lifting scoops or similar devices which lift the material a certain distance and let it fall, thereby intensifying the mixing action. Gravity mixers can sometimes be suitably employed for the mixing of materials which must not be subjected to the severer mechanical stresses exerted by the mixing elements of positive-action mixers, but they are hardly suitable for mixing sticky or highly viscous materials or for solids differing greatly in physical character and therefore difficult to mix.

Fig. 3 (p. 499) shows a so-called double-helical mixer (belonging to class 3) in which mixing is performed by two helical "ribbons" in concentric arrangement, one a right-hand and the other a left-hand screw, so that the material is moved back and forth in the container. This type is used for mixing powders or thin pastes. In Fig. 4 a turbodisperser is illustrated (class 4); it comprises a centrifugal turbine impeller which rotates at high speed and produces flow patterns whose direction and intensity depend on the shape of the container and on the shape and speed of the impeller. A kneader (class 2) has two sets of paddle blades rotating in opposite directions; in the machine shown in Fig. 5 the two bladed shafts not only rotate at high speed on their own axes but also perform a combined rotary motion about their common center (planetary kneader). The blades are disposed in a helical pattern around the shafts and are staggered in relation to one another, thus producing intensive radial and axial counterflow in the material. Such mixers are more particularly suitable for mixing materials of a viscous, sticky or plastic consistency. Another type of kneader is illustrated in Fig. 6; it has two Z-shaped kneading blades rotating in opposite directions on horizontal shafts at different speeds. The bottom of the container is shaped like a divided trough, with a raised ridge along the center line. These kneaders are especially suitable for highly viscous materials.

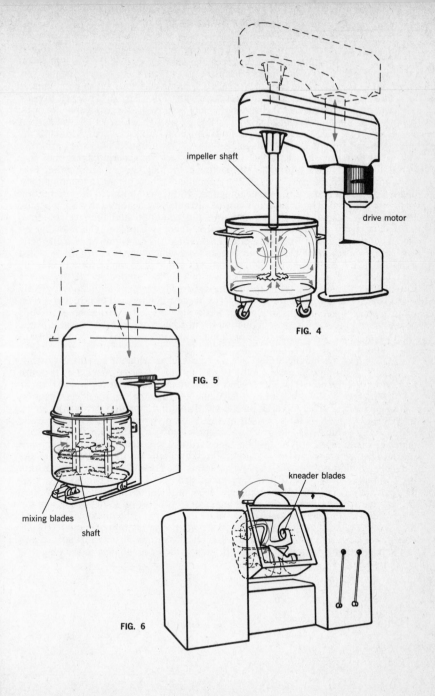

impeller shaft

drive motor

FIG. 4

FIG. 5

mixing blades

shaft

kneader blades

FIG. 6

Iron (or steel) can be conceived as consisting of numerous randomly disposed magnetic units, or domains, which cancel one another so that the piece of metal as a whole exhibits no magnetic polarity (Fig. 1). When the iron is magnetized, these domains become aligned in the same direction and thus act in combination to produce overall magnetic properties: the iron has thus become a magnet (Fig. 2). This can be done by placing the iron in the field of force of an existing magnet or by placing it within a coil of insulated wire (Fig. 3) through which an electric current is passed; in the latter case the coil with its iron core forms an electromagnet. If the core is of soft iron, it loses its magnetism almost immediately after the current in the coil is switched off—i.e., when the electromagnet is de-energized. On the other hand, steel will retain a substantial proportion of the magnetism it acquires and thus form a permanent magnet, in which the magnetic domains persist in retaining their aligned orientation after the external magnetic field which produced this orientation has been removed. The orientation can be disrupted by heating the steel, whereby the magnetic domains revert to their random condition and the steel becomes partly or wholly demagnetized. The electromagnet is based on the fact that an electric current passing through a circular conductor (Fig. 4) produces a magnetic field—i.e., is surrounded by magnetic lines of force which together form the so-called magnetic flux. A coil consisting of many turns of wire (Fig. 5) can be conceived as the superposition of a corresponding number of circular conductors. If the coil is provided with an iron core (Fig. 6), the flux density (or magnetic induction) is greatly increased. This is due to the property of ferromagnetism possessed by iron and certain other metals. A ferromagnetic material has a high magnetic permeability (μ), this being the ratio of the magnetic induction (B) in a piece of magnetic material to the external magnetic field strength (H) producing the induction. The permeability of air and nonmagnetic materials is unity.

Fig. 5 shows the magnetic field set up by a coil in which a current is flowing. Externally its properties are generally similar to those of a bar magnet, with south and north pole respectively. The highest field strength occurs at the center (Fig. 7). As already stated, the presence of a ferromagnetic material (in particular, iron) in the magnetic field gives rise to magnetic induction in that material, which is linked to the field strength of the coil itself by the relation $B = \mu H$, where μ denotes the permeability (which has the order of magnitude of 10^3 to 10^4 for ferromagnetic materials). The permeability may be conceived as the criterion for the increase in the number of magnetic lines of force brought about by the orientation of the magnetic domains in the iron. In practical terms, the presence of an iron core within the coil makes the magnetic field very much stronger. The permeability varies for different values of the magnetic induction, even for the same material (Fig. 8, p. 505). The magnetic-field strength of the coil increases in proportion to the strength of the current flowing through it; as a result, more and more of the domains in the iron core become aligned until finally they are all oriented in the same direction (Fig. 2). The iron is then said to have become magnetically saturated. Any further increase in the magnetic-field strength (H) will produce little or no change in the magnetic induction (B).

(more)

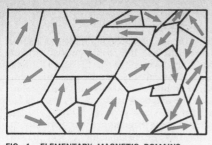

FIG. 1 ELEMENTARY MAGNETIC DOMAINS
IN A POLYCRYSTALLINE MATERIAL
WHICH BEHAVE LIKE TINY MAGNETS
WITH RANDOMLY ORIENTED FIELD
DIRECTIONS (ARROWED)

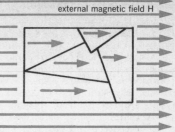

FIG. 2 FERROMAGNETIC SATURATION:
ALL THE ELEMENTARY DOMAINS
ARE NOW ORIENTED PARALLEL
TO THE EXTERNAL MAGNETIC
FIELD

external magnetic field H

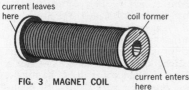

current leaves here

coil former

FIG. 3 MAGNET COIL

current enters here

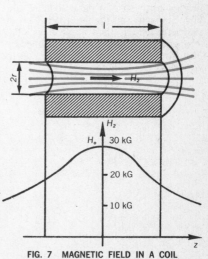

FIG. 4 MAGNETIC FIELD
OF A CIRCULAR CONDUCTOR

S   N

FIG. 5 MAGNETIC FLUX (LINES OF FORCE)
IN A COIL THROUGH WHICH
A CURRENT IS FLOWING

l

$2r$

H_2

H_2

H_o 30 kG

20 kG

10 kG

z

FIG. 7 MAGNETIC FIELD IN A COIL

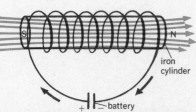

S   N

iron cylinder

battery

FIG. 6 ELECTROMAGNET WITH IRON CORE

Small electromagnets are used in a wide variety of electrical equipment. Familiar examples presented in Vol. I include the electric bell, the relay, the loudspeaker, the telephone, etc. There are innumerable other types of apparatus and electrical machinery in which electromagnets play an essential role: measuring equipment, television tubes, switch gear, remote-control equipment, tape recorders, signaling devices, telecommunication equipment, etc. The magnetic lenses in electron microscopes are basically electromagnets. Electric motors and generators are, of course, very important applications of electromagnets. Lifting magnets are used for handling scrap iron, steel plates, etc.; such magnets may have lifting capacities of several tons. Powerful electromagnets play an important part in various branches of research— e.g., in cyclotrons and similar equipment.

To produce a powerful magnetic field, the electromagnet should have a core which resembles as closely as possible a ring interrupted only by a narrow gap (Fig. 10); it is within this gap that the powerful field is developed, especially if the two poles are so shaped as to produce a concentration of the lines of force (Fig. 11). Examples of electromagnets embodying this principle and designed to produce high field strengths are illustrated in Figs. 12 and 13. The unit of magnetic-field strength is the oersted, this being the force in dynes which acts on a unit magnetic pole at any point in a magnetic field; the unit of magnetic flux is the weber; the gauss is the unit of magnetic induction. To obtain high field strengths in large volumes of space, it is necessary to dispense with the iron core: a large and long coil with a large number of windings and a powerful electric current will produce a strong and fairly homogeneous magnetic field in its interior. Theoretically it would be possible to increase the field strength to any desired value, but in actual practice the heat evolved in the windings causes major difficulties. Water cooling is usually employed for large magnets: the water circulates through pipes embedded in the windings, or the wires themselves are hollow and conduct both electricity and water.

An interesting application of the electromagnet in research is in the production of extremely low temperatures close to absolute zero ($-273°$ C) by the adiabatic demagnetization of a paramagnetic salt. The specimen to be cooled is embedded in such a salt, and the latter is cooled in liquid helium (about $-269°$ C). A strong magnetic field is then applied, which serves to orient all the magnetic dipoles (elementary "atomic magnets" formed by the orbiting electrons) in the salt, releasing heat from this into the helium, so that the temperature of the salt drops and comes very close to absolute zero.

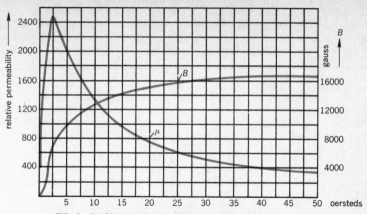

FIG. 8 DIAGRAM SHOWING HOW THE PERMEABILITY
AND THE MAGNETIC INDUCTION B
ARE RELATED TO THE FIELD STRENGTH H

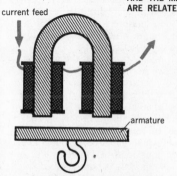

FIG. 9 HORSESHOE ELECTROMAGNET
WITH ARMATURE

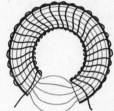

FIG. 10
MAGNETIC LINES OF FORCE
IN AN INTERRUPTED MAGNETIC
CIRCUIT

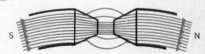

FIG. 11 TAPERED POLE SHOES

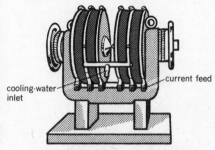

FIG. 12 WATER-COOLED FARADAY-TYPE
ELECTROMAGNET

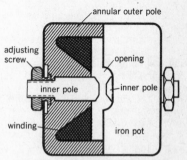

FIG. 13 A MODERN ELECTROMAGNET
(DOUBLE-POT TYPE)

DIRECT-CURRENT MACHINES

A direct-current (or DC) machine is either an electric motor fed by a DC power supply or a generator (dynamo) which produces direct current. The motor and the generator are essentially identical in construction: in general, a DC machine can be used for either purpose. Fig. 1 schematically illustrates such a machine. It comprises the stator, which (in this example) has two magnetic poles for producing the magnetic field, and the rotor (usually called the armature in a DC machine) which rotates between the poles. In very small motors the field may be produced by a permanent magnet; otherwise electromagnets are used, the stator poles being provided with windings (known as field windings or exciting windings) through which current flows. The armature also has a winding consisting of conductors (coils of wire) disposed in grooves formed in the armature core, the latter being composed of sheet-steel laminations. The ends of the armature coils are each connected to one of the insulated copper segments of the commutator. In a generator the current for the external circuit is collected from the commutator by brushes (spring-loaded contact pieces, usually of carbon). Conversely, the current for driving a motor is applied to the armature through the brushes and commutator; at the same time, part of the current is used to energize the field windings. The commutator is mounted on the armature shaft and rotates with it. Fig. 2 schematically shows the "developed" armature winding and the commutator segments, which in reality are of course arranged in a cylindrical shape. In this example the armature has a so-called lap winding. Motors usually have a wave winding, however, in which the winding does not overlap its previous course in loops, but follows a wavelike course.

When current is passed through the armature winding of a motor, the magnetic fields of armature and stator strive to place themselves parallel to each other. As a result, the armature develops a torque (turning moment) about its shaft. The magnitude of the torque is proportional to the strength of the magnetic field and of the current. Several types of DC motor can be distinguished according to the manner of connection of the armature and field windings respectively. In the *shunt motor* the field winding is connected in parallel with the armature and is thus energized by a current of constant voltage, so that the magnetic field is constant. When load is applied to the motor, the speed decreases, but not considerably. Thus this type of motor has a fairly constant speed at all loads and is especially suitable for driving machine tools, lifts, etc. In the *series motor* the field and armature windings are connected in series, so that the strength of field is dependent on the motor load and varies with the armature current. Such motors develop a high torque at starting and run at a speed depending on the load; they are suitable for cranes, traction, etc., because of the high starting torque and their flexibility of operation. Under no-load conditions the speed may become dangerously high, however. The third form of DC motor is the *compound motor*, which has, in addition to the shunt field winding (in parallel with the armature), a series winding which reinforces the field and gives a fairly high starting torque while retaining the speed-limiting properties of the shunt winding. It is used for driving machine tools, presses, shearing machines, etc. Its characteristic properties are a compromise between those of the two foregoing types of motor.

The DC generator is the reverse of the motor in that the armature is rotated by an external source of power, so that the armature conductors intersect the lines of force of the magnetic field of the stator. The principle of the generator has already been dealt with on page 322. In the *separately excited generator* (Fig. 3) the field winding (or exciting winding) is supplied with current from an independent source

(more)

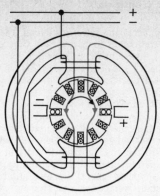

FIG. 1 DC GENERATOR

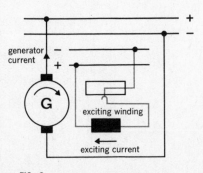

FIG. 2 LAP WINDING
ON ARMATURE

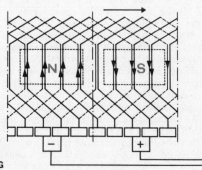

FIG. 3

at constant voltage. With increasing (electrical) load, the voltage at the output terminals of the generator undergoes some decrease. If necessary, this decrease can be compensated, and the voltage thus kept constant, by appropriately increasing the excitation voltage applied to the field winding. This type of generator is suitable for supplying current to an electrical system of constant voltage—e.g., for the charging of storage batteries.

In analogy with those DC motors already discussed, there are three main types of *self-excited generator*: shunt, series and compound. In all these machines a portion of the current produced by the generator itself is used for energizing the field windings. In the *shunt generator* the field winding is in parallel with the armature winding (Fig. 4). With increasing load there is a greater decrease in voltage than in the case of the separately excited generator; nevertheless, at not too high loads the decrease is only gradual, so that this type of generator is also suitable—in conjunction with a shunt rheostat—for constant-voltage systems. However, with increasing load the voltage falls off more rapidly, and above a certain maximum load the characteristic curve even becomes retrograde, with the voltage decreasing to zero at short-circuit. The shunt generator is the commonest type; it is used in power stations, electro-chemical works, etc.—i.e., in situations where no frequent and no excessively large load variations occur; it can also be used for the charging of storage batteries. The *series generator* has the field winding connected in series with the armature winding (Fig. 5). The voltage becomes higher as the load increases. This type of generator is seldom used: only as an auxiliary device for loss compensation in long-distance power transmission systems or in constant-current distribution systems in which the current, not the voltage, is kept constant. (Normally, a constant-voltage system is used, in which the voltage remains constant at all currents and all consuming appliances—lamps, motors, etc.—are connected in parallel across the mains.) The *compound generator* has a shunt winding (in parallel with the armature) and, in addition, a series field winding to compensate for the drop in voltage with increasing load. With appropriate design of the respective windings it is possible to obtain a voltage which remains virtually constant at all loads (Fig. 6). The compound generator is used for systems in which the voltage remains constant without need for adjustment to compensate for frequent and large variations of load, e.g., in rolling mills.

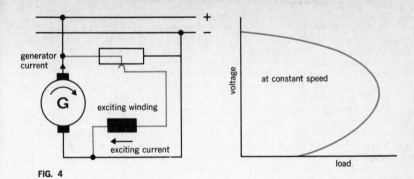

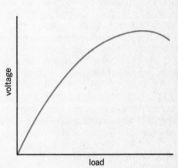

FIG. 4

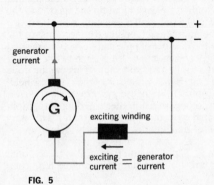

generator current

exciting winding

exciting current = generator current

FIG. 5

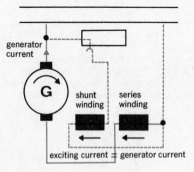

generator current

shunt winding   series winding

exciting current = generator current

FIG. 6

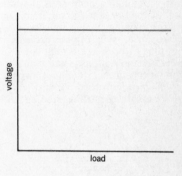

If the armature winding of a DC generator is connected to continuous slip rings instead of to a segmented commutator, an alternating current (AC) is collected from these rings by the brushes (see page 322). There may be two slip rings, connected to two opposite points of the rotor winding, or three slip rings, connected to three equidistant points; in the latter case a three-phase alternating current is obtained. To avoid the technical difficulty of collecting a high-intensity current from slip rings, the form of design now usually adopted for an alternator (AC generator) is that of the "revolving field alternator." In this machine the armature winding, in which the output current is generated, is on the stator (the stationary frame of the machine), while the field system revolves on the rotor, the latter being called the magnet wheel in this type of machine. The direct current for excitation—i.e., for energizing the winding of the magnet wheel—is supplied through brushes and slip rings; as it is a current of much lower strength than the output current of the generator, the technical problem is greatly reduced.

The voltage produced by the AC generator depends on the strength of the excitation current, the speed of rotation, and the number of pairs of poles on the magnet wheel. The frequency of the voltage is also directly dependent on the speed and the number of pole pairs. Figs. 1 and 2 show a two-pole and a four-pole rotor (i.e., with one and two pole pairs respectively). The number of tappings (connections) of the stator winding increases in proportion to the number of pole pairs. Since it is desired to maintain a certain frequency (usually 50 or 60 cycles/sec.), it is not possible to vary the speed of rotation to adjust the voltage to the required value. This can be done only by varying the excitation current, i.e., varying the intensity of the magnetic field.

Fig. 4 is a sectional drawing of a two-pole three-phase *synchronous alternator*, with DC excitation and a fixed frequency, as envisaged in the preceding paragraph. Fig. 3 is a diagram of the voltage generated by a three-phase alternator with star-connected stator winding. It is this type of machine that is most widely used for generation of electricity. The driving power is provided by steam, water or gas turbines or by internal-combustion engines. Depending on the speed of these prime movers, two-pole or four-pole generators are used to obtain the required frequency of 50 or 60 cycles/sec.

If a synchronous alternator is run up to its normal speed (synchronous speed)—i.e., the speed corresponding to the frequency of the circuit—and connected to an existing power-distribution system, it will continue to run as a *synchronous motor* on removal of the external driving power. Such an AC motor, which is similar in construction to a synchronous alternator, always needs an external agency—e.g., a small auxiliary motor—to run it up to speed before being put on to load. It will then continue running at constant speed, "in step" with the frequency, at any load up to a certain overload; if this is exceeded, the motor will fall "out of step" and stop. This ability to run at constant speed, irrespective of the load, makes it very useful for certain applications. Small synchronous motors are used to drive electric clocks, timing mechanisms, etc. The drawback of the synchronous AC motor is that it is not self-starting; it needs an auxiliary starting motor or the rotor must be provided with an auxiliary starting winding; in either case the starting torque is poor.

(more)

FIG. 1

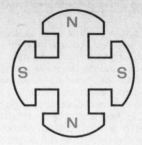

FIG. 2

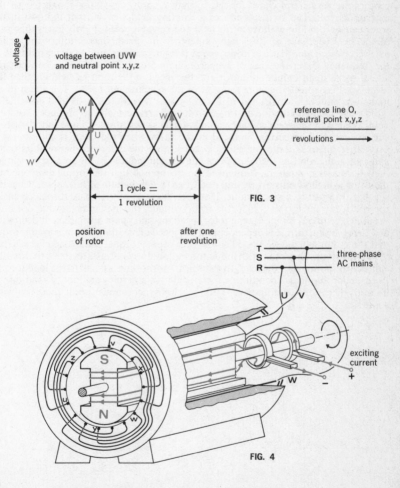

voltage

voltage between UVW
and neutral point x,y,z

reference line O,
neutral point x,y,z

revolutions ⟶

$$\frac{1 \text{ cycle}}{1 \text{ revolution}} =$$

position
of rotor

after one
revolution

FIG. 3

three-phase
AC mains

exciting
current

FIG. 4

A three-phase winding such as the stator of a three-phase synchronous motor produces a so-called rotating magnetic field which has an effect similar to that of a mechanically rotated magnet system. In the machine shown in Fig. 4 (p. 511), conceived as a motor, the rotor consists of an iron core provided with peripheral slots in which the "winding" is disposed in the form of longitudinal copper or aluminum bars connected to rings at each end forming a short-circuited winding without any external connection through slip rings. This is known as a squirrel-cage rotor (Fig. 1). The motion of the rotating field in relation to the rotor bars causes a voltage to be induced in them. This voltage, and the current associated with it, depends on the strength of the magnetic field and on the speed of the field relative to the bars. The squirrel-cage motor is a particular type of *induction motor*. Such motors are in general characterized by having only the stator connected to the external circuit. The stator produces a rotating field. The current induced in the rotor causes it to follow the rotating field and run at a speed slightly lower than that of the field. Because of this slight difference in speed, these motors are called *asynchronous motors*: i.e., the rotor speed is asynchronous, not an exact multiple of the frequency (as in the case of synchronous motors). The difference between the actual speed of an induction motor and the speed of the rotating field is called the "slip" and is expressed as a percentage of the synchronous speed. The stator of an induction motor is usually fed with three-phase current, but it is also possible to build single-phase induction motors with special arrangements to produce a rotating field. When the *squirrel-cage induction motor* is switched on, a large initial rush of current occurs. The current induced in the rotor when the latter is still at rest is very large; then, as the rotor gains speed, so that the difference between its speed and that of the rotating field decreases, the current induced in the rotor correspondingly decreases in strength. It might thus be supposed that the torque developed by the rotor is highest on starting and decreases to a minimum at full speed, when the slip has dropped to a small value. However, because of the phase displacement of the rotor current in relation to the rotor voltage, the maximum torque does not occur at starting. With increasing rotor speed the said phase displacement decreases, and the resulting torque is dependent on the speed of rotation. The maximum torque that the motor can develop is called the breakdown torque; if the load exceeds this, the motor stops. The rotor of an induction motor always adjusts its speed to a certain value corresponding to the load, so that the motor torque is equal to the load torque. Even under "no-load" conditions there is always a certain amount of load due to friction; the rotor therefore never quite reaches synchronous speed; there is always some slip.

(more)

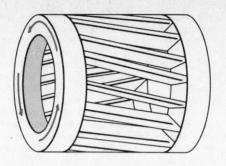

FIG. 1   SQUIRREL-CAGE ROTOR

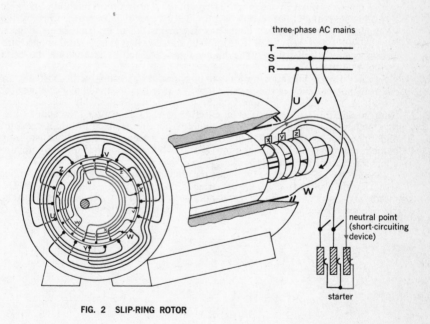

FIG. 2   SLIP-RING ROTOR

The phase displacement between rotor current and magnetic field can be modified by varying the resistance of the rotor winding. In this way the torque characteristic can be altered, so that the starting current can be limited while a reasonably high starting torque is nevertheless obtained. This is not possible with a squirrel-cage induction motor, however, as the rotor winding is unsuitable for this kind of control. Instead, it is necessary to have a wound rotor, i.e., a rotor provided with a winding connected to slip rings (Fig. 2, page 513), enabling variable resistors (rheostats) to be connected in series to the winding. These so-called starting resistors may be regarded as an extension of the winding. For starting the motor the resistors are temporarily connected into the rotor circuit, to give good starting conditions. Then, when the motor has attained its normal working speed, the resistors are disconnected and no longer perform any function; the brushes are lifted off the slip rings in order to reduce wear. The slip rings are short-circuited and the rotor then functions essentially in the manner of a squirrel-cage rotor. This type of motor is called a *slip-ring induction motor*. To reduce the starting current, any motor may initially be connected to a voltage which is lower than the nominal voltage. This can be done by means of a variable resistor connected in series with the stator winding; the resistance is reduced as the motor gains speed; when working speed is reached, the resistor is bypassed and performs no further function. Another method of reducing the voltage at starting is by using an autotransformer starter whereby a reduced voltage from a transformer is applied, in one or more steps, to the stator. A third method, applicable to three-phase induction motors, is provided by star-delta control, whereby the stator winding is temporarily connected in star instead of in delta. The voltage is thereby reduced in the ratio of $1/\sqrt{3}$, and the starting current is reduced to one-third of that drawn at full voltage. The changeover is effected by means of a star-delta switch.

Speed control of induction motors can be obtained by various methods: frequency control (the frequency of the power supply is varied by means of a special device; this alters the synchronous speed of the revolving magnetic field); pole-changing control (alteration of the number of effective poles by regrouping the stator coils; this method enables the speed to be changed stepwise in fixed ratios; it does not allow continuous variation); addition of rotor or secondary resistance (slip-ring induction motor); stator voltage control. The two first-mentioned methods are also applicable to synchronous motors.

A third type of AC motor (besides the induction motor and the synchronous motor) is the *commutator motor*, constructed on the same general lines as the DC motor and used in household appliances such as vacuum cleaners, food mixers, etc. They are sometimes called universal motors because they can be used on DC or AC.

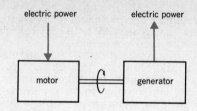

FIG. 1

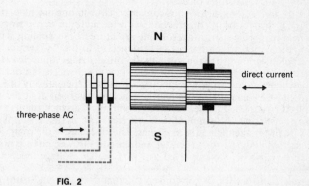

FIG. 2

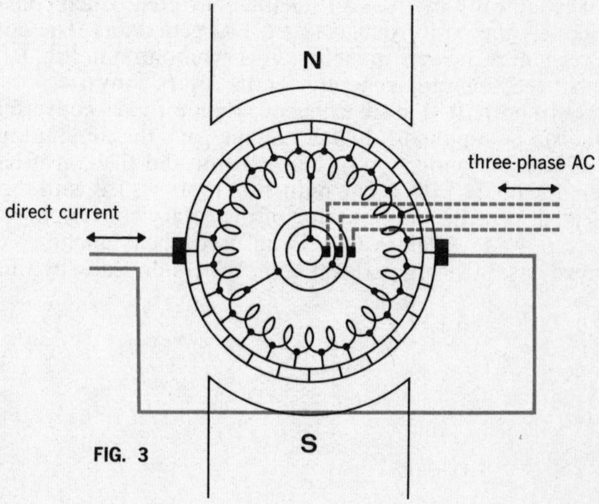

FIG. 3

CONVERTERS

Converter is a general term applicable to a rotary machine or combination of machines whose function is to convert AC to DC or vice versa, or which is used for AC phase conversion or frequency conversion.

The *motor-generator* is a combination of an electric motor and a generator connected to the same shaft. The motor, fed with one type of current, drives the generator, which produces current of the desired type. For instance, alternating current can be converted into direct current. The principle is indicated in Fig. 1 (page 515). An advantage of the motor-generator is that the input and the output current can be independently controlled. Against this, however, the overall efficiency is relatively low, being the product of the efficiencies of the two machines individually and therefore lower than either of them.

The *rotary converter* (or synchronous converter) is a rotating machine, resembling a multipolar DC generator in construction, with the addition of slip rings. It has a stationary field-magnet system and a rotating armature whose winding is connected to a commutator upon which are brushes connected to the DC terminals. At the opposite end of the armature to the commutator are slip rings connected to tappings in the armature winding; the brushes on these rings are connected to the AC terminals. The machine and the armature winding are shown schematically in Figs. 2 and 3 (page 515); in this particular example the machine has been designed for three-phase AC. If AC is supplied to the converter, it will run as a synchronous motor, and DC can be taken from the commutator. Alternatively, if DC is supplied to the machine, AC can be taken from the slip rings. The efficiency of the rotary converter is higher than that of the motor-generator and the cost is lower, but it is not possible —unless special methods are employed—to vary the DC voltage independently of the AC voltage.

The *cascade converter* (or motor-converter) comprises a wound-rotor induction motor mounted on the same shaft as a rotary converter. Suitable points of their windings are interconnected. The set runs at half the synchronous speed of the induction motor, and the slip energy from the rotor of the motor is fed to the armature of the converter and converted to DC. The other part of the energy input to the motor is transmitted mechanically by the shaft to the converter, which converts it to DC output (in doing this the converter acts as a DC generator). The combination of electrical and mechanical power transmission is symbolized in Fig. 1. This machine is intermediate between a motor-generator and a rotary converter, combining some of the advantages of both. It is more expensive than a rotary converter, but has the advantage that it can be supplied at higher voltages, and the commutating difficulties are less severe. The interconnection of motor rotor and the converter armature is illustrated schematically in Fig. 2. The induction motor is fed with three-phase AC, and DC is collected from the commutator of the rotary converter.

Converters are nowadays relatively seldom used, their function—conversion of AC to DC in most cases—being performed more economically by silicon rectifiers.

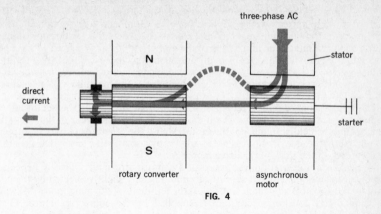

three-phase AC

N

stator

direct
current

S

starter

rotary converter

asynchronous
motor

FIG. 4

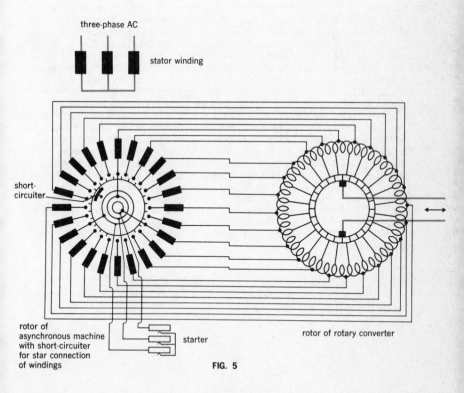

three-phase AC

stator winding

short-
circuiter

rotor of
asynchronous machine
with short-circuiter
for star connection
of windings

starter

rotor of rotary converter

FIG. 5

WARD-LEONARD CONTROL

This is a control system whereby the speed of a DC motor (see page 506) can be very accurately varied by the application of a variable voltage obtained from a motor-generator by shunt regulation. This system provides a wide range of speed control in both directions of rotation of the motor and low power losses. It is expensive and used only for large motors driving such machinery as rolling mills, cranes, winding engines, rotary printing presses, etc. A typical Ward-Leonard control circuit is shown in Fig. 1. The main motor M, which drives the machinery, is a separately excited DC motor which is fed by the continually running control generator G of the Ward-Leonard motor-generator set. The drive motor A of this set (usually an induction motor) is fed with current from the mains. The DC for the excitation (energizing the field windings) of the motor M and the generator G may be supplied by an external source or, if this is not available, by a small self-exciting DC generator E mounted on the shaft of the drive motor A.

With constant excitation, the speed of a shunt motor is very largely dependent on the voltage of the current supplied to it and is only slightly affected by the magnitude of the load. The voltage applied to the terminals of the main motor M (which in general has constant excitation) is supplied by the control generator G. By means of the field rheostat F, the excitation of the control generator can be so adjusted that the generator produces a voltage that can cause the main motor to run at the desired speed. With this rheostat it is thus possible to obtain "infinitely variable" speed control and also to reverse the direction of rotation of the main motors, this being done by reversing the direction of the field current. With Ward-Leonard control the speed of the main motor can be accurately varied to any value from zero to nominal speed, without involving any appreciable loss of power. Since the motor speed is moreover inversely proportional to the field current, a further speed increase can be obtained by field weakening, though the torque will thereby be reduced. A further advantage is that when the speed is reduced, the momentum of the driven machinery will drive the motor, which will thus act temporarily as a dynamo and supply energy back to the mains (regenerative braking: energy recuperation). A drawback of the Ward-Leonard system is its cost, this being due to the fact that, in addition to the main motor M that drives the machinery, it is necessary to provide the drive motor A and the control generator G, both of which have to be of practically the same power as M; the small generator E is also an extra item. The main motor M can be installed some considerable distance away from the Ward-Leonard motor-generator set, as only electrical connections are required. Fig. 2 shows the various main units of this control system. The black arrows indicate the energy-flow direction during regenerative braking. The motor-generator is sometimes replaced by a rectifier whose output can be varied continuously from zero.

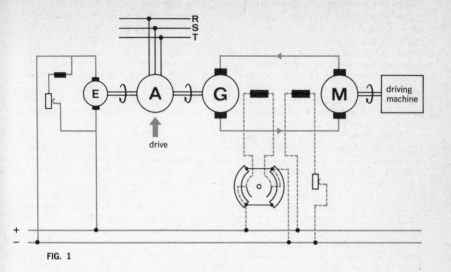

FIG. 1

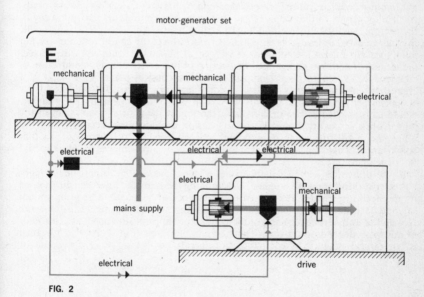

FIG. 2

519

The two methods available for the transmission and distribution of electric power are: underground insulated cables; bare conductors suspended at a safe height aboveground. The latter method—i.e., overhead lines—is generally adopted for high-voltage long-distance transmission, because the cost is lower than for corresponding buried cables, especially at higher voltages.

Before an overhead transmission line is built, a survey of the proposed route must be carried out. The general route may be selected from maps, approximating as closely as possible to the straight line connecting the beginning and end of the route, but avoiding natural or man-made obstacles. If possible, the alignment should avoid woods and forests; it should also not pass close to towns or villages, having regard to their probable future expansion.

The overhead line must not be located too close to railway overhead electrification lines, or to telephone and telegraph lines, because of the danger of induction. Also, it must be routed sufficiently far away from airfields, artillery training ranges, etc. Major terrain obstacles, such as extensive marshy areas, unstable hillsides presenting a landslide hazard, and mining subsidence areas, must likewise be avoided.

Once the route has been decided, a final detailed survey must be made. Ground levels are determined at frequent intervals for the purpose of establishing the longitudinal profile along the route. Part of such a profile is shown in Fig. 1. It shows all the significant topographical features along the route and is plotted to a greatly exaggerated vertical scale in relation to the horizontal scale—e.g., 1:250 vertically and 1:1250 horizontally. The purpose of the profile is to enable the positions and heights of the supports to be determined, so as to conform to requirements of minimum ground clearance and special arrangements at crossings of railways, roads, etc. The positions of the supports and the curve representing the maximum sag of the bottom conductor (corresponding to highest temperature) are plotted. As a rule, in order to avoid unfavorable conditions of vibration of the conductors due to wind load, successive spans are not made exactly equal. When the conductors are erected, the tension is equal in all spans of the section of line concerned. On completion of erection, the conductors are clamped at each suspension point. As a result of a change in temperature the condition of equal tension in all spans would no longer be satisfied if the suspension points were rigid. A fall in temperature, for example, would cause a greater rise in conductor tension in the shorter spans than in the longer ones. In reality, however, the suspension insulators are flexible enough to allow the conductor some longitudinal (in-line) freedom of movement, so that equalization of tension in adjacent spans is automatically achieved despite rise and fall of temperature.

Fig. 2 illustrates a typical lattice steel tower ("pylon") for supporting a high-voltage overhead transmission line. Determining the requisite tower height is based on the need to provide sufficient ground clearance of the conductors at maximum sag and with due regard to conductor deviation caused by swinging in the wind. Horizontal and vertical spacing of the conductors must ensure adequate clearance at midspan, depending on span length, sag and voltage. Fig. 2 shows a so-called wire-clearance diagram based on these considerations. The sag curve of a conductor

(more)

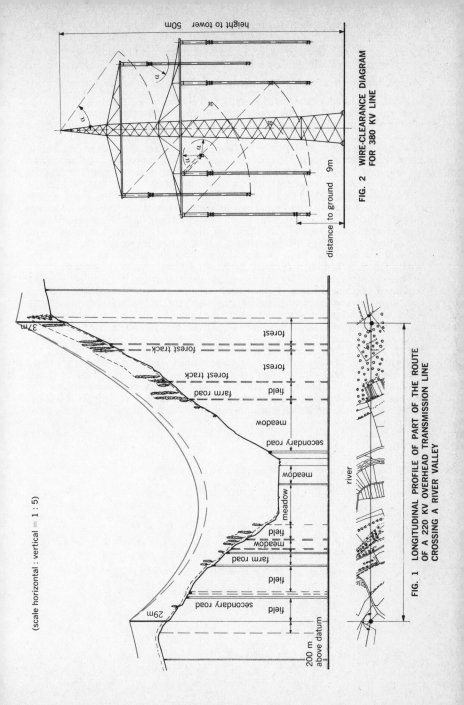

(scale horizontal : vertical = 1 : 5)

FIG. 1   LONGITUDINAL PROFILE OF PART OF THE ROUTE
OF A 220 KV OVERHEAD TRANSMISSION LINE
CROSSING A RIVER VALLEY

FIG. 2   WIRE-CLEARANCE DIAGRAM
FOR 380 KV LINE

is theoretically a so-called catenary, but can, for spans up to about 1600 ft. (500 m), be adequately approximated by a parabola. Thus the sag at midspan can be calculated from the formula $f = \dfrac{WL^2}{8T}$, where W is the weight of the conductor per unit length, L is the span, and T is the tension (Fig. 3).

Of primary importance to the electrical design of an overhead transmission line are the requisite values of voltage drop and power loss. For example, voltage loss must not exceed 6% for low voltage, 10% for high voltage, and 15% for extra-high-voltage transmission lines. Power loss should not exceed 7%. The conductors in present-day use are mainly of three types: steel-cored aluminum (Fig. 4), hard-drawn aluminum, and hard-drawn copper; the first-mentioned type is used for the majority of extra-high-voltage transmission lines. Corrosion of conductors may be a serious problem in aggressive industrial or marine atmospheres. Conductors consist of three or more wires stranded together. They are insulated from the supports, which are earthed. Various types of insulators are employed, such as the suspension type (Fig. 5), as employed for 220 kv lines. Insulators are so designed as to minimize current leakage and prevent surface flashover in rain or fog or under conditions of atmospheric pollution. Widely employed insulating materials are porcelain and toughened glass, in conjunction with fittings of galvanized cast iron or steel.

The supports used for carrying an overhead transmission line may be of wood, reinforced concrete, or steel. Lattice steel towers consist of painted or galvanized steel members and are used mainly for high-voltage and extra-high-voltage lines. Rolled-steel sections or tubular members are used for these structures. Aluminum alloys are also used to a certain extent as construction materials for transmission-line supports; they have the advantages of low weight and high resistance to corrosion, but they are more expensive than steel. From the functional point of view the following types of support are to be distinguished: intermediate supports (most supports are of this kind; the conductors are supported on suspension or pin type insulators); angle supports (at line deviations; the conductors are attached to tension insulators); section supports (these are provided with tension insulators and serve to limit the length of continuous line); terminal supports (at each end of the line).

(more)

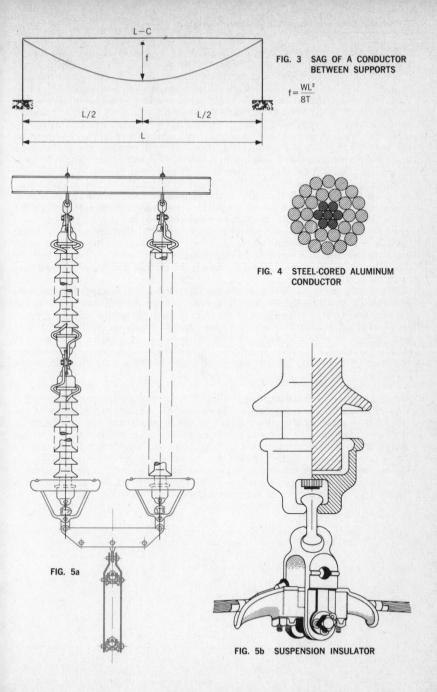

FIG. 3 SAG OF A CONDUCTOR BETWEEN SUPPORTS

$$f = \frac{WL^2}{8T}$$

FIG. 4 STEEL-CORED ALUMINUM CONDUCTOR

FIG. 5a

FIG. 5b SUSPENSION INSULATOR

The foundations for overhead transmission line supports depend on the type of support and on the nature of the soil. Wooden-pole supports and light concrete poles are often simply installed in excavated or bored holes. Broad-base lattice steel towers for high-voltage and extra-high-voltage lines generally have concrete foundations (Fig. 6). These may be so-called uplift and compression foundations (Fig. 6b), which have to resist upward and downward forces, produced by wind pressure, conductor tension, dead weight, etc. If the soil has very poor structural properties, it may be necessary to use pile foundations. Steel or concrete piles may be employed for the purpose. In firm soil, the foundations can be formed by placing concrete and reinforcement in bored holes, which may be widened at the bottom to form enlarged footings. The corner members of the towers are embedded in the concrete. Overvoltages in transmission lines may be due to internal causes (switching operations and faults); the line insulation and clearances are so contrived as to be able to withstand these overvoltages. External overvoltages are caused by lightning strokes on the line. The line conductors are shielded against lightning by means of overhead earth wires suitably earthed at the supports. There is usually a single earth wire running centrally, above the power conductors; sometimes two earth wires are installed. The supports themselves are also earthed. Fig. 7 shows how the potential due to a fault current is dissipated in the earth at the base of a broad-base tower.

In British practice, high-voltage transmission lines carry voltages ranging from 66 kv to 132 kv; extra-high-voltage lines carry voltages from 220 kv to 380 kv; in all cases the power is transmitted in the form of three-phase alternating current at 50 cycles/sec. The cost of an overhead line depends largely on conductor size and voltage. It is necessary to select the most economical combination of the two. In various countries the trend is toward the construction of much-higher-voltage transmission lines, e.g., 750 kv. These voltages result in greater economy in the transmission of electric power over long distances.

Conductor erection: The conductors are supplied wound on drums, which are installed in position at the end of a section. They are mounted to revolve freely and are controlled by a braking device. The conductors are usually hauled along the ground by a winch or sometimes a tractor. As the conductors pass each support, hauling is stopped, and they are fitted through running blocks, which are then hoisted up to the crossarms of the supports. Alternatively, the conductor may be run out clear of the ground, i.e., pulled through pulleys suspended overhead (Fig. 8). This is done mainly to avoid causing damage to the conductors. Finally, after suitable adjustment and check of the amounts of sag, the conductors are clamped at each suspension point (a conductor suspension clamp is shown in Fig. 5b).

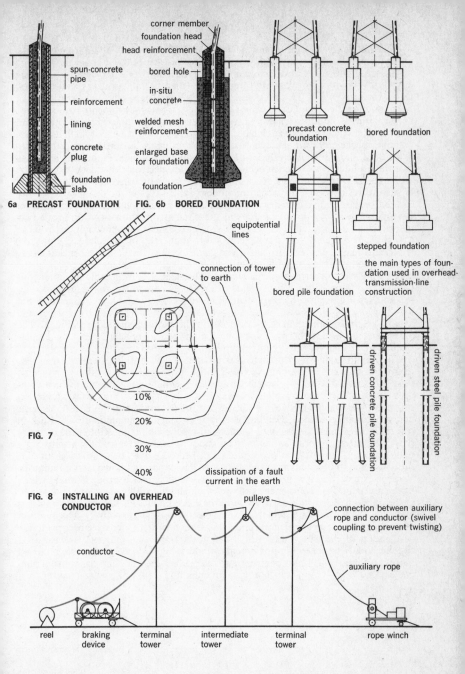

spun-concrete pipe

reinforcement

lining

concrete plug

foundation slab

6a PRECAST FOUNDATION

corner member
foundation head
head reinforcement

bored hole

in-situ concrete

welded mesh reinforcement

enlarged base for foundation

foundation

FIG. 6b BORED FOUNDATION

precast concrete foundation

bored foundation

stepped foundation

the main types of foundation used in overhead-transmission-line construction

bored pile foundation

driven concrete pile foundation

driven steel pile foundation

equipotential lines

connection of tower to earth

10%

20%

30%

40%

FIG. 7

dissipation of a fault current in the earth

FIG. 8 INSTALLING AN OVERHEAD CONDUCTOR

pulleys

connection between auxiliary rope and conductor (swivel coupling to prevent twisting)

conductor

auxiliary rope

reel

braking device

terminal tower

intermediate tower

terminal tower

rope winch

High-frequency electromagnetic energy, i.e., high-frequency alternating currents, in the form of electromagnetic waves or displacement currents (see page 528) require no material medium for their propagation. Because of this property, standing electrical oscillations of very high frequencies can be established in metal cavities; this is the principle of the cavity resonator (see Vol. I, page 70). In general, this is a space bounded by conducting walls in which electromagnetic energy can be stored as oscillations (standing waves) whose frequency is determined by the shape and size of the cavity. The resonator has a lower frequency limit below which it will not respond. If one end wall of a cylindrical cavity resonator is removed and the cylindrical wall is extended, the tubular hollow conductor obtained in this way is known as a wave guide. It is so named because it can guide electromagnetic waves coaxially with its axis. More specifically it is a hollow conducting tube used for the propagation of such waves (Fig. 1). The oscillation nodes of the waves in the radial direction are located on concentric cylindrical surfaces which are concentrated closer together toward the wall of the wave guide; in the circular direction they are located on planes which are spaced at equal angular distances and pass through the cylinder axis (Fig. 2). At very high frequencies (from about 100 megacycles/sec. onwards) the electromagnetic fields in the wave guide do not break down, because they alternate so rapidly that no charge exchange at all can take place in the wall, as the conducting electrons travel too slowly for this. Hence considerable instantaneous current may flow between two points of the wall. Slots in the wall present no obstacle; they are bridged by displacement currents. The longitudinal edges of a slot function like the plates of a condenser, whereas the transverse edges (which extend axially, i.e., in the direction of the currents in the wall) act inductively (Fig. 3). The displacement currents between the longitudinal edges are surrounded by circular magnetic flux lines (or lines of force) similar to those around a straight conventional conductor through which a current is passing. These flux lines emerge from the cavity and give rise to circular electric flux lines, which in turn produce a magnetic flux, and so on. In this way an electromagnetic radiation is generated which withdraws energy from the wave guide. A slot in a wave guide (Fig. 3) thus functions similarly to an electric dipole (see Vol. I, page 70), except that in this case a magnetic field is primarily created. For this reason it is called a magnetic dipole. Both in respect of its geometric form and its mode of functioning it constitutes the counterpart of the electric dipole (rather in the manner of the positive and the negative of a photograph). The radiation properties described here are put to practical use for the construction of antennae (or aerials), such devices being known as slot radiators. The waves that occur in wave guides are classified into two groups: waves which have only an electric component and waves which have only a magnetic component in the axial direction of the wave guide; these are known as TM waves (transverse magnetic waves) and TE waves (transverse electric waves) respectively. The numbers of circular and radial nodal surfaces (in this sequence) are added as subscripts to the designation of the wave; thus TM_{12} denotes a transverse magnetic wave with 1 circular nodal surface and 2 radial nodal cylinders (Fig. 4). Wave guides are employed more particularly in radio engineering, where they are used for the transmission of electrical energy from transmitter to antenna and from antenna to receiver.

FIG. 1 TRANSFORMATION OF CAVITY RESONATOR INTO
WAVEGUIDE

FIG. 2 NODAL PLANES
(TWO CYLINDRICAL,
ONE PLANE)

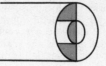

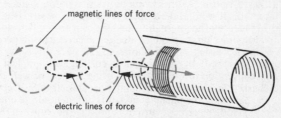

FIG. 3 RADIATION OF ELECTROMAGNETIC ENERGY
THROUGH A SLOT IN A WAVEGUIDE

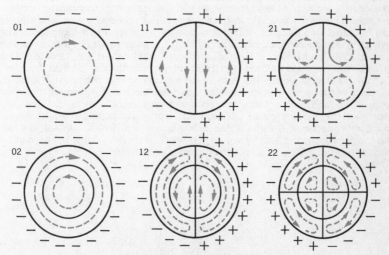

FIG. 4 NODAL PLANES AND MAGNETIC LINES OF FORCE
OF VARIOUS WAVE TYPES IN CYLINDRICAL
WAVEGUIDES

DISPLACEMENT CURRENT

If a charged capacitor (or condenser) is connected into a direct current circuit, it causes an interruption of the flow of current (Fig. 1). On the other hand, if the same circuit is fed with alternating current, the capacitor is alternately charged and discharged—i.e., the alternating current continues to flow (Fig. 2); the electric field between the capacitor plates, whose intensity and direction vary with time, thus constitutes an immaterial link in the transmission of electric energy through the circuit. This energy flow of alternating direction between the plates behaves like an electric current; it is associated with circular magnetic flux lines (or lines of force) whose direction of rotation varies synchronously with the electric field (Fig. 3). This transmission of electric energy by the action of an "immaterial" electromagnetic field instead of by the action of charge carriers such as electrons or ions is called displacement current. In actual practice, the plates of a capacitor are separated not by an air gap but by a dielectric (an insulating material) through which induction can take place, i.e., allowing magnetic or electric lines of force to pass, for example: mica, glass, plastics, etc. (Fig. 4). The alternating electromagnetic field causes a displacement of the positive in relation to the negative charges in the atoms or of the dielectric (Fig. 5b); normally these charges (positive atomic nuclei surrounded by electrons) are so disposed that the atoms are outwardly neutral (Fig. 5a).

Under the influence of an alternating field the charges oscillate about their centers of neutrality. Moving charges produce an electric current; with periodically oscillating charges an alternating current is formed, and since this is due to displacement of electric charges, the term "displacement current" was applied to it. However, although this conception of the current as being associated with the displacement of electric charges in a material medium (the dielectric) is a convenient aid to visualizing the nature of this phenomenon, it is not an accurate one. This is evident from the fact that a displacement current is produced even when there is a vacuum —i.e., no material of any kind—between the capacitor plates. The presence of a suitable dielectric does, however, play a part in that it intensifies the transfer of electromagnetic energy in the alternating field. This intensification is expressed by the dielectric constant (or permittivity), which is the ratio of the electric displacement produced in a particular medium to the electric force producing it as compared with the ratio for a vacuum; it indicates how many times higher a capacitor with that particular dielectric can be charged than a similar capacitor with a vacuum between its plates; i.e., it is the ratio of the capacitances for those two respective cases. The oscillating motion of charges extracts energy from the electric field, however; this is more evident as the frequency of the alternating field increases. The inertia resistance of the charge carriers (which transfer the electric charge to the dielectric) opposes the high accelerations and decelerations associated with very high frequencies of the alternating field. Consequently, increased damping of the oscillations occurs, more particularly with substances having a high dielectric constant, in which case the displacement current becomes very small. If the dielectric is removed and a vacuum substituted for it, the current increases in intensity.

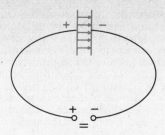

FIG. 1   CAPACITOR IN DC CIRCUIT

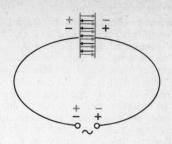

FIG. 2   CAPACITOR IN AC CIRCUIT

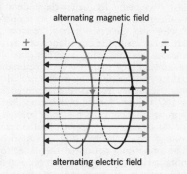

alternating magnetic field

alternating electric field

FIG. 3   LINES OF FORCE IN
ALTERNATING ELECTROMAGNETIC FIELD

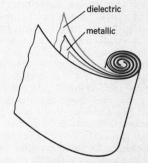

dielectric

metallic

FIG. 4   CYLINDRICAL CAPACITOR
COMPOSED OF SHEETS OF
METAL AND INSULATING MATERIAL

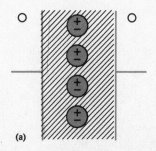

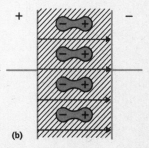

(a)

(b)

FIG. 5   ATOMS OF THE DIELECTRIC IN THE NEUTRAL
CONDITION (a) AND IN AN ELECTRIC FIELD (b)

When a current I flows through a series of resistors (Fig. 1), the voltage E between two points is proportional to I and the resistance R that the current encounters; this is expressed by the formula: $E = I \cdot R$ (Ohm's law). In this way a voltage can be divided into two or more lower voltages in any desired ratio. If I is an alternating current of frequency ω, an ordinary (ohmic) resistor has no effect on the frequency; but if the circuit includes a capacitor with a capacitance C, the capacitive reactance thereof is $R = \omega/C$, and if the circuit comprises an inductor (a coil) with inductance L, the inductive reactance due to this is $R = \omega L$. Because of these effects, voltage dividers with inductors and capacitors will divide a high-frequency alternating current in a different ratio from a low-frequency one. Examples of the application of this principle are afforded by the high-pass and the low-pass filter (Fig. 2). To calculate the impedance, or total apparent resistance, of an AC circuit with ohmic resistors, inductors and capacitors, the principle indicated in Fig. 3 is applied. Impedance is a complex property comprising the ohmic resistance of the circuit, the inductive reactance (due to the inductance of the circuit), and the capacitive reactance (due to the capacitance), the last two being frequency-dependent as indicated above. To explain the underlying theory would be outside the scope of this article. At a certain frequency the inductor and the resistor have resistances which are exactly equal but of opposite effect. As a result, the impedance of a series oscillating circuit (Fig. 4a) attains a minimum at this frequency, whereas that of a parallel oscillating circuit (Fig. 4b) attains a maximum. This is known as the resonance frequency ($\omega = \sqrt{LC}$). In high-frequency electronic engineering such oscillating circuits are utilized for filtering out a particular frequency from a frequency mixture.

Electronic equipment requires a certain DC voltage for its operation; this is supplied by a so-called power unit which appropriately converts the main voltage—for example, 240 volts AC. Fig. 5 shows how the AC voltage supplied by the transformer is converted into a pulsating DC voltage whose "alternating" components are suppressed by low-pass filter stages. The smoothing action of the capacitors presents an analogy with that of elastic containers (balloons) in a compressed-air pipeline, whereby the initially existing pulsations in air pressure are smoothed away as a result of the equalizing action of the balloons.

(more)

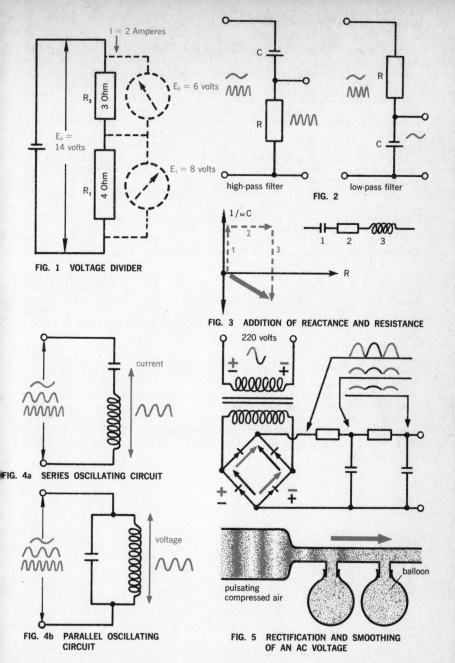

I = 2 Amperes

R₂ 3 Ohm

E₂ = 6 volts

E₀ = 14 volts

R₁ 4 Ohm

E₁ = 8 volts

FIG. 1   VOLTAGE DIVIDER

C

R

high-pass filter

R

C

low-pass filter

FIG. 2

$1/\omega C$

1   2   3

R

FIG. 3   ADDITION OF REACTANCE AND RESISTANCE

current

FIG. 4a   SERIES OSCILLATING CIRCUIT

voltage

**FIG. 4b   PARALLEL OSCILLATING
CIRCUIT**

220 volts

+   −

−   +

+

−

+

pulsating
compressed air

balloon

**FIG. 5   RECTIFICATION AND SMOOTHING
OF AN AC VOLTAGE**

531

Fig. 6 shows a commonly adopted circuit for an amplifier tube (or valve). The input AC voltage applied to the grid of the tube causes the anode current I to be increased and decreased in the same rhythm (see also Vol. I, page 74). The current I flows through the anode resistor R_a and the cathode resistor R_k and produces a voltage drop in both. The higher the value of R_a, the higher becomes the output AC voltage. The value of R_k (and therefore the voltage drop in R_k) is so chosen that the grid remains negative in relation to the cathode. The AC voltage produced by the variations of I in R_k is often undesirable. (It should be borne in mind that the current I is controlled by the difference between grid voltage and cathode voltage.)

In the circuit known as a cathode follower (Fig. 7) there is no anode resistor; the output voltage is taken across the cathode resistor R_k. Voltage variations at the grid produce anode (and therefore cathode) current variations in phase, so that the cathode potential rises and falls in sympathy with the grid voltage. This arrangement provides a high input impedance and low output impedance. The voltage amplification of the cathode follower is equal to unity (or less). There is no phase inversion between input and output with a resistive load.

A certain proportion of the output voltage of an amplifier can be so fed back to the input that the input voltage is in part compensated. This so-called negative feedback at first merely causes some reduction of the amplification. Now, in every amplifier there occur deviations from exact proportionality between input and output voltage (linearity). The actual voltage of an amplifier may be conceived as the sum of a truly linear output voltage and an error voltage. The feedback adds part of the error voltage to the input voltage, so that a correcting voltage is produced at the output of the amplifier. The residual error in the output voltage is reduced to such an extent that it just suffices to produce the correcting voltage. In a case where a very constant DC voltage is required, the arrangement known as a stabilizing circuit is employed (Fig. 9). The voltage is regulated, for example, by reference to a glow-discharge lamp (such as a neon lamp) in which there is practically no voltage drop—i.e., the voltage across the electrodes remains constant irrespective of the strength of the current discharged through the lamp. A fixed proportion of the voltage to be stabilized is compared with the glow-lamp voltage, and the difference is fed to the input of the amplifier. The output voltage of the differential amplifier controls the grid of a tube (or valve) which determines the current in the voltage divider and thus keeps the voltage E_2 constant (apart from a residual error) even when load variations occur.

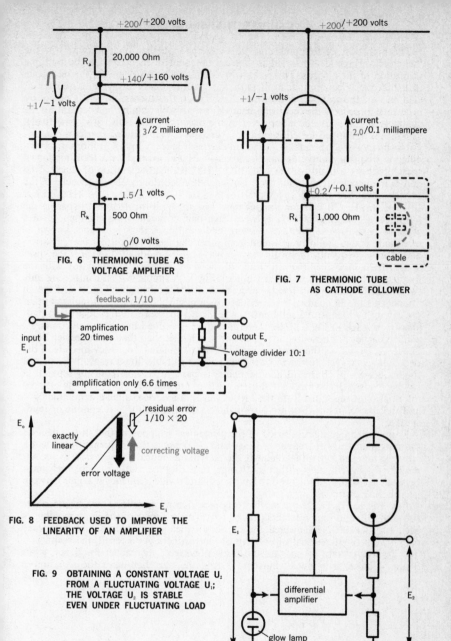

+200/+200 volts

R_a 20,000 Ohm

+140/+160 volts

+1/−1 volts

current 3/2 milliampere

1.5/1 volts

R_k 500 Ohm

0/0 volts

FIG. 6 THERMIONIC TUBE AS VOLTAGE AMPLIFIER

+200/+200 volts

+1/−1 volts

current 2,0/0.1 milliampere

+0.2/+0.1 volts

R_k 1,000 Ohm

cable

FIG. 7 THERMIONIC TUBE AS CATHODE FOLLOWER

feedback 1/10

amplification 20 times

input E_i

output E_o

voltage divider 10:1

amplification only 6.6 times

E_o

residual error 1/10 × 20

exactly linear

correcting voltage

error voltage

E_i

FIG. 8 FEEDBACK USED TO IMPROVE THE LINEARITY OF AN AMPLIFIER

FIG. 9 OBTAINING A CONSTANT VOLTAGE U_2 FROM A FLUCTUATING VOLTAGE U_1; THE VOLTAGE U_2 IS STABLE EVEN UNDER FLUCTUATING LOAD

E_1

E_2

differential amplifier

glow lamp

533

The high cost of telecommunication systems—whether on the "wire" or the "wireless" principle—soon led to the development of methods for the multiple utilization of the circuits. This is known as channeling and denotes the technique of using one telecommunication path for simultaneously carrying a number of channels for the transmission of messages. To do this successfully, it is obviously necessary to possess the technical means of suitably combining and separating the channels at the start and at the end of the telecommunication path respectively. The means employed for effecting the separation is the so-called *carrier frequency*. Each channel corresponds to a different carrier frequency or, to be more precise, a different frequency range or band of frequencies. For example, the transmission of speech requires a band width of 3400 Hz ("Hz" is the abbreviation for "hertz," the unit of frequency equal to one cycle per second). Accordingly, the frequencies of the carrier waves are spaced at intervals of 4 kHz (= 4000 Hz)—e.g., 4 kHz, 8 kHz, 12 kHz, 16 kHz, 20 kHz for a six-channel system. With this frequency spacing the intermediate range of 600 Hz between each pair of neighboring channels ensures effective separation of the transmissions and suppresses the disturbance known as cross talk, due to spillover of sound from other channels. The terms "carrier wave" and "carrier frequency" indicate that the wave can be modulated to carry the lower speech frequency—i.e., the amplitude of the wave is varied in sympathy with the frequency of the sounds to be transmitted (Fig. 1). A simple system comprising only two channels will be considered in order to illustrate the principle (Fig. 2). The speech frequency band in channel 1 is produced by a microphone and is modulated upon the carrier by means of a modulator, whence it is passed to the transmission path (drawn in black in Fig. 2). The same procedure is applied in the case of channel 2 with its carrier 2, whose frequency, however, is higher than that of the carrier 1 (red in Fig. 2). By way of the transmission path the modulated frequency mixture of the two channels travels to the receiving end, where suitable filters respectively permit the frequencies of channel 1 and channel 2 to pass. After this frequency separation has been effected, the modulated carrier waves are demodulated—i.e., the original information is separated from the carrier waves in a device called a demodulator—and the speech frequencies are turned into audible speech in a telephone or loudspeaker.

From the foregoing discussion of the principle it is apparent that, if f_{max} denotes the maximum carrier frequency range and Δf denotes the frequency band width required for each individual channel, then it will theoretically be possible to accommodate $f_{max}/\Delta f$ channels within that range. In reality the number of channels must be smaller, because of the need to have adequately wide frequency margins between neighboring channels.

Fig. 3 is a diagram of a so-called ring modulator, comprising two transformers (U1 and U2) and four rectifiers (tracks 1–4). To the center of the inner winding of each transformer is connected the carrier frequency CF, whose high frequency causes very rapid reversal of polarity of the rectifiers. As a result, the rectifiers 1 and 2 allow current to pass during one half-wave of the carrier frequency, while 3 and 4 block the flow of current; during the next half-wave the situation is

(more)

**FIG. 1   MODULATION OF A HIGH-FREQUENCY CARRIER OSCILLATION (CF)
BY A LOW-FREQUENCY SPEECH OSCILLATION (LF)**

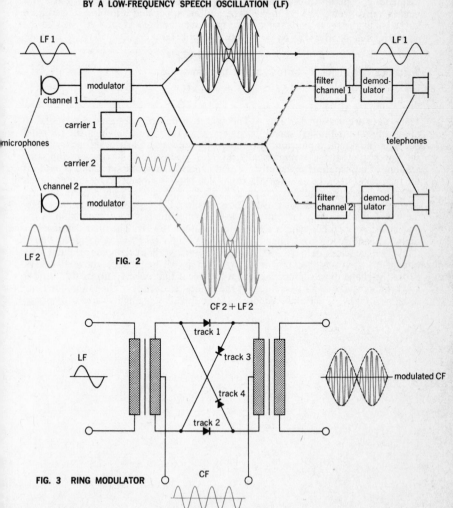

FIG. 2

FIG. 3   RING MODULATOR

reversed. The low-frequency current LF is applied to the input winding of the first transformer and, in consequence of the cross connections in the circuit, undergoes a phase displacement of 180 degrees. On the output side, a carrier frequency whose amplitude is modulated to the low frequency emerges from the ring modulator. On the right in Fig. 3 the low frequency is represented by the enveloping curve (dotted) of the high-frequency amplitudes.

A mathematical analysis shows that the two frequencies (LF and CF) fed to the modulator are present in it as a sum and a difference of the frequencies. The sum is referred to as the upper side frequency f_s, and the difference as the lower side frequency f_d. Speech transmission requires not just a single frequency, but a frequency range from 0.3 kHz to 3.4 kHz, and an upper and a lower side band are formed. Let f_c denote the carrier wave frequency and let Δf_s and Δf_d denote the respective side bands; then $\Delta f_s = f_c + (0.3 \rightarrow 3.4 \text{ kHz})$

$$\Delta f_d = f_c - (0.3 \rightarrow 3.4 \text{ kHz})$$

With a carrier frequency of 16 kHz, the side bands are:

$$\Delta f_s = 16.3 \rightarrow 19.4 \text{ kHz}$$
$$\Delta f_d = 12.6 \rightarrow 15.7 \text{ kHz}$$

These data are indicated in Fig. 4. The apexes of the triangles each correspond to the position of the lowest speech frequency (0.3 kHz). The triangle in the LF band occupies the standard position. The upper side band is similarly positioned, whereas the lower side band is symmetrically reversed. Each side band contains the same amount of information as the other. For this reason only one side band is used for the transmission of the message; the other side band is suppressed by suitable filters before it enters the transmission path. A filter which allows a particular frequency range to pass is known as a band-pass filter. Fig. 5 shows the circuit arrangement and the damping curve of a band-pass filter for the upper side band in the transmission of speech with a carrier frequency of 16 kHz. The filter consists of a high-pass and a low-pass filter. The high-pass filter (H P) is so designed that it allows frequencies above 16.3 kHz to pass almost without damping; the low-pass filter (L P) is so designed that it strongly damps all frequencies above 19.4 kHz. From the shape of the damping curve it is evident that only the upper side band is let through by the band-pass filter: at the highest frequency of the lower side band (in this example 15.7 kHz) the band-pass filter already has a high degree of damping.

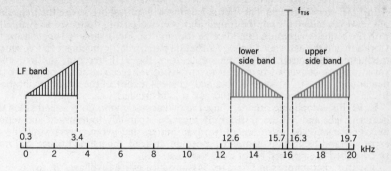

FIG. 4   MODULATION OF A SPEECH BAND WITH CARRIER
FREQUENCY $f_{T16} = 16$ kHz
(LF AND UPPER SIDE BAND IN STANDARD POSITION
LOWER SIDE BAND IN REVERSED POSITION)

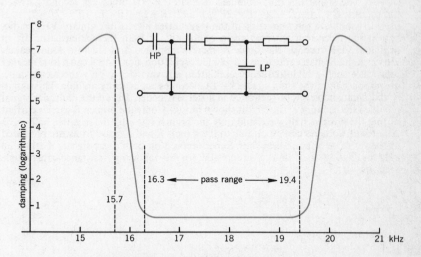

FIG. 5   BAND-PASS FILTER FOR UPPER SIDE BAND OF
CARRIER FREQUENCY $f_{T16} = 16$ kHz (CIRCUIT
DIAGRAM AND DAMPING CURVE)

In VHF (very high frequency) broadcasting with ultrashort waves, the frequency of the waves radiated from the transmitter is increased and decreased in sympathy with the acoustic vibration that must be transmitted. Fig. 1 shows a simple acoustic vibration and the associated voltage-variation pattern at the antenna of a frequency-modulated transmitter. In the demodulator of the VHF receiver, the frequency variations are converted back into variations of voltage; i.e., at the output of the demodulator an AC voltage is obtained which is a copy of the acoustic vibration. It can be amplified and used to energize a loudspeaker.

In VHF stereophonic broadcasting, the two acoustic vibrations intended for the hearer's right and left ear respectively must be separately transmitted and reproduced. Transmission by means of two transmitters and two receivers would be an expensive and technically complex procedure. Instead, the method which will now be described is employed (pilot frequency method).

The two microphones in Fig. 2 (p. 541) produce two AC voltages R and L, from which first the sum signal $(R+L)$ and the difference signal $(R-L)$ are formed. In the arrangement in Fig. 2 this is achieved by means of transformers; other circuits may be used in practice.

Instead of using the AC voltage direct from the microphone, a so-called multiplex signal is transmitted. This signal is composed as follows (Fig. 3):

The amplitude of a 38 kHz oscillation (for example) is varied proportionally to the signal $(R-L)$ just as in medium-frequency broadcasting (amplitude modulation). As a result, an AC voltage (C) is produced, which varies the 38 kHz frequency between the values $(+1) \cdot (R-L) + T$ and $(-1) \cdot (R-L) - T$, where T denotes a constant 38 kHz frequency (the carrier). The AC voltage (C) is added to the signal $(R+L)$. The signal (D) thus obtained oscillates 38,000 times per second between the values $(R+L)+(R-L)+T = 2R+T$ and $(R+L)-(R-L)-T = 2L-T$. It is used to modulate the frequency of the transmitter just as in ordinary VHF radio, and at the receiver it emerges again as the output from the demodulator. In an ordinary VHF receiver the signal is merely amplified and fed to a loudspeaker. However, no human ear can hear a 38 kHz oscillation, nor indeed can a loudspeaker reproduce such a high-frequency oscillation anyway. Hence only the mean value, corresponding to the sum signal $(R+L)$, becomes acoustically audible. The signals R and L can, however, be separated in a stereo decoder. Thus if the multiplex signal (D) is fed to the input of the circuit shown in Fig. 4, only the branch R carries current during the positive half-wave, and only the branch L during the negative half-wave. As a result, voltages corresponding to the signals R and L respectively are produced in the resistors of these branches. Superimposed on both these signals is still a 38 kHz oscillation, but this is suppressed in the following resistance-capacitance sections.

(more)

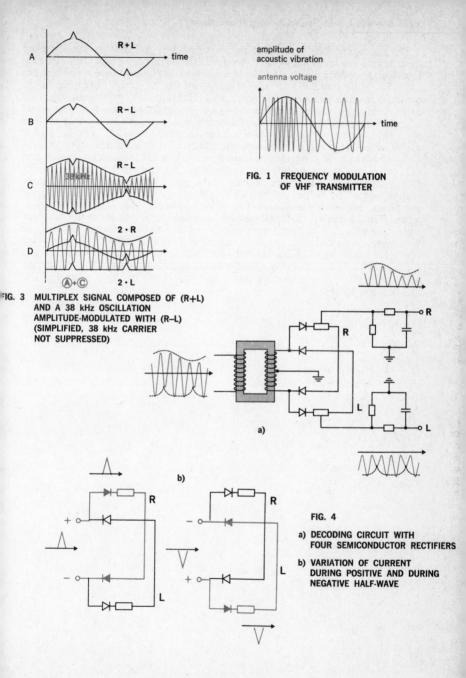

A R+L → time

B R-L

C R-L 38kHz

D 2·R 2·L

Ⓐ+Ⓒ

FIG. 3 MULTIPLEX SIGNAL COMPOSED OF (R+L)
AND A 38 kHz OSCILLATION
AMPLITUDE-MODULATED WITH (R-L)
(SIMPLIFIED, 38 kHz CARRIER
NOT SUPPRESSED)

amplitude of
acoustic vibration

antenna voltage

→ time

FIG. 1 FREQUENCY MODULATION
OF VHF TRANSMITTER

FIG. 4

a) DECODING CIRCUIT WITH
FOUR SEMICONDUCTOR RECTIFIERS

b) VARIATION OF CURRENT
DURING POSITIVE AND DURING
NEGATIVE HALF-WAVE

539

The procedure that is adopted in actual practice differs from the foregoing description: The multiplex signal (D) contains the frequencies arising from the sum signal, up to approximately 15 kHz, and additionally the frequency range between $38 - 15 = 23$ kHz and $38 + 15 = 53$ kHz arising from the signal $(R - L)$(Fig. 5a). In the upper frequency range the frequency of 38 kHz is especially prominent, inasmuch as it occurs also when the difference signal $(R - L)$ temporarily vanishes (sound source in the middle position). Accurate transmission of such a frequency mixture by a VHF transmitter would require a large frequency deviation, i.e., a large peak difference between the instantaneous frequency of the modulated wave and the carrier frequency. It is possible to manage with a smaller frequency deviation and yet obtain the same quality of transmission by completely suppressing the 38 kHz carrier and, instead, transmitting a constant 19 kHz "pilot frequency" along with the signal. In the decoder this 19 kHz frequency is filtered out by a resonant circuit and nonlinearly amplified. The 38 kHz harmonic that is thus produced is again filtered out in a 38 kHz resonant circuit, amplified, and added to the multiplex signal as a substitute for the suppressed carrier. The further process of decoding can then again be performed in the same manner as described above for the simplified multiplex signal (D). Fig. 6 is a block diagram of a stereo decoder. In actual practice the signal $(R + L)$ is added only after the decoding circuit shown in Fig. 4, since the splitting up into a positive and a negative half-wave concerns only the signal $(R - L)$ oscillating at 38 kHz.

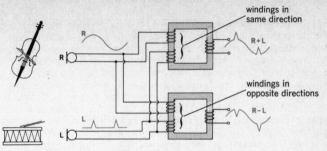

FIG. 2 OBTAINING THE SIGNALS (R+L) AND (R-L) FROM THE OUTPUT VOLTAGES OF TWO MICROPHONES R AND L WITH THE AID OF TWO TRANSFORMERS

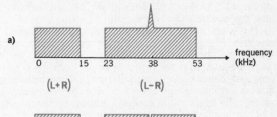

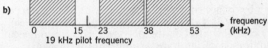

FIG. 5 FREQUENCY DISTRIBUTION IN THE MULTIPLEX SIGNAL
   a) 38 kHz CARRIER IS PRESENT
   b) 38 kHz CARRIER SUPPRESSED AND REPLACED BY kHz PILOT FREQUENCY

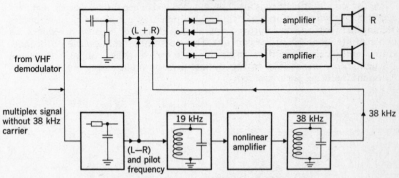

FIG. 6 BLOCK DIAGRAM OF A STEREO DECODER WITH LOW-FREQUENCY AMPLIFIERS CONNECTED IN SERIES BEHIND IT

TIME DIVISION MULTIPLEXING

Like carrier signaling (see page 534), time-division multiplexing is a technique for the transmission of two or more signals over a common path, using different time intervals for the different signals. However, instead of employing different frequency ranges (speech bands) transmitted by the modulation of progressively higher carrier frequencies in successive channels, time-division multiplexing telescopes the various speech signals together, as it were, by availing itself of the relative slowness of response of the human sense of hearing, which fails to detect gaps of up to 25 milliseconds' (25/1000 sec.) duration in the transmission and which is even able to compensate for deficiencies in continuity or clarity. Because of this, it is possible to pack a large number of "time channels," each carrying a message (i.e., a separate conversation), into this time lag. Within this time lag the two speakers conducting a conversation along one of the time channels are interconnected for brief periods of 0.5 microseconds (1/2,000,000 sec.) by means of electronic gate circuits operating at a repetition frequency of 10 kHz (10,000 cycles/sec.). Thus 250 connections are established within the duration of the time lag, each connection lasting only for 0.125 millisecond. Thus about 2000 time channels can be packed into the time lag. Multiple use of the transmission path is thus achieved by telescoping together a number of impulse sequences, each associated with a particular speech connection (Fig. 1). The speed of the impulse sequence is measured in bits/sec. or kbits/sec., a "bit" being a unit of information content (see Vol. I, page 298). Based on the time-division multiplexing principle is the TASI system (an abbreviation of "time assignment speech interpolation"), developed more particularly for use with undersea telephone cables, as it enables more channels to be transmitted than is economically feasible with carrier-signaling techniques. Fig. 2 schematically shows this system applied to a cable which carries 36 channels and through which 72 lines can be connected. Speech detectors ascertain whether a particular channel is engaged or free. As soon as more than 36 speech connections are simultaneously required, a central control device, operating with electronic gate circuits, initiates multiple utilization of the transmission path by time-division multiplexing. The control operations, for which a separate channel is reserved (channel 37 in Fig. 2), must likewise take place within the time lag. Faulty transmission is liable to occur only if it takes the speech detectors more than 100 milliseconds to find a free channel; in that case a syllable of speech (average duration 120 milliseconds) will be lost.

Time-division multiplexing is also coming into use in general telephone engineering, as it cuts down the amount of technical equipment by getting rid of the elaborate switching network and in improved operational reliability because it reduces the number of junctions (points of contact). The most important application of time-division multiplexing is in data processing, however, in which impulse sequences play such an important role. In Fig. 3, time-division multiplexing and frequency-division multiplexing are compared.

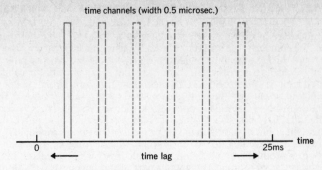

FIG. 1 MULTIPLE UTILIZATION OF A TRANSMISSION PATH BY SIX TIME CHANNELS (WIDTH NOT TO SCALE)

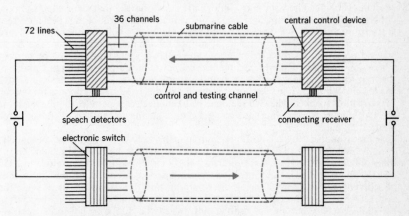

FIG. 2 SCHEMATIC DIAGRAM TO SHOW HOW THE TASI SYSTEM FUNCTIONS

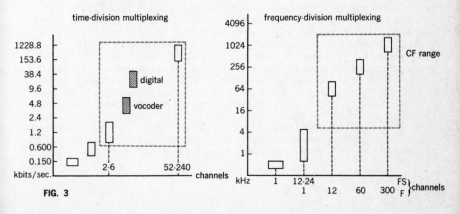

FIG. 3

The procedure in making a telephone call is as follows. The caller picks up the receiver; the contact U1 (Fig. 1) closes and bypasses the capacitor C. Now a direct current flows through the line and serves to prepare the equipment at the telephone exchange for dealing with the call. When the dial tone is heard, the caller can dial the desired number. When he does this, the contact nsa short-circuits the microphone and telephone circuit until the dial has returned to its original position. During the return motion the contact nsi is opened and closed as many times as corresponds to the digit that has been dialed. When the receiver is resting in its cradle, the contact nsr bridges the contact nsi. When the caller speaks into the microphone M, the latter modulates the above-mentioned current or—in the case of an electrodynamic microphone—produces an alternating speech current which oscillates in sympathy with the acoustic vibrations. The greater part of this current travels along the line to the other person's instrument. A small proportion flows through the short circuit consisting of the winding 1 of the telephone transformer and the resistor 4. Most of the speech current also flows through a winding 2 of the transformer before reaching the line. The windings 1 and 2 are so connected that their effects almost entirely cancel each other out. Hence only a small proportion of the speech current is transmitted to the winding 3, so that the caller's own voice is only faintly heard in his own receiver. This arrangement suppresses the disturbing influence of background noises in the room from where the caller is speaking. The speech current coming in from the distant instrument flows through the two windings 1 and 2 of the transformer in the same direction and is therefore passed without loss of strength to the third winding. The telephone connected to this winding reproduces the speech of the distant speaker. The two rectifiers (GL) connected in parallel to the telephone serve to eliminate crackling and contact noises arising in the line. W denotes the alternating-current bell.

Behind the dial (Fig. 2) of the telephone instrument is a small gear mechanism with a contactor. When a number is dialed, the spiral spring is tensioned. When the dial is released, the spring force returns it to its initial position. This return motion has to be performed at a controlled constant speed, which is ensured by means of a centrifugal governor. During the return motion the interrupter arms open the contact nsi as many times as corresponds to the digit dialed. The cam on the shaft of the dial actuates the contact nsa/nsr only when the dial is in its initial position (position of rest) and for two-thirds of the distance between the initial position and the finger hole "1."

Some telephone systems have push-button dialing by means of 10 or 12 buttons. When a button is depressed, a transistor generator sends two audio frequencies—out of a possible eight such frequencies—through the line to the telephone exchange (Fig. 3). The circuit (Fig. 4) functions as follows: When the caller picks up the receiver, the contact GU is closed, but the transistor is short-circuited by the contact r. When one of the push buttons is depressed, this contact r opens and activates the transistor. The contact a short-circuits the microphone-telephone circuit, so that no interfering tones can be transmitted by the microphone during the transmission of the signals. At the same time, the resonant circuit capacitors C3 and C4 are each connected, by respective contacts, to a tapping of the transformers U2 and U3 respectively. While the contacts u and r are actuated by every button, each button closes a different combination of two resonant circuit contacts. The bell and speech circuit arrangements are similar to those in the rotating dial system already described.

(more)

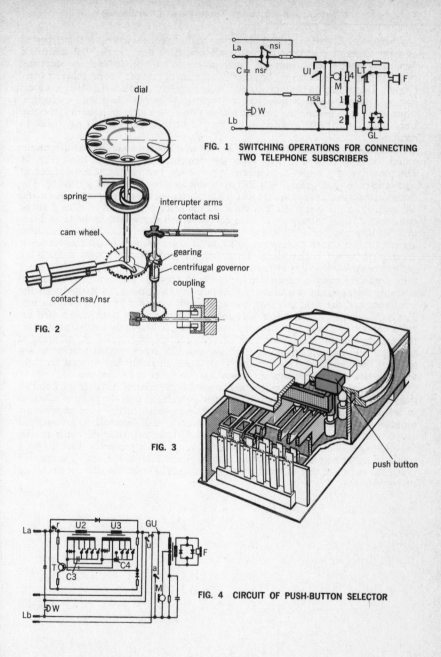

dial

FIG. 1 SWITCHING OPERATIONS FOR CONNECTING
TWO TELEPHONE SUBSCRIBERS

spring

interrupter arms

contact nsi

cam wheel

gearing

centrifugal governor

coupling

contact nsa/nsr

FIG. 2

FIG. 3

push button

FIG. 4 CIRCUIT OF PUSH-BUTTON SELECTOR

545

When the caller dials a number, the circuit to the telephone exchange is interrupted as many times as corresponds to that number. Each digit on the dial controls a "selection stage" comprising an electromagnetic multiple-contact (or stepping) switch called the "selector." The latter has one input and can connect this to many outputs. Depending on the digit dialed, the first selector connects the caller to a local line or a long-distance (trunk) line, and further selectors—at distant exchanges, when trunk calls are made—connect one section of line after another in successive steps until the caller is connected to the telephone of the persons he wishes to speak to.

The selector equipment of the type known as a "noble metal uniselector motor switch" comprises a motor, an adjusting element, and a contact bank (Fig. 1). The motor is of a somewhat unusual type: it has two electromagnets placed at right angles to each other and a Z-shaped soft-iron rotor without a winding. The rotor shaft actuates two contacts which alternately switch the magnet coils on and off; each time, the coil which has attracted the rotor is switched off. The Z shape ensures that the attracting force of the energized coil always acts in the same direction. With each half revolution of the rotor the contact arms of the adjusting element are rotated a distance corresponding to one contact position of the contact bank. To stop the motor, both coils are simultaneously energized by a relay, causing the rotor to be arrested in an intermediate position between the magnet poles; the contact arms are then located accurately at the center of a contact position. This method of stopping the rotation is a distinctive feature of this selector, as it eliminates the vibrations and the wear and tear associated with the older forms of construction. For local connections, selectors with four pairs of arms are used; those used for trunk connections have eight pairs. Two of the arms (or four in the eight-arm selector) move along the contact positions and serve to control the running and stopping of the rotor and transfer certain control currents. The other two (or four) arms are lifted off the contacts during rotation. Only when the motor has stopped are these arms pressed (by electromagnetic action) against the contact positions (Figs. 2 and 3).

Another important device is the "noble metal rapid-contact relay" (Figs. 2 and 3), used in the construction of so-called couplers, which can take the place of selectors in telephony and are becoming increasingly important. In this type of relay the number of moving parts is reduced to a minimum. Five of these relays are combined into a unit. There are no separate mechanical parts for actuating the contacts. The movable springs of the four or six contacts of a relay are pressed against the fixed opposite contact springs directly by the magnetic field of the coil. The moving masses are so small that the contacts close within 2/1000 second after the operating coil is energized.

(more)

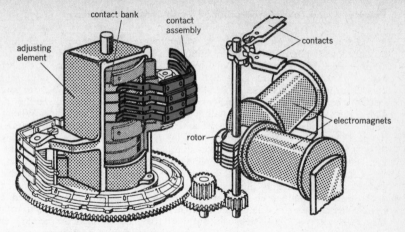

FIG. 1 UNISELECTOR MOTOR SWITCH

contact bank

contact assembly

adjusting element

contacts

electromagnets

rotor

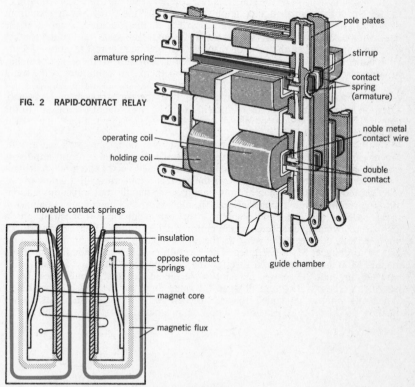

pole plates

stirrup

contact spring (armature)

armature spring

FIG. 2 RAPID-CONTACT RELAY

noble metal contact wire

operating coil

holding coil

double contact

movable contact springs

insulation

opposite contact springs

magnet core

magnetic flux

guide chamber

FIG. 3 DIAGRAM SHOWING MODE OF FUNCTIONING OF RAPID-CONTACT RELAY WITH SUPERIMPOSED MAGNETIC FLUXES

The number-selection procedure in a telephone exchange can most suitably be explained with reference to the selection of a two-digit number. Formerly two-motion selectors were used (see Vol. I, page 112), in which the number was selected from a square bank comprising 10×10 contacts. In the modern "noble metal uniselector motor switch" (see page 546), however, all the 100 contacts are arranged side by side in a semicircle. Hence when the digit corresponding to the "tens" is dialed, the selector must, at each dialing impulse, jump over ten contacts. When the "ones" digit is dialed, the selector proceeds step by step, i.e., from one contact to the next. At the first current impulse the relay A energizes the delay-action relay V (Fig. 1), which in turn closes the motor circuit. The selector cannot yet rotate, however, since both motor magnets are energized through the zero position contact nr. When the relay A is de-energized at the end of the current impulse, the rotor is released, and the selector rotates to the first main rest position, where the first "ten" of the dialed number commences. Now when a second current impulse is emitted, the changeover contact a de-energizes the magnet M_2, and the selector begins to rotate. It completes its first "tens" step, which brings it to the beginning of the second "ten." The cam contacts nr and zr on the selector play an important part in controlling the selector movements. The main rest contact hr marks the end and also the beginning of a "ten." The intermediate rest contact zr serves to stop the motion of the selector if the number-dialing impulses are emitted too slowly. As appears from the—greatly simplified— circuit diagram in Fig. 1, the selector is stopped at the intermediate rest position ZR (always the sixth step within each "ten") if the contact a is still reversed when the selector reaches this intermediate rest position. On the other hand, if the relay A has already become de-energized when the selector reaches the intermediate rest position, it does not stop there, but continues to the main rest position. Here it is held by the action of the reset contact a and the main rest contact. When the next current impulse is emitted, the same procedure is repeated. After the last impulse of a series of current impulses has brought the selector to the relevant main rest, the control relay V is de-energized; selection of the "ones" then commences.

This latter operation, in which the selector moves along one "step" at a time, can be explained with reference to Fig. 2. When the first dialing current impulse is emitted, the relay A causes the delay-action control relay V to be energized. A contact v energizes the motor circuit. Through the contact a (in series with the low-ohmic relay winding D) both magnets are now energized, so that the motor remains stationary. The second winding D is energized through another contact v and a contact a. In that case the relay D does not respond, as the current flows in the opposite direction through the winding connected in the motor circuit. At the end of the first current impulse, with release of the relay A, the motor holding current circuit is broken. The motor magnet M_1 is energized. The selection performs only one step, since the magnet M_1 remains energized through the rest contact a and the contact d, notwithstanding that the contact m_1 is open. At the next current impulse the relay D responds because the current flows in the same direction through both windings. When this impulse has ended, the selector can perform its next step, but is once again stopped by the relay D because now the magnet M_2 remains energized.

(more)

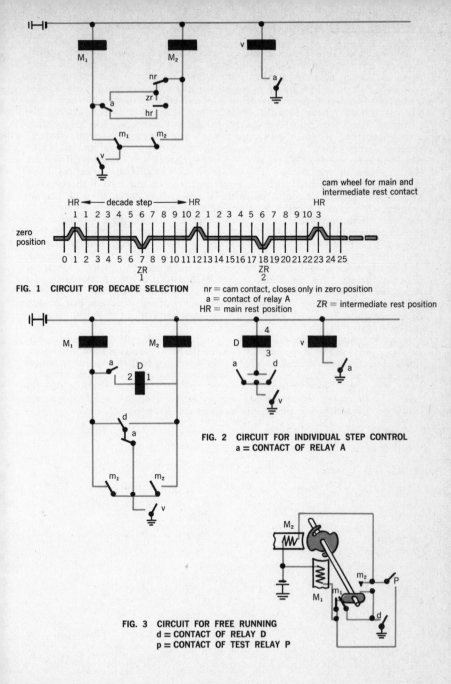

FIG. 1 CIRCUIT FOR DECADE SELECTION

cam wheel for main and intermediate rest contact

nr = cam contact, closes only in zero position
a = contact of relay A
HR = main rest position
ZR = intermediate rest position

FIG. 2 CIRCUIT FOR INDIVIDUAL STEP CONTROL
a = CONTACT OF RELAY A

FIG. 3 CIRCUIT FOR FREE RUNNING
d = CONTACT OF RELAY D
p = CONTACT OF TEST RELAY P

As a result of the advances in switching techniques described in the foregoing, there are already a number of experimental telephone exchanges operating with electronic switch gear. In principle, any relay having two functional positions ("on" and "off") can be replaced by a system of electronic components (e.g., diodes, transistors) in which a flow of electrons is permitted to pass or is blocked. These technical advances have, broadly speaking, been reflected in developments in the telephone system itself, including the direct dialing of numbers in foreign countries by any subscriber. A feature which has not changed, however, is the "star-shaped" pattern, as exemplified by the accompanying map showing central and main telephone exchanges in the Federal Republic of Germany (West Germany). There are eight central exchanges, to each of which are connected eight main exchanges. These in turn are each connected to eight junction exchanges, and each of these has up to eight end exchanges connected to it. In cases where the number of junction exchanges or end exchanges is larger than the maximum that can be connected to the main exchange or the junction exchange concerned, these last-mentioned exchanges have to be doubled. This may be the case in the vicinity of major cities, where large numbers of telephone subscribers are concentrated in a relatively limited area. The "star" arrangement has advantages and disadvantages. An advantage is that the expensive long-distance selection equipment need be installed only at the center of the star. The disadvantage associated with it is that, in the event of a breakdown or technical trouble of any kind in these important exchanges, large sections of the telephone system are immediately put out of action. For gradual enlargement and extension of a telephone system the "star" arrangement provides the least expensive solution.

However, really economical operation can be achieved only by sufficient intermeshing of the telephone system so as to fulfill the requirements of telephone operation and reliability. Instead of having to pass all calls through the central exchanges it might well be possible in certain cases to bypass these and, for example, employ a more economical cross connection between two main exchanges. Now that the "star" system has been completed in Germany, further development for the moment consists in the addition of such connections as and when the need arises.

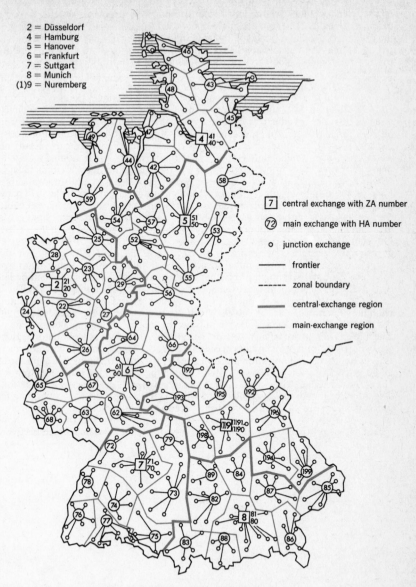

2 = Düsseldorf
4 = Hamburg
5 = Hanover
6 = Frankfurt
7 = Suttgart
8 = Munich
(1)9 = Nuremberg

7   central exchange with ZA number

72   main exchange with HA number

o   junction exchange

———   frontier

-----   zonal boundary

———   central-exchange region

———   main-exchange region

Central and main-exchange telephone regions for
the Federal Republic of Germany, not including Berlin

The telephone lines are the most expensive part of a telephone system. Within a particular locality they connect all the individual subscribers to the exchanges, and the exchanges are interconnected by trunk lines. In the early days of the telephone, overhead lines were usual; nowadays, where these are used at all, they are confined to short final lengths of line to individual subscribers. Overhead lines have numerous disadvantages: stresses to which they are subjected by the action of wind, ice formation, etc., make it necessary to employ hard copper or bronze wires up to 3 mm in diameter, whereas 0.6 mm diameter would be sufficient to convey the electric current, i.e., such lines are wasteful in material, most of this being needed for mechanical strength. Besides, overhead telephone wires are exposed to inductive action from adjacent high-voltage equipment (particularly if it is faulty) and from atmospheric electricity (especially thunderstorms). In the course of time most of the overhead lines were replaced by cables, thus getting rid of the unsightly congestion of wires that disfigured some cities in prewar days. In a cable a large number of circuits can be accommodated, each of which comprises two wires, called a "pair." In local cables the pairs are usually insulated from one another by dry cellulose in the form of paper pulp or paper tape wrapped around the wires; in long-distance (or trunk) cables they are insulated with plastics, e.g., polyvinyl chloride or polyethylene. Since cables are, with few exceptions, laid underground, they are thus protected from the adverse influences affecting overhead telephone lines. In addition, the cable can be given appropriate strength and resistance to aggressive chemical influences. For this purpose it is provided with a metal sheath (lead or aluminum), an armoring of steel tape, and an outer wrapping of bitumen-impregnated jute. Such cables can be laid directly in the ground. In local telephone systems the cables may be installed in underground multiple-way ducts (Fig. 1) which extend between cable-jointing chambers (manholes). These cables, which are not armored, are threaded through the ducts and are connected to one another by means of jointing sleeves (Fig. 3). The cable is composed of eight quads, i.e., groups of four wires (conductors); there are two pairs forming two circuits in each such group. The object of this arrangement is to prevent cross talk (unwanted sound due to capacitive and inductive action from adjacent circuits). There are two forms in which the wires in the cable are stranded, i.e., twisted together: "star-quad formation" (Fig. 4a) and "multiple-twin formation" (Fig. 4b). Five quads are assembled into a basic group of conductors containing ten pairs of wires (Fig. 5). Such basic groups are in turn assembled into main groups, which are disposed in several layers within the protective sheath. Fig. 6 shows typical cross-sectional arrangements for a 300-pair and a 1500-pair cable respectively.

(more)

multiple-way duct

FIG. 1   CABLE DUCT

FIG. 2   JOINTING CHAMBER

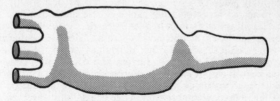

**FIG. 3   LEAD JOINTING SLEEVE AT BRANCHING OF ONE
MULTIPAIRED CABLE INTO THREE SMALLER CABLES**

A
B

St
DM

**FIG. 4   STAR-QUAD FORMATION (St) AND MULTIPLE-TWIN
FORMATION (DM)**

**FIG. 5   BASIC GROUP CONTAINING
PAIRS OF CONDUCTORS**

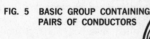

quad

**FIG. 6   COMPOSITION OF MULTI-
PAIRED CABLES COMPRISING
MAIN GROUPS AND BASIC
GROUPS OF CONDUCTORS**

300-pair cable

50
DA

main group with
five basic groups
(50 pairs)

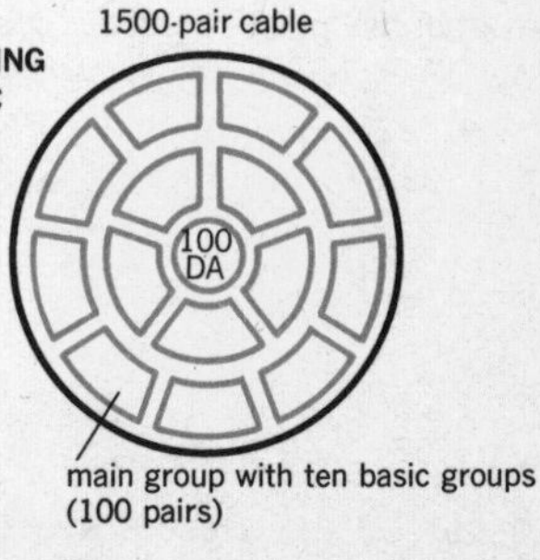

1500-pair cable

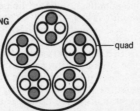

100
DA

main group with ten basic groups
(100 pairs)

553

The telephone cables extend between distribution boxes and dividing boxes; the individual subscribers' lines are connected to the latter. In the telephone exchange the cables are split up into separate groups of conductors in multiple cable joints and connected to a main distributing frame (Fig. 7). The branching of a cable into individual lines in a telephone system is shown schematically in Fig. 8.

In local systems each subscriber is connected to the exchange by a pair of individual copper wires. For long-distance cables, however, multiplex carrier techniques are being increasingly adopted (see page 534); time-division multiplexing is more particularly preferred for submarine cables. These methods make new demands upon the cables; the use of high-frequency carrier waves (e.g., 240 kHz for 300 speech channels) increases the likelihood of cross talk between adjacent conductors. Attempts to overcome these and other difficulties led to the development of the so-called coaxial cable. The two conductors of a coaxial unit consist of a hollow cylinder and a central wire respectively (Fig. 9), separated by a distance of 2 to 5 mm (0.08 to 0.4 inch). Sometimes a cable is composed of a number of such units. Alternatively, a cable may comprise one large coaxial unit together with a number of ordinary quads grouped around it (Fig. 10). A coaxial cable of this kind can carry as many as 960 speech channels or it can be used to carry the wide frequency band of a television transmission. At higher carrier frequencies the electromagnetic energy tends to "become detached" from the metal conductor and to be propagated into space as waves (a property which, at still higher frequencies, is utilized in the wave guide: see page 526). For this reason it is essential to ensure that the coaxial unit is geometrically accurate, i.e., the axial wire must be precisely central within the cylindrical tube surrounding it. If this condition is not satisfied, the carrier waves will be reflected back from the cylinder wall, causing "ghost" images in the case of a television transmission. This requirement of geometric precision, which is essential to ensure good electromagnetic performance of the line, must be fulfilled without undue sacrifice of flexibility of the cable. It should, for example, be possible to bend it to a radius of about 4 ft. ($1\frac{1}{4}$ m) so that it can be wound on a reel. To obtain this flexibility, the cylindrical tube of the coaxial cable is made of coiled copper tape, the central wire being maintained in position by means of spacer discs made of trolitul (an insulating material) or by means of a helix made of styroflex (Fig. 9). Despite progress in cable-manufacturing technology, higher carrier frequencies undergo considerable damping. Thus, about 99.9% of the input energy is lost over a distance of five or six miles, so that amplification becomes necessary. These intermediate amplifiers, called repeaters, are now often installed underground, completely buried, or provided with access shafts for servicing (Fig. 11). Submarine telephone cables have to be of much stronger construction than cables for normal use on land, as they are subjected to high tensile forces (especially at the time of laying) and very high pressures when installed in great depths of water. For this reason, coaxial cables for submarine use do not contain a cavity between central wire and cylindrical tube, but are filled solid with a suitable insulating material (gutta-percha, plastic), besides containing steel "armor" wires for additional strength.

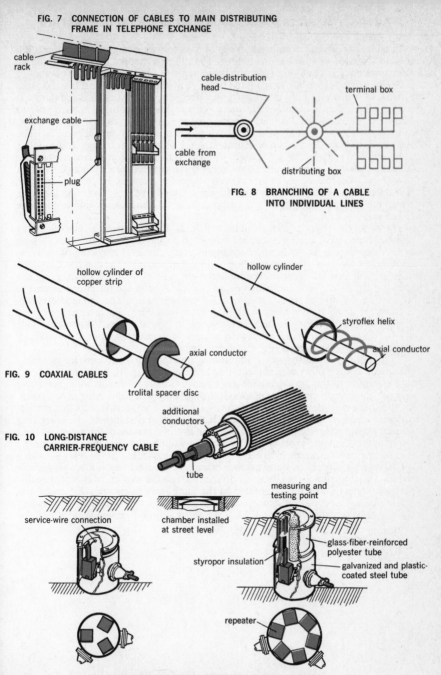

FIG. 7 CONNECTION OF CABLES TO MAIN DISTRIBUTING FRAME IN TELEPHONE EXCHANGE

cable rack

exchange cable

plug

cable-distribution head

cable from exchange

terminal box

distributing box

FIG. 8 BRANCHING OF A CABLE INTO INDIVIDUAL LINES

hollow cylinder of copper strip

axial conductor

FIG. 9 COAXIAL CABLES

trolital spacer disc

hollow cylinder

styroflex helix

axial conductor

FIG. 10 LONG-DISTANCE CARRIER-FREQUENCY CABLE

additional conductors

tube

service-wire connection

chamber installed at street level

measuring and testing point

styropor insulation

glass-fiber-reinforced polyester tube

galvanized and plastic-coated steel tube

repeater

FIG. 11 REPEATERS FOR MULTIPLE-TUBE COAXIAL CABLES IN UNDERGROUND CHAMBERS (UP TO SIX REPEATERS PER CHAMBER)

AUTOMATIC MESSAGE ACCOUNTING AND CIRCUIT TESTING

In telephony, the automatic recording of data for preparing the subscriber's bill is closely related to automatic switching. The simplest method consists in using an electromechanical counter, called a message register, which records the number of calls made by a subscriber. In modern telephone systems, however, a more elaborate procedure, known as multiunit registration, according to the distance and duration of the call, is usually employed. In some systems this procedure is applied only to long-distance calls, while local calls are charged at a flat rate, irrespective of duration; in others, all calls are charged on the basis of the number of "message units." The registration of calls is based on electrical impulses which actuate relays. The impulses are continually fed to the call-metering device during the conversation; they are fed in quicker succession according as the call distance is longer. In practice, a certain number of "distance zones" is introduced—nine, for example. For each zone there is a different standard impulse rate, the time interval between two impulses (the zone time) being the length of time corresponding to one message unit in the zone concerned.

The impulses are emitted by a call timer (Fig. 1). This comprises an electric motor which drives a shaft at constant speed. Mounted on the shaft are cam wheels which actuate contacts and thus produce the impulse sequences. The spacing of the cams is different on the wheels corresponding to the different zones. The difference between normal tariff and the cheaper evening and night tariff is obtained by changing the speed of the motor. The metering devices are installed in the telephone exchange, where they are read at regular intervals; the number of message units indicated by each meter provides the basis for billing each subscriber. Each metering device comprises several drums on which the numerals for "units," "tens," "hundreds," etc., are marked and which are rotated by means of a ratchet-and-pawl mechanism actuated by electromagnets (Fig. 2). The meter dial is shown in Fig. 3.

Every telephone selector has, in addition to the wipers (contact arms) for switching the speech circuit, other wipers for testing and control (Fig. 4). The wiper C is used for "engaged" testing (or testing for free circuit). Connected to this wiper is the test relay P; the speech circuit is connected through only when this relay is energized; for this there must be a sufficiently high voltage at the contact that the wiper reaches on completion of the dialing operation. If this voltage is sufficient, the energizing current can flow through the relay P. As soon as the armature of this relay has been attracted, its contacts short-circuit part of the winding, so that the voltage at the selector contact decreases. While P is energized, the winding of the relay C (which is connected to the selector contact and belongs to the next selection stage) and the reduced winding of P form a "voltage divider." When the wiper C of another selector reaches a contact which is connected to the same relay C, its relay P receives too low a voltage and cannot become energized. The control circuit of this second selector tests whether the relay P has responded within a certain time; if not, the "engaged" signal is transmitted to the subscriber.

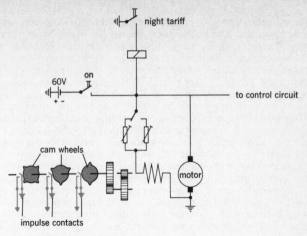

FIG. 1   DIAGRAM OF A CALL TIMER

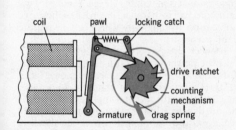

FIG. 2   RATCHET-AND-PAWL COUNTING DEVICE

FIG. 3

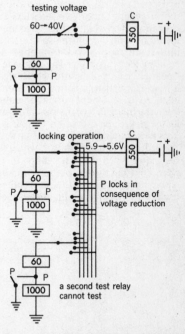

FIG. 4   TESTING AND LOCKING PROCEDURE FOR SELECTORS

CURRENT SUPPLY FOR TELECOMMUNICATION SYSTEMS

The suitable supply of current is an essential factor for the proper functioning of a telephone system. The public electricity supply mains as such are not a suitable source of current, as the permissible tolerances with regard to voltage and the occasional power failures would seriously interfere with proper telephonic communication. Besides, the requirements as to the type of current, voltage, current strength, and frequency vary greatly. What is primarily essential in telephony is a completely undisturbed supply of direct current; the lines and equipment are designed for 60 volts, whereas the public mains generally supply alternating current at a substantially higher voltage. The alternating main current is transformed down, rectified, and used to charge a storage battery of sufficient capacity to supply the telephone system with current for at least six hours in the event of a supply failure of the main (Fig. 1). In about 97% of the cases this is a sufficient length of time to put right the fault in the main and thus resume the recharging of the battery. Otherwise a standby generating set, consisting of a diesel engine coupled to a generator, must temporarily supply charging current.

Certain types of telecommunication equipment require an alternating-current supply. Even in such cases the main's alternating current is first converted into direct current—because this can be stored in a storage battery which gives a constant supply of current independent of the mains and possible troubles affecting them—and then converted back into alternating current. This last-mentioned conversion can be done by means of a motor-generator, i.e., a DC motor driving an AC generator, or by means of an inverter (also known as an inverted rectifier) (Figs. 2 and 3). Two motor-generators or two inverters are connected in parallel, in order to improve operational reliability. For telecommunication installations which are in remote locations with no on-the-spot attendant personnel—e.g., directional radio transmitters—it may be advantageous to connect the power supply directly with a standby supply system. The requisite electric power for operating the equipment is obtained from a generator whose shaft is coupled to that of an electric motor (with a flywheel) which in turn is connected, through an induction coupling, to a diesel engine (Fig. 4). Under normal operating conditions the motor, supplied with current from the main, drives the generator and the flywheel. In the event of a power failure in the main, the induction coupling engages, and the momentum of the flywheel provides sufficient mechanical power to start the diesel engine. The latter runs at the same speed as the motor, which has now been automatically disconnected from the main. When the main supply is restored, the diesel engine is mechanically disconnected and the motor is electrically reconnected to the main. An automatic fault-reporting system brings these events to the attention of a central monitoring station, usually some considerable distance away.

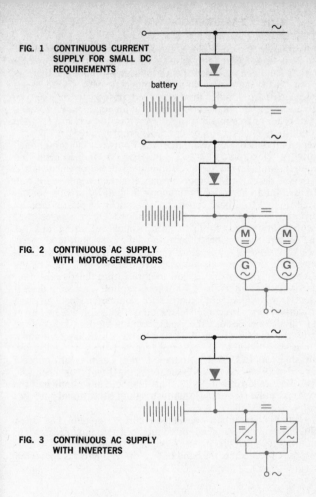

FIG. 1 CONTINUOUS CURRENT SUPPLY FOR SMALL DC REQUIREMENTS

battery

FIG. 2 CONTINUOUS AC SUPPLY WITH MOTOR-GENERATORS

FIG. 3 CONTINUOUS AC SUPPLY WITH INVERTERS

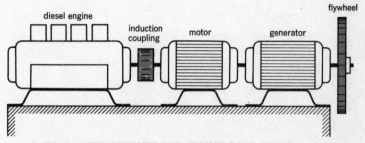

diesel engine   induction coupling   motor   generator   flywheel

FIG. 4 MOTOR-GENERATOR WITH STANDBY DIESEL ENGINE

Video tape recording (magnetic picture recording) is the "visual" counterpart of audio tape recording. It is the technique of storing video signals by means of a tape magnetized along its length in accordance with the signals impressed upon it. The tape, consisting of a plastic material coated with magnetic oxide, is fed past a recording head, and the signals are stored in the magnetized oxide particles. This type of recording has the important advantage of being ready for immediate playback without development or other processing. For this reason it is now extensively used for the recording of both monochrome and color television programs.

An early forerunner of tape-recording technique was Poulsen's "telegraphone" (1898) for the recording of telegraphic messages and, later on, of speech on thin steel wire as the magnetic storage medium. A subsequent development was the "magnetophone," a device developed by the German electrical-engineering firm of AEG in 1930 and based on an invention by Pleumer. It used iron oxide powder as the storage material applied to a base consisting of a paper or plastic tape. In 1953 the Radio Corporation of America (RCA) developed Olson's video tape recorder, operating with a tape speed of 6 m/sec., longitudinal recording (Fig. 1a), with a fixed recording or reproducing head. Because of the very high tape speed, this apparatus was not very suitable for practical purposes. In the Ampex machine developed by Ginsburg in 1956 the signals are recorded transversely (Fig. 1b), the tape speed being 38 cm/sec. This apparatus had four recording heads (and four reproducing heads) positioned at intervals of 90 degrees around a rotating disc. It was the first really serviceable video tape recorder for commercial use, giving a picture of such good quality that a prerecorded television transmission was no longer distinguishable from a "live" one. Besides, the tape speed was similar to that of the audio tape recorders used in broadcasting. With appropriate ancillary equipment this video recorder was also suitable for color-picture recording for television (NTSC system used in America). In 1959 the Japanese firm of Toshiba introduced a recorder with a tape speed of 19 cm/sec. and oblique recording (Fig. 1c) on a helically guided tape by means of a single rotating head. This machine has about half the resolving power of the Ampex and is suitable for industrial and educational purposes and private use.

Fig. 2 schematically shows the Ampex video tape-recording system. The video signal frequency-modulates a carrier-frequency oscillation of approximately 50 MHz with a deviation from -0.9 to $+2.1$ MHz—i.e., giving a frequency band of 49.1 to 52.1 MHz. For the purpose of recording, the band of 49.1–50–52.1 MHz is converted

(more)

FIG. 1

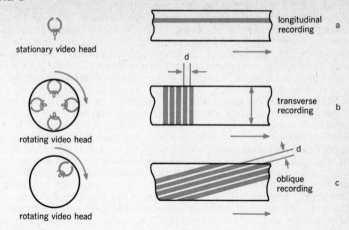

longitudinal recording — a

transverse recording — b

oblique recording — c

stationary video head

rotating video head

rotating video head

FIG. 2 SIMPLIFIED BLOCK DIAGRAM OF A VIDEO TAPE-RECORDING SYSTEM OPERATED ON THE FOUR-HEAD AMPEX PRINCIPLE

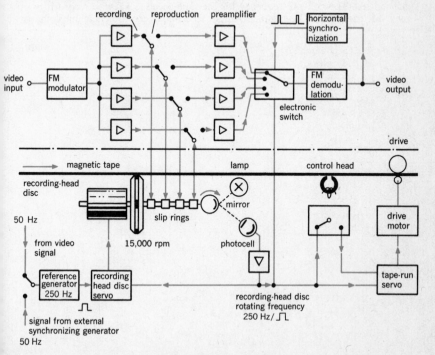

to 6.3–7.2–9.3 MHz and is fed simultaneously to the four rotating recording heads (Fig. 3). A picture line of 64 microseconds' duration is recorded on a track length of 2.4 mm—i.e., at a speed of 38.1 m/sec. The tracks are 0.25 mm wide and are 0.131 mm apart; a complete 625-line picture is recorded on a 15.2 mm length of the 50.8 mm wide tape. One revolution of the recording-head assembly corresponds to a tape length of 4×0.381 = approx. 1.6 mm; for a speed of 250 revolutions per second the tape travels at a speed of 38.1 cm/sec. Reproduction of the recording is done by means of the same head assembly. It converts the magnetic recording into frequency-modulated signal voltages. These are converted to 49–52 MHz and demodulated to produce a video signal. In contrast with the recording process, the four reproducing heads are *consecutively* connected to the demodulator in order to obviate the overlaps between the individual heads during recording. For recording, the tape speed and the rotational speed of the heads need not fulfill very exacting requirements of uniformity. On the other hand, for reproduction the head and the tape must be synchronized with an accuracy of 0.1 microsecond (or 0.15 micron). For this purpose the tape speed is controlled by means of a control frequency corresponding to the actual recording-head speed and recorded on the tape along with the picture, and the speed of rotation of the heads is controlled by the frame-synchronizing impulse of the video signal. Additional electronic control devices compensate for inaccuracies in the positioning of the four heads in relation to one another. The reproduction of color video signals calls for something like 100 times greater accuracy even than for monochrome. The audio signal, i.e., the sound accompanying the picture, is recorded on a 1 mm wide track at the edge of the tape. The video and audio signals are exactly synchronized. The control frequency for the tape speed is likewise recorded on a 1 mm wide edge track. The complete Ampex video-tape recording machine is shown in Fig. 4.

(more)

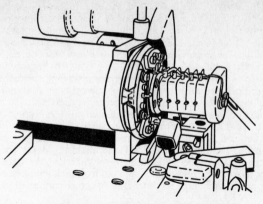

FIG. 3 AMPEX MACHINE (RECORDING-HEAD DISC WITH
MAGNETIC RECORDING HEADS AT 90-DEGREE
SPACING AND COLLECTOR SPRINGS FOR HEAD
CONTACTS ON THE ROTATING SHAFT)

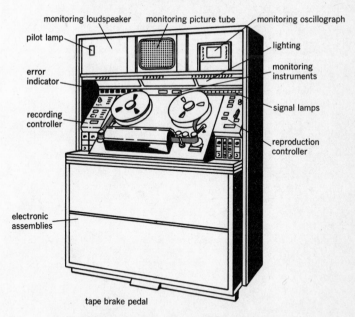

FIG. 4 FOUR-HEAD VIDEO TAPE-RECORDING MACHINE

Because of the convenience associated with the immediate readiness for reproduction of the recorded pictures, video tape recording is used not only in television
broadcasting but also for industrial and educational purposes, as well as for amateur
use in the home. Somewhat simpler machines operating with a single recording
head are employed in these cases, where rather less exacting requirements as to
picture quality are applied. A machine of this type which records the signals in
tracks inclined at an angle of about 4 degrees is shown schematically in Fig. 5.
The tape is wrapped around a tilted tape-guide cylinder containing one rotating
head. One inclined track records half a picture ($\frac{1}{50}$ second); the jump from one
track to the next occurs within the blanking interval—i.e., during the fraction of a
second when there is no picture on the video screen—and is therefore not seen. As
the speed of the head is only about 20 m/sec., the resolving power is only about
half that of the four-head machine (2.5 MHz as against 5 MHz). The tape speed has
been reduced to 19.05 cm/sec. without adversely affecting the signal-to-noise ratio
to a significant degree. The audio track and control track are likewise located at the
edge of the tape. The video tape recorder (Fig. 6) can be connected to a television set.

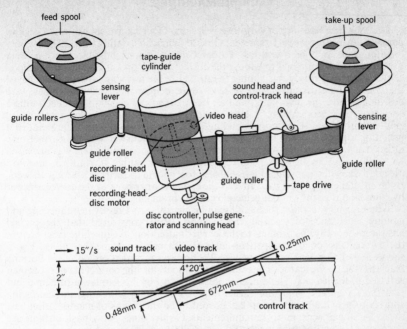

FIG. 5 DIAGRAM OF VIDEO RECORDING IN OBLIQUE TRACKS ON MAGNETIC TAPE; THE TRACKS ARE FORMED BY GUIDING THE TAPE ON A RECORDING-HEAD DISC WITH INCLINED AXIS

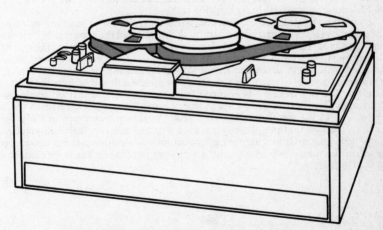

FIG. 6 VIDEO TAPE RECORDER FOR INDUSTRIAL USE

Teaching machines are of various types, some of which are shown in the accompanying illustrations. Fig. 1 represents Pressey's machine dating from 1927. Its only function was to pose questions. To each question four possible answers were presented; the learner had to choose the correct one by pressing a button (multiple-choice method). Pressey was aiming to rationalize the teaching process. He believed that material inducements and rewards were helpful. His early machines were designed to give the deserving pupil a piece of candy or chewing gum each time a correct answer was chosen. A later device of his was the so-called punchboard (Fig. 2) with holes—each corresponding to an answer—into which the point of a pencil or a pricker could be inserted, but only if the answer was the correct one. Other machines were devised in which the choice of possible answers was increased to twenty—e.g., in Briggs's "subject matter trainer" (Fig. 3). Another inventor, Norman Crowder, considered it useful also to evaluate the learner's wrong answers. In his machine (Fig. 4) he tried to lead the erring learner back to the correct answer by roundabout paths (on the principle called "branching").

Skinner, who was more successful than his predecessor Pressey in publicizing and gaining acceptance of teaching by machine (1954), considered that the correct answer must always be brought to the learner's attention: the machine must present the correct answer for comparison after the learner has made his choice. Fig. 5 shows one of Skinner's machines, with the answers recorded on a disc. In Porter's machine (Fig. 6) the learner's answer was fed into a slot; the correct answer was then revealed to the learner. The machine illustrated in Fig. 7 is similar in principle, but is worked by a knob which the learner rotates. For dealing with programs consisting entirely of text matter the knob has proved to be the means of manipulation best suited to human sensomotoric requirements. On the other hand, push buttons are preferable for audiovisual machines. Fig. 8 illustrates a special teaching machine which instructs the learner in the operation of a piece of apparatus and shows the correct manipulations on a television screen. In some machines the correct answer is not directly given; instead, the progress of the program is made dependent on the learner's giving the correct answer. Some machines of this type are shown in Figs. 9–12, including a digital computer (Fig. 11).

From the viewpoint of educational psychology, the use of teaching machines is based on concepts of behavioristic learning theories and the principle of "conditioning" as envisaged by Pavlov in his famous experiments on animals at the end of the last century. Teaching machines have indeed proved successful in many branches of education. Their advantage lies in skillful breakdown and organization of the course into a series of steps. Responsibility for progress devolves upon the individual student. The object of all programed teaching is to let the learner teach himself "privately," at his own pace, without fear of being "trapped" by a human teacher's questions. At the same time, he is compelled to master every step in the program actively; he is at all times directly involved, whereas in an ordinary school class the pupil gets only a limited amount of opportunity to demonstrate his knowledge or grasp of the subject matter actively; for most of the time he has to be content with a passive listening role.

(more)

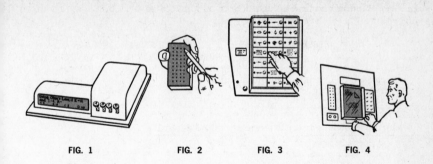

FIG. 1   FIG. 2   FIG. 3   FIG. 4

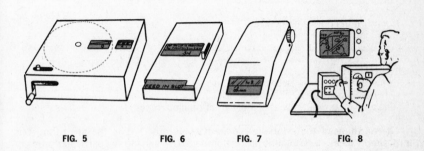

FIG. 5   FIG. 6   FIG. 7   FIG. 8

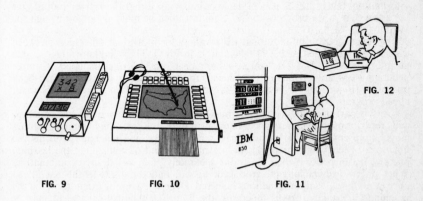

FIG. 9   FIG. 10   FIG. 11

FIG. 12

In the teaching program, the subject matter must be so organized that the learner can easily follow the flow of instruction, as represented by a flow diagram or an algorithm:

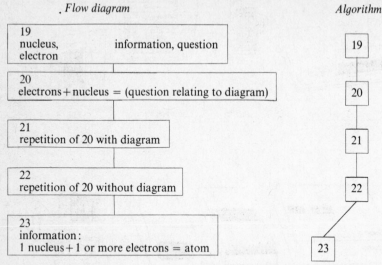

(The above example is part of a program on electronics.)

The meaning of the symbolic representation embodied in the algorithm (this term simply means a systematic process or method of solving any particular problem in a number of steps conforming to well-defined rules) is that the machine presents the learner with a series of "learning elements" (LE): LE 19 to 21 give and repeat information, but in LE 22 it is of decisive importance that the learner indeed give the answer; if so, he proceeds to LE 23; if not, then he must again go through the routine from LE 19 to 22.

The incorporation of additional LE to allow for individual differences between learners is exemplified by LE 119, 94 and 95 in Fig. 1. These are known as "simple program steps." In Fig. 2 a "simple subsequence" to LE 101 is included. If simple program steps are frequently introduced, the procedure is called a "wash-back program" (Fig. 3), and if a system of subsequences is adopted, it is called a "wash-ahead program" (Fig. 4).

Fig. 5 is part of an algorithm that makes use of explanatory steps, subsequences and repetitions ("complex wash-back program"). Repetitions, jumps, detours may be presented to the learner. In general, the teaching machine performs the functions of teaching (imparting the information) and of controlling the learning process. It can perform these functions separately or in any desired combination. When programed teaching is introduced, it is not a good thing to let the machine at once take charge of all the functions involved. Depending on the extent to which the machine undertakes various functions, the following types of equipment may be distinguished:

(more)

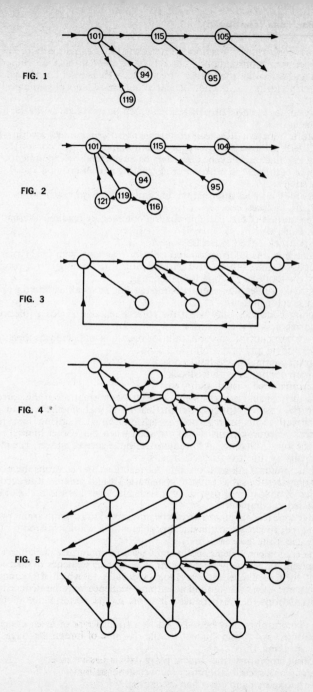

FIG. 1

FIG. 2

FIG. 3

FIG. 4

FIG. 5

1. The machine performs the function of instructing (giving information). The other functions (repetition, confirmation of correctness of answers, etc.) are embodied in the program and are under the learner's free control. He can set his own pace.

2. The machine performs the function of instructing and reduces physiological strain in various ways:

(a) elimination of the need to turn pages, reduction or optimization of the number of eye movements, etc.;

(b) sensomotoric function (the learner turns knobs); sometimes several sensory paths are utilized: acoustic plus visual presentation (push-button control); instruction given by the machine may in some cases be augmented by experiments.

3. The machine performs the function of instructing and of giving a visual presentation of the subject:

(a) a tape recorder supplies information, in conjunction with visual presentation of tables, diagrams, pictures, etc.;

(b) visual presentation of text (i.e., information obtained by reading) alternates with pictures, diagrams, etc.;

(c) taped sound track linked to a slide projector.

4. The machine presents information and calls for reactions from the learner in the form of a reply or some other active participation; the running of the program is stopped until the required reaction is forthcoming:

(a) the program continues after every reaction (writing or speaking into the machine, or pressing a selector button);

(b) the program continues only when the correct reaction is given (the machine checks the correctness of the reaction).

5. The machine presents information, calls for reactions and confirms their correctness (or otherwise) in any of the following ways:

(a) the program continues, without feedback;

(b) the program continues, with feedback;

(c) a signal is presented (visual, audio, etc.).

5(a) controls the learning process without explicit confirmation of the correctness of replies. In the case of 5(b) and 5(c) the learner is told whether or not he has answered correctly; this is regarded as advantageous in giving him positive encouragement. Encouragement is also provided when the learner himself is made to compare the answer that he has formulated with the correct answer. The program may achieve this by direct or indirect means.

6. The machine presents information, calls for reactions and confirms their correctness; it also prepares the subject matter systematically and presents it in accordance with the rules of educational psychology; such machines are more elaborate and are to be classed as computers:

(a) a computer-type teaching machine can be programed to present variant problems of the same type; it can set arithmetical problems to a whole classroom of pupils and speedily check the answers;

(b) electronic equipment is being increasingly used in organizing the subject matter itself and presenting it in a systematic form; the correct sequence of instruction is mapped out in accordance with psychological rules to which the computer is adjusted; in some cases the optimum learning sequence may be deduced by the computer itself from the "experience" it gains in the performance of teaching programs.

7. Language laboratories may be regarded as a special type of teaching machines whose main object is to train students in the oral use of foreign languages. There are, broadly speaking, three types:

(a) the listening laboratory (the student plays only a passive role);

(b) the listening and speech laboratory (active participation);

(c) the listening, speech and recording laboratory.

(more)

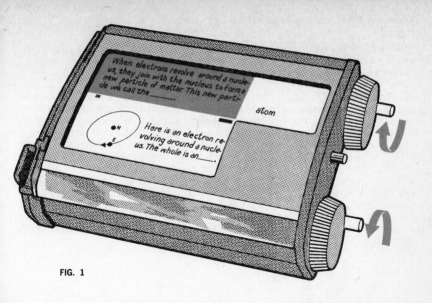

When electrons revolve around a nucleus, they join with the nucleus to form a
new particle of matter. This new particle we call the _____.

Here is an electron re-
volving around a nucle-
us. The whole is an _____.

atom

FIG. 1

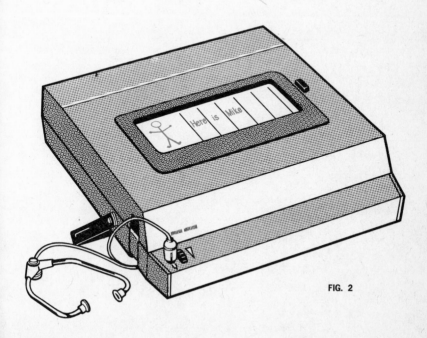

Here is Mike

FIG. 2

Programed instruction by machines aims to give the student immediate confirmation of the correctness of his answers to the questions and problems presented to him and thus to reinforce the acquisition of each step by knowledge of progress. However, the machine should not take complete control of the teaching process to the exclusion of all initiative on the learner's part. He should always be allowed a certain amount of independence of decision or choice—i.e., a controlled degree of freedom. The teaching machine has a psychological effect in stimulating the learner's interest: he feels a desire to manipulate it and is thus stimulated to learn.

The illustrations on pp. 571 and 573 show some teaching machines which have been developed in Germany. Fig. 1 (p. 571) shows the "Promentaboy," a machine which presents written information in simple easy-to-understand form, whereby much of the physiological strain is taken out of the reading-and-learning process. The second machine is the "Probiton," developed by Zielinski and Schöler, which gives audiovisual instruction (Figs. 1–3, p. 573): spoken and read text, pictures and sound illustrations together provide the optimum learning conditions. Finally, Correll's teaching machine (Fig. 2, p. 571) is an example of an audiovisual device for teaching very young children to read.

FIG. 1
MACHINE READER
FOR TEACHING

headphone socket

a   b   c

a) volume control
b) forward run
   and selector keys
c) synchronizing
   buttons

FIG. 2
MACHINE WITH COVER
REMOVED, SHOWING
PROGRAM STRIP

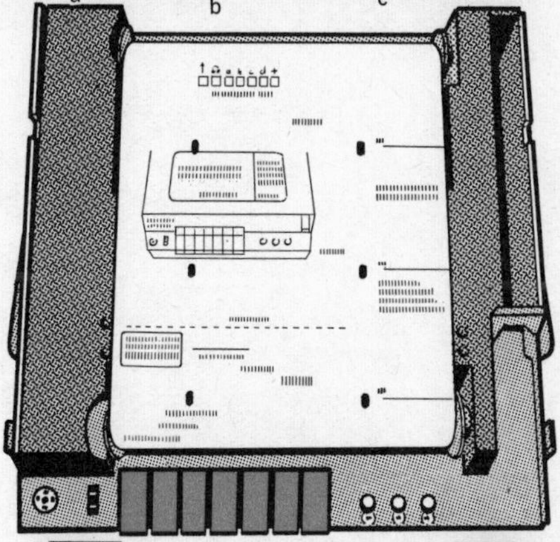

FIG. 3
MACHINE WITH COVER
REMOVED AND NO
PROGRAM STRIP

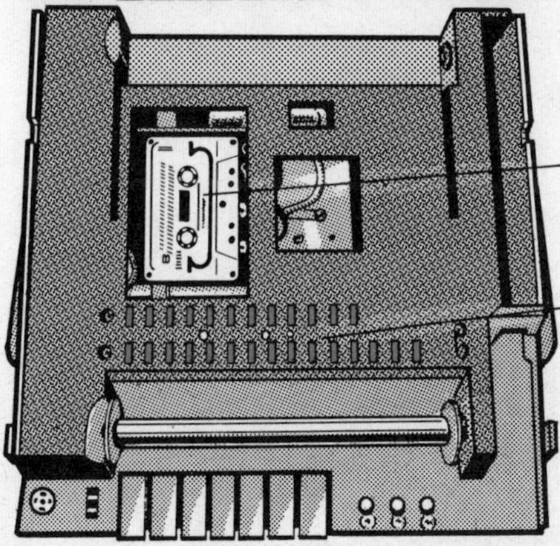

tape cassette

contact strips

INDEX

[References to illustrations are in italics.]

Ablating reentry shield, 362, 370
Ablation cooling, 362, 364
Abrasive blasting, 102
Absolute pressure, 490
Absorbers, 48, *49. See also* Shock absorber
Absorption tower, 50, *51*
Acceleration lanes, 250, 252
Acceleration pickup, 482
Accelerator pedal, 256, *257*
Accelerometer, 374, *375*
Access shafts, 64
Accordion, 452, *453*
Accumulators, 340
Acid cooler, 50, *51*
Acme screw thread, 156, *157*
Acoustic effect, 456
Acoustic insulation, 52
Acoustic mines (explosives), 234, *235*
Action (piano), 448
Activated antiroll tanks, 360, *361*
Active acoustic torpedo, 238
Active stabilizer, 358, 360, *361*
Active charcoal, 406
Addendum circle, 164, *165*
Addition circuits, 292
Additive mixing, 464
Additive principle, 288
Adiabatic compression, 344
Adiabatic demagnetizing, 50
Advanced gas-cooled nuclear reactor, 30, 32, *33*
Advancing longwall working, 66
Advancing system, 66
Aerial, 526
Aerofoil, 330, *331*, 332, *333*, 334, *335*
Aerosols, 20, *21*
Aerothermodynamic duct engine, 344, *345*
Aerotopography, 60
Agitators, 56
Ailerons, 330, 332, *333*, 340
Aiming, 222
Aiming devices, 222, *223*, 224, *225*
Air-circulating equipment, 404, *405*
Air compressor, 204
Air-conditioning
  in incubators, 404, *405*
  in mines, 72
  in oxygen tents, 404, *405*
  in space suits, 372
Air-cooled engine, 320, *321*
Air crossings, 72
Air-cushion bearings, 402, *403*
Air cylinders, 442, *443*
Air drier, 338
Air-extraction shafts, 72
Air-flask section (torpedo), 236, *237*

Air locks, 72
Air-release valve, 54, *55*
Air-relief valves, 152, *153*
Air turbine drill, 402, *403*
Aircraft construction, 328, 329
Aircraft hydraulic power systems, 340, *341*, 342, *343*
Airfields, 346
Airflow sheets, 72, *73*
Airfoil, 330, *331*, 332, *333*, 334, *335*
Albis variable filter, 460, *461*
Alcohols, 42
Al-Fin process, 114
Algorithm, 568
Alkazide method, 40, *41*
Allwall artificial kidney, *423*, 424
Alternating current, 322, 326
Alternating current machines, 510, *511*, 512, *513*, 514, *515*
Alternating electric field, 398, 414
Altitude tinting, 299
Alum, 48
Alumina, 92
Aluminum
  in Al-Fin process, 114
  alloys of, 328
  casting of, 98, *101*
  dies of, 108
  in galvanizing, 144
  processing of, 92, *93*
  in space suits, 372
  thread-rolling in, 160
  ultrasonic soldering of, 138
Aluminum chloride, 144
Aluminum oxide, 40
  abrasive of, 128
  in thermite welding, 136
  in thermite pressure welding, 132
Aluminum potassium sulphate, 48
Aluminum powder fuel, 364
Amatol, 226
America's cup, 348
American screw thread, 156
Ammonia
  catalytic oxidation of, 50
  gas mask for, 406
  as refrigerating agent, 68
Ammonia process, 50, 51
Ammonia synthesis, 40, *41*, 44
Ammoniacal copper chloride solution, 40
Ammonium nitrate, 364
Ammonium perchlorate, 364
Ammonium sulphate fertilizers, 48
Ammunition, 226, *227*, 228, *229*
Ampex video tape recorder, 560, 562, *563*
Amphibious tanks, 240, 242, *243*
Amplification, 400, 468, 554
Amplifier, 450, 454, *455*

Amplifier system, 472, *473*
Amplifier tube, 532, *533*
Amplitude modulation, 538
Anaerobic fermentation, 430, 438, 440
Analogue-digital converter, 390
Anchor bolt, 76, *77*
Angle of friction, 156, 157
Angular contact bearings, 171, *172*
Annealins, 114
Annular bit, 62
Annular measuring value, 154, *155*
Annular tube, 398, *399*, 490
Annulus, 206, *207*
Anode current, 366
Antenna, 526
Antiaircraft gun carriers, 240, *241*, 244
Antiaircraft guns, 218, *219*, 222, 224, *225*
Antiaircraft shell, 230
Anticathode, 398, *399*
Anticoagulins, 422
Antifriction bearings, 170, *171*, 172, *173*, 174, *175*, 176, *177*
Antiknock gasoline, 268
Antilogarithm, 484
Antimony, 94
Antiparallel linkage, 200, *201*
Antipersonnel mines, 234
Antiroll tanks, 358, *359*, 360, *361*
Antisubmarine rocket, 236, *237*
Antitank mines, 234
Anvil, 118, *119*
Applied geophysics, 60
Apollo spacecraft, 368, 370, 372, 382–383, *384–385*
Aqualung, 442, *443*
Arbor, 186
Arc furnace, 90, *91*
Arc pressure welding, 130
Arc welding, 130, 134, *135*
Archimedean screw, 422
Argonarc process, 136
Armature, *332*, 324, *325*, 506, *507*, 510
Arming device, 234, *235*
Armor-piercing shell, 228, *229*, 230
Armored machine-gun carriers, 240, *241*
Armoring (tank), 242
Arresting mechanism, 214, *215*, 216, *217*
Arsenic, 94
Articulated tank vehicle, 54
Artificial heart, 408, *409*, 410, *411*, 412, *413*, 414, *415*, 416, *417*
Artificial heart valves, 408, 410, *411*
Artificial kidney, 418, *419*, 420, *421*, 422, *423*, 424, *425*

Artillery, 218, *219*
  self-propelled, 240, *241*
Asphalt oil fuel, 364
Asroc, 236, *237*
Assault tank, 240, *241*
Astronavigation, 374
Asynchronous motor, 512
Athodyd engine, 344, *345*
Atlas intercontinental ballistic
  rocket, 368, *369*
Atmospheric humidity, 404
Atmospheric pressure, 490
Atomic bomb, 22
Atomic-hydrogen welding, 136,
  137
Atomic-powered artificial heart,
  414
Atomic reactors. *See* nuclear
  reactors
Atomizer, 404, *405*
Audiovisual instruction, *571*, 572
Auditory nerve, 400, *401*
Auditory ossicles, 400, *401*
Auditory receptor cells, 400
Auger, 440, *441*
Autoclave, 92
Automated packaging, 426, *427*,
  428, *429*
Automatic choke, 256
Automatic clutch, 196, *197*
Automatic forging machine, 118
Automatic lathes, 174
Automatic waterer, 432
Automatic welding, 130
Automation in mining, 76, *77*
Automobile, 256, *257*, 258, *259*
Automobile generator, 322, *323*,
  324, *325*, 326, *327*
Autotransformer starter, 514
Auxiliary hammers, 118, *119*
Auxiliary lift aircraft, 346
Auxiliary radiator, 320, *321*
Auxiliary tanks, *353*, 354
Axial-cylinder pump, 340
Axial-flow pumps, 362
Azides, 52
Azo compounds, 52

Backfilling, 74
Background noise, 472
Backward extrusion, 122, *123*
Bacterial filter, 404, *405*
Bag-type artificial heart, 416, *417*
Balance wheel, 216
Ball bearings, 170, *171*, 172, 173,
  402, *403*
Ball catch, 446, *447*
Ball joint, 190, *191*
Ball valve, 408, 410, *411*
Ballast tank, *353*, 354
Ballistic trajectory, 226, *227*
Ballistics, 224
Balls, 170, *171*, 172, *173*
Band-pass filter, 460, *461*
Band saws, 126
Bar magnet, *502*
Barite, 48
Barrel-shaped rollers, 172
Base fuse, 230

Bass strings, 448
Bastard file, 126
Battery, 326
Battery compartment, 236
Battery-powered electric motors,
  354
Bauxite, 92, *93*
Bayer process, 92, 93
Beading, 140, *141*
Beam-and-crank mechanism, 202,
  203
Beam-splitting device, 290
Bearings
  air-cushion, 402, *403*
  applications, 176, *177*
  casting, *115*, 116
  for high-speed turbine drill,
    402, *403*
  manufacture, 174, *175*
  materials, 174
  production, 116
Beat-frequency oscillator, 458,
  459
Beater shaft, 438
Beef cattle, 432
Bell-crank lever, 448, *449*
Bell die, 130
Bellows, 452, *453*
Belt conveyors, 64, 76, 88
Belt drive, 180, *181*, 210
Belt feeder, *81*
Belt pulleys, 438
Benardos's arc-welding process,
  134, *135*
Bench shears, 140
Bending, 56, 140
Bending molding, 56
Bessemer continuous casting, 98,
  99
Betatron, 398, *399*
Bevel gear, 162, *163*
Bevel-gear planing machine, 168,
  169
Bevel-gear transmission, 206, *207*
Bicycle chain drive, 210
Bielliptical orbit, 378, *379*
Bilge keel, 358, *359*
Billet, 96, 98, *99*, 100, 101
Binding for skis, 444, *445*, 446,
  447
Bipropellant system, 362
Black phosphorus, 46
Blanking, 140
Blasting techniques, 76
Blast furnace, 84, *85*
Blowpipe soldering, 138, *139*
Blown extrusion, 58
Blown flap, 334, *335*
Boiling-water nuclear reactor, 27,
  28, *29*
Boltzmann constant, 22
Bone-conduction receiver, 400
Booster coil, 24, *25*
Booster rocket, 368
Borax, 138
Bore of engines (cubic capacity),
  264, *265*, 282, 304, *305*
Bore-safe fuses, 250
Bored foundation, 524, *525*
Borehole, 60, 62

Boring mills, 174
Bottom-pouring casting, 96, *97*
Boundary layer, 330, 332, *333*
Bourdon-tube pressure gauge,
  490, *491*
Bourrelet, 226, *227*
Bowman's capsule, 418, *419*
Box molding, 102, *103*
Brake mean effective pressure,
  262, 266
Brakes, 258, *259*
Branching, 566
Brass, 110, 138, 174
Brass sheet, 174
Brazing, 138, *139*
Breathing bag, 442
Breech block, 218, 220, *221*
Breeder nuclear reactor, 38, *39*
Bremsstrahlung, 22, 23, 24, *25*
Bridge (piano), 448
Bridge (roads), 248
Bridge (submarine), 352, *353*
Bridge-laying tank, 240, *243*
Briede–de Lavaud process, 112,
  113
Briggs subject matter trainer, 566,
  567
Briggsian logarithm, 484
Brine, 68
Broaching, 126, *127*
Bronze bearings, 116
Building industry, foam plastics
  in, 52
Build-up welding, 130, 306
Bubble chamber, 476, *477*
Bulkheads, 352
Bullion (lead), 84
Bumping table, 78
Burnout, 368
Bushes, 326
Bushing, 176
  casting of, 112, 114
Butane, 20, 42, *43*
Butt welds, 134
Butterfly valve, 154, *155*
Buttress screw thread, 156, *157*

Cable sheathing extrusion, 124
Cadastral maps, 296
Cadmium, 90
Cage, 70, *71*, 170, 174, 206, 207
Cage guides, 70
Cage winding, 70
Calcine, 86
Calcining, 94
Calcining kiln, *93*
Calcium phosphate, 46
Calder Hall nuclear reactor, 30,
  31
Calendering rolls, 428
Caliber, 228
Call timer, 556, *557*
Cam lever, 488, *489*
Cam mechanism, 198, *199*, 212,
  213
Camber, 334
Camshaft, 276, *277*, 283, *284*, *285*,
  286, *287*, 312, *313*
Canister (gas mask), 407

Cannon. *See* Guns
Capacitive gauge, 492
Capacitive resistance, 530
Capacitors, 458, 528, 529
Capsule-type pressure gauge, 490, *491*
Carbon, 46
Carbon brush, 322, *323*
Carbon dioxide, 26, 30, 32, 52
Carbon monoxide, 40, 44, 406
  from coal-dust explosions, 74
  conversion of, 42, 44, *45*
Carburetor, 256, 272, *273*, 308, *309*
Cardan gears, 488, *489*
Cardan joint, 190, *191*
Cardan shaft, 256, *259*
Cargo storage, 358
Carriage, 178, *179*
Carrier frequency, 534, *535*
Carrier transmission, 534, *535*, 536, *537*
Carrier wave, 294, 390
Cartography, 296–303
Cartridge case, 218, 226, *227*
Cartridge case extraction system, 220
Cascade converter, 516, 517
Case-hardening steel, 174
Casing (submarine), 354
Cast iron
  bearing cages of, 174
  dies of, 108
  molds of, 112
  pipe casting of, 112
Cast steel, 68, 242
Castable propellant, 364
Casting, 56, 96, *97*, 328
Casting molds, 148
Catalyst, 40, 42, 50
Catch lock, 214, *215*
Caterpillar mold, 100, *101*
Cathode follerer, 532, *533*
Cattle bar, 432, *433*, 434, *435*
Caustic soda, 442
Caving, 64, 66
Cavity resonator, 526
Celestial equator, 396
Celestial guidance system, 374
Cellophane, 420
Celluloid molds, 148
Cellulose tetranitrate, 420
Cement-sand molding, 102
Cement slurry, 68
Cementation, 68
Cemented carbide, 126, 128
Central traffic signals, 250, 252
Center lathe, 178, *179*
Center punch, 128
Centerless grinding, 128, *129*
Centrifugal casting, 100, 112, *113*, 114
Centrifugal clutch, 196, *197*
Centrifugal governor, 544, *545*
Centrifugal impeller mixer, 500
Centrifugal pumps, 362, 434
Ceramic pickup, 472
Cermets, 116
Cesium vapor, 366
Chain conveyor, 76

Changeover valves, 152, 442, *443*
Change-speed gearbox, 206, *207*, 242, *243*
Channeling, 534
Characteristic of logarithm, 484
Charge, 218
Charts, 296
Chaser, 158
Chassis, 242, *243*
Check valve, 154, *155*
Chemical consolidation, 102
Chemical foaming agents, 52, 53
Chemical reaction welding, 130
Chile saltpeter, 50
Chimney effect, 72, *73*
Chiseling, 126, *127*
Chloride salts, 94
Chlorides, 94
Chlorinated hydrocarbons, 20
Chlorine gas, 94
Choke, 256, *257*
Chrominance signal, 290, 292, 294
Chromium, 150
Chromium oxide, 40, 42, 44
Chromium steel, 122, 174
Chuck, 178
Churning, 52, *53*
CIE chromaticity diagram, 288, *289*
Circlips, 176
Circuit testing, 556, *557*
Circular milling, 128
Circular pitch, 164, *165*
Circular saws, 126, 127
Circulatory system, human, 408, *409*, 410, 412
Clamping sleeves, 402
Classifying ores, 78
Clavichord, 448
Claw clutch, 194, *195*
Clipping, 140
Climate-control system, 404, 405
Clock making, 106
Clockwork drives, 216
Clockwork fuse, 230
Closed-circuit SCUBA system, 442
Closed-circuit wind tunnel, 336, 337
Closed die forging, 120, *121*
Closed-jet wind tunnel, 336
Closed molds, 102
Clotting, 422
Cloud chamber, 476, *477*
Cloud track, 476, *477*
Cloverleaf intersection, 254, *255*
Clutch, 188, 194, *195*, 196, *197*, 318, *319*
Clutch-and-brake steering, 242, *243*
Clutch pedal, 256, *257*
Coal cutters, 76
Coal dust explosions, 74
Coal mining, 65, 66
Coating, 58, 134, 146, 147, 148, *149*, 150, *151*
Coating knife, 58, *59*
Coaxial cable, 554, *555*
Cocking mechanism, 216, *217*
Coil-spring clutch, 318, *319*

Coil-spring coupling, 192, *193*
Coiling, 140, *141*
Coke, 84
Cold-chamber pressure die casting, 110, *111*
Cold drive system, 366
Cold extrusion, 122, *123*
Cold-forming processes, 118
Cold pressing, 116, *117*
Cold pressure welding, 130, 132
Collectors, 82
Collision trajectory, 380
Collodion, 420
Color coder, 291
Color television, 288, *289*, 290, *291*, 292, *293*, 294, *295*
Color television camera, 294
Color triangle, 288, *289*
Combination fuse, 230
Combination shell, 230
Combined harvester, 438
Combustion chamber, 268, 270, *271*, 306, *362*, *363*, 364, *365*
Comets, 376, 392
Command module, 383, *384–385*
Comminuter, 434
Comminuting effect, 498
Comminution, 78, *81*, 89, 116
Communication engineering, 456
Communications satellites, 386, 390, *391*
Commutator, 322, *323*, 326, 506, 510
Commutator motor, 514
Compatible color television system, 291
Complementary colors, 288, *289*
Composite barrel, 220, 221
Composite casting, 112, 114, *115*
Composite solid fuel, 364
Compound generator, 508
Compound motor, 506
Compound slide, 178
Compressed air, 354, 402
Compressed-air cylinders, 442, *443*
Compression deformation, 118
Compression foundation, 524, *525*
Compression molding, 56, *57*
Compression ratio, 266, 268, 270, 306
Compression ring, 280, 281
Compression stroke, 260
Concentrate, *80*, 84, 86, 88, 94
Conditioning, 566
Conductor suspension clamp, *523*, 524
Conductors, 522, *523*
Cone clutch, 194, 195
Cone-and-friction-ring variable speed drive, 182, *183*
Cone-pulley transmission system, 180, *181*
Cone type speed-control friction drive, 208, *209*
Conic sections, 376, 378
Conical pulleys, 182, *183*
Connecting arm, 198, *199*
Connecting rod, 204, 260, *261*

Connector valve, 342, *343*
Conning tower, 352, *353*
Consumer packaging, 426, 428
Contact-electrode welding, 132
Contact horns, 234, *235*
Contact mines, 234, 235
Contact pressure, 208
Contact process, 48, *49*
Contact reactor, 50, 51
Continuous casting, 98, *99*, 100, *101*
Continuous conveying systems, 76
Contour lines, *298–300*
Control gyroscope, 358, 360
Control valves, 340
Convergence adjustment, 292
Convergent-divergent nozzle, 338, *339*
Conversion process, 40
Converter, 86, *87*, 516, *517*
Converter tower, 48, *49*
Conveying systems, 76
Conveyor, 76
Conveyor rollers, 100
Coolant in nuclear reactors
  carbon dioxide gas as, 26, 30, 32
  heavy water as, 26, 28
  helium as, 32, 34
  sodium as, 36, 38
  water as, 26, 28
Coordinated control system, 250
Copper, 86, *87*
  continuous casting of, 98
  dies of, 108
  solder of, 138
  ore, 82, 94
  production of, 86, *87*
Copper pyrites, 84
Copper-tungsten production, 116
Copper-zinc alloy, 138
Cordite propellant, 222
Core, 60, 62, 102, 103
Core barrel, 62
Cork Harbour Water Club, 348
Corona, 392
Corti's organ, 400
Cosine oscillations, 468
Cosmic rays, 372
Cottrell dust, 46
Cottrell-type electrostatic dust precipitator, 46, *47*
Cotton wool, 52
Counterboring, 128, *129*
Countercoil mechanism, 220, *221*
Countercurrent classifier, 78, 80, *81*
Countercurrent principle, 422
Counterflow principle, 500
Countersinking, 128
Counterweight, 70
Counting devices, 214, 216
Coupler, 198, *199*, 200, *201*, 546
Coupling, 188, *189*, 190, *191*, 192, *193*
Covered-arc welding, 134, 136, *137*
Cow shed, 432, *433*, 434, *435*
Crank, 200, *201*
Crank mechanism, 198, *199*, 200, 212, 316

Crankshaft, 204, 260, *261*, 284, 304, *305*, 318
Crawler-mounted power drills, 76
Crawler tracks, 242, 244, 245
Cristofori, Bartolommeo, 448
Croning's shell molding process, 104, *105*
Cross-peen hammer, 118
Cross-shaped bar assembly, 204, *205*
Cross slide, 198
Cross stringing, 448, *449*
Cross talk, 534, 552
Crossed-axis helical gears, 162, *163*
Crude oil, 42
Cruising Club of America, 350
Crusher cylinder, 218, *219*
Crushing and screening plant, 78, *81*
Cryolite, 92
Crystal pickup, 470, *471*, 472
Cubic capacity, 264, *264*
Cumberland Fleet, 348
Cupping, 140, *141*
Cupric oxide, 406
Curb alignment, 250
Curing, 102, 104, 364
Current coil, *325*, 326
Current supply (telecommunications systems), 558, *559*
Cursor, 486, *487*
Cutter bar, 440, *441*
Cutter blade, 438
Cutter-loading machine, 76
Cutting discs, 126
Cutting knife, 440
Cutting torch, 98
Cyanide copper baths, 150
Cycle (electric), 322
Cylinder, 260, 261
  capacity of, 262
  charging of, 274
  head of, 310, *311*
  -pressure diagram, 260, *263*
Cylinder-type forage harvester, 440
Cylindrical-piston meter, 494, *495*
Cylindrical roller bearings, *171*, 172

Dairy cows, 432–35
Damper, piano, 448, *449*
Damping action
  in coupling, 192, *193*
  drive shaft and, 256
Damping moment of a ship, 358
Dashpot, 154, *155*
DC generator, 506, *507*, 508
Dead roasting, 94
Dead smooth file, 126
Debismuthizing, 84, *85*
Deceleration (navigation), 374
Deceleration lanes, 250, 252
Decibels, 456
Dedendum circle, 164, *165*
Deep drawing, 142, *143*
Deep-drawing steel strip, 174

Deep drill holes, 62
Deep-drilling engineering, 62, *63*
Deflecting roller, 100
Deformability, 118, 124
Delayed action fuse, 230
Delta-wing aircraft, 330, *331*
Demand regulator, 442, *443*
Demodulation, 534, 538, 562
Dentistry, 402
Depressors, 82
Depth control, 238
Depth indicator, 70
Depth-regulating mechanism, 236
Desilvering, 84, *85*
Detergents, phosphorus in, 46
Detonating cap, *231*, 232, *233*
Detonation, *267*, 268
Detonator, 234, *235*
Deuterium, 22, *23*, 24
Deuterium oxide (heavy water), 22, 26
Deuterons, *23*
Dial mechanism (fluid measurement), 496, *497*
Dialing (telephones), 548
Dialyser, 420
Dialysis, 422
Diametral clearance, 176
Diametral pitch, 164
Diamond, 126, 402, *403*
Diaphragm-type pressure gauge, 490, *491*
Diaphragm valve, 152, *153*
Diastole, 410
Dichroic mirror, 294
Die, 58, 108, *109*, 110, *111*, 116, 117, *117*, 120, *124*, *125*, 130
Die casting, 108, *109*, 110, *111*
Die plate, 158
Die shaping, 58
Dielectric, 528, *529*
Dielectric constant, 528
Diesel engine, 242, 278, 354
Diesel fuel, transportation of, 54
Differential construction, 328
Differential gauge, 490
Differential gear, 256, 258, *259*
Differential mechanism, 166
Differential piston, 152, *153*
Differential-pressure piston, 340
Diffuser, 338
Diffusion, 420
Diffusion cloud chamber, 476
Dip brazing, 138, *139*
"Dip" of ore deposit, 60, *61*, 64
Direct circuit, 28
Direct circuit boiling-water reactor, *27*, 28
Direct current, 558
Direct-current generator, 322, *323*, 324, 326
Direct-current machines, 506, *507*, 508, 509
Direct drive, 206
Direct extrusion, 124, *125*
Direction indicators, 258
Director system, 224, *225*
Disc brakes, 258, *259*
Disc clutch, 194, *195*
Disc-shaped single cutter, 160, *161*

Disc-type flexible coupling, 192, *193*
Disc-type speed-control friction drive, 208, *209*
Dispatch packaging, 426
Dispenser, aerosol, 20, *21*
Displacement (submarine), 352
Displacement currents, 526, 528, *529*
Displacement pump, 340
Displacement-type meters, 494
Distilling furnace, 88, *89*
Distortion factor, 468, *469*, 470
Distributor, 256
Division (slide rule), 486, *487*
Diurnal motion, 394
Dividing head, 166
Divining rod, 60
Docking (spacecraft), 380, 382, 383, *384–385*
Doctor knife, 58, *59*
Double-acting clutch, 195, *196*
Double-acting pump, 434, *435*
Double-action accordion, 452
Double-cut file, 126
Double-diaphragm valve, 152, *153*
Double helical gear, 162, *163*
Double-helical mixer, *499*, 500
Double-row bearing, 170, *171*, 172, *173*
Double-thrust bearing, 170
Draft (mining), 72
Drafting machine, 200, *201*
Drag, 224, 330, 332, 334
Drag-link mechanism, 200, *201*
Dragon class (yachts), 350
Draw-key transmission, 206, *207*
Drawing, 56
Drawing process, 56
Dressing, 80
Dressing of ores, 78–83
Drill bit, 402
Drill rod, 62, *63*
Drilling, 76, 128, *129*
Drilling derrick, 62, *63*
Drilling mud, 62, 68
Drilling platform 62, *63*
Drip-feed oil lubrication, 176
Drive pin, 216
Drive shaft, 182, *183*, 190, 256 *259*
Driving band, 218, 226, *227*
Driving cone, 182
Driving-key transmission system, 184
Drop arm, 258, *259*
Drop forging, 116, 120
Drop keel, 348, *349*
Drop shaft method, 68, *69*
Drop-weight brake, 70
Drop-weight device, 216, *217*
Dross, 84
Drum brakes, 258
Drum separator, 82, *83*
Drum winding, 70
Dry-cell battery, 400
Dry galvanizing, 144, *145*
Dry-sand molding, 102
Dry-sump lubrication, 280, *281*

Ducted fan, 346
Ductility, 118
Dunlop process, 52, *53*
Duration, 456
Dust mask, 406, *407*
Dusting, 74, *75*
Dwight-Lloyd sintering machine, 94, *95*
Dyes, nitric acid in manufacture of, 50
Dynamic balancing, 318
Dynamic pressure, 330
Dynamic similarity, 336
Dynamo, 322, *323*, 324, *325*, 326, *327*, 506, *507*

Ear trumpet, 400
Eardrum, 400, *401*
Early Bird communications satellite, 386
Earth-orbiting satellite, 370
Earth satellites, 338
Earth wire, 524
Eccentric crank linkage, 204, *205*
Echo communication satellite, 386
Echo effect, 462, *463*
Eclipses, 392
Ecliptic, 392, 394
Economy-type forage harvester, 438, *439*
Eddy, 332, *333*
Efferent arteriole, 418, *419*
Eiffel wind tunnel, 336, *337*
Eight-cylinder car, 264
Electric current, 322, *323*
Electric dipole, 526
Electric filters, 48, *49*, 454, *455*
Electric furnace, 46, *47*
Electric power, nuclear reactor output of, 26, 28, 30, 32, 34
Electric pressure gauge, 492, *493*
Electric zinc furnace, 90, *91*
Electrical amplification, 400
Electrical-contact materials, 116
Electrical exploration methods, 60, *61*
Electrical hearing aid, 400, *401*
Electrical measurement technology, 456
Electrical resistance soldering, 138 *139*
Electrical-resistance welding, 130
Electrically propelled torpedo, 238
Electroacoustics, 456
Electrocleaning, 150
Electrodeposition, 148, *149*
Electroforming, 148, *149*
Electrogalvanizing, 144, 150, *151*
Electrohydraulic drive, 222, *223*
Electrolysis, 86, *87*, 88, *89*, 90
Electrolytes, 420
Electrolytic cell, 92, 93
Electrolytic refining cell, 84, 86, *87*
Electrolytic tin-plating process, 150, *151*
Electromagnet, 324, 342, 454, 502, *503*, 504, *505*

Electromagnetic artificial heart, 413, 414
Electromagnetic drum separator, 82, *83*
Electromagnetic pickup, 470
Electromagnetic waves, 526
Electromechanical generator, 460
Electromechanical musical instruments, 456
Electron, 398, 478
Electron accelerator, 398
Electron beam, 388, 398
Electron gun, 366, 392
Electron-volt, 478
Electronic circuits, 530, *531*, 532, *535*
Electronic filters, 456, 460, *461*
Electronic music, 456, *457*, 458, *459*, 460, *461*, 462, *463*, 464, *465*, 466, *467*
Electronic music studio, 456, *457*
Electronic organ, 454, *455*
Electric-resistance strain gauge, 480, *481*, 482, *483*
Electronic switchgear, 550
Electronic teaching equipment, 570, 572, *573*
Electrophonic musical instruments, 456, 460
Electroplating, 150, *151*
Electroslag process, 134, 135
Electrostatic dust precipitators, 46 *47*
Electrostatic fields, 366
Electrostatic precipitator, 46, *47* 48, *49*
Elevating, 222
Elevation (aiming), 244, *247*
Elevator, 332, *333*, 340
Elevator grip brake, 214, *217*
Elevators (airplane), 334, *335*
Elliptic trammel, 204, *205*
Elliptical orbits, 376, 378
Embossing, 58, 140, *141*
Embossing dies, 148
Enamel (tooth), 402, *403*
End-burning rocket, 364, *365*
End play, 176
"Endless" products, 58
Energy
  nuclear fusion release of, 22, *23*, 24, *25*
  release in stars, 22
Engine, 256, *259*
  cubic capacity (bore) of, 282, 304, *305*
  cylinders in, 282, *283*
  efficiency of, 274, *275*, 276, *277*, 280, *281*
  fan of, 280, 281
  flexibility of, 266, 276, 310, 312
  friction in, 280, 282
  gas flow, 308, *309*
  life of, 304
  mechanical efficiency of, 280, *281*
  oscillation system in, 274, *275*
  performance of, 260, *261*, 262, *263*
  power loss in, 280
  power output of, 266

resonance in, 274, 276
speed of, 282, *283*, 284, *285*, 286, *287*, 314, 316, 318
tuning of, 304, *305*, 306, *307*, 308, *309*, 310, *311*, 312, *313*, 314, *315*, 316, *317*, 318, *319*, 320, *321*
turbulence in, 268
Engine lathe, 128, *179*
English Standard screw thread, 156
Ensilage, 430, *431*
Epsom salts, 48
Equal temperamental tuning, 450
Equatorial bulge, 396
Equilibrium, human, 400
Equinox, 396
Erard, Sebastian, 448
Escape velocity, 383
Escape wheel, 216, *217*
Escapement, 216, *217*
Escapement action, 441, *449*
Etching technique, 328
Evaporator, 50, *51*
Evolution, 414
Exact tuning, 450
Excitation circuit, 510
Exciting winding, 324, *325*, 326, *327*
Exhaust air, 72
Exhaust-driven turbosuper-charger, 278
Exhaust duct, 272, *273*
Exhaust manifold, 312
Exhaust pipe, 312
Exhaust stroke, 260
Exhaust valve, 268, 270, *271*, 276, 310, *311*
Expendable mold, 102, 106, *107*
Experimental aerodynamics, 336
Exploder mechanism, 238
Explosion welding, 130, 132, *133*
Explosions
of coal dust, 74
in mines, 74
Explosive
nitric acid in manufacture of, 50
sulphuric acid use in, 48
Explosive bolts, 368
Explosive filling, 226
Explosive powder, 218
Expressways, 248
External gear teeth, 162, *163*
External gem cutting, 168, *169*
External thread cutting, 158, *159*
Extracorporeal heart, 414
Extruders, 56, 428
Extruding, 328
Extrusion, 57, 58, *59*, 116, 122–125
cold, 122–*123*
hot, 124, *125*

Face of tooth, 164, *165*
Facepiece, 406, *407*
Facing, 128
Fail-safe principle, 328
Fairing, 332, *335*

Fans, 72, 73, 336, *337*
Fast breed nuclear reactor, 38
Feed, 184
Feed hopper, *81*
Feed shaft, 178, 184, *185*
Feed troughs, 432
Feedback, 532, *533*
Feedback oscillator, 454, *455*
Felt washers, 176
Felt wool, 52
Female-mold process, 56, *59*, 426, *427*
Fences (on airplane wings), 332, *333*
Fermentation, 430, 438, 440
Ferromagnetic ores, 82, 502
Ferrophosphorus, 46
Ferrosoferric oxide, 82
Fertilizers, 50
Fiberboard, 56
Field gun, 222
Field windings, 506
Fighter aircraft, 346
File disks, 174
Filing, 126, *127*
Filler metal, 130, 134, 138
Filters, 116, 530, *531*
Fin-stabilized projectile, 228, *229*
Fin stabilizing system, 360, *361*
Finger board (accordion), 452, 453
Finger pump artificial heart, 414, *415*
Finn Dinghy (yacht class), 350
Fire-central device, 222, *223*
Fire ladder, 210, *211*
Fire refining, 86
Fireclay, 108
Firecracker welding, 136, *137*
Firedamp, 74
Firing mechanism, 234, *235*
Firing pin, 218, 220
Firing pressure, 260, 306
Firing table, 224
Fit of bearings, 176
5.5 m class (yacht), 350
Fixed ammunition, 226
Fixed keel, 348, *349*
Fixed pulley, 210, *211*
Flanged coupling, 188, *189*
Flanging, 140, *141*
Flank of tooth, 164, *165*
Flap valve, 154
Flaps, *333*, 334, *335*, 340, 342, *343*
Flash, 120, *121*, 174
Flash roasting, 94, *95*
Flat color reproduction, 299–303
Flat riser aircraft, 346
Flat-trajectory guns, 218, *219*
Flexible coupling, 188, 192, *193*
Float lever, 54, *55*
Float-type pressure gauge, 492, *493*
Floating-center coupling, 188, *189*
Flotation, *80*, 82, *83*, 86, 88
Flotation agents, 82
Flow-control valves, 340
Flow forming, 142, *143*

Flow measurement, 494, *495*, 496, *497*
Flow mixer, 500
Flue dusts, 88
Fluid drive, 182, *183*
Fluidized bed, 94, *95*
Fluidized bed kiln, 48
Fluidized-solids reactor, 94, *95*
Fluosolids-roasting furnace, 94, *95*
Flux, 134, 136, 138, 144, *145*, 502
Flying Dutchman, 350
Flyovers, 248
Flywheel, 318, *319*
Flywheel-type forage harvester, 440
Foam plastics, 52, *53*
Foam rubber, 52
Foaming agent, 52, *53*, 104
Folding, 140
Follower, 212, *213*
Foot brake, 258
Forage, 430
Forage harvester, 438, *439*, 440, *441*
Fording tanks, 240, 242, *243*
Forge welding, 130
Forging, 118–121
Forging dies, 120, *121*
Forging furnace, 118
Forging hammers, 118, *119*
Forging press, 118
Forging roller, 118, *121*
Forging tools, 118, *119*
Formaldehyde condensation products, 52
Formed cutter, 166, *167*
Forward extrusion, 122, *123*
Four-bar linkage, 200, *201*, 202, 204
Four-color process, 299
Four-cylinder car, 264
Four-cylinder radial-type engine, 238, *239*
Four-stroke engine, 260, 262
Free phase system, 454
Free-reed accordion, 452, *453*
Freewheeling clutch, 196, *197*
Freezing process, 68, *69*
Frequency, 510
Frequency amplifiers, 292
Frequency converter, 460, *461*
Frequency dividers, 454
Frequency-division multiplexing, 542, *543*
Frequency modulation, 538, *539*
Frequency response, 468
Fretz-Moon process, 130
Friction, 156, 332, 370
Friction brake, 214, *217*
Friction clutch, 194, *195*, 242, 438
Friction-drive mechanisms, 208, *209*
Friction-gear mechanism, 198
Friction linings, 196
Friction plates, 196
Friction ring, 182, *183*
Friction welding, 132, *133*
Frictional heating, 74, 126, 370
Front-axle assembly, 258, *259*

Frother, *80*, 82
Fuel elements, 28, 30, 32, 34, 38
Fuel-injection system, 274, *275*, 278, *279*
Fuel tanks, *353*, 354
Full-track tank, 240
Fuming sulphuric acid, 48, *49*
Fundamental tone, 458
Fuse, 226, *227*, 228, 229, 230, *231*, 232, *233*, 234
Fuse bond, 146
Fuselage, 328
Fusion, 22
Fusion reactors, 22
Fusion welding, 130, 134–137

GALCIT propellant, 364
Galena, 84, 88, 90
Galvanic contact potential differences, 60
Galvanizing, 144, *145*
Gangue, 66, 82, 86
Gas cap, 74, *75*
Gas-cooled nuclear reactor, 30, *31*
Gas cooler, 50, *51*
Gas flow, 310
Gas generator, 362
Gas mask, 406, *407*
Gas preheater, 50, *51*
Gas pressure, 362
Gas pressure-operated hydraulic accumulators, 340
Gas pressure welding, 130, *131*
Gas propellant, 20
Gas separator, 54, *55*, 496, *497*
Gas-shielded-arc welding, 134, 136
Gas-tight diaphragm, 338, *339*
Gas turbine engine, 346
Gas washer, 40
Gas welding, 130, 134, *135*
Gauge factor, 480
Gauss, 504
Gear casting, 112
Gear cutting, 166, *167*, 168, *169*
Gear drive, 180, *181*
Gear hydraulic motor, 340
Gear lever, 256, *257*
Gear mechanism, 198, *199*, 206, *207*
Gear pump, 340
Gear ratio, 164, 206
Gear teeth, 162, *163*, 164, *165*, 198
Gear tooth grinding machine, 168, *169*
Gear wheel, 162
Gearboxes, 176, 180, *181*, 206, 256, *259*
Geared clutch, 194, *195*
Gears, 162, *163*, 164, *165*, 206, *207*
Gemini spacecraft, 370, 380, *381*
Generator, 322, *323*, 324, *325*, 326, *327*, 454, *455*, 456
Geophysical exploration, 60, *61*
Geophysics, applied, 60
Glaser bubble chamber, 476, *477*
Glass fiber, 370
Glass wool, 52

Glauber's salt, 48
Globoid, 206
Glomerulus, 418, *419*
Glow tube, 458
Gold, 150, 372
Grade-separated intersection, 241, 250, *251*, *253*, 254, *255*
Grand piano, 448, *449*
Graphite
    for dies, 108
    for electric conductivity, 148
    for lubrication, 124
    for molds, 98
    as nuclear reactor moderator, 30, 32, 36
Gravimetric exploration, 60, *61*
Gravity die casting, 108, *109*
Gravity mixer, *499*, 500
Grease grooves, 176
Grease lubrication, 176
Green fodder, 430
Green-sand molding, 102
Grinding, 102, 128, *129*, 150, 168, *169*, 174
Grinding wheels, 128
Grip brake, 214, *217*
Grip-roller clutch, 196, *197*
Grip-roller locking device, 214, *215*
Grit blasting, 144
Ground mines, 234
Ground tracking station, 386, *387*, 388, 390, *391*
Group casting, 96, *97*
Grout, 68
Gudgeon bearings, 358
Gudgeon pin, 316, *317*
Guide rollers, *99*, 100
Gun barrel, 218, *219*
Gun barrel casting, 112
Gun chamber, 218
Gun mount, 222
Guns, 218, *219*, 220, *221*, 222, *223*, 224, *225*
Gutta-percha molds, 148
Gypsum, 48
Gypsum plaster, 104, 108, 116, 148
Gyratory crusher, *81*
Gyroscope, 236, 238, *239*, 246, *247*, 358, 359, 374, 375
Gyrostabilizer, 358, *359*

Haber-Bosch process, 40, *41*
Half-track tank, 240
Hammer forging, 118, *119*, 120, *121*
Hammer (piano), 448, *449*, 450
Hand brake, 256, 258
Hand forging tools, 118, 119
Hand width, 390
Handicap, 350
Hardinge countercurrent classifier, 78, *81*
Hard-metal tool production, 116
Hard soldering, 138, *139*
Hard surfacing, 130
Harmonic series, 450
Harmonic vibrations, 450, 454

Harpsichord, 448
Haulage, 76, *77*
Haulage roads, 64, *65*
Haymaking machine, 436, *437*
Hayracks, 432
Hazelett process, 100, *101*
Head prism, 356, *357*
Headframe, 70
Headings, 64
Headstock, 178
Hearing aids, 400, *401*
Heart, 408, 409
    See also Artificial heart
Heart-lung machine, 410, 414
Heart transplant, 410
Heart valves, 408, 410, *411*
Hearth furnace, 84
Heat exchanger, 48, *49*, 356
Heat-sealed packaging, 428, *429*
Heating jacket, 56
Heavy water
    as nuclear reactor coolant, 26, 28
    as nuclear reactor moderator, 26, 28
Heavy water (deuterium oxide), 22, 26
Heliarc process, 136
Helical gear, 162, *163*
Helical gear cutting, 166, 168
Helical spring, 60, *61*
Helical teeth cutting, 166, *167*
Helicopter, 346
Helium, 362, 372
    as nuclear reactor coolant, 32, 34
Helix, 156, *157*
Hemispherical combustion chamber, 284
Hemodialysis, *419*, 420
Heparin, 412, 420
Herreshoff furnace, 94, *95*
Herringbone gear, 162, *163*
Heterogeneous solid fuel, 364
High-angle guns, 218, *219*
High-carbon steel, 128
High explosive shell, 226, *227*, 228, *229*
High fidelity, 468, *469*, 470, *471*, 472, 473, *473*, 474, *475*, 476, *477*
High-intensity magnetic separators, 82
High-lift truck, 210, *211*
High-pass filter, 460, *461*, 530, *531*, 536
High-precision bearings, 402
High-speed steels, 128
High-speed subsonic wind tunnel, 336, *337*, 338
High-speed turbine drill, 402, *403*
High temperature nuclear reactor, 32, 34, *35*
Hirudin, 420, 422
Hobbing process, 166, *167*
Hohmann transfer orbit, 378, *379*
Hoistways, 70, *71*
Holding furnace, 110
Homing equipment, 236
Homing torpedo, 238

Homogeneity, in mixing, 498
Hooke's law, 480
Hopcalite, 406
Horizontal centrifugal casting, 112
Horizontal cold-chamber pressure die casting, 110, *111*
Horizontal continuous casting, 98, *99*
Hot-chamber pressure die casting, 110, *111*
Hot-dip galvanizing, 144, *145*
Hot extrusion, 124, *125*
Hot-forming process, 118–121
Hot sawing machines, 126
Hot sealing, 428
Hovering, 346, 352
Howitzer-carrying tank, 240, *241*
Howitzers, 218, 244
Hue, definition of, 288
Hull (tank), 242, *243*
Human waste regeneration, 372, *373*
Hydraulic accumulators, 340
Hydraulic adjusting cylinder, 338, *339*
Hydraulic agitators, *433*, 434
Hydraulic braking cylinders, 220
Hydraulic chock, 76, *77*
Hydraulic cylinder, 340, *341*
Hydraulic elevating gear, 222, *223*
Hydraulic fluid, 222, 342
Hydraulic motor, 340
Hydraulic power systems, 340, *341*, 342, *343*
Hydraulic power transmission, 340
Hydraulic presses, 124
Hydraulic pump, 340
Hydrocarbons, manufacture of, 42
Hydrochloric acid, for pickling, 144
Hydroforming, 142
Hydrogen bomb, 22
Hydrogen peroxide, 362
Hydrogen sulphide, 40
Hydromechanical separation, 80
Hydrometallurgical processes, 86
Hydrophilic ores, 82
Hydrophobic ores, 82
Hydroplanes, 352, *353*, 354
Hydropneumatic suspension system, 244
Hydrostatic arming device, 234, *235*
Hydrous ferrous sulphate, 48
Hyperbolic orbits, 376, 378
Hypersonic wind tunnels 336, 338
Hypoid gear, 162
Hypsometric coloring, 299

Ignition coil, 256, 320
Ignition switch, 256, 257
Ignition system, 320
Illuminating shell, 228, *229*
Image orthicon tube, 290, 294

Immersion composite casting, 114, *115*
Impact breaker, *80*
Impact extrusion, 122
Impact fuse, 230, *231*, 232, *233*
Impact molding, 56
Impedance, 530
Impeller blades, 440, *441*, 498
Impeller casting, 112
Imperial Smelting process, 90, *91*
Impermeability, 426
Impinging spray fuel injection, 362
In-the-ear hearing aid, 400, *401*
Incubator, 404, *405*
Indexing (gear cutting), 166, 168
Indicator diagram, 260, *263*
Indirect circuit, 28
Induced drag, 330
Induction, 322, 323, 502, 520
Induction heating, 118
Induction motor, 512
Induction soldering, 138, *139*
Induction welding, 132, *133*
Inductive resistance, 530
Inert-gas metal-arc process, 136, *137*
Inert-gas tungsten-arc process, 136, 137
Inertial navigation systems, 356, 374, *375*
Infinitely variable speed control, 208, *209*
Influence mines, 234, *235*
Infrared devices, 246
Ingot casting, 96, *97*
Ingots, 96, *97*
Inhaling tube, 442
Injection fuel system (rockets), 362, *363*
Injection molding, 56, *57*
Injection nozzle, 278, *279*
Inlet pipe, 308
Inlet valve, 276, 310, *311*
Inlet-valve guide, 310, *311*
Inner ear, 400, *401*
Insert hearing aid, 400
Insulation of cables, 552
Insulators, 522, *523*
Intake slide, 260
Integral construction, 328, *329*
Integrated semiconductor circuit, 400
Integration (navigation), 374
Intercontinental missile, 374
Interference drag, 332
Intermittently operated tunnel, 338, *339*
Internal broach, 126, *127*
Internal-combustion engine, 204
  performance of, 260, *261*, 262, *263*
  in tanks, 242
  valve central mechanism in, 198
  valve gear of, 212, *213*
Internal-expanding shoe clutch, 194, *195*
Internal friction, 192
Internal gear cutting, 168, *169*
Internal gear teeth, 162, *163*

Internal thread cutting, 158, *159*
Internal thread milling, 160, *161*
Internally toothed annulus, 206, 207
International Yacht Racing Union, 350
Interplanetary space flight, 366, 370, 372
Interrupted projection, *296*–*297*
Intersections, 248, *249*, 250, *251*, 252, *253*, 254, *255*
Inverted extrusion, 124, *125*
Inverter, 558, *559*
Investment molding, 106, *107*
Involute curve, *163*, 164
Involute tooth, 164, *165*
Involution, 484
Ion-drive rocket, 366, *367*, 378
Ionization, 22, 24, 398, 476
Ionization chamber, 366, *367*
Ions, 476, 478, *479*
Iron
  continuous casting of, 98, 100
  galvanizing of, 144, *145*
Iron ores, 82
Iron oxide, 40, 560
Iron oxide catalyst, 44
Iron pyrites, 48, *49*, 84
Iron vitriol, 48
Ironstone mining, 76
Irrigated electrostatic precipitator, 48, *49*
Islands, traffic, 250, 252
Isobutane, in aerosol spray dispensers, 20
Isostatic pressing, 116

Jaw coupling, 188, *189*
Jet flap, 334
Jig, 78
Junctions, 248, *249*, 250, *251*, 252, *253*, 254, *255*

Kaufman system (rocket propulsion), 366, *367*
Kepler's laws of planetary motion, 376
Kerosene, 362, 368
"Key fossils," 60
Keyboard, 448, 452, *453*, 454, *455*
Kick maneuvers, 378
Kidney, 418, *419*, 420
  See also Artificial Kidney
Kidney transplant, 418
Kneader, 56, 500, *501*
Knee-type horizontal milling machine, 186, *187*
Knocking, *267*, 268
Knuckle screw thread, 156, *157*
Koepe winding system, 70, *71*

Labyrinth washers, 176
Lacquers, nitric acid in manufacture of, 50
Lactic acid, 430
Laminar airflow, 332, *333*

Laminated plastics, 56
Laminated sheeting, 428, *429*
Laminates, 56, 428
Lamination, 58
Land-based guns, 218, *219*
Land mines, 234
Landing flap, 334, *335*
Landing gear, 342, *343*
Language laboratory, 570
Lap winding, 506, *507*
Latex, 52
Lathe, 158, 166, 178, *179*, 180, *181*, 198
Lattice steel towers, 521, *522*
Launch window, 378
Launching frame for torpedos, 236, *237*
Launching rocket, 386, *387*
Layer tinting, 299
Layout of intersections, 250, *251* 252, *253*, 254, *255*
Lazy tongs, 202
Lead, *85*
Lead bronze in bearings, 114
Lead-chamber process, 48
Lead ore, 82, 84, 94
Lead pipe extrusion, 124
Lead processing, 90, *91*
Lead screw, 158, 178
Leading, 88, *89*, 90
Leaf springs, 258
Leather seals, 176
Left-turning traffic, 250, 252
Lettering of maps, 299, 302, *303*
Letterpress, 299
Levels, 64, *65*
Lever, 200, *201*
Lever-and-ratchet mechanism, 200, *201*
Lever system (piano), 448, *449*
Lift, 330, 332, 334
Lift-and-force pump, 434, *435*
Lifting tackle, 210
Lincoln milling machine, 186
Linear accelerator, 478, *479*
Linkage, 200, *201*, 202, *203*
Liquid manure, *433*, 434, *435*
Liquid propellant rocket engines, 362, *363*
Lithography, 299
Livestock fattening, 432
Loader mining machines, 76, *77*
Loaders, 76, *77*
Loading valve, 342, *343*
Local variometer, 60
Locking mechanisms, 214, *215*, 216, *217*
Loco-hauled trains, 64
Locomotives in mines, 76
Logarithm, 484, *485*
Long thread milling, 160, *161*
Longitudinal turning, 126, 128
Longwall coal mining, 76
Longwall working, *65*, 66, *67*
Loose-housing system (cowshed), 432
Lost-wax casting process, 106, *107*
Loudness, 450, 456
Loudspeakers, 456, 472, 474, *475*

Low-frequency generator, 454
Low-pass filter, 460, *461*, 530, *531*, 536
Low-speed wind tunnel, 336, *337* 338
Lubrication, 176
Luminance signal, 291, 292, 294
Luna 3 moon probe, 388
Lunar module, 382, 384–385
Lunar Orbiter program, 374, 388
Lunar probe, 378
Lunik, 386

Mach number, 330, 336
Machine guns, 244
Machining, 56, 126, 166, *167*
McPherson suspension unit, 258
Macromixing, 498
Macromolecules, 398
Magnesium alloy, 328
Magnetic bottle, 24, *25*
Magnetic clutch, 194, *195*
Magnetic dipole, 526
Magnetic field, 22, 23, 322, 323, 398, *399*, 502, *503*
Magnetic induction, 502
Magnetic mines, 234, *235*
Magnetic mirrors, 24
Magnetic permeability, 502, *505*
Magnetic pumping, 24
Magnetic recorders, 456
Magnetic separation, 82, *83*
Magnetic tape recorder, 462
Magnetization, 502, 503
Magneto bearings, 172
Magnetometric method, 60
Magnetron, 366
Main bearings, 283, *284*
Major road, 248
Male-mold process, 56, 58, *59*, 426, *427*
Malleability, 118
Maltese cross mechanism, 214
Mammoth pumps, 68
Mandrel, 120, *121*, 124, *125*
Manganese dioxide, 406
Manganese ores, 82
Manganese-silicon steel, 174
Manifold, 272
Manometer, 154
Mantissa, 484, 486
Manual choke, 256
Manual haulage, 76
Manual welding, 130
Manure disposal, 432, *433*, 434, *435*
Manure gutter, 432
Map graticule, 298
Map legend, 303
Map making, 296–303
Marforming, 142, *143*
Mariner space probe, 386
Mascher's integral-casting process, 114, *115*
Maser amplifier, 386, 390

Matchboxes, 46
Matte, 86, *87*
Mean chord, 330, *331*
Mean effective pressure, 260, *263*, 266, *267*, 308, *309*
Measuring bridge, 482, *483*
Mechanical agitation, *433*, 434
Mechanical efficiency, 262
Mechanical fuse, 230
Mechanical power transmission, 198, *199*
Medical betatron, 398, *399*
Mercast process, 106
Mercury, 106
Mercury-float gauge, 492, *493*
Mercury space capsule, 372, *373*
Meshing gears, 206, *207*
Mesopause, 370
Message accounting, 556, *557*
Message register, 556
Metal fatigue, 328
Metal spraying, 144, 146, *147*
Metal-tower silo, 430, *431*
Metalliferous ore preparation, 78–83
Metallizing, 146, *147*
Metering, 54, *55*
Meteorites, 372, 392
Methane, 74, *75*
Methanol synthesis, 42
Methyl alcohol, 42
Metric screw thread, 156
Micromixing, 498
Microphone, 400, *401*
Microwaves, 386
Middle ear, 400, *401*
Milage counter, 214, 216, 218
Military aircraft, 330, 342
Milking room, 432
Milky Way, 392, 394
Milling, 128, 129, 160, *161*
Milling cutter, 212, *213*
Milling machines, 128, 166, 186 *187*, 198
Mine cars, 76
Mine gas, 74, *75*
Mine ventilation, 72, *73*
Mineral-dressing process, 78–83
Mineral oil, 342
Mineral prospecting, 60, 61
Minerals, 60
Mines (explosives), 234, *235*
Minimum-energy transfer orbit, 378
Mining, 64, 76
    automation in, 76, *77*
    mechanization in, 76, *77*
Minor road, 248
Minuteman missile, 364
Missile, 338
Missile-launching submarine, 356
Miter gears, 162
Mixers, 498, *499*, 500, *501*
Mixing drums, 56
Mixing milk, 498
Mixing (music), 464
Mixing (physical materials), 498, *499*, 500, *501*
Mobile rocket launcher, 240, *241*

Moderator in nuclear reactors
 graphite as, 30, 32, 36
 heavy water as, 26, 28
 water as, 26
 zirconium hydride as, 36
Modulating wave, 294
Modulation, 456, 534, 538, 544, 562
Modulator, 458, 460
Modulus of elasticity, 480
Moeller artificial kidney, 422, *423*
Molding, 56, *57*, 96, *97*, 98, *99*, 100, *101*, 102, *103*, 104, *105*, 106, *107*, 108, 148, *149*
Molding box, 102, *103*, 104, *105*
Molding sand, 104
Molten bead process, 428
Molten metal soldering, 138, *139*
Molten-slag bath, 134, *135*
Molten zinc bath, 144, *145*
Moltopren, 52
Molybdenum, 116, 362
Monaural pickup, 472
Monaural reproduction, 466
Monobloc barrel, 220, *221*
Monochrome television receiver, 291
Moon landing, 368, 372, 382–383, *384–385*
Moored mines, 234, *235*
Moore's process, 112, *113*
Mortar projectile, 228, *229*
Mortars (guns), 218
Motion-picture equipment, 214
Motor-converter, 516, *517*
Motor-generator, 516, 518, *519*, 558, 559
Motorcycle chain drive, 210
Mouthpiece, 442, *443*
Mower, 436
Muff coupling, 188, *189*
Muffler, 312
Multiblade grab, 68
Multifuel tanks, 242
Multilevel intersection, 248, 250 *251*, *253*, 254, *255*
Multiplaten presses, 56
Multiple-carburetor assemblies, 272, 308, 309
Multiple-disc clutch, 180, 194, *195*
Multiple-hearth furnaces, 86, 88, 94, *95*
Multiple-immersion casting, 96, *97*
Multiple pantograph, 202, *203*
Multiple refining system, 86
Multiple-spindle automatic lathe, 174
Multiple thread, screw, 156
Multiple-twin formation, 552, *553*
Multiplex principle, 454
Multiplex signal 538, *539*, 540, *541*
Multiplexing, 542, *543*
Multiplication (slide rule), 486, *487*
Multiplicative mixing, 460
Multiribbed cutter, 160

Multistage axial-flow compressor, 338
Multistage rocket, 368, *369*
Multiunit registration, 556
Multivibrator, 458, *459*
Music, *see* Electronic music and *specific musical instruments*
"Musique concrete," 466
Muzzle brake, 220, *223*
Muzzle velocity, 218, 228

N-nitroso compounds, 52
Naperian logarithms, 484
Naphthalene, 42
Natural frequency, 316, 358
Natural gas, 42, *43*
 drilling for, 62
 exploration for, 60, *61*
Natural-potential exploration, 60, *61*
Natural rubber, 52
Natural scale, 450
Nautilus, U.S.S., 354
Naval guns, 218, *219*, 222, *223*
Naval mines, 234, *235*
Navigation, 374, *375*
Nebulae, 394
Needle roller bearings. *171*. 172
Negative feedback, 532
Negative-mold process, 56, *59*
Nephrons, 418, *419*
Neutral buoyancy, 352
New Jersey process, 90, *91*
New York Yacht Club, 348
Nichols-Freeman flash roaster, 94, *95*
Nickel ore, 94
Nickel-silver, solder of, 136
Nimbus weather satellite, 386
Ninth harmonic, 450
Nitrates, 50
Nitric acid, 50, 51
Nitriding bath, 318
Nitrogen, 362, 372
 in aerosol spray dispensers, 20
 in foam plastic preparation, 52
Nodes, 450
Noise
 in electronic music, 456
 maser amplifier and, 386
Noise generator, 458
Nonferrous metals, 96
Nonmagnetic steel, 174
Nonreturn valve, 154, 340, *341*. 342, *343*
Nonswept short-span wing, 330, *331*
Norton gearbox, 184, *185*
Nose fuse, 230, *231*
Notation (electronic music), 466, *467*
Notch effect, 310, 318
NTSC system, 290–294
Nuclear energy production, 22
Nuclear fusion, 22 *23*, 24, *25*
 temperatures in, 24
 See also Nuclear reactors *and specific types of reactors*
Nuclear-fusion processes, 22, *23*

Nuclear-powered submarine, 354
Nuclear reactors, 26–39
 advanced gas-cooled type of, 30, 32, *33*
 breeder type of, 38, 39
 boiling-water type of, *27*, 28, *29*
 Calder Hall type of, 30, *31*
 coolant in, 26, 28, 30, 32, 34, 36, 38
 electric power output of, 26, 28, 30, 32, 34
 fast breed type of, 38
 gas-cooled type of, 30, *31*
 high-temperature type of, 32, 34, *35*
 moderator in 26, 28, 30, 32, 36
 pressure-tube type of, 26, *27*
 pressurized-water type of, 26, *27*
 pyrolytically precipitated carbon in, 32
 sodium type of. 36, *37*
 superheated-steam type of, 28, *29*
 thermal breeder type of, 38
 thermal output of, 26, 28, 30, 34
 thermal type of, 26–29
 thorium-uranium cycle in, 38
 uranium-plutonium cycle in, 38

Ocean bed mapping, 299, *299*
Octane filter, 460
Oersted, 504
Offshore drilling, 62. *63*
Ogive, 226, *227*
Ohm's law, 530
Oil, exploration for, 60, *61*
Oil circulation lubrication, 176
Oil cooler, 320
Oil flow, 320, *321*
Oil-level lubrication, 176
Oil lubrication, 176
Oil mist lubrication, 176, 402
Oil scraper ring, 280, *281*
Oil-splash lubrication, 176
Oleum, 48, *49*
Olson's video tape recorder, 560
Open-arc welding, 134
Open circuit SCUBA system, 442
Open-circuit wind tunnel, 336, *337*
Open-die forging, 120
Open-jet wind tunnel, 336, *337*
Open molds, 102
Open orbits, 378
Open sand molding, 102
Orbits, 376, *377*, 378, *379*, 380, *381*
Ores, preparation of, 78–83
Organ, 454, *455*
Oscillating circuit. 530, *531*
Oscillating crank mechanism, 204, *205*, 488
Oscillating-piston principle, 54
Oscillating pump artificial heart, 414, *415*
Oscillation, 454, *455*, 456, 468, *469*, 528, 530

Oscillation tube, 272, *273*, 278
Oscillographs, 60, 218
Osmosis, 420
Outer ear, 400, *401*
Oval gear meter, 54, *55*
Oval-runner meter, 494, *495*, 496, *497*
Oval window (in ear), 400
Overboring, 304
Overhead camshafts, *283*, 284, *285*, 286, *287*, 315
Overhead transmission lines, 520, *521*, 522, *523*, 524, *525*
Overpasses, 248
Over-stringing, 448, *449*
Overtones, 450, 454
Overvoltage, 524
Overwind-prevention device, 70
Oxidation tower, 50, 51
Oxide ceramic materials, 126
Oxides, 78
Oxyacetylene welding, 134, *135*
Oxygen, 372
  liquids, 362
Oxygen tent, 404, *405*

Packaging
  aerosol pressurized, 20, *21*
  technology of, 426, *427*, 428, *429*
Packing material, 52
Padding, 52
Paddle agitator, *433*, 434
Paddle mixer, *499*, 500
PAL system, 290, 294
Paleography, 60
Pallets, 216
Pantograph, 202, *203*, 488, *489*
Parabolic antennas, 388, 390
Parabolic orbits, 376, 378
Parallel linkage, 200, *201*, 202, *203*
Parallel oscillating circuit, 530, *531*
Paramagnetic ores, 82
Parking bake, 258
Parking orbit, 380, 382
Partial roasting, 94
Particle accelerator, 478, *479*
Passenger lift grip brake, 214, *217*
Passive acoustic torpedo, 238
Passive stabilizer, 358, *359*
Pastel colors, 288, *289*
Pattern, 102
Pawl, 198, *199*, 214, *215*, 216, *217*
Pelletizing, 46, *47*
Pelton wheel, 402
Pendulum, 216
Pentane, 52
Petroleum refining, 48
Percussion drilling, 62
Periflex coupling, 192, *193*
Period (electricity), 322
Periodic-reverse process, 150
Periodic time, 376, 380, 390
Periscope, 246, 356, *357*
Permanent-magnet alloys, 116
Permanent mold, 102
Permeability, 426
Permittivity, 528

Personnel carrier, 240, *241*
Petrography, 60
Petrol, 42
Petroleum drilling, 62
Phase alternation line, 294
Phase displacement, 468, *469*
Phase response, 468
Phenol, 42
Phenolic plastic, 56
Phonograph record masters, 148
Phosphate, 122, 420
Phosphor dots, 292, *293*
Phosphoric acid, 46
Phosphorus, 46, *47*
Photoelectric sound converter, 462
Photogeology, 60
Photographic film, 388
Photomodular, 462, *463*
Piano tuning, 450, *451*
Pianoforte, 448, *449*
Pickling, 150, *151*
Pickup, 60
Pickup instruments, 60, *61*
Picture scanner, 462
Piercing, 140
Piezoelectric artificial heart, 414
Piezoelectric effect, 218, 470
Piezoelectric gauge, 492
Piezoelectric stereo pickup, 470, *471*
Piezoelectric transducer, 416
Piggyback multistage rocket, 368, *369*
Pile foundation, 524, *525*
Pillaring, 64
Pillaring method of mining, 66
Pilot frequency method, 538
Pin joints, 200, 204
"Pinch effect," 24, *25*, 472
Pinion, 162, 206
Pinion-cutter process, 168, *169*
Pinking, *267*, 268
Pipe casting, 112
Pipe manufacture, 132, *133*
Piston, 198, 260, *261*, 276
Piston and cylinder, 204
Piston ring, 280, 281
  casting of, 112
Piston stroke, 304, *305*
Piston-type pressure gauge, 490, *491*
Piston valve, 340, *341*
Pit molding, 102, *103*
Pitch (musical), 450, 456
Pitch circle of gears, 164, *165*
Pitch point, 164
Pitching (of ships), 358
PIV drive, 182, *183*
PIV gear, 210
Pivots, 204
Plain milling machine, 186
Plain wing flap, 334, *335*
Planet wheel, 206, *207*
Planet-wheel motor, 494, *495*
Planetarium, 392, *393*, 394, *395*, 396, *397*
Planetary friction drive, 208, *209*
Planetary gear systems, 206, *207*, 212

Planetary orbits, 276
Planing, 126, *127*
Plasma, 22, *23*, 24, *25*
Plaster molding, 104
Plastic
  in ablating reentry shield, 370
  for bearing cages, 174
  for gears, 162
  for molds, 148
  nitric acid in manufacture of, 50
  as packaging, 426, 428, *429*
  in precision casting, 106
  processing of, 56, *57*, 58, *59*
Plasticizers, 46
Platinum-nickel catalysts, 42
Plate clutch, 194, *195*
Platinum catalyst, 50, *51*
Plumbicon tubes, 294
Plunger hydraulic motor, 340
Plunger pump, 340, *341*, 434
Plutonium, 38, 239
Pneumatic recuperator, 220, *221*
Pneumatic agitation, 433, *434*
Pneumatic bearings, 402, *403*
Pocket hearing aid, 400, *401*
Point fuse, 230, *231*
Poison gas, 406
Polaris missile, 356, 364
Polaroid film, 388
Polybutadiene fuel, 364
Polymer plastic fuel, 364
Polymerization, 364
Polymers, 56
Polystyrene, 52, *53*
Polyurethane foam plastics, 52
Polyurethane fuels, 364
Polyvinyl chloride, 52, *53*
Pony haulage, 76
Porter's machine, 566, *567*
Porting, 310
Positive-action mixer, 499, 500
Positive drive, 210
Positive infinitely variable drive, 182, *183*
Positive infinitely variable gear, 210
Positive-mold process, 56, 58, *59*
Potassium, 40
Potassium nitrate, 48, 50
Potassium oxide, 40
Potassium perchlorate, 364
Potential wall, 22, *23*
Potentiometer, 374, *375*
Pouring gate, 102, *103*, 106
Pouring gate system, 108, *109*
Pouring ladle, *113*
Powder metallurgy, 116, *117*
Powder-train fuse, 230
Power manure loader, 432, *433*
Power-operated valves, 152–155
Power output, 262, 264
Powerline route survey, 520, 521
Prandtl wind tunnel, 336, *337*
Precessional motion, 396
Precision casting, 106, *107*
Precision tuning, 208
Premature babies, 404
Preparation of ores, 78–83
Preprinted lettering, 299

Pressey's machine, 566, *567*
Pressure casting, 112, *113*
Pressure compensator, 430, *431*
Pressure die casting, 110, *111*
Pressure drag, 330
Pressure gauge, 442, 490, *491*, 492, *493*
Pressure hull, 352, *353*
Pressure line, 164
Pressure measurement, 490, *491*, 492, *493*
Pressure mine, 234, *235*
Pressure-reducing valve, 340, 442
Pressure rollers, 130, *131*, 132, *133*
Pressure sintering process, 90
Pressure thread rolling, 160, *161*
Pressure-tube nuclear reactor, 26, *27*
Pressure vessel, 26
Pressure welding, 130, *131*
Pressurized-water nuclear reactor, 26, *27*
Prestressed concrete, 30
Primary colors, 288, *289*
Primary tone, 458
Primary water system, 356
Primer, 218, 226, *229*
Printing plates, 299
Process color reproduction, 299
Producer gas, 40
Programmed orbit, 386
Projectiles, 218, 226, *227*, 228, *229*
Projection (maps), 297
Projection welding, *131*, 132
Projector (planetarium), 392, *393*, 394, *395*, 396, *397*
Propane, 20
Propellant charge, 226, *227*
Propellant gases, 20, 220, 222
Propellant pumps, 362, *363*
Propellant systems (rocket engine), 362, *363*, 364, *365*, 366, *367*
Propeller, 236
Propeller casting, 112
Propeller mixer, 500
Propeller pumps, 434, *435*
Propeller shafts, 190
Propeller VTOL aircraft, 346
Prospecting, 60, *61*
Protective face masks, 406, *407*
Protein metabolism, 418
Protons, 476, 478
Proximity fuse, 230, *231*
Pulley drive systems, 210, *211*
Pulley mechanism, 198, *199*
Pulp, 82
Pump, 340
Pump rotor, 182, *183*
Punchboard, 566, *567*
Punching, 140
Purity adjustment, 292
Push-button dialing, 544, *545*
Pylons, *521*, 522
Pyrotechnics, 228, *229*

Quadrature modulation, 294
Quadric-crank mechanism, 200, *201*

Quads, 552
Quantum-mechanical tunnel effect, 22
Quartz crystal, 218
Quartz fiber, 370
Quenching effect, 108
Quick-action air-relief valves, 152, *153*
Quills, 448

Rabble arms, 94, *95*
Races, 170, 172
Racing engine, 276, *277*, 282, 286
Rack, 162, *163*, 164, *165*
Rack-and-pinion gearing, 162, *163*, 164, *165*
Rack-cutter process, 168, *169*
Radar, 246
Radial bearings, 170, *171*, *172*
Radial-cylinder pump, 340
Radiation dose, 398
Radiator, 320, *321*
Radio receiver tuning mechanism, 208, *209*
Radiotelephonic communication, 390
Radius vector, 376
Railway wheel casting, 112
Raking forks, 436
Ram extrusion, 58, *59*
Ramjet engine, 344, *345*
Ramp generator, 458, *459*
Ranger space probe, 386, 388
Raschig rings, 48
Ratchet mechanism, 198, *199*, 214, *215*, 216, *217*
Ratchet-and-pawl counting device, 556, *557*
Ratchet wheel, 198, *199*
Reaction force, 374
Reactor, 26, 94
Reactor core, 26, 32, 34, 36, 38
Reactor fuel, 26
Rear-axle housing, 258
Reciprocals, 486
Reciprocating engine, 236
Reciprocating pump, 340
Recoil, 220, 222, 244
Recoilless gun, 244, *245*
Record player, 470, *471*
Record-player drive, 208
Recording head, 560
Rectangular tones, 456, 458, *459*
Rectifier, 458
Recuperator, 220, *221*
Red mud (bauxite residue), 92
Red phosphorus, 46
Reducing agent, 40
Reed oscillators, 454
Reeds, 452, *453*
Reentry, 362, 370, *371*
Reentry trajectory, 370, 371
Refinery gas, 42, *43*
Refining, 84–93
Refining furnace, *85*
Reforming, 42
Refrigeration, 40, 68
Regatta, 348, 350
Regenerative chambers, 88

Regenerative principle, 362
Registers, 452, 453
Regulator, 324, *325*, 326
Relay communication satellite, 386
Release binding, 444, *445*, 446, *447*
Relief valve, 340
Rendezvous procedure, 380, 381, 382, 383, *384–385*
Repeaters, 554, *555*
Repetition action (piano), 448
Reproducing heads, 462, *463*, 464, *465*
Resin adhesives, 328
Resistance, 530, *531*
Resistance butt welding, 130, *131*
Resistance-capacitance generators, 458
Resistance noise, 458
Resistance pressure gauge, 492
Resistance pressure welding, 130, *131*, 132, *133*
Resistors, 458
Resonance, 192, 474
Resonance accelerator, 478, *479*
Resonance frequency, 530
Respirator, 442
Retort furnace, 88, 89
Retractable table, 98, *101*
Retreating longwall working, 66
Retro-rocket motors, 382
Reverberation device, 462, *463*
Reverberation plate, 462
Reverberation signal, 454
Reverberatory furnace, 84, 85, 86, 87
Reversing gear, 184, *185*, 206, *207*
Revolving field alternator, 510
Revolving furnaces, 86
Revolving turret, 244, *247*
Reynolds number, 336
Rifle striker-cocking mechanism, 216, *217*
Rifling, 218, 224, 225
Right-hand rule, 322, *323*
Rigid coupling, 188
Ring-balance pressure gauge, 490, *491*
Ring modular, 534, *535*
Riser, 102, *103*
Riveting, 328
Road tank vehicles, 54, *55*
Roads, 248, *249*, 250, *251*, 252, *253*, 254, *255*
Roasted blende, 90
Roasting, 84, 86, 88, 94
Roasting furnaces, *95*, 99
Rocker arm, 198, 200, *201*, 283, 284, *285*, 312, *313*
Rocket armament, 240
Rocket-propelled missiles, 338
Rocket-propelled torpedo, 236
Rocket propulsion, 344, *345*; 362, *363*, 364, *365*, 366, *367*
Rockwell scale, 174
Roll forming, 140, *141*
Roll-off, 330
Roller bearings, 170, *171*, 172, *173*, 402

Roller bit, 62, 63
Roller blind, 214, *215*
Roller grizzly, *81*
Roller tappet, 212
Rollers, 170, *171*, 172, *173*
Rolling (machining), 116
Rolling (of ships), 358.
Rolling mills, 118
Roof lagging, 76
Room-and-pillar coal mining, 66
Rotary casting process, 100, *101*
Rotary converter, 516, *517*
Rotary cutter, 166
Rotary drilling, 62, *63*
Rotary-drum artificial kidney, 421, *422*
Rotary island, *251*, 252
Rotary haymaking machine, 436, *437*
Rotary kiln, 46, *47*, 48, *49*, 88, 92, 94, *95*
Rotary process, 88
Rotary pump, 340
Rotary pump artificial heart, 414, *415*
Rotary valves, 340
Rotating band, 226, *227*
Rotating cutter, 186
Rotating file disk, 74
Rotating magnetic field, 512
Rotor, 506, 510
Rotor chamber, 402
Rotor motor, 494, *495*
Rotor winding, 510
Rough sheet, 56, 58, *59*
Roundabout, *251*, 252
Royal Ocean Racing Club, 350
Rudders, 236, *237*, 332, *333*, 340
Runner, 96, *97*, 102, *103*
Running gear, 244, *245*

Saddle (lathe), 158, 178, *179*, 186, *187*
Safety binding (skis), 444, *445*, 446, *447*
Safety lamp, 74, *75*
Sail-driven boats, 348, *349*, 350, *351*
St. Joseph Lead Co. process, 90, *91*
Salt-bath displacement process, 114, *115*
Salt-bath soldering, 138, *139*
Saltpeter, 48, 50
Sand casting, 102, *103*
Sand-lined molds, 112, *113*
Sandblasting, 150
Sandwich structures, 328, 329
Sandwich-type artificial kidney, 422, *423*
Saponification, 150
Satellites, 338, 376, 386, *387*, 388, *389*, 390, *391*
Saturated colors, 288, *289*
Saturation, definition of, 288
Saturn rocket, 368, *369*
Sawing, 126, *127*
Sawtooth generator, 458, *459*
Sawtooth tones, 456, 458, *459*

Schnorkel, 354, 356, *357*
Schooner rigging, 348
Score (electronic music), 466, *467*
Scout tank, 240, *241*
Scraping, 150
Screening, 80
Screw conveyor, 90, *91*, 432
Screw cutting, 158, *159*
Screw die, 158, *159*
Screw extrusion, 58, *59*
Screw mechanism, 198, *199*
Screw pump artificial heart, 414, *415*, 434
Screw spindle, 436
Screw tap, 158, *159*
Screw thread, 128, 156, *157*, 160, *161*
Screw-thread cutting, 128, *129*
SCUBA, 442, *443*
Sea mapping, 299, *299*
Sealing systems, 176
Seam welding, *131*, 132
Seaming, 140, *141*
SECAM system, 290, 294
Second-cut file, 126
Secondary water system, 356
Sedimentary rocks, 60
Segregation (traffic), 250, *251*, 252, *253*, 254, *255*
Seismic exploration, 60, *61*
Selection stage, 546
Selector valve, 342, *343*, 340
Selectors (telephone), 546, 556, 557
Self-aligning bearings, 170, *171*, 172, *173*
Self-aspirating engine, 278
Self-excitation, 324
Self-exciting generator, 508
Self-feeder (animal feeding), 432
Self-propelled gun, 240, *241*, 244
Sellers screw thread, 156
Semi-integral construction, 328, *329*
Semicircular canals, 400, *401*
Semicloverleaf intersection, *253*, 254
Semiconductor components, 454
Semiconductor diodes, 326, *327*
Semifixed ammunition, 226
Semipermanent molds, 108
Semipermeable membrane, 420, 422, 424
Semitone intervals, 450
Sendzimir process, 144, *145*
Separate lift engines, 346
Separate-loading ammunition, 218, 222
Separately excited generator, *507*, 508
Separation (airflow), 332, *333*
Separation (ores), 78, 80
Sequential transmission system, 288
Series generator, 508
Series motor, 506
Series oscillating circuit, 530, *531*
Series staging, 368, *369*
Serrated coupling, 188, *189*

Service module, 383, *384–385*
Setoff button (piano), 448, *449*
Settling classifier, 78, *83*
Seventh harmonic, 450
Shading, 296, *296*
Shadow boundry (moon), 378
Shadow-mask tube, 292, *293*
Shaft furnace, 84, *85*, 90, *91*
Shaft lining, 68
Shaft sinking, 68, *69*
Shaking tables, 78
Shaping, 126
Shear-bolt coupling, 196, *197*
Shearing, 140
Sheet-metal work, 140–143
Sheeting calender, 58, *59*
Shelf-testing rig, 226, *227*
Shell molding, 104, *105*
Shells, 226, *227*, 228, *229*
Shelved roasting kiln, 48
Sherardizing, 144
Shielded-arc welding, 136, *137*
Ship stabilizing, 358, *359*, 360, *361*
Shock absorber, 188, 192
Shock-wave wind tunnel, 338, *339*
Shock waves, 330, 344, 370
Short-thread milling, 160, *161*
Shovel-type mobile loaders, 76, *77*
Shower-head type fuel injection, 362
Shrunk-on heat-sealed packaging, 428, *429*
Shunt generator, 508
Side gear unit (tank), 242, *245*
Side bands, 536, *537*
Side-thrust rockets, 368
Sidereal day, 394
Siderite, 82
Sighting device, 222, *223*
Silage, 430, *431*, 432
Silbermann, Gottfried, 448
Silica, 46
Silicic acid esters, 342
Silicon carbide abrasive, 128
Silos, 430, *431*
Silver, 138, 150
Simultaneous transmission system, 288, 290, *290–291*
Sine oscillation, 468
Sine-wave generator, 458
Single action accordion, 452
Single-cut file, 126
Single-point cutting tool, 158, *159*
Single-pointer dial mechanism, 496, *497*
Single-row bearings, 170, *171*, 172
Single-stage rocket, 366
Single thread screw, 156
Single-thrust bearing, 170
Sinkable barge, 62, *63*
Sinkable drilling platform, 62, *63*
Sintered metals, 174
Sintering, 46, 58, 84, 88, 116, *117*
Sintering belt, 94, *95*
Sinusoidal cam drive, 204

Sinusoidal tones, 456, 458
Six-cylinder car, 264
Sizing, 78, 80
Skin diving, 442
Skinner's machine, 566, *567*
Skip, 70, *71*
Skip winding, 70
Skis, 444, *445*, 446
Slag, 84, 86, *87*
Slavjanov's method, 134, *135*
Sledgehammers, 118, *119*
Sleeve casting, 112
Slide rule, 484, *485*, 486, *487*
Slider, 204, *205*
Slider-crank linkage, 204, *205*
Sliding gear system, 184
Slip, 116, 206, 208
Slip casting process, 116, *117*
Slip clutch, 196, *197*
Slip ring, 322, *323*, 326, *327*, 510
Slip-ring induction motor, 514
Slip-ring rotor, 512, *513*
Slitting saw, *127*
Slot-and-crank drive, 204, *205*
Slot radiator, 526
Slotted flap, 334, *335*
Slotting machine, *127*, 168, *169*
Sluice valve, 154, *155*
Slurry, 106
Slush casting, 108
Smelting works, 78
Smooth file, 126
Snorkel, 356, *357*
Snort, 356, *357*
Soda lime, 442
Soda water glass, 46
Sodium, 36, *37*, 38, 286, 287
Sodium aluminate, 92
Sodium type hollow-stem exhaust
    valves, 286, *287*, 310
Soft soldering, 138, *139*
Solar battery, 386, *387*
Solar day, 394
Solar radiation, 372
Soldering, 138, *139*
Soldering iron, 138, *139*
Sole holder, 446, *447*
Solenoid valve, 152, *153*
Solid-propellant rocket engine,
    364, 365
Sonar, 356
Sonotrode, 428, *429*
Soot sludge, 42, *43*
Sorting (ores), 78
Sound amplifier, 400, 454, *455*
Sound mixtures, 456
Soundboard, 448, 452, 453
Sounds, 456
Southern Yacht Club, 348
Space capsule, 372, 373
Space flight, 366
Space-flight orbits, 376, *377*
Space probes, 386, *387*, 388, *389*,
    390, *391*
Space suit, 372, *373*
Spangles, 144
Spark-ignition engine, 260
Spark plug, 268, 270, 320
Speed control, 514, 518, *519*
Speed ratio, 164

Sphalerite, 84, 88, 90
Spigot-and-socket cast-iron pipes,
    100
Spigot pipe casting, 112
Spindle-drive motor, 186
Spinet, 448
Spinning, 142, *143*
Spiral bevel gears, 162
Spiral gears, 206, *207*
Splash plate fuel injection, 362
Splices, 464
Split flap, 334, *335*
Split-type muff coupling, 188, *189*
Spoilers, *333*, 334
Sponges, 52
Spontaneous combustion, 74
Spontaneous ignition, 260
Spot welding, *131*, 132
Spray dispenser, 20, *21*
Spray gun, 146, *147*
Spraying, 146
Spread-coating machines, 58
Spring catch, 444, *445*
Spring-loaded bush, 446, *447*
Spring-loaded hydraulic
    accumulators, 340
Spring-operated recuperator, 220,
    221
Spring-plate clutch, 318, *319*
Spring retaining rings, 176
Spring-tube pressure gauge, 490,
    491
Sprocket, 210
Spud-leg pontoon, 62, *63*
Spur gear, 162, *163*
Spur-gear teeth cutting, 166
Square flatter hammer, 118, *119*
Square screw thread, 156, *157*
Square-wave generator, 458, *459*
Square-wave tones, 456, 458, *459*
Squirrel-cage rotor, 512, *513*
Stability movement, 358
Stabilizer, 358, *359*, 360, *361*
Stabilizing circuit, 532
Stabilizing system, 246
Stabilizing tanks, 358, *359*
Stagnation point (airflow), 332,
    333
Stainless steel, 30, 32, 36, 38, 174,
    328, 402
Stall-barn system, 432, *433*
Stall-warning device, 332, *333*
Stalling angle, 332
Stampings, 120, 328
Standard-shift car, 256, *257*
Standard zinc refining process, 88,
    89
Star class, 350
Star clusters, 394
Star-delta control, 514
Star-quad formation, 552, *553*
Star-tracking guidance system,
    374
Star-wheel mechanism, 214, 216,
    217
Starting resistor, 514
Starting sheets, 86
Starting torque, 506, 510, 512,
    514

Stationary molds, 98, 99, 100,
    101
Stator, 506
Stearin molds, 148
Steel, 174
    in Al-Fin process, 114
    armor plates of, 242
    continuous casting of, 98, 100
    dies of, 108
    forging of, 118
    galvanizing of, 144, *145*
    hot extrusion of, 124
    molds of, 112
Steel alloys, 118
Steel-apron conveyors, 76
Steel-band coupling, 192, *193*
Steering gear, *257*, 258, *259*
Steering-wheel lock, 256
Stein, Johann Andreas, 448
Steinway action, 448, *449*
Stellar-monitored inertial guid-
    ance, 374
Stellar time, 394
Stellarator torus, 24, *25*
Stellite, 310
Stenosis, 408
Stepped roller grizzly, *81*
Stereo-decoder, 540, *541*
Stereophonic broadcasting, 538,
    539, 540, *541*
Stereophonic pickup, 472
Stereophonic reproduction, 466
Stick-up lettering, 299
Stock, 156
Stone-dust barrier, 74, *75*
Stop pin, 446, *447*
Stoping mining, 64, 72
Storage batteries, *353*, 354, 400
Stowing, 74
Straight-line link mechanism, 488,
    489
Straight milling, 128
Strain gauge, 480, *481*, 482, *483*
Strain hardening, 118
Streamline flow, 330
Stretch forming, 56, 142
Stretching rolls, 428
Strike of an ore deposit, 60, *61*,
    64
Striker-cocking mechanism, 216,
    217
Stringers, 328
Ströder washer, 46
Stroke-bore ration, 282, 284
Stylus, 472
Styropor, 52
Subcarrier wave, 294
Submarine, 352, *353*, 354, *355*,
    356, *357*
Submarine rocket, 236, *237*
Submerged-arc welding, 136, *137*
Subroc, 236, *237*
Subsonic combustion, 344
Subtraction circuits, 292
Sulphates, 48, 94, 420
Sulphide ores, 86
Sulphides, 78, 82, 94
Sulphidic lead-zinc ores, 82
Sulphur compounds, 42
Sulphur dioxide, 48, *49*, 94

Sulphuric acid, 48, *49*
  in copper refining, 86
  for electrolytic cleaning, 150
  in nitric acid preparation, 50
  in ore roasting, 94
  for pickling, 144
Sun wheel, 206, *207*
Supercharging, 278, *279*
Superheated steam nuclear
  reactor, 28, *29*
Superphosphate fertilizers, 48
Supersonic aircraft, 328
Supersonic speeds,'330
Supersonic wind tunnel, 336, 338
Surface-friction drag, 330
Surface-to-air missile, 364
Surfacing (submarine), 354
Suspension insulator, 522, *523*
Suspension system, 248, 259
Sustainer motor, 368
Swash plate, 340, *341*
Swash-plate agitator, *433*, 434
Swept-wing aircraft, 322, 330
Swing-wing aircraft, 330, *331*
Swirl-type spray fuel injection,
  362
Switch contact, 454
Swiveling jet, 346, *347*
Synchronous alternator, 510, *511*
Synchronous converter, 516, *517*
Synchronous motor, 510
Synchronous satellites, 390
Syncom, 386
Synthesis gas, 42, *43*
Synthetic ammonia process, 40,
  41, 42
Synthetic fibers
  nitric acid in manufacture of,
    50
  sulphuric acid in manufacture
    of, 48
Synthetic resin, 370
Synthetic rubber, 52
Synthetic-rubber seals, 176
Systole, 410

T-junction, 248, *249*, 252
T-tail, 332, *335*
Table concentrator, 78, 80, *81*
Tabling, 78, 80, *81*
Tail blades, 236, *237*
Tail section, 236
Tail sitters (VTOL aircraft), 346
Tail units (aircraft), 328
Tailings, *80*, 82
Tailstock, 178
Tandem staging, 368, 369
Tangent-elevation values, 224,
  225
Tank (military), 240, *241*, 242,
  243, 244, *245*, 246, *247*
Tank trucks, 54, *55*
Tank wagon, 434, *435*
Tape recorder, 560, *561*, 562, *563*,
  564, *565*
Tape recorder drive, 208
Tape recordings, 464, 465
Tape-speed control, 462
Taper roller bearings, 172, *173*

Taper socket, 178
Tapered socket, 186
Tappet, *283*, 284, *285*, 312
Tar, 42
TASI System, 542, *543*
Teaching machines, 566, *567*, 568,
  569, 570, *571*, 572, *573*
Teaching program, 570, *571*, 572
Telemetry antennae, 390
Telephone cables, 552, *553*, 554,
  555
Telephone-exchange, 544–551
Telephony, 544, *545*, 546, *547*,
  548, *549*, 550, *551*
Telephoto photography, 388
Telescopes, 246
Television camera, 388, *389*, 390
Television transmission, 390
Telstar, 386, 390
Temperature
  in nuclear fusion reactors, 24,
    30, 32, 34
  in "town-gas" detoxification, 44
  underground, 72
Tempered scale, 450
Template, 148
Template molding, 102, *103*
Tenor strings, 448
Tensioning roller, 210
Thermal breeder nuclear reactor,
  38
Thermal insulation, 52
Thermal nuclear reactor, 26–29
Thermal power, 26, 28, 30, 34
Thermal reduction, 88, 90
Thermionic cathode, 532
Thermionic tube, 532, *533*
Thermite pressure welding, 132
Thermite welding, 136, *137*
Thermoforming, 426, *427*
Thermonuclear fusion, 22, *23*, 24,
  25
Thermonuclear reactions, 22
Thermoplastics, 56
Thermosetting plastics, 56
Thermostat, 420, *421*
Thorium, 38, 232
Thorium-uranium cycle, 38
Thread-milling cutter, 160, *161*
Thread-rolling machine, 160, *161*
Three-layer electrolysis, 92, *93*
Three-phase alternating current,
  510
Three-phase generator, 326, *327*
Three-way valve, 152
Thrombosis, 412
Throttle, 256
Throttling effect, 308, 316
Throttle valves, 340
Thrust, 364
Thrust bearings, 170, 172, *173*
Thrust chamber, 364, *365*
Thyratron, 458
Tilting meter, 494, *495*
Tilting-mold casting, 96, 97
Timbre, 456, 458
Time channels, 542
Time division multiplexing, 542,
  543
Time fuse, 230

Time regulator, 464
Tin, 144
Tin-plating, 150
Tiros satellite, 386
Tissue rejection, 410, 412, 418
Titanium, 284, 314, *315*, 316
Titanium alloy, 326
TNT, 226
Toggle action, 446
Tone, 450, *451*
Tone control, 400, *401*
Tone mixtures, 456
Tone-pitch regulator, 464
Tool steels, 122, 126
Toolbox, 200, *201*
Tooth enamel, 402, *403*
Toothed coupling, 188, *189*
Toothed roll crusher, *81*
Top carriage, 222
Top slide, 178, *179*
Torpedoes, 236, *237*, 238, *239*,
  356
Torque, 142, 206, 260, 262, 266,
  282, 506, 510, 512, 514
Torque transmission, 188
Torricellian vacuum, 490
Torsion-bar suspension, 244
Tough pitch, 86
"Town gas" detoxification, 42, 44,
  45
Toxemia, 418
Tracer ammunition, 226
Track drive wheels, 244, *245*
Track rollers, 244, 245
Traction wheel, 208
Tractor takeoff shaft, 434, 438,
  440
Traffic circle, *251*, 252, 254
Traffic streams, 248, *249*, 250,
  251, 252, *253*, 254, *255*
Train of gears, 160
Trajectories, 226, *227*, 376, *377*,
  378, *379*, 380, *381*
Transfer orbits, 378
Transistors, 454
Transition point (airflow), 332,
  333
Transmission lines, 520, *521*, 522,
  523, 524, *525*
Transmission ratio, 164, 182, 206,
  208, 210, 312
Transonic wind tunnel, 336, 338
Transparency, 426
Transverse electric waves, 526
Transverse magnetic waves, 526
Transverse turning, 128
Traveling grate, 88, 94
Traversing, 244, *247*
Traversing gear, 222, *223*
Trebel mechanism, 452
Tree pattern, 106, *107*
Tricolor tube, 292, *292*
Trigonometric functions, 484
Trimming tab, 334, *335*
Trimming tanks, *353*, 354
Tritium, 22, *23*, 24
Trunk cable, 552
Trunnions, 222
Tubbing, 68
"Tubbing" shaft lining, 68

Tube, 124, 130, 132
Tube manufacture, 132, *133*
Tube production, 98
Tubs, 70
Tumbler gear, 184
Tumbling, 102, 144, 174
Tumors, 398
Tuner, 292
Tungsten, 116, 362, 366
Tungsten carbide, 62, 128
    production of tools, of, 116
Tuning (piano), 450, *451*
Tuning pins, 448, *449*
Tunneling, 64, *65*
Tunnels (underpasses), 248
Turbine, 236
Turbine drill, 402, *403*
Turbine mixer, 500
Turbodisperser, 500, *501*
Turbodrill, 62
Turbogenerators, 356
Turbojet, 344
Turbojet VTOL aircraft, 346
Turbopump supply system, 362,
    363
Turbulence, 310, *311*
Turbulent air flow, 332, *333*
Turning, 126, *127*, 128
Turning mills, 174
Turret lathes, 174
Turret-mounted naval guns, 222
Twin carburetors, 272, *273*
Twin-coil artificial kidney, 424,
    425
Twist drill, 128
Two-circuit boiling water reactor,
    28, *29*
Two-motion selector, 548
Two-pointer dial mechanism, 496,
    497
Two-stroke engine, 262

U-tube manometer, 490, *491*
Ultrafiltration, 422
Ultrasonic combustion, 344
Ultrasonic soldering, 138
Ultrasonic welding, 130, 132, *133*,
    428, *429*
Underground drilling, 76
Underground exploration, 60, *61*
Underpasses, 248
Undersea telephone cables, 542
Underwater breathing apparatus,
    442, 443
Ungine-Séjournet extrusion
    method, 124
Uniselector motor switch, 546,
    547, 548
Universal coupling, 190, *191*
Universal milling machine, 186
Universal motor, 514
Universal shaft, 190, *191*, 256,
    258, *259*, 436, *437*, 438
Unmanned satellites, 386, *387*
Uphill pouring casting, 96, *97*
Uplift foundation, 524, *525*
Uranium, 22, 26, 30
Uranium dioxide, 26, 30
Uranium oxide, 32

Uranium-plutonium cycle, 38
Uranium-thorium carbide, 32
Urea, 52, *53*, 420
Urea-formaldehyde foams, 52, *53*
Uremia, 418
Ureter, 418, *419*
Urinary bladder, 418, *419*
Urine, 418

V-belt, 210, 438
V-shaped screw thread, 156, *157*
V2 rockets, 362, 374
Vacuum dezincing plant, 84, *85*
Vacuum die casting, 110
Vacuum jacket, 398, *399*
Vacuum molding, 55, 426, *427*
Vacuum packaging, 428, *429*
Vacuum shaping process, 56, 58,
    59
Vacuum vessel, 338, *339*
Valve control mechanism, 198
Valve flutter, 316
Valve gear, 212, *213*
Valve lift, 312, *313*, 314
Valve mechanism, *283*, 284, *285*,
    314, *315*
Valve noise, 458
Valve-opening periods, 276
Valve-operated artificial heart,
    414, *415*
Valve-seal rings, 310
Valve timing, 276, *277*, 312
Valves, 152–155, 310, *311*, 340
Vanadium pentoxide catalyst, 48
Vane hydraulic motor, 340
Vane pump, 340
Variable capacitors, 454
Variable-geometry wing, 330, *331*
Variable-speed direct-current
    motors, 182
Variable-speed mechanical drive,
    182, *183*
Variable-stroke pump, 340
Ventilation, 72
Ventilation doors, 72
Ventilation of mines, 64, 72, *73*
Vertical agitator, *433*, 434
Vertical centrifugal casting, 112,
    113
Vertical cold-chamber pressure
    die casting, 110, *111*
Vertical continuous casting, 98,
    99
Vertical milling machine, 186
Vertical-shaft furnace, 84, *85*
Vertical stabilizer, 332
Vertical take-off-and-landing
    aircraft, 346, *347*
VHF stereophonic broadcasting,
    538, *539*, 540, *541*
Vibrato device, 454, 458
Video tape recording, 560, *561*,
    562, *563*, 564, *565*
Vidicon tubes, 294, 388, 390
Vinyl chloride, 20
Virginals, 448
Vise, 198
Vocoder, 462
Voltage coil, 324, *325*

Voltage divider, 530, *531*
Voltage loss, 522
Volume control, 400, 454
Volume-metering equipment, 54,
    55, *494*
Volumetric efficiency, 272, *273*,
    276, 278, 282, 308, 312
VT fuse, 230
VTOL aircraft, 346, *347*

Ward-Leonard control, 70, 222,
    223, 518, *519*
Warhead, 236, *237*
Wash-ahead program, 568
Washback program, 568
Washing tower, 48, *49*
Watch-spring steel, 452
Water
    as ingot coolant, 96, *97*
    as nuclear reactor coolant, 26,
    28
    as nuclear reactor moderator,
    26, 28
Water atomizer, 404, *405*
Water-ballast tanks, *353*, 354
Water-cooled casting wheel, 100,
    101
Water-cooled copper shoes, 134,
    135
Water-cooled engine, 320, *321*
Water-cooled mold, 98, *99*, *101*,
    112, *113*
Water-cooled turbine drill, 402,
    403
Water evaporator, 404, *405*
Water gas, 40
Water gauge, 490
Water glass, 46, *47*, 108
Water purification, 372, *373*
Waveguide, 478, 526, *527*
Wax molds, 148
Weather satellites, 386, 388, *389*
Weber, 504
Wedge-shaped combustion
    chamber, 270, *271*
Welding, 56, 130–137
Wet ball mill, *81*
Wet classifier, *81*
Wet galvanizing, 144, *145*
Wet-mill concentration, 78, *80*, *83*
Wet precipitator, 48, *49*
Wheatstone bridge, 482, *483*
Wheel sintering process, 58
Whirl sintering, 58
White noise, 456, 458
White phosphorus, 46
Whitford screw thread, 156, *157*
Wick oil lubrication, 176
Wilson cloud chamber, 476, *477*
Wind tunnel, 336, *337*, 338, *339*
Winding, 70, *71*
Winding gears, 79
Winding ropes, 70, *71*
Winding shaft, 72, 76
Winding-speed regulator, 70
Windrow, 438, 440
Windshield projectile, 228, *229*
Wing geometry, 328, 330, *331*
Wire brushing, 150

Wire-clearance diagram, 521, *522*
Wire-guided torpedo, 238, *239*
Work hardening, 118
Work spindle, 178, 180, 184, *185,*
186, 187
Work table, 198, 204
Work table lathe, 186, *187*
Worm drive, 154, *155*
Worm gear, 162, *163*, 186, 206,
207
Worm wheel, 166, 178
Wound rotor, 514
Wrest plank, 448

X-rays, 398

Y junction, 248
Yachts, 348, *349*, 350, *351*
Yacht clubs, 348, 350
Yacht Racing Association, 348
Young's modulus, 480

Zeiss planetarium, 392, *393*
Zerener's method, 134, *135*
Zeta (nuclear-fusion research), 24,
25
Zinc, 88–91

in bearings, 114
for galvanizing, 144, *145*
Zinc alloy casting, 110
Zinc blende, 88
Zinc-carbon electric cell, 234, *235*
Zinc chloride, 144
Zinc ores, 82, 94
Zinc oxide catalyst, 42
Zinc sulphide (sphalerite), 84, 88,
90
Zircaloy, 26
Zirconium, 36
Zirconium hybride, 36